MW01378078

KENNESAW COLLEGE
CONSTITUTION
WIS DOM
JUSTICE
MOD ERATION
1963
UNIVERSITY SYSTEM OF GEORGIA

The Photochemistry of Atmospheres

Earth, the Other Planets, and Comets

The Photochemistry of Atmospheres

Earth, the Other Planets, and Comets

Edited by

JOEL S. LEVINE

Atmospheric Sciences Division
NASA Langley Research Center
Hampton, Virginia

1985

ACADEMIC PRESS, INC.

(Harcourt Brace Jovanovich, Publishers)

Orlando San Diego New York London
Toronto Montreal Sydney Tokyo

ACADEMIC PRESS, INC.
Orlando, Florida 32887

United Kingdom Edition published by
ACADEMIC PRESS INC. (LONDON) LTD.
24–28 Oval Road, London NW1 7DX

Library of Congress Cataloging in Publication Data

Main entry under title:

The Photochemistry of atmospheres.

Includes index.
1. Atmospheric chemistry. 2. Photochemistry.
3. Atmosphere, Upper. 4. Planets--Atmospheres.
I. Levine, Joel S.
QC879.6.P48 1985 551.5'11 84-12356
ISBN 0-12-444920-4 (alk. paper)

PRINTED IN THE UNITED STATES OF AMERICA

85 86 87 88 9 8 7 6 5 4 3 2 1

To my wife, Arlene, and daughter, Lisa,
for their love, understanding, support, and patience

Contents

II
The Other Planets

6. The Photochemistry of the Atmosphere of Venus
RONALD G. PRINN

7. The Photochemistry of the Atmosphere of Mars
CHARLES A. BARTH

8. The Photochemistry of the Atmospheres of the Outer Planets and Their Satellites
DARRELL F. STROBEL

III
Comets

9. The Photochemistry of Comets
WALTER F. HUEBNER

IV
Appendixes

Contributors

Numbers in parentheses indicate the pages on which the authors' contributions begin.

CHARLES A. BARTH (337), Department of Astrophysical, Planetary, and Atmospheric Sciences, and Laboratory for Atmospheric and Space Physics, University of Colorado, Boulder, Colorado 80309

T. E. GRAEDEL (39), AT&T Bell Laboratories, Murray Hill, New Jersey 07974

WALTER F. HUEBNER (437), Theoretical Division, Los Alamos National Laboratory, Los Alamos, New Mexico 87545

WILLIAM R. KUHN (129), Department of Atmospheric and Oceanic Science, University of Michigan, Ann Arbor, Michigan 48109

JOEL S. LEVINE (3), Atmospheric Sciences Division, NASA Langley Research Center, Hampton, Virginia 23665

RONALD G. PRINN (281), Department of Earth, Atmospheric, and Planetary Sciences, Massachusetts Institute of Technology, Cambridge, Massachusetts 02139

DARRELL F. STROBEL (393), Department of Earth and Planetary Sciences, The Johns Hopkins University, Baltimore, Maryland 21218

DOUGLAS G. TORR (165), Center for Atmospheric and Space Sciences, Utah State University, Logan, Utah 84322

RICHARD P. TURCO (77), R & D Associates, Marina del Rey, California 90295

Preface

This book is about the photochemical and chemical processes in atmospheres—the atmosphere of our planet, past, present, and future; the atmospheres of the other planets and their satellites; and comets. Beginning in the 1960s, there has been an explosion in our knowledge and understanding of the photochemical and chemical processes in atmospheres. These insights have resulted from a new awareness and national as well as international concern about the effects of anthropogenic perturbations on the composition and photochemistry of the atmosphere and on climate. Stories concerning increasing levels of air pollution and acid precipitation, the possible inadvertent depletion of stratospheric ozone, and changes in atmospheric carbon dioxide and climate continually appear on the front pages of our newspapers and on the nightly news. In addition, the United States and the Soviet Union have embarked on a vigorous program of planetary exploration, resulting in close visits to all of the planets known to the ancients—Mercury, Venus, Mars, Jupiter, and Saturn. Close-up photographs of these planets and their satellites, once restricted to the pages of scientific journals, now regularly appear in the popular press and on television.

Prior to this volume, no single book has treated the subject of the photochemistry of atmospheres in its entirety. The present volume considers the subject of atmospheric photochemistry over large temporal and spatial scales. On a temporal scale, the composition and photochemistry of our atmosphere over its 4.6 billion year history are outlined, with particular emphasis on the strong coupling between atmospheric evolution and photochemistry and the origin and evolution of life. The composition and photochemistry of the present atmosphere (troposphere, stratosphere, and upper atmosphere) are described, including the implications and consequences of present and future anthropogenic perturbations. In addition, the coupling involving photochemistry, composition, and climate is considered in a separate chapter. On a spatial scale, the subject material ranges from the photochemistry of the hot, massive atmosphere of Venus to that of the frozen gases in cometary nuclei at the very edge of the solar system, which are heated by solar radiation, forming diffuse atmospheres around comets. Of such vastness are the temporal and spatial scales of atmospheric photochemistry covered in this volume. General topics in atmospheric photochemistry, such as composition and structure, the transfer of incoming solar radiation, the principles governing the rates of photochemical and chemical processes, the role of eddy and molecular transport, and the continuity–transport equation used in theoretical numerical modeling studies, are discussed throughout the volume.

Each chapter was written by a scientist active in that area of atmospheric research. Each of the nine chapters begins with a historical introduction and background, followed by a detailed discussion of the relevant photochemical and chemical processes, and concludes with comments on the directions for future research. The book includes a series of tables in each chapter, as well as appendixes at the end of the volume, that contain detailed information on the structure and composition of the atmospheric region and details of the photochemical and chemical processes, parameters, and rate constants discussed. We have tried to summarize the present state of knowledge in each chapter. The number of references was limited by space considerations, and hence, it was not possible to include comprehensive reference lists for each topic. This is especially true for some research cited in figures from various reviews that are reproduced in this volume.

We believe that this volume will serve two very diverse audiences. It can be used in advanced-level courses in the atmospheric and planetary sciences and will be of value to the larger general audience interested in learning more about current atmospheric/climatic environmental problems, their causes and consequences, as well as the recent discoveries concerning the atmospheres of our neighboring worlds.

As will become evident in the following pages, much of the new information and understanding of the photochemistry of the atmospheres of the Earth, the other planets, their satellites, and comets has resulted both directly and indirectly from the research programs and space missions supported by the National Aeronautics and Space Administration (NASA) at our nation's colleges and universities, government and industrial laboratories, and the NASA field centers. It is therefore most fitting for the preparation of this volume to coincide with the Silver Anniversary of NASA.

On a personal note, it is a pleasure to thank colleagues (too numerous to identify here) for their guidance and insight in my own studies of the photochemistry of atmospheres, some of which is discussed in this volume. It is also a pleasure to acknowledge the continued support and counsel of Dr. James D. Lawrence, Jr., Chief of the Atmospheric Sciences Division at the NASA Langley Research Center. I also thank Lisa K. Levine for her skill and good humor in the typing and retyping of portions of this volume.

Introduction

The Photochemistry of Atmospheres

Atmospheric photochemistry treats the interaction of atmospheric gases with incoming solar radiation (usually solar X rays, and ultraviolet and visible radiation). Photochemical processes initiate most of the chemistry in the atmospheres of the planets, their satellites, and comets. The absorption of solar radiation (photons) by atmospheric gases leads to the photoionization and photodissociation of atmospheric species. Photoionization, which in general requires more energetic photons (shorter-wavelength radiation) than photodissociation, results in the production of positively charged molecules and atoms, and electrons, usually in the outer regions of atmospheres (where the incoming solar radiation is more intense and of greater energy). Photodissociation involves the breakup of a molecule into its constituent molecules, radicals, or atoms. The absorption and scattering properties of an atmosphere determine the intensity and spectral energy distribution of the incoming solar radiation as it traverses the atmosphere. Atmospheric gases selectively absorb incoming solar X rays and ultraviolet and visible radiation (this selective absorption initiates the photoionization and photodissociation reactions.) Clouds, haze layers, and aerosols in the atmosphere both absorb and scatter incoming solar radiation. Since the transfer of incoming solar radiation through the atmosphere occurs in the vertical direction, the photochemical and kinetic chemical processes initiated by its absorption also vary in the vertical direction (i.e., with altitude). Hence, atmospheric photochemical and chemical processes vary with altitude as a function of the intensity and spectral energy distribution of the incoming solar radiation, and the chemical composition of the atmosphere.

Photochemical processes differ from planet to planet as a result of varying atmospheric composition and the level of solar radiation incident on the atmosphere. In general, chemical reactions between major atmospheric constituents occur at very slow rates. Most atmospheric chemical reactions are initiated by chemically active photodissociation products. Hence, to a large extent, the kinetic chemistry of atmospheres is controlled by chemically active photodissociation products at trace levels [parts per million by volume (ppmv) or less]. Throughout this volume, the photochemistry and chemistry of atmospheres and their variation with altitude will be considered. In the case of the Earth's atmosphere, the photochemistry and chemistry of the early atmosphere and its evolution over geological time, the troposphere (surface to 10 km), stratosphere (10–50 km), and upper atmosphere (>50 km) are considered in separate chapters. The structure, composition, photochemistry and chemistry of the lower and upper regions of the atmospheres of Venus, Mars, and the outer planets and their satellites are also treated in separate chapters.

Atmospheric Environmental Concerns and Planetary Exploration

Our understanding of the photochemistry/chemistry of atmospheres has advanced significantly since the early 1960s, primarily as a result of two unrelated activities: (1) a new awareness and concern of the role of anthropogenic activities on atmospheric composition and climate and (2) a vigorous program of planetary exploration. The present and future effects of anthropogenic activities on atmospheric composition and climate have become environmental problems of great national and international concern. These problems include the general deterioration of the quality of the air we breathe, the increase of air pollution and acid precipitation throughout the world, the possible inadvertent depletion of ozone (O_3) in the stratosphere (which shields the surface of the Earth from lethal solar ultraviolet radiation), and the effects of increasing atmospheric levels of carbon dioxide (CO_2) and other trace greenhouse gases on climate. These atmospheric and climatic environmental problems have several things in common: they are initiated by trace-level atmospheric gases produced or perturbed by anthropogenic activities, and most of these anthropogenic gases are transformed by various atmospheric photochemical and chemical processes.

During the very time that these atmospheric environmental problems were being identified and studied, the United States and the Soviet Union embarked on a vigorous program of planetary exploration. There has begun exploration of all of the planets known to the ancients—Mercury, Venus, Mars, Jupiter, and Saturn—by a series of planetary flybys, orbiters, and landers. This "Golden Age of Planetary Exploration" has significantly expanded our knowledge and understanding of the composition, structure, and photochemistry/chemistry of the atmospheres of our neighboring worlds.

During this same period, an active program of Earth-orbiting spacecraft has provided new information about the structure and composition of the Earth's upper atmosphere. This has been a period of rapid and significant advance in our understanding of the photochemistry/chemistry of the Earth's atmosphere and the atmospheres of the other planets and their satellites.

Throughout the chapters in this volume we shall find the same gaseous species occurring in widely differing atmospheric and planetary environments. For example, sulfuric acid (H_2SO_4) is a major component of both acid precipitation on Earth and of the clouds on Venus; O_3 is found both in the Earth's stratosphere and in the lower atmosphere of Mars; CO_2, the major constituent of the atmospheres of Mars and Venus, is also a trace, but increasing, component of the Earth's atmosphere, resulting from the burning of fossil fuels. Atmospheric species initiate widely different chemical transformations in different atmospheric regions. For example, in the troposphere, nitrogen dioxide (NO_2) is the precursor for the photochemical production of O_3, whereas in the stratosphere, it leads to the chemical destruction of O_3.

The new awareness and understanding of the photochemistry of atmospheres prompted this volume. In the following nine chapters and two appendixes, we have attempted to summarize the current understanding of atmospheric photochemistry, the causes and consequences of atmospheric and climatic environmental problems, and the excitement and sense of discovery of planetary exploration. Throughout the volume there are discussions on the general principles of atmospheric photochemistry that apply to the specific processes described in each chapter. For example, the general principles governing the rates of photochemical and chemical processes are discussed in Chapters 1 and 2; atmospheric ionization processes are covered in Chapter 5; the pressure and density structures of atmospheres are covered in Chapter 7; atmospheric species transport by eddy and molecular diffusion is covered in Chapter 6; and the coupled continuity–transport equation is described in Chapters 1 and 8 for eddy diffusion and molecular diffusion, respectively.

Earth

The Early Atmosphere

The origin, early history, composition, photochemistry, and chemical evolution of the Earth's atmosphere are discussed in Chapter 1. A new picture of the composition of the prebiological paleoatmosphere (i.e., the early atmosphere prior to life), based in large part on photochemical considerations, has emerged. This new picture suggests that the early atmosphere may have been composed of molecular nitrogen (N_2), CO_2, and water vapor (H_2O), instead of the earlier idea that it consisted of a mixture of ammonia (NH_3), methane (CH_4), and molecular hydrogen (H_2). The H_2O, CO_2, and N_2 that we now believe most probably constituted the Earth's early atmosphere were originally trapped in the upper layers of the planet's interior and were subsequently released, forming the atmosphere as a result of *volatile outgassing.* (It is generally believed that the atmospheres of Earth, Venus, and Mars formed as a result of the outgassing of volatiles originally trapped in the upper layers of the solid planet during the late stages of planetary accretion. In contrast, the atmospheres of the outer planets may be the gaseous remnants of the solar nebula gas cloud that condensed to form the solar system some 4.6 billion years ago.) Most of the outgassed H_2O condensed and then precipitated out of the atmosphere, forming the oceans. The bulk of the outgassed CO_2 left the atmosphere, first by dissolution in the oceans and then ultimately by incorporation in sedimentary carbonate rocks. Most of the outgassed N_2, a relatively chemically inert gas, remained in the atmosphere, where it accumulated to become the major constituent. The composition of the early atmosphere and its chemical evolution over geological time are closely related to the origin and evolution of life. It is thought that the constituents of the early atmosphere, energized by solar ultraviolet radiation and atmospheric lightning, abiotically formed organic molecules of increasing complexity that eventually evolved into the first living organisms. Over geological time, the composition, and hence, the photochemistry of the atmosphere changed as a result of several time-varying geochemical and biological processes, including volatile outgassing, volcanic emissions, gravitational escape of light gases, formation of the oceans and sedimentary carbonate rocks, and the emergence and evolution of life.

The Troposphere

Chapter 2 considers the photochemistry/chemistry of the troposphere, the lowest region of the atmosphere, which extends from the surface to 10 to 15 km, depending on latitude. The troposphere contains about 80–85% of the total mass of the atmosphere. This chapter covers the interplay between photochemistry/chemistry and meteorology on a local, regional, and global scale; atmospheric photochemistry in liquid and aerosol phases; species concentrations and budgets; and some of the environmental implications and consequences of tropospheric photochemistry, including acid precipitation, photochemical smog, the degradation of atmospheric visibility, and corrosion by atmospheric species. The major gases in the troposphere (N_2 and O_2) are chemically relatively inert. Most of the photochemistry in the troposphere is initiated by reactions of the hydroxyl radical (OH), with trace gases produced by biogenic processes [e.g., CH_4, NH_3, nitric oxide (NO), and hydrogen sulfide (H_2S)] and anthropogenic activities [e.g., carbon monoxide (CO), NO, sulfur dioxide (SO_2) and nonmethane hydrocarbons]. The hydroxyl radical is produced by the reaction of excited atomic oxygen [$O(^1D)$] with H_2O. Excited atomic oxygen is formed by the photodissociation of O_3, which is chemically produced by the combination of atomic oxygen (O) and O_2. The source of oxygen atoms needed for the formation of O_3 in the troposphere is the photodissociation of NO_2 by visible solar radiation (<420 nm). Nitrogen dioxide is re-formed by the reaction of O_3 with NO, produced by high-temperature combustion, atmospheric lightning, and biogenic activity.

The relationships between OH, $O(^1D)$, O_3, O, NO_2, and NO briefly outlined above are good examples of the strong photochemical and chemical coupling between trace gases in the atmosphere. The chemistry that leads to the formation of acid precipitation is another example of chemical transformations in the atmosphere. Sulfuric acid, the major component of acid precipitation, and nitric acid (HNO_3), its fastest increasing component, are chemically produced from SO_2 and NO, respectively. Anthropogenic activities are significant sources of both SO_2 and NO. Studies and reports by the U.S. National Academy of Sciences (NAS) and the U.S. Environmental Protection Agency (EPA) have considered the origin, chemical transformations, transport, deposition, and environmental consequences of acid precipitation. A major thrust of tropospheric photochemistry is the assessment of the role and importance of anthropogenic activities on the composition of the troposphere, particularly its future composition. Specific environmental concerns unique to the photochemistry of the troposphere include the deterioration of air quality, the increase of air pollution and acid precipitation, and the increase of trace gases in the troposphere that affect climate via their greenhouse effect [CO_2, O_3, CH_4, and nitrous oxide (N_2O)].

The Stratosphere

The photochemistry/chemistry of the stratosphere is considered in Chapter 3. The stratosphere extends from the top of the troposphere to ~ 50 km and contains almost all of the remaining mass of the atmosphere. About 90% of the total atmospheric O_3 is found in the stratosphere (with most of the remainder in the troposphere). The absorption of solar ultraviolet (200–300 nm) radiation by stratospheric O_3 shields the surface of the Earth from this lethal radiation. Much of the photochemistry of the stratosphere centers on the production and destruction of O_3, particularly the possible inadvertent depletion of O_3 by gases resulting from anthropogenic activities. Possible anthropogenic perturbations to stratospheric O_3 have been identified as a problem of national and international concern. The possible inadvertent depletion of stratospheric O_3 and its effects on climate and on life at the Earth's surface have been the subjects of continuing studies and reports by the World Meteorological Organization (WMO), NAS, and the U.S. National Aeronautics and Space Administration (NASA). The photochemical destruction of stratospheric O_3 results from a series of catalytic cycles involving the odd-hydrogen species HO_x [atomic hydrogen (H), OH, and the hydroperoxyl radical (HO_2)], the odd-nitrogen species NO_x (NO and NO_2), and the odd-chlorine species, ClO_x [atomic chlorine (Cl) and chlorine oxide (ClO)]. The odd-nitrogen and odd-chlorine species are ultimately produced or controlled by anthropogenic activities.

Climate

The interactions between tropospheric and stratospheric trace gas concentrations, photochemistry, and climate are considered in Chapter 4. Water vapor, CO_2, O_3, CH_4, and N_2O absorb and then reradiate Earth-emitted infrared radiation, resulting in a greenhouse temperature enhancement of the surface and troposphere and a cooling in the stratosphere. This chapter covers the radiative properties of these greenhouse gases and radiative climate modeling as well as the other parameters that control and/or regulate the climate of our planet (e.g., albedo, clouds, the solar constant, atmospheric circulation, and atmosphere–ocean exchange of energy). Atmospheric levels of CO_2, CH_4, and N_2O appear to be increasing, probably due to anthropogenic perturbations. Increased levels of these greenhouse gases may affect the future climate. Possible climate change resulting from anthropogenically produced greenhouse gases, most notably CO_2, has been the subject of studies and reports by the WMO, NAS, and EPA.

The Upper Atmosphere

The photochemistry/chemistry of the upper atmosphere, which includes the mesosphere (50–85 km), thermosphere (85–500 km), and exosphere (>500 km), is covered in Chapter 5. The structure, composition, and photochemistry/chemistry of the ionosphere (the region of positively charged molecules and atoms, and electrons, between 60 and 400 km) and magnetosphere (the region of charged particles contained by the Earth's magnetic field) and the gravitational escape of light gases (hydrogen and helium) from the exosphere are also discussed in this chapter. The exosphere eventually merges into the interplanetary medium. Much of the new information about the photochemistry/chemistry, structure, and composition of the upper atmosphere resulted from experiments aboard the NASA Earth-orbiting *Atmosphere Explorer*, *Dynamics Explorer*, and *Solar Mesosphere Explorer* satellites.

The Other Planets

Venus

Beginning with Chapter 6, which covers the photochemistry/chemistry of the atmosphere of Venus, we turn our attention away from the Earth's atmosphere. (For all practical purposes, the planet Mercury is devoid of an atmosphere and, hence, is not considered in this volume.) The atmosphere of Venus is very massive (surface pressure of 90 atm) and very hot (surface temperature of 750 K). The surface is perpetually covered by very thick clouds of H_2SO_4. The atmosphere consists predominantly of CO_2 (~95%), with small amounts of N_2 (3.5%) and argon (Ar) (1.5%), and trace amounts of H_2O, SO_2, CO, H_2, and various sulfur and halogen species. A major photochemical/chemical question concerning the atmosphere of Venus is the stability of CO_2 against photochemical destruction and the very low atmospheric levels of CO and atomic and molecular oxygen, the photolytic products of CO_2. This chapter also covers the chemical reactions between atmospheric gases and the minerals constituting the very hot surface of Venus. Much of our information about the structure, composition, and photochemistry of the atmosphere of Venus resulted from experiments aboard the NASA *Mariner 5* and *10* flybys and *Pioneer Venus* orbiter and entry probes (*Mariner 5* was launched 14 June 1967 and encountered Venus on 19 October 1967; *Mariner 10* was launched 3 November 1973 and encountered Venus on 5 February 1974; *Pioneer Venus* was launched 20 May 1978, the orbiter achieved Venus orbit on 4 December 1978, and the probes entered the atmosphere on 9 December 1978), and the U.S.S.R. *Venera* series of spacecraft (through the 1960s, 1970s, and 1980s).

Mars

The photochemistry/chemistry of the atmosphere of Mars is discussed in Chapter 7. Mars, like the atmosphere of Venus, is predominantly CO_2 (95.3%), with small amounts of N_2 (2.7%), Ar (1.6%), O_2 (0.13%), CO (0.07%), H_2O (0.03%), and O_3 (0.03 ppmv). Water vapor and O_3 concentrations in the Martian atmosphere are variable with season and latitude. The amount of atmospheric H_2O is controlled by the temperature of the surface and atmosphere. Ozone is present when the atmosphere is cold and dry. Measured

O_3 over the northern polar cap was found to maximize in winter, decrease during spring, and disappear during summer. Measurements indicated that as H_2O increased, O_3 decreased. The concentration and variability of O_3 on Mars can be explained in terms of the presence of H_2O and the odd-hydrogen species that result from the photolysis of H_2O and control the photochemical destruction of O_3. In contrast to the thick, hot atmosphere of Venus, the atmosphere of Mars is thin (surface pressure of 0.006 atm) and cold (mean surface temperature of $\sim$220 K). The annual sublimation and precipitation of CO_2 out of and into the polar caps produce a planetwide pressure variation of 2.4 mbars, a pressure change of 37% compared to the mean atmospheric pressure of Mars of 6.36 mbars. Atoms of hydrogen, oxygen, and nitrogen escape from the exosphere of Mars. The escape flux of hydrogen to oxygen atoms is in the ratio 2:1. If the escape flux of hydrogen and oxygen atoms has always operated at the present rate, then over its history Mars has lost the equivalent of 2.5 m of liquid H_2O over its entire surface. There is evidence that the climate of Mars was different in the past from what it is today. High-resolution spacecraft photographs indicate the presence of small and large runoff channels, tributary networks, and examples of widespread fluid erosion. Much of the information about the structure, composition, and photochemistry of the atmosphere of Mars came as a result of the experiments aboard NASA Mars probes, including the *Mariner 4* flyby (launched 28 November 1964, encountered Mars on 14 July 1965), the *Mariner 6* and *7* flybys (launched 25 February and 27 March 1969, respectively; encountered Mars on 31 July and 5 August 1969, respectively), the *Mariner 9* orbiter (launched 30 May 1971, went into Mars orbit on 13 November 1971), and the twin *Viking* orbiters and landers (*Viking 1* was launched 20 August 1975, went into orbit 19 June 1976, and landed 20 July 1976; *Viking 2* was launched 9 September 1975, went into orbit 7 July 1976, and landed 3 September 1976).

The Outer Planets

The photochemistry/chemistry of the atmospheres of the outer planets (Jupiter, Saturn, Uranus, and Neptune) and their satellites is discussed in Chapter 8. In contrast to the terrestrial planets (Mercury, Venus, Earth, and Mars), the outer planets are more massive (15–320 Earth masses), larger (4–11 Earth radii), and possess multiple satellite and ring systems. Their massive atmospheres are composed of molecular hydrogen and helium and compounds of carbon, nitrogen, and oxygen, primarily present in the form of saturated hydrides (CH_4, NH_3, and H_2O) at approximately the solar ratios of carbon, nitrogen, and oxygen. The composition of these atmospheres suggests that they are remnants of the primordial solar nebula gas cloud that condensed to form the solar system. These atmospheres also contain heavy hydrocarbons resulting from the photochemistry of CH_4. Titan, the largest satellite of Saturn, has an appreciable atmosphere (surface pressure of $\sim$1.5 atm) of N_2 (76–98%), CH_4, Ar, neon (Ne), CO_2, CO, and heavy hydrocarbons. Io, one of the Galilean satellites of Jupiter, was discovered to be volcanically very active, with an atmosphere of SO_2 and its photolysis products. Much of our information about the structure and composition of the atmospheres of Jupiter, Saturn, Titan, and Io resulted from the experiments aboard the NASA *Pioneer 10* and *11* and *Voyager 1* and *2* flybys (*Pioneer 10* was launched 3 March 1972 and encountered Jupiter on 4 December 1973; *Pioneer 11* was launched 6 April 1973 and encountered Jupiter on 3 Dec 1974 and Saturn on 1 September 1979; *Voyager 1* was launched 5 September 1977 and encountered Jupiter on 5 March 1979 and Saturn on 12 November 1980; *Voyager 2* was launched 20 August 1977 and encountered Jupiter on 9 July 1979 and Saturn on 26 August 1981). *Voyager 2* will continue its exploration of the outer planets with flybys of Uranus (January 1986) and Neptune (August 1989). In 1988, the NASA *Galileo* spacecraft, consisting of an orbiter and an entry probe, is scheduled to continue the exploration of the atmosphere of Jupiter and the Galilean satellites.

Comets

We leave the photochemistry/chemistry of planetary atmospheres and consider the photochemistry of comets in Chapter 9. Comets contain varying amounts of frozen volatiles that vaporize when heated by the Sun. As comets approach the Sun, the frozen nucleus slowly vaporizes, releasing gases that form the coma and tail. With the exception of the light gases (atomic hydrogen and helium), planetary atmospheres are gravitationally bound in contrast to cometary comae. Cometary atmospheres are generated by the evaporation of frozen gases in the nucleus, which escape almost instantaneously with supersonic speed. A comet is practically all atmosphere. The photochemistry/chemistry of planetary atmospheres is very nearly a steady-state problem dominated by reactions involving radical species formed via photochemical processes, whereas the photochemistry/chemistry of comets is strongly time dependent and dominated by ion–neutral molecule reactions. Many volatiles and their photochemical products have been detected in the coma and tail, including H_2O, OH, atomic oxygen, NH_3, various sulfur species, and other molecular ions. Studies of the composition and photochemistry of comets may provide new insights into our understanding of the early history of the solar system. Interest in cometary research is increasing with the return of

Halley's Comet in 1985–1986. Six different cometary probes are presently planned to investigate Halley's Comet and its environment. The International *Sun–Earth Explorer-3* (*ISEE-3*) satellite will fly through the tail of Halley's Comet after first passing through the tail of Comet P/Giacobini–Zinner in September 1985. Two Japanese and two U.S.S.R. space probes will explore Halley's Comet and its environment. But the best equipped spacecraft to Halley's Comet is the European Space Agency's *Giotto* mission.

In addition to advances in our understanding of atmospheric photochemistry resulting from concerns about the effects of anthropogenic activities on the atmosphere and climate, and from the exploration of the upper atmosphere and planetary atmospheres by spacecraft, considerable progress has been made in the area of numerical modeling of photochemical and chemical processes in atmospheres. The photochemical model has become a standard tool in the theoretical investigation of photochemical processes in atmospheres. These models calculate the vertical distribution of atmospheric species by the simultaneous solution of species-coupled continuity–transport equations, one coupled equation for each atmospheric species under consideration. The species-coupled continuity–transport equation and its application to atmospheric photochemistry are discussed in Chapter 1.

The volume includes a series of tables containing information on various atmospheric photochemical and chemical processes and information on the optical properties of the atmosphere that control the transfer of incoming solar radiation through the atmosphere, initiating photochemical processes. Appendix I lists the various photodissociation and photoionization reactions, products, rates, and energy thresholds for the reactions. Appendix II lists the various chemical reactions, products, and rates involving neutral atmospheric species. The reactions listed in these appendixes indicate where in the text the reaction is discussed, for easy cross-reference. The reactions, products, and rates for ionospheric processes are included in the tables and appendix in Chapter 5, which treats ionospheric photochemistry and chemistry.

I

Earth

1

The Photochemistry of the Early Atmosphere

JOEL S. LEVINE
Atmospheric Sciences Division
NASA Langley Research Center
Hampton, Virginia

I. Introduction

This chapter deals with the composition and photochemistry of the early atmosphere and the evolution of the atmosphere over geological time. The story began some 4.6 billion years ago with the formation of the Earth and its atmosphere. Over geological time, the composition of the atmosphere has changed significantly as a result of the outgassing of volatiles originally trapped in the interior of our planet, the gravitational escape of light gases, the

ISBN 0-12-444920-4

geochemical cycling of gases between the atmosphere, the oceans, and the surface, the emergence of life, and various atmospheric photochemical and chemical processes.

The study of the composition of the early atmosphere draws on many disciplines other than atmospheric chemistry. These disciplines include astronomy, biology, biochemistry, geology, geochemistry, meteorology, and oceanography. The subject is truly interdisciplinary in nature.

Atmospheric evolution is countinuing today on a time scale of decades, or less, rather than on millions or billions of years. Today the composition and photochemistry of the atmosphere, as well as the climate, are changing, due primarily to the input of gases from various anthropogenic activities, for example, fossil fuel burning, industrial and manufacturing activities, and certain agricultural practices. Present-day changes in atmospheric composition are of local, national, and international concern. These changes include the general deterioration of the quality of the air we breathe, enhanced levels of air pollution and acid precipitation, increasing levels of carbon dioxide (CO_2) and possible climate change, and the possible inadvertent depletion of ozone (O_3) in the upper atmosphere.

As we go further back in time, the geological record becomes less certain and eventually nonexistent. At the same time, widely varying interpretations of the scant geological record are possible. Studies of the composition and photochemistry of the early atmosphere cannot be attempted without theoretical model calculations. Therefore, it is very important to recognize the very speculative nature of these studies, which are very model and assumption dependent. We have restricted this chapter to theoretical studies of the composition and photochemistry of the early atmosphere based on photochemical models and have omitted the literature dealing with laboratory kinetic studies of the photochemistry of early atmospheric gas mixtures and studies of the early geological record, which are worthy of chapters of their own.

The origin and evolution of the atmosphere are intimately related to the origin and evolution of life on our planet. [See Table I for some of the milestones in the origin and evolution of the atmosphere and life (Cloud, 1983).] It is generally believed that the simple molecules in the early atmosphere energized by atmospheric lightning and solar ultraviolet (UV) radiation abiotically formed complex organic molecules via atmospheric reactions. These molecules then rained out of the atmosphere into ponds, shallow seas, and the oceans, where through aqueous solution chemical reactions, they formed organic molecules of greater complexity, and eventually life itself (Miller, 1953; Schwartz, 1981; Schlesinger and Miller, 1983a,b). Once life formed and photosynthetic organisms evolved, molecular oxygen (O_2) built up in the atmosphere. Accompanying the accumulation of O_2 in the

Table I

HISTORY OF THE EARTH, ITS ATMOSPHERE, AND LIFE[a]

Event	Years ago[b]
Formation of the Sun, the Earth and its atmosphere(?)	4.6 B
Oldest known sedimentary rocks	3.8 B
Origin of life(?)	3.8 B
Oldest stromatolites	3.5 B
Microbial fossils (proalgae?)	2.8 B
Atmospheric O_2 reaches 1%	2.0 B
Gunflint blue-green algae	2.0 B
Oldest eukaryotic cells	1.4 B
Atmospheric O_2 reaches 7%	670 M
First known metazoans	670 M
Atmospheric O_2 reaches 10%	550 M
Atmospheric O_3 shields surface from solar UV radiation	550 M
First hard-shelled animals	550 M
Atmospheric O_2 reaches 100%	400 M
Large fishes and first land plants	400 M

[a] Adapted from Cloud (1983).
[b] B ≡ Billion, M ≡ million.

atmosphere was the evolution of O_3, which is photochemically produced from O_2. Increasing levels of atmospheric O_3 began to shield the Earth's surface from lethal solar UV radiation. This shielding by O_3 eventually permitted life to leave the safety of the oceans and go ashore for the first time (Berkner and Marshall, 1965). As these examples illustrate, there was strong coupling between the evolution of the atmosphere and the evolution of life.

II. Structure of the Atmosphere

The Earth's atmosphere extends out to several thousand kilometers above the surface, where it eventually merges with interplanetary space. The total mass of the atmosphere (5.1×10^{21} g) is very small compared to the mass of the oceans (1.39×10^{24} g) and the mass of the Earth (5.98×10^{27} g) (Walker, 1977). The composition of the atmosphere at the Earth's surface is given in Table II. Electrically neutral molecules account for almost the entire mass of the atmosphere. Superimposed on the neutral atmosphere are distinct regions of electrically charged particles (electrons and positively charged atoms and molecules), however, resulting from the ionization of neutral gases by high-energy (X-ray and UV) solar radiation and cosmic rays.

The neutral atmosphere is subdivided into distinct regions defined by the temperature gradient within the region (Fig. 1) (Levine and Graedel, 1981).

Table II

COMPOSITION OF THE PRESENT ATMOSPHERE

	Surface concentration[a]	Source
Major and minor Gases		
Nitrogen (N_2)	78.08%	Volcanic, biogenic
Oxygen (O_2)	20.95%	Biogenic
Argon (Ar)	0.93%	Radiogenic
Water vapor (H_2O)	Variable, up to 4%	Volcanic, evaporation
Carbon dioxide (CO_2)	0.034%	Volcanic, biogenic, anthropogenic
Trace gases		
Oxygen species		
Ozone (O_3)	10–100 ppbv	Photochemical
Atomic oxygen (O) (ground state)	10^3 cm^{-3}	Photochemical
Atomic oxygen [$O(^1D)$] (excited state)	10^{-2} cm^{-3}	Photochemical
Hydrogen species		
Hydrogen (H_2)	0.5 ppmv	Photochemical, biogenic
Hydrogen peroxide (H_2O_2)	10^9 cm^{-3}	Photochemical
Hydroperoxyl radical (HO_2)	10^8 cm^{-3}	Photochemical
Hydroxyl radical (OH)	10^6 cm^{-3}	Photochemical
Atomic hydrogen (H)	1 cm^{-3}	Photochemical
Nitrogen species		
Nitrous oxide (N_2O)	330 ppbv	Biogenic, anthropogenic
Ammonia (NH_3)	0.1–1 ppbv	Biogenic, anthropogenic
Nitric acid (HNO_3)	50–1000 pptv	Photochemical
Hydrogen cyanide (HCN)	~200 pptv	Anthropogenic(?)
Nitrogen dioxide (NO_2)	10–300 pptv	Photochemical
Nitric oxide (NO)	5–100 pptv	Anthropogenic, biogenic, lightning, photochemical
Nitrogen trioxide (NO_3)	100 pptv	Photochemical
Peroxyacetylnitrate ($CH_3CO_3NO_2$)	50 pptv	Photochemical
Dinitrogen pentoxide (N_2O_5)	1 pptv	Photochemical
Pernitric acid (HO_2NO_2)	0.5 pptv	Photochemical
Nitrous acid (HNO_3)	0.1 pptv	Photochemical
Nitrogen aerosols		
Ammonium nitrate (NH_4NO_3)	~100 pptv	Photochemical
Ammonium chloride (NH_4Cl)	~0.1 pptv	Photochemical
Ammonium sulfate [$(NH_4)_2SO_4$]	~0.1 pptv(?)	Photochemical

Table II (*cont.*)

	Surface concentration[a]	Source
Carbon species		
Methane (CH_4)	1.7 ppmv	Biogenic, anthropogenic
Carbon monoxide (CO)	70–200 ppbv (N hemis.) 40–60 ppbv (S hemis.)	Anthropogenic, biogenic, photochemical
Formaldehyde (H_2CO)	0.1 ppbv	Photochemical
Methylhydroperoxide (CH_3OOH)	10^{11} cm^{-3}	Photochemical
Methylperoxyl radical (CH_3O_2)	10^{8} cm^{-3}	Photochemical
Methyl radical (CH_3)	10^{-1} cm^{-3}	Photochemical
Sulfur species		
Carbonyl sulfide (COS)	0.5 ppbv	Volcanic, anthropogenic
Dimethyl sulfide [$(CH_3)_2S$]	0.4 ppbv	Biogenic
Hydrogen sulfide (H_2S)	0.2 ppbv	Biogenic, anthropogenic
Sulfur dioxide (SO_2)	0.2 ppbv	Volcanic, anthropogenic, photochemical
Dimethyl disulfide [$(CH_3)_2S_2$]	100 pptv	Biogenic
Carbon disulfide (CS_2)	50 pptv	Volcanic, anthropogenic
Sulfuric acid (H_2SO_4)	20 pptv	Photochemical
Sulfurous acid (H_2SO_3)	20 pptv	Photochemical
Sulfoxyl radical (SO)	10^{3} cm^{-3}	Photochemical
Thiohydroxyl radical (HS)	1 cm^{-3}	Photochemical
Sulfur trioxide (SO_3)	10^{-2} cm^{-3}	Photochemical
Halogen species		
Hydrogen chloride (HCl)	1 ppbv	Sea salt, volcanic
Methyl chloride (CH_3Cl)	0.5 ppbv	Biogenic, anthropogenic
Methyl bromide (CH_3Br)	10 pptv	Biogenic, anthropogenic
Methyl iodide (CH_3I)	1 pptv	Biogenic, anthropogenic
Noble gases (chemically inert)		
Neon (Ne)	18 ppmv	Volcanic
Helium (He)	5.2 ppmv	Radiogenic
Krypton (Kr)	1 ppmv	Radiogenic
Xenon (Xe)	90 ppbv	Radiogenic

[a] Species concentrations are given percentage by volume, or in terms of surface mixing ratio, parts per million by volume (ppmv $\equiv 10^{-6}$), parts per billion by volume (ppbv $\equiv 10^{-9}$), parts per trillion by volume (pptv $\equiv 10^{-12}$), or in terms of surface number density (cm^{-3}). The species mixing ratio is defined as the ratio of the number density of the species to the total atmospheric number density (2.55×10^{19} molec cm^{-3}). There is some uncertainty in the concentrations of species at the ppbv level or less. The species concentrations given in molec cm^{-3} are generally based on photochemical calculations, and species concentrations in mixing ratios are generally based on measurements.

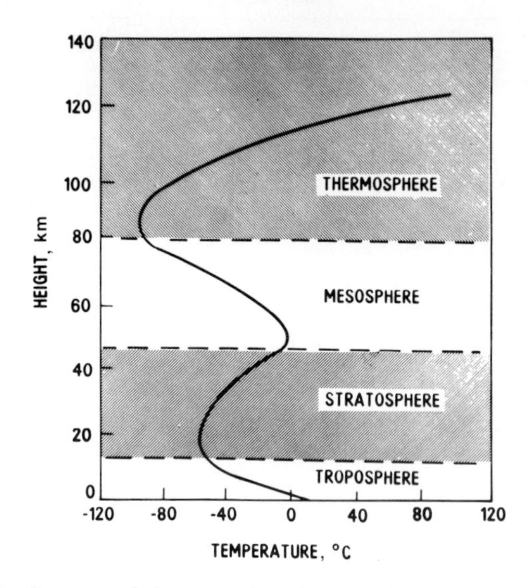

Fig. 1. Structure of the atmosphere. From Levine and Graedel (1981).

The variation of temperature with altitude throughout the atmosphere is shown in Fig. 1. Most of the solar extreme UV radiation ($\lambda < 100$ nm) is absorbed in the thermosphere above 100 km. Solar UV radiation between 100 and 200 nm is absorbed between about 50 and 80 km, and radiation between 200 and 300 nm is absorbed by O_3 within the stratosphere. Most of the solar visible and infrared (IR) radiation ($\lambda > 300$ nm) traverses the entire atmosphere and reaches the surface.

About 80–85% of the total mass of the atmosphere resides in the troposphere. The troposphere extends from the surface to ~10 km at high latitudes and to ~15 km in the tropics. The troposphere is in direct contact with the biosphere, and hence, regulates or modulates the transfer of gases and particulates into and out of the biosphere. Almost all of the water vapor (H_2O) in the atmosphere is found in the troposphere, where its distribution is controlled by the evaporation–precipitation cycle. The troposphere is a region of active vertical motion resulting from its negative temperature gradient. Upward vertical motions lead to the formation of clouds and precipitation. The vertical distribution of most gases in the atmosphere is controlled by the combined effects of photochemical/chemical processes and vertical transport by large-scale eddy motions. In theoretical one-dimensional photochemical calculations of species profiles, vertical eddy transport is parameterized using a vertical eddy diffusion coefficient profile. The climate of our planet is largely controlled by tropospheric phenomena, including the concentration and distribution of greenhouse absorbing gases, clouds, atmospheric aerosols, the global circulation of tropospheric air, and the exchange of energy between the troposphere, the oceans, and the surface. The photochemistry/chemistry of the troposphere is discussed in Chapter 2, and global climate in Chapter 4.

Most of the remaining mass of the atmosphere is found in the stratosphere, directly above the troposphere and extending to ~50 km. The stratosphere is a region of positive temperature gradient. The positive temperature gradient results from the absorption and subsequent heating of solar UV radiation (200–300 nm) by O_3 in the stratosphere. About 90% of the O_3 in the atmosphere resides in the stratosphere, with the remainder found in the troposphere. Anthropogenic perturbations to stratospheric O_3 may affect the levels of solar UV radiation reaching the surface and the climate of our planet. The photochemistry/chemistry of the stratosphere is discussed in Chapter 3.

Above the stratosphere are the regions of the mesosphere (50–85 km), thermosphere (85–500 km), and exosphere (an isothermal region beginning at ~500 km). The photochemistry of these regions is initiated by the absorption of extreme UV radiation ($\lambda < 200$ nm) and X rays from the Sun and by cosmic radiation. The chemical composition of the atmosphere is fairly uniform up to ~100 km (with the exceptions of H_2O and O_3), although the total atmospheric number density (molecules per cubic centimeter) decreases exponentially with altitude. Above ~100 km, the vertical distribution of various gases is controlled by photochemical/chemical processes and transport by molecular diffusion, as opposed to mixing by eddy diffusion in the lower atmosphere. The exosphere is the region where the molecular mean free path exceeds the Earth's radius and light gases, such as atomic hydrogen and helium, can gravitationally escape into space. The photochemistry/chemistry of the mesosphere, thermosphere, and exosphere is discussed in Chapter 5.

The regions of electrically charged particles superimposed on the neutral atmosphere are found in distinct layers, collectively called the ionosphere. The ionospheric regions are the D layer (below ~90 km), the E layer (90–120 km), the F-1 layer (a daytime feature centered at ~150 km), and the F-2 layer (200–400 km). The photochemistry/chemistry of the ionosphere is also described in Chapter 5.

III. Formation of the Earth

It is generally believed that our planetary system—the Sun, the Earth, the other planets, their satellites, comets, and meteors—condensed out of the solar nebula, an interstellar cloud of gas and dust, ~4.6 billion years ago. The chemical composition of the solar nebula most probably reflected the cosmic abundance of the elements (Table III). Volatiles, elements that were either

Table III

COSMIC ABUNDANCE OF THE ELEMENTS[a]

Element	Abundance[b]	Element	Abundance[b]
$_{1}H$	2.6×10^{10}	$_{44}Ru$	1.6
$_{2}He$	2.1×10^{9}	$_{45}Rh$	0.33
$_{3}Li$	45	$_{46}Pd$	1.5
$_{4}Be$	0.69	$_{47}Ag$	0.5
$_{5}B$	6.2	$_{48}Cd$	2.12
$_{6}C$	1.35×10^{7}	$_{49}In$	2.217
$_{7}N$	2.44×10^{6}	$_{50}Sn$	4.22
$_{8}O$	2.36×10^{7}	$_{51}Sb$	0.381
$_{9}F$	3630	$_{52}Te$	6.76
$_{10}Ne$	2.36×10^{6}	$_{53}I$	1.41
$_{11}Na$	6.32×10^{4}	$_{54}Xe$	7.10
$_{12}Mg$	1.050×10^{6}	$_{55}Cs$	0.367
$_{13}Al$	8.51×10^{4}	$_{56}Ba$	4.7
$_{14}Si$	1.00×10^{6}	$_{57}La$	0.36
$_{15}P$	1.27×10^{4}	$_{58}Ce$	1.17
$_{16}S$	5.06×10^{5}	$_{59}Pr$	0.17
$_{17}Cl$	1970	$_{60}Nd$	0.77
$_{18}Ar$	2.28×10^{5}	$_{62}Sm$	0.23
$_{19}K$	3240	$_{63}Eu$	0.091
$_{20}Ca$	7.36×10^{4}	$_{64}Gd$	0.34
$_{21}Sc$	33	$_{65}Tb$	0.052
$_{22}Ti$	2300	$_{66}Dy$	0.36
$_{23}V$	900	$_{67}Ho$	0.090
$_{24}Cr$	1.24×10^{4}	$_{68}Er$	0.22
$_{25}Mn$	8800	$_{69}Tm$	0.035
$_{26}Fe$	8.90×10^{5}	$_{70}Yb$	0.21
$_{27}Co$	2300	$_{71}Lu$	0.035
$_{28}Ni$	4.57×10^{4}	$_{72}Hf$	0.16
$_{29}Cu$	919	$_{73}Ta$	0.022
$_{30}Zn$	1500	$_{74}W$	0.16
$_{31}Ga$	45.5	$_{75}Re$	0.055
$_{32}Ge$	126	$_{76}Os$	0.71
$_{33}As$	7.2	$_{77}Ir$	0.43
$_{34}Se$	70.1	$_{78}Pt$	1.13
$_{35}Br$	20.6	$_{79}Au$	0.20
$_{36}Kr$	64.4	$_{80}Hg$	0.75
$_{37}Rb$	5.95	$_{81}Tl$	0.182
$_{38}Sr$	58.4	$_{82}Pb$	2.90
$_{39}Y$	4.6	$_{83}Bi$	0.164
$_{40}Zr$	30	$_{90}Th$	0.034
$_{41}Nb$	1.15	$_{92}U$	0.0234
$_{42}Mo$	2.52		

[a] From Cameron (1968). Copyright 1968 Pergamon Press.

[b] Abundance normalized to silicon (Si) = 1.00×10^{6}.

gaseous or that formed gaseous compounds at the relatively low temperature of the solar nebula, were the overwhelming constituents. The major volatile forming elements were hydrogen, by far the major constituent, followed by helium, oxygen, nitrogen, and carbon (see Table III). Less abundant in the solar nebula, but key elements in the planetary formation process, were the refractory elements, such as silicon, magnesium, iron, nickel, and aluminum, which formed solid compounds at the relatively low temperature of the solar nebula. Through the processes of coalescence and accretion of the refractory elements and their compounds, the terrestrial planets (Mercury, Venus, Earth, and Mars) formed. The terrestrial planets may have grown to their full mass in as little as 10 million years.

Theoretical calculations indicate that the rate of accretion, accompanied by collisional heating, was high enough to melt the Earth as it was forming. This heating resulted in the segregation or differentiation of the interior into a core composed of iron surrounded by an iron-free mantle of silicates. Volatiles incorporated in a late-accreting, low-temperature condensate may have formed as a veneer surrounding the newly formed Earth. This volatile-rich veneer resembled the chemical composition of carbonaceous chondritic meteorites, which contain relatively large amounts of H_2O and other volatiles. This scenario which has the iron migrating to the core, surrounded by an iron-free silicate mantle at the time of its formation, is the *inhomogeneous accretion model* (Turekian and Clark, 1969; Walker, 1976).

When the iron migrated to the core and Earth became geologically differentiated is of critical importance in determining the composition of the early atmosphere. The inhomogeneous accretion model has gained in acceptance at the expense of the *homogeneous accretion model.* The homogeneous accretion model predicted that the Earth formed initially as a cold and chemically homogeneous body, with free iron uniformly distributed throughout the planet's interior. According to this model, sometime after it formed, the Earth began to heat due to radiogenic decay. The heating caused the Earth to become molten, at which time the iron migrated to the core. In the homogeneous accretion model, the Earth was heated and the iron migrated to the core sometime after its formation (during the first billion years). In contrast, the inhomogeneous accretion model predicts that the Earth formed as a hot, differentiated body, with the iron already in the core.

According to both the inhomogeneous and homogeneous models for the formation of the planets, the atmospheres of the terrestrial planets (Earth, Venus, and Mars) formed as a result of volatile outgassing, the release of trapped volatiles from the solid planet. However, the absence (as predicted by the inhomogeneous model) or presence (as predicted by the homogeneous model) of iron in the mantle determined the oxidation state and composition of the outgassed volatiles (Holland, 1962), as discussed in the next section.

Some volatile outgassing may have also been associated with the impact heating during the accumulation of the late-accreting veneer, resulting in an instantaneous formation of the atmosphere, coincident with the final stages of the formation of the planets. It has also been suggested that a thick atmosphere of molecular hydrogen (H_2) and helium, the overwhelming constituents of the solar nebula, may have surrounded the early Earth. This primitive remnant atmosphere would have dissipated very quickly, however, if it ever existed at all, due to the rapid gravitational escape of hydrogen and helium. The collapse of the solar nebula and the formation of the Earth and its early atmosphere have been discussed in more detail by Turekian and Clark (1969), Walker (1976, 1977), and Canuto *et al.* (1983).

IV. Composition of the Prebiological Atmosphere

In this section, we shall consider the oxidation state and composition of the early prebiological atmosphere (the atmosphere prior to life). Both highly and mildly reducing gas mixtures for the early atmosphere will be considered. A highly reducing atmosphere is one containing more methane (CH_4) than CO_2, more ammonia (NH_3) than molecular nitrogen (N_2), and more H_2 than H_2O. For many years, it was assumed that the early atmosphere was strongly reducing (Hart, 1979). The more recent picture envisions the early atmosphere as a mildly reducing mixture of CO_2, N_2, and H_2O, with trace amounts of H_2. It is interesting to note that many of the molecules believed to be constituents of the early atmosphere (CO_2, H_2O, NH_3, and CH_4) absorb Earth-emitted IR radiation and may have compensated for the reduced total luminosity of the young Sun ($\sim 75\%$ of the present value) in determining the climate of the Earth (Hart, 1978; Kuhn and Atreya, 1979).

The changing ideas concerning the composition of the early atmosphere will be briefly discussed. The early ideas about the composition of the prebiological atmosphere were strongly influenced by several factors, including spectroscopic studies of the chemical composition of Jupiter, by laboratory experiments on the abiotic production of the complex organic molecules needed for life, and by early ideas of the formation and composition of the Earth.

Early spectroscopic studies indicated the presence of large amounts of H_2, CH_4, and NH_3 in the atmosphere of Jupiter (see Chapter 8). It is generally believed that unlike the atmospheres of the terrestrial planets (Earth, Venus, and Mars), which formed as a result of volatile outgassing, the atmospheres of Jupiter and the outer planets are captured remnants of the solar nebula that originally condensed to form the solar system. It was thought the atmosphere of Jupiter had not undergone significant evolution over geological time due to

Table IV
VOLATILES IN THE ATMOSPHERE, OCEANS, AND SEDIMENTARY ROCKS[a]

	H_2O (g)	CO_2 (g)	N_2 (g)
Atmosphere	1.7(19)[b]	2.45(18)	3.87(21)
Oceans	1.4(24)	1.38(20)	2.18(19)
Sedimentary shell	1.5(23)	2.42(23)	1.0(21)
	1.6(24)	2.4(23)	4.9(21)
"Excess" volatiles[c]	1.6(24)	2.3(23)	4.9(21)

[a] From Walker (1977).
[b] $1.7(19) \equiv 1.7 \times 10^{19}$.
[c] "Excess" volatiles are those volatiles not contributed by the weathering of igneous rocks to produce sedimentary rocks and, hence, represent the volatiles originally trapped in the interior and released via outgassing.

its great mass and low exospheric temperature, resulting in a low gravitational escape rate. Since Jupiter's atmosphere contains large amounts of H_2, CH_4, and NH_3 at the present, it was reasoned that the early Earth should have also contained large quantities of these gases. Early laboratory experiments on chemical evolution using Jupiter-like mixtures of CH_4, NH_3, and H_2 exposed to lightning and solar UV radiation resulted in the production of the complex organic molecules needed for the origin of life (Miller, 1953). Finally, if iron were uniformly distributed throughout the mantle, as predicted by the earlier homogeneous accretion model, the outgassed volatiles would have been strongly reducing compounds of CH_4, NH_3, and H_2 (Holland, 1962). The inhomogeneous accretion model, however, which is presently favored, predicts volatile outgassing of H_2O, CO_2, and N_2 if the iron were already out of the mantle and in the core prior to the start of outgassing (Holland, 1962).

There is very little question but that the Earth outgassed tremendous quantities of H_2O, CO_2, and N_2 over geological time (Table IV). The question is whether outgassing of CH_4, NH_3, and H_2 occurred for a short while prior to the longer period of extensive outgassing of H_2O, CO_2, and N_2. According to the inhomogeneous accretion model, the answer is *no*. The early Earth never outgassed significant amounts of CH_4, NH_3, and H_2, since the iron was already in to the core prior to the start of volatile outgassing. Furthermore, more recent chemical evolution experiments indicate that complex organic molecules needed for the origin of life can be synthesized in mildly reduced mixtures (see Section VI). In addition to these findings in geochemistry and biochemistry, theoretical calculations indicate that an early atmosphere of CH_4 and NH_3 would have been very unstable against photochemical/

Table V
AVERAGE COMPOSITION OF HAWAIIAN VOLCANIC GASES[a]

Gas	Vol %
H_2O	79.31
CO_2	11.61
SO_2	6.48
N_2	1.29
H_2	0.58
CO	0.37
S_2	0.24
Cl_2	0.05
Ar	0.04

[a] From Walker (1977).

chemical processes and rain out (in the case of NH_3), and hence, such an atmosphere would have been very short-lived, if it ever existed at all. (From a photochemical point of view, both H_2 and N_2 may be considered as chemically inert gases.) More about this in the next section.

For the reasons just outlined, we conclude that the early prebiological atmosphere resulted from volatile outgassing and consisted primarily of H_2O, CO_2, and N_2, with trace amounts of H_2. A somewhat related question concerns the time period over which volatile outgassing and the formation of the atmosphere took place. The two extreme scenarios are (1) that outgassing was very rapid and was completed in the Earth's very early history, or (2) that outgassing was gradual and continued for several billion years. Studies in geology, geochemistry, and paleoclimate modeling favor the first case (Walker, 1977).

We shall consider the photochemistry of early atmospheric mixtures of CH_4 and NH_3, and H_2O, CO_2, and N_2, in the following sections, but first point out that the latter mixture is not unlike that emitted by present-day volcanoes (Table V). It is important to note that no O_2, the second most abundant constituent of the atmosphere (21% by volume), is released via volcanic activity. Clearly, the composition of the present atmosphere (Table II) bears little resemblance to the composition of the early prebiological atmosphere. The bulk of the H_2O that outgassed from the interior condensed out of the atmosphere, forming the Earth's vast oceans. Only small amounts of H_2O remain in the atmosphere, with almost all of it confined to the troposphere. At the surface, the H_2O vapor concentration is variable, ranging from a fraction of a percent to a maximum of $\sim 4\%$ by volume. At the top of the troposphere, H_2O has a mixing ratio in the parts per million by volume (ppmv) range.* Most of the CO_2 that outgassed over the Earth's history formed sedimentary carbonate rocks [calcite, $CaCO_3$, and dolomite, $CaMg(CO_3)_2$] after dissolution in the ocean. The mixing ratio of CO_2 in the present atmosphere is ~340 ppmv. It has been estimated that the preindustrial (around the year 1860) level of atmospheric CO_2 was ~280 ppmv, with the increase in atmospheric CO_2 attributable to the burning of coal. For every CO_2 molecule presently in the atmosphere, there are almost 10^5 CO_2 molecules incorporated as carbonates in sedimentary rocks.

All of the carbon presently in sedimentary rocks outgassed from the interior and was at one time in the atmosphere in the form of CO_2. Hence, the early atmosphere may have contained orders of magnitude more CO_2 than it presently contains (Hart, 1978). Molecular nitrogen is basically chemically inert, and most of the outgassed nitrogen accumulated in the atmosphere over geological time to become the most abundant constituent (78% by volume). Argon (isotope 40), a chemically inert gas resulting from the radiogenic decay of potassium (isotope 40) in the crust, built up over geological time to become the third most abundant permanent constituent of the atmosphere (1% by volume), after N_2 and O_2. The present atmospheric compositions of Venus and Mars most probably reflect different fates of the outgassed volatiles, particularly H_2O and CO_2 (see Chapters 6 and 7). In the absence of liquid H_2O on these planets, the bulk of the outgassed CO_2 remained in the atmosphere, where it accumulated to become the overwhelming atmospheric constituent (~95% by volume in the atmospheres of both Venus and Mars).

V. Photochemistry of the Prebiological Paleoatmosphere

In this section we shall study the photochemistry of the prebiological paleoatmosphere and consider two different possible compositions: (1) the classical view of a strongly reducing paleoatmosphere of CH_4, NH_3, and H_2, and (2) the more recent view of a mildly reducing atmosphere of H_2O, CO_2, and N_2. As already noted, from a photochemical point of view, both H_2 and N_2 may be treated as chemically inert species in the lower atmosphere. Both

* The concentration of major atmospheric constituents is expressed in terms of percentage by volume. The concentration of minor or trace atmospheric constituents are usually given in terms of mixing ratio, a dimensionless quantity. The species mixing ratio is defined as the ratio of the number density (molecules per cubic centimeter) of the species in question to the total atmospheric number density (molecules per cubic centimeter). A mixing ratio of 10^{-2} is parts per hundred by volume (pphv), or simply percentage by volume, 10^{-3} is parts per thousand by volume (ppthv), 10^{-6} is parts per million by volume (ppmv), 10^{-9} is parts per billion by volume (ppbv), and 10^{-12} is parts per trillion by volume (pptv).

gases are stable against photochemical and chemical processes and exhibit a near-constant mixing ratio throughout the lower and middle atmosphere. Before we consider the photochemistry and chemistry unique to these two mixtures, we shall briefly discuss some basic ideas common to all of the photochemical and chemical processes to be discussed throughout this volume.

A. Photochemical and Chemical Processes

Atmospheric gases are transformed from one species to another by photochemical and chemical processes. A photochemical process is initiated by the absorption of solar photons that possess enough energy either to break the molecular bond, resulting in a photodissociation reaction, or cause the molecule or atom to loss one or more electrons, resulting in a photoionization process. In general, photoionization reactions require more energy and, hence, are more important in the upper atmosphere, where the incoming solar radiation is more energetic and intense. Photodissociation and photoionization processes can be represented by reactions (1) and (2), respectively,

$$\mathrm{AB} + h\nu \longrightarrow \mathrm{A} + \mathrm{B} \tag{1}$$

$$\mathrm{C} + h\nu \longrightarrow \mathrm{C}^{+} + e^{-} \tag{2}$$

where h is Planck's constant and ν the frequency of the incident solar radiation supplying the energy needed to either dissociate or ionize the atmospheric species in question. For each photochemical reaction we shall indicate the wavelength of solar energy needed to initiate the process. In reaction (1), molecule AB has undergone photodissociation, resulting in the production of two new species (either molecules, radicals, or atoms), A and B. In reaction (2), species C (either a molecule or atom) has undergone photoionization, forming a positively charged ion (C^{+}) and an electron (e^{-}). Photoionization reactions produce electrically charged atmospheric regions of ions and electrons in the upper atmospheres of planets and will not be discussed any further here (see Chapter 5 for a general discussion of ionospheres).

We can estimate the characteristic atmospheric lifetime t_{AB} (sec) of a species AB against photochemical destruction [reaction (1)] as

$$t_{\mathrm{AB}} = 1/J_{\mathrm{AB}}, \tag{3}$$

where J_{AB} (sec^{-1}) is the photolytic destruction rate of species AB and can be determined via

$$J_{\mathrm{AB}} = \int_{\lambda_1}^{\lambda_2} I_{\infty}(\lambda)\sigma_{\mathrm{AB}} e^{-\tau(i)\sec\theta}, \tag{4}$$

where $I_\infty(\lambda)$ is the flux of solar radiation incident at the top of the atmosphere (photons cm^{-2} sec^{-1}), σ_{AB} the absorption cross section (cm^2) for species AB, λ_1 and λ_2 represent the wavelength range (nm) over which species will undergo photolysis, θ is the solar zenith angle, and $\tau(i)$ the optical depth (dimensionless quantity) due to the various gases that absorb solar radiation in the spectral range λ_1 to λ_2. The optical depth $\tau(i)$ is represented as

$$\tau(i) = \sum N_i \sigma_i, \tag{5}$$

where N_i (molec cm^{-2}) is the column density through the atmosphere of the absorbing species above a given altitude z, and σ_i the absorption cross section

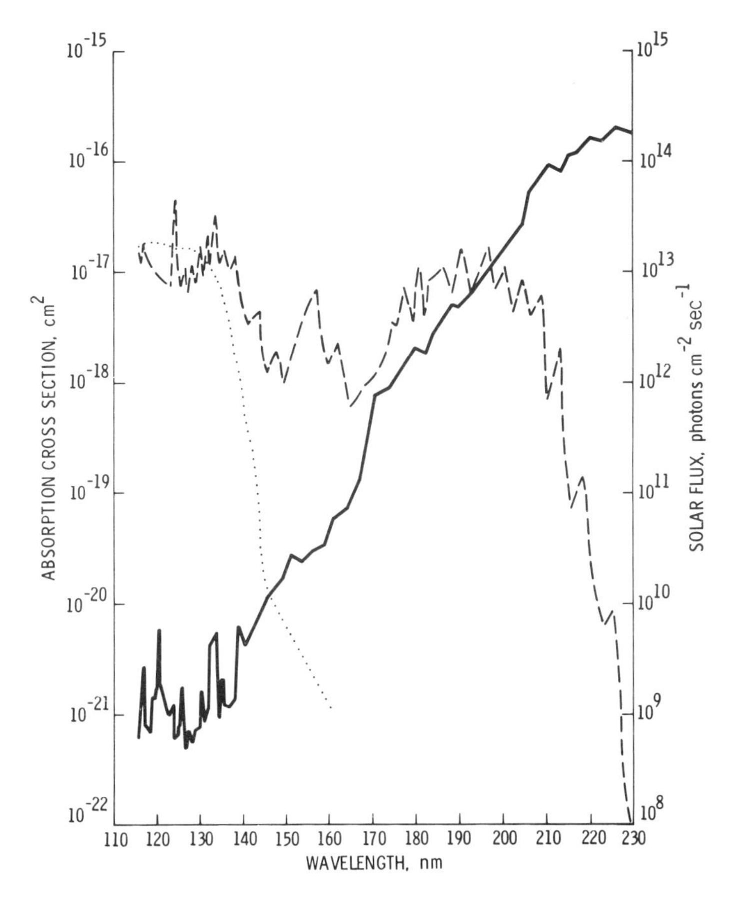

Fig. 2. Absorption cross sections corresponding to the photolysis of methane (···, CH_4) and ammonia (– – –, NH_3). Also shown is the solar flux (——, 110–230 nm) incident on top of the atmosphere. From Levine *et al.* (1982).

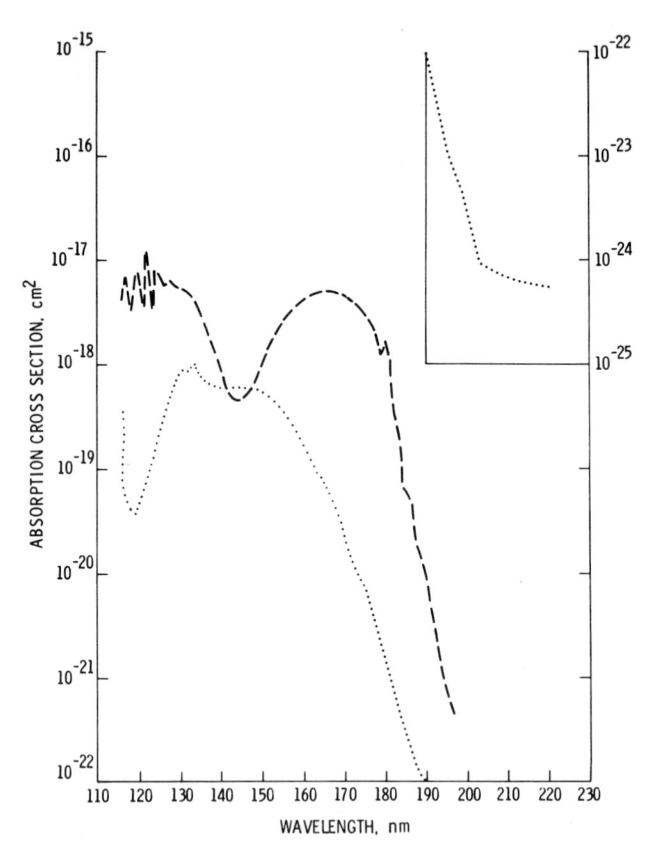

Fig. 3. Absorption cross sections corresponding to the photolysis of carbon dioxide (···, CO_2) and water vapor (---, H_2O). From Levine *et al.* (1982).

of the absorbing species. The absorption cross sections of the species of interest to the early atmosphere (i.e., CH_4 and NH_3, CO_2 and H_2O, and O_2 and O_3) are given in Figs. 2 to 4. These figures also show the variation of solar UV radiation with wavelength at the top of the atmosphere $[I_\infty(\lambda)]$. The column density N_i can be expressed as

$$N_i = \int_z^\infty n_i \, dz, \tag{6}$$

where n_i is the number density (molec cm^{-3}) of the absorbing species at an altitude z, and ∞ represents the top of the atmosphere.

A chemical process transforms one species to another species via the following reaction:

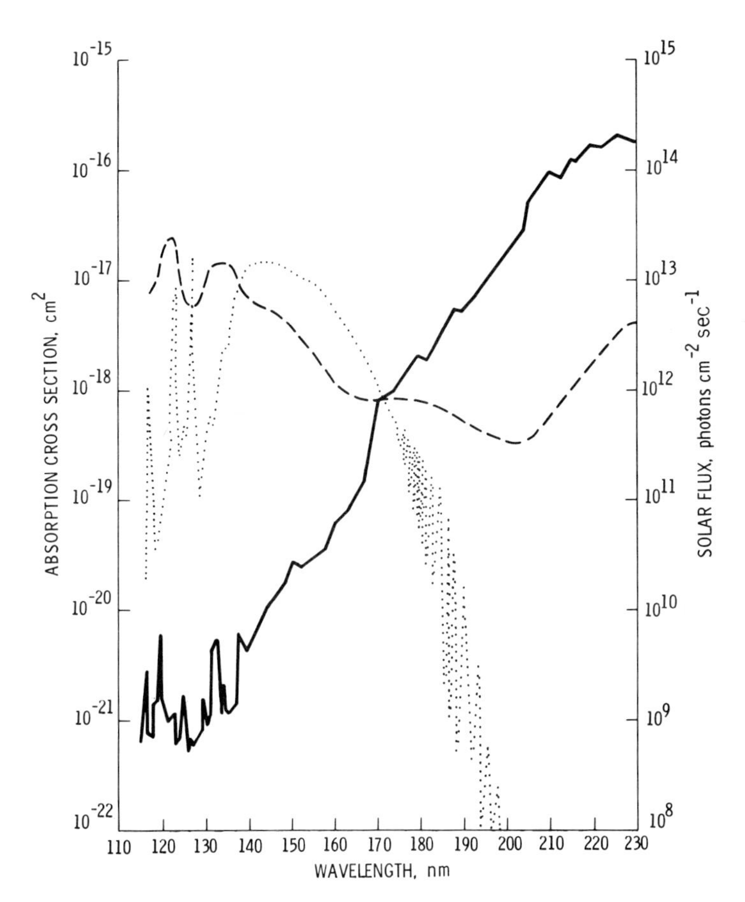

Fig. 4. Absorption cross sections corresponding to the ultraviolet photolysis of molecular oxygen (···, O_2) and ozone (---, O_3). Also shown is the solar flux (——, 110–230 nm) incident on the top of the atmosphere. From Levine *et al.* (1982).

$$DE + FG \xrightarrow{K} DF + EG \tag{7}$$

where K is the kinetic reaction rate with units of cm^3 molec^{-1} sec^{-1} for two-body kinetic reactions and units of cm^6 molec^{-1} sec^{-1} for three-body kinetic reactions. The characteristic atmospheric lifetime t_{DE} (sec) of species DE against destruction by reaction with species FG can be estimated as

$$t_{DE} = 1/K[FG], \tag{8}$$

where [FG] is the concentration (molec cm^{-3}) of reacting species FG, and K the kinetic reaction rate that controls the rate at which the reaction proceeds.

B. The Coupled Continuity–Transport Equation

Knowledge of the photochemical and chemical processes that lead to the transformation of one atmospheric species to another can be used theoretically to calculate the vertical distribution of each species throughout the atmosphere. This calculation is based on the solution of the coupled continuity–transport equation for each atmospheric species. Since one atmospheric constitutent may be a source of another constituent, the calculation involves the simultaneous solution of a series of coupled continuity–transport equations, one for each atmospheric constituent. The solution of the one-dimensional species-coupled continuity–transport equation leads to the vertical distribution of the species under study, by considering the various photochemical/chemical production and loss terms as well as the effects of vertical eddy diffusion on the distribution of the species.

The general form of the coupled continuity–transport equation for any atmospheric constituent, designated by the subscript i, may be represented by

$$\frac{\partial n_i}{\partial t} + \frac{\partial \phi_i}{\partial z} = P_i(n_j) - L_i(n_j)Mf_i, \tag{9}$$

where n_i is the number density of species i (molec cm^{-3}), t time (sec), ϕ_i the vertical eddy flux of the species i (molec cm^{-2} sec^{-1}), z the vertical distance or altitude (cm), $P_i(n_j)$ represents the volume production rate (molec cm^{-3} sec^{-1}) of the various photochemical/chemical processes that lead to the production of species i (which involves species other than species i, represented collectively as j), $L_i(n_j)$ the volume destruction rate (molec cm^{-3} sec^{-1}) of the various photochemical/chemical processes that lead to the destruction of species i (which again involves species other than species i), M is the total number density of the atmosphere (molec cm^{-3}) (which decreases exponentially with altitude z), and f_i the mixing ratio (a dimensionless quantity) of species i. The species mixing ratio is defined as

$$f_i = n_i/M. \tag{10}$$

The vertical flux of species i, ϕ_i, can be represented by

$$\phi_i = -K_z M \frac{\partial f_i}{\partial z}, \tag{11}$$

where K_z is the vertical eddy diffusion coefficient (cm^2 sec^{-1}), a prescribed parameter that is a measure of vertical mixing by atmospheric eddy motions. The same vertical eddy diffusion coefficient profile is used to calculate the vertical mixing of all species, since it is based on mean atmospheric vertical motions. Above ~100 km, however, the distribution of atmospheric gases is controlled by molecular diffusion rather than by eddy diffusion, and the molecular diffusion coefficient, which is unique for each gas, replaces the eddy diffusion coefficient. In this chapter, however, we are considering the photochemistry/chemistry of the early troposphere and stratosphere and, hence, only need to include the effects of vertical eddy diffusion.

For steady-state or time-dependent ($t = 0$) calculations, the general form of the species-coupled continuity–transport equation reduces to the following expression [with Eq. (11) substituted in Eq. (9)]:

$$\frac{\partial}{\partial z}\left(K_z M \frac{\partial f_i}{\partial z}\right) = -P_i(n_j) + L_i(n_j) M f_i. \tag{12}$$

For short-lived atmospheric species, whose vertical distributions are controlled by photochemical and chemical processes rather than by atmospheric transport, the transport portion of Eq. (12) (the left side of the equation) may be neglected. In this case, the species is said to be in *photochemical equilibrium*, and its vertical distribution may be determined by equating the production and destruction terms on the right side of Eq. (12). The mixing ratio of a species in photochemical equilibrium may be expressed as

$$f_i = P_i(n_j)/L_i(n_j)M. \tag{13}$$

The system of equations represented by Eqs. (12) and (13) (for short-lived species), one for each atmospheric species, is solved numerically by computer. The solution of Eq. (12) is subject to lower and upper boundary conditions. Usually these boundary conditions take the form of either the species mixing ratio or an upward or downward species flux at the lower and upper boundaries. Typical flux boundary conditions include the upward flux of biogenically produced or volcanically emitted species at the surface, the downward flux of species deposited at the surface, or the flux of gravitationally escaping light species at the top of the atmosphere. Initial "first guess" species profiles are needed to start the iterative numerical calculation. The final calculated solution is not sensitive to the first-guess approximation. Developments in numerical solutions to large systems of differential equations and rapid advances in high-speed computer technology have made numerical solution of species continuity–transport equations a routine technique in studying the photochemistry and chemistry of atmospheres. The results of this technique of numerically modeling the chemistry of atmospheres is discussed throughout this volume.

C. Ammonia and Methane

A consequence of the inhomogeneous accretion model is that the Earth never outgassed significant quantities of CH_4, NH_3, H_2, or other highly reduced species. Biogenic production of CH_4 and NH_3, the overwhelming source of these gases in the present atmosphere, was not operable at this early period in the history of the Earth. Other possible sources of CH_4 and NH_3 include the influx of cometary material (Oró, 1961; and see Chapter 9) and the localized catalytic fixation of atmospheric N_2 to form NH_3 in the presence of naturally occurring titanium dioxide in desert sands (Henderson-Sellers and Schwartz, 1980). Although there is no evidence to suggest that these various sources of CH_4 and NH_3 were adequate to produce an early atmosphere with significant levels of CH_4, NH_3, and other highly reduced hydrogen compounds, some still favor such a mixture for the composition of the prebiological early atmosphere (e.g., see Hart, 1979). But what about the presence of even trace quantities of CH_4 and NH_3 in the early atmosphere? Several studies have considered the photochemistry and chemistry of NH_3 and CH_4 in the early atmosphere, and from them it was concluded that even small amounts of these gases would have been destroyed very rapidly. Ammonia and CH_4 are destroyed by direct photolysis by solar radiation [reactions (14) and (15)] and by chemical reaction with the hydroxyl radical (OH) [reactions (16) and (17)]. The OH radical needed in reactions (16) and

$$NH_3 + h\nu \longrightarrow NH_2 + H \qquad \lambda \leqslant 230\ \text{nm} \tag{14}$$

$$CH_4 + h\nu \longrightarrow CH_2 + H_2 \qquad \lambda \leqslant 145\ \text{nm} \tag{15}$$

$$NH_3 + OH \longrightarrow NH_2 + H_2O \tag{16}$$

$$CH_4 + OH \longrightarrow CH_3 + H_2O \tag{17}$$

$$H_2O + h\nu \longrightarrow OH + H \qquad \lambda \leqslant 240\ \text{nm} \tag{18}$$

$$H_2O + O(^1D) \longrightarrow 2\ OH \tag{19}$$

(17) is formed from H_2O via direct photolysis [reaction (18)] and its reaction with excited atomic oxygen [$O(^1D)$] [reaction (19)].

Photochemical calculations by Kuhn and Atreya (1979) indicate that the lifetime of NH_3 against direct photolysis [reaction (14)] is very short. They found that the lifetime of NH_3 ranges from less than a day for a mixing ratio of 10^{-8} to $\sim 10^4$ days for a mixing ratio of 10^{-4}. They did not assess the lifetime of NH_3 against reaction with OH [reaction (16)] but concluded that it was probably too small to affect their results. In addition to the photochemical and chemical loss of NH_3 via reactions (14) and (16), NH_3, being very water soluble, is readily rained out of the atmosphere. The mean lifetime of NH_3 against loss due to rainout in the present atmosphere is ~ 10 days (Levine *et al.*, 1982).

Levine *et al.* (1982) investigated the lifetime of CH_4 against photochemical [reaction (15)] and chemical [reaction (17)] destruction. Their calculations indicated that photolysis of CH_4 is an efficient destruction mechanism in the upper atmosphere. Above 100 km, CH_4 has a lifetime against photolysis of only a few days. Their calculations indicated that other gases, however, particularly H_2O, shield CH_4 from photolytic destruction in the lower atmosphere. Photochemical calculations by Levine *et al.* (1982) showed that reaction with OH is the major destruction mechanism for CH_4 in the lower atmosphere, however, resulting in a lifetime for even trace quantities of CH_4 (mixing ratio of 10^{-6}) of ~50 years. Hence, these photochemical studies indicate that in the absence of a continuous source, even trace quantities of NH_3 and CH_4 were extremely short-lived in the early atmosphere, if they existed at all.

Theoretical calculations of the distributions of NH_3 and CH_4 in the prebiological troposphere for different prescribed upward surface fluxes $\phi(NH_3)$ and $\phi(CH_4)$, ranging from 10^9 to 10^{12} molec cm^{-2} sec^{-1}, are shown in Figs. 5 and 6, respectively. As already mentioned, sources to produce significant levels of NH_3 and CH_4 in the prebiological early atmosphere have not been identified. Arbitrary surface fluxes were assumed in these calculations to show the sensitivity of calculated profiles to surface fluxes. These calculations include the photochemical/chemical destruction mechanisms shown in Eqs. (14) to (17) as well as the subsequent chemistry involving the amine (NH_2) and methyl (CH_3) radicals, including their reactions with hydrogen, leading to the reformation of NH_3 and CH_4, respectively. The NH_3 profiles (Fig. 5) decrease rapidly with altitude, showing the combined effects of rain out and the photolysis of NH_3. The CH_4 profiles (Fig. 6) are constant throughout the troposphere, reflecting the mean atmospheric lifetime of CH_4, which is significantly greater than the vertical eddy mixing time (50 years versus several months).

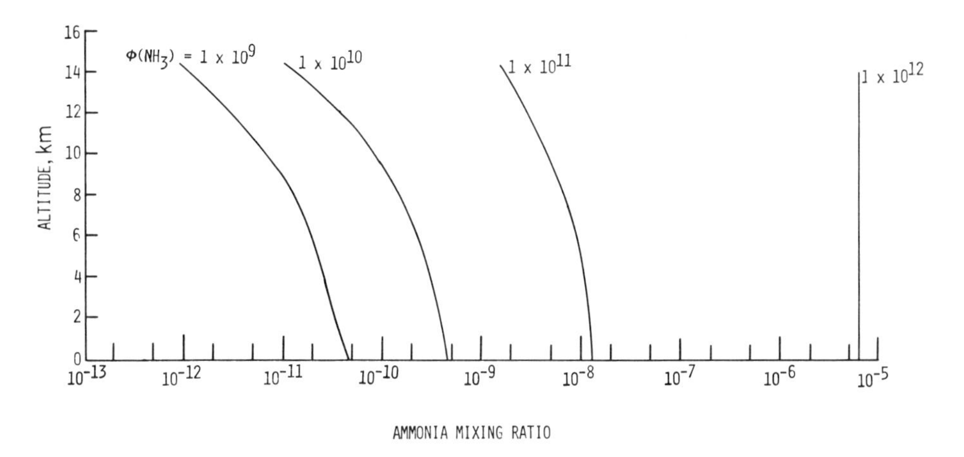

Fig. 5. Vertical distribution of NH_3 in the prebiological paleoatmosphere calculated for different surface fluxes of NH_3.

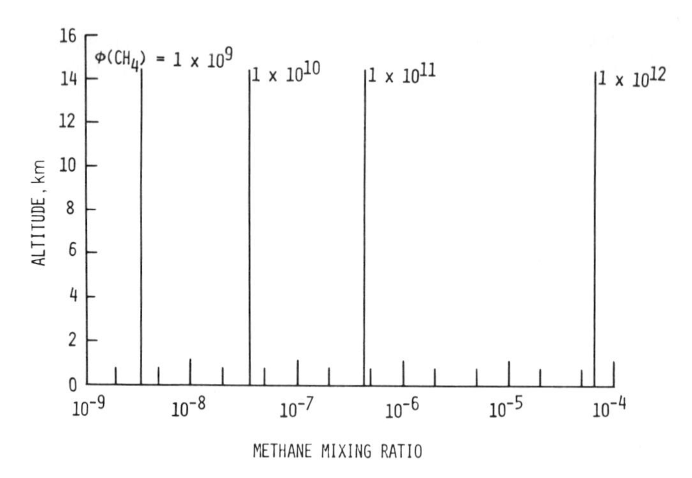

Fig. 6. Vertical distribution of CH_4 in the prebiological paleoatmosphere calculated for different surface fluxes of CH_4.

D. Carbon Dioxide and Water Vapor

We know that the Earth outgassed significant quantities of H_2O, CO_2, and N_2 (Table IV). Here we shall consider the photochemistry and chemistry of an early atmospheric mixture of H_2O, CO_2, and N_2. The photodissociation of H_2O and CO_2 resulted in a prebiotic source of O_2 in the early atmosphere, as well as in the production of formaldehyde (H_2CO) and hydrogen cyanide (HCN), both key species in chemical evolution and the origin of life.

E. Oxygen

It is generally agreed that the atmosphere was transformed from a mildly reducing mixture to the present oxidizing atmosphere by the evolution of O_2, produced as a by-product of photosynthetic activity. Clearly, photosynthetic activity, which may have started as early as 3 billion years ago, was the major source of atmospheric O_2. However, the photolysis of H_2O and CO_2 represented an early prebiotic source of O_2. The photochemical production of prebiotic O_2 was initiated by the photolysis of H_2O vapor [see Eq. (18)]. A small percentage ($< 10\%$) of the atomic hydrogen produced by the photolysis of H_2O will eventually escape into space. At present, the hydrogen escape flux is 10^8 hydrogen atoms cm^{-2} sec^{-1}. There is no evidence to suggest that this escape was less in the early history of the atmosphere; in fact, it may have been significantly greater. If it were not for the gravitational escape, eventually hydrogen would combine with OH, reforming H_2O. The OH produced in reactions (18) and (19) forms atomic oxygen (O) via the following reaction:

$$OH + OH \longrightarrow O + H_2O \tag{20}$$

The photolysis of CO_2 is another source of O:

$$CO_2 + h\nu \longrightarrow CO + O \qquad \lambda \leqslant 230 \text{ nm} \tag{21}$$

The atomic oxygen formed in reactions (20) and (21) combines in a three-body reaction to form O_2:

$$O + O + M \longrightarrow O_2 + M \tag{22}$$

where M is any molecule needed to absorb the excess energy and momentum of the reaction. Molecular oxygen is also formed via the reaction of O and OH:

$$O + OH \longrightarrow O_2 + H \tag{23}$$

Some of the prebiological O_2 photochemically produced via reactions (18) to (23) was lost via the the oxidation of minerals exposed to the atmosphere during the course of weathering. Prebiological O_2 was also lost by direct photolysis [reaction (24)] and by reaction with H_2 [reaction (25)], which led to the reformation of H_2O:

$$O_2 + h\nu \longrightarrow O + O \qquad \lambda \leqslant 242 \text{ nm} \tag{24}$$

$$2H_2 + O_2 \longrightarrow 2H_2O \qquad \text{(net cycle)} \tag{25}$$

Levels of O_2 in the prebiological atmosphere were very sensitive to atmospheric levels of H_2O, CO_2, and H_2 and to the flux of incoming solar radiation, which initiates the photolysis of H_2O, CO_2, and O_2. There is reason to believe that several of these parameters may have varied significantly over geological time (Hart, 1978; Canuto *et al.*, 1982, 1983). The H_2O distribution in the troposphere is controlled by its saturation vapor pressure, which is regulated by the tropospheric temperature profile. There is no reason to believe that the tropospheric temperature profile was significantly different in the early history of the Earth (Walker, 1977; Kasting and Walker, 1981). Even though very large amounts of H_2O may have outgassed in the early history of our planet, it seems unlikely that the early atmosphere contained significantly more H_2O than it presently contains, since any outgassed H_2O in excess of its saturation vapor pressure would have simply condensed into cloud droplets and then precipitated out of the atmosphere, forming the oceans. Carbon dioxide is a different story. Prior to the formation of carbonates, the early atmosphere may have contained significantly higher levels of CO_2, perhaps orders of magnitude more CO_2 than presently found in the atmosphere (Hart, 1978). To assess the importance of CO_2 on prebiotic levels of O_2, we have performed calculations for the preindustrial level (280 ppmv) and for 100 times

Table VI

SOLAR ULTRA VIOLET FLUX AS A FUNCTION OF AGE[a]

Age (years)	Ultraviolet enhancement
10^6	10^4
10^7	500
5×10^7	100
10^8	32
5×10^8	8
10^9	4
5×10^9	1

[a] From Canuto *et al.* (1982). Reprinted by permission from *Nature*. Copyright 1982 Macmillan Journals Ltd.

that value, a mixing ratio of 0.028 (2.8% by volume). While very little is known about levels of H_2 in the early atmosphere, previous studies suggested an H_2 mixing ratio of between 1.7×10^{-5} (Kasting and Walker, 1981) and 10^{-3} (Pinto *et al.*, 1980). (In our calculations, however, we used an H_2 mixing ratio as high as 10^{-1} to assess its effect on O_2 levels.) To assess the importance of H_2 on prebiotic levels of O_2, we performed calculations over a wide range of H_2 concentrations.

Measurements of young, Sun-like stars obtained with the *International Ultraviolet Explorer* (*IUE*) satellite suggest that during its "T Tauri" phase the young Sun may have emitted considerably more UV radiation than it presently emits (Canuto *et al.*, 1982, 1983; Zahnle and Walker, 1982), although the total visible luminosity of the young Sun was only 75% of its present value (Hart, 1978). The variation in UV radiation emitted by a Sun-like star over its history is summarized in Table VI (Canuto *et al.*, 1982). To assess the role of enhanced solar UV radiation on prebiotic levels of O_2, we performed calculations over a wide range of UV levels.

The results of these calculations are summarized in Figs. 7 to 10. Figure 7 shows the vertical distribution of prebiotic O_2 for three diffferent sets of parameters. Profile B represents the "standard" case: $CO_2 = 1$ (preindustrial level = 280 ppmv), $H_2 = 1.7 \times 10^{-5}$, and solar UV = 1 PAL (present atmospheric level). Profile A represents a combination of parameters to give a minimum O_2 profile ($CO_2 = 1$, H_2 mixing ratio = 10^{-1}, and UV = 1), while profile C represents a combination of parameters to give a maximum O_2 profile [$CO_2 = 100$, H_2 mixing ratio = 1×10^{-6}, and UV = 100 PAL). All three calculations exhibit a similar distribution with altitude. The O_2 maximum occurs above 40 km, the region of maximum photolysis of CO_2,

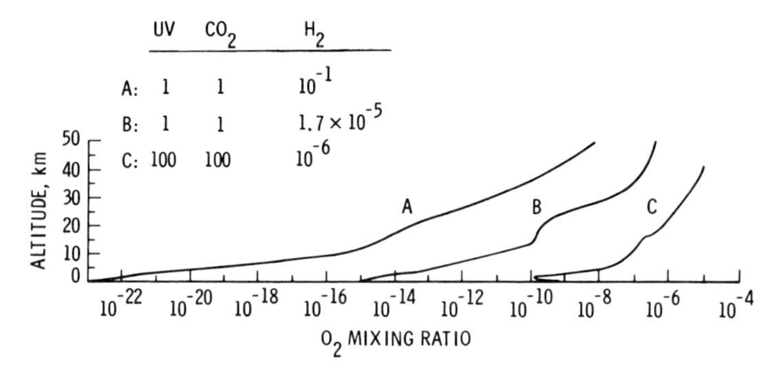

Fig. 7. Vertical distribution of O_2 in the prebiological paleoatmosphere for three different sets of assumed values for CO_2, H_2, and solar ultraviolet.

and the O_2 minimum occurs close to the ground, away from the region of maximum O_2 production.

The variation of the surface O_2 mixing ratio as a function of CO_2 and H_2 for different solar UV levels is shown in Figs. 8 to 10. Figure 8 gives the surface O_2 mixing ratio as a function of $CO_2(CO_2 = 1$, 10, and 100) and $H_2(H_2 = 10^{-6}$ to $10^{-1})$ for solar UV = 1. Figures 9 and 10 give similar calculations for solar UV = 10 and 100, respectively.

It is instructive to consider the individual reactions that lead to the photochemical production of atomic oxygen. The O production reactions include, in addition to reactions (20), (21), and (24), reaction (26):

$$HO_2 + H \longrightarrow O + H_2O \tag{26}$$

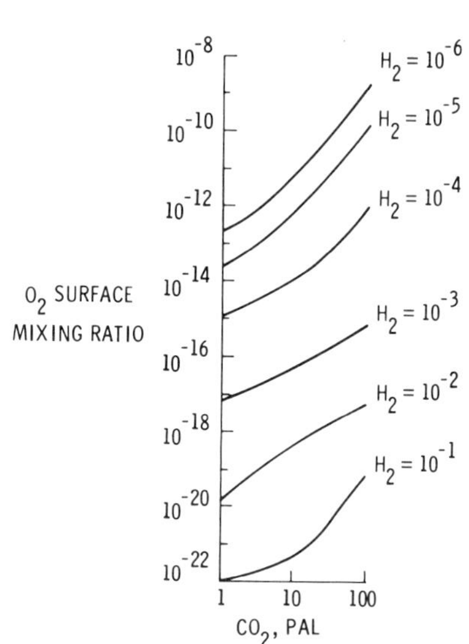

Fig. 8. Variation of surface O_2 mixing ratio as a function of CO_2 and H_2 for solar UV = 1.

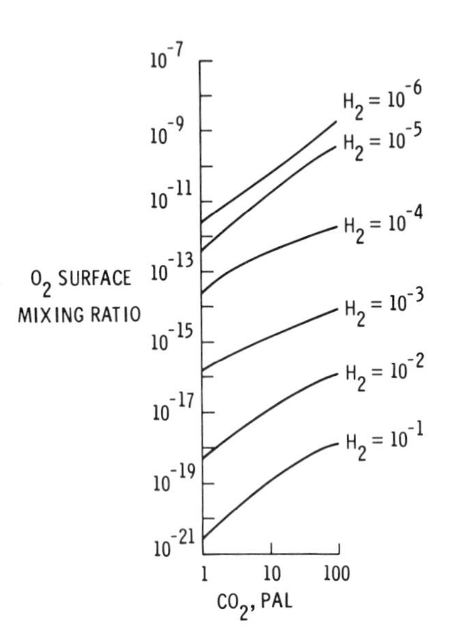

Fig. 9. Variation of surface O_2 mixing ratio as a function of CO_2 and H_2 for solar UV = 10.

Reaction (24), the photolysis of O_2, is not a *net* source of O, since O_2 was initially formed by the combination of O [reactions (22) and (23)]. Similarly, reaction (26) is not a real source of O, since the hydroperoxyl radical (HO_2) itself was formed by O_2 via the reaction

$$H + O_2 + M \longrightarrow HO_2 + M \tag{27}$$

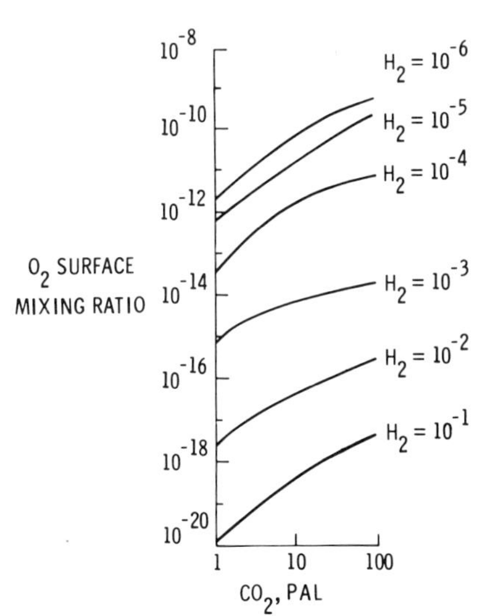

Fig. 10. Variation of surface O_2 mixing ratio as a function of CO_2 and H_2 for solar UV = 100.

The only reactions leading to the *net* production of O are reactions (20) and (21). The variation of the production rates of O via these four reactions [(20), (21), (24), and (26)] as a function of altitude are shown in Figs. 11 to 13. Also plotted is the total O production rate (the sum of the individual reactions leading to O production) as a function of altitude. The photolysis rate for H_2O [reaction (18)], while not a direct source of O, is also shown for comparison with the CO_2 photolysis rate. These three sets of calculations were performed for the following conditions:

Fig. 11: $CO_2 = 280$ ppmv, $H_2 = 17$ ppmv, solar UV = 1
Fig. 12: $CO_2 = 28000$ ppmv or 2.8%, $H_2 = 17$ ppmv, solar UV = 1
Fig. 13: $CO_2 = 28000$ ppmv, $H_2 = 17$ ppmv, solar UV = 100

For all of these calculations, we see that the photolysis of CO_2 is the dominant source of O. Furthermore, the photolysis rate of CO_2 exceeds that of H_2O at all altitudes, with the exception of 0 to 10 km in Fig. 11.

These calculations indicate that the photolysis of CO_2 and H_2O led to the photochemical production of trace amounts of O_2 in the prebiological atmosphere. These levels of O_2 may have affected the chemical evolution process in the prebiological paleoatmosphere. However, it must be

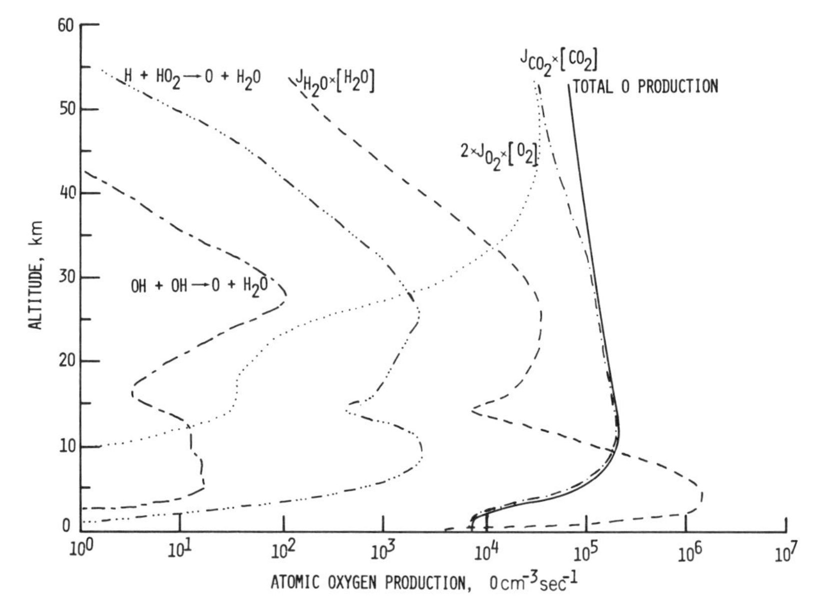

Fig. 11. Vertical distribution of the production rate of atomic oxygen (O) for $CO_2 =$ 280 ppmv, $H_2 = 17$ ppmv, and solar UV = 1. Also shown is the photolysis rate of H_2O.

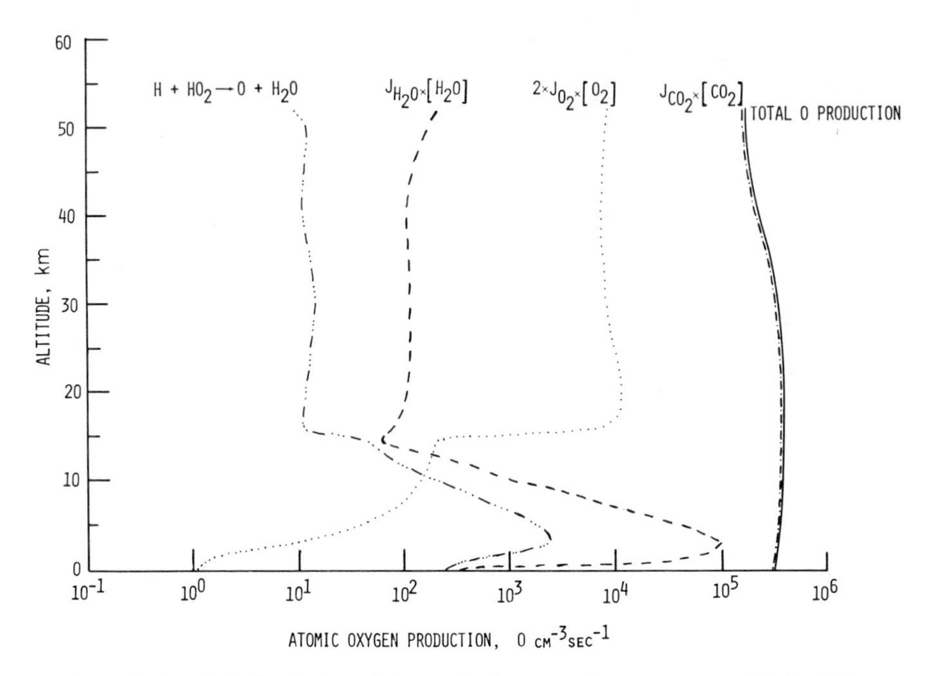

Fig. 12. Vertical distribution of the production rate of atomic oxygen (O) for CO_2 = 28,000 ppmv, H_2 = 17 ppmv, and solar UV = 1. Also shown is the photolysis rate of H_2O.

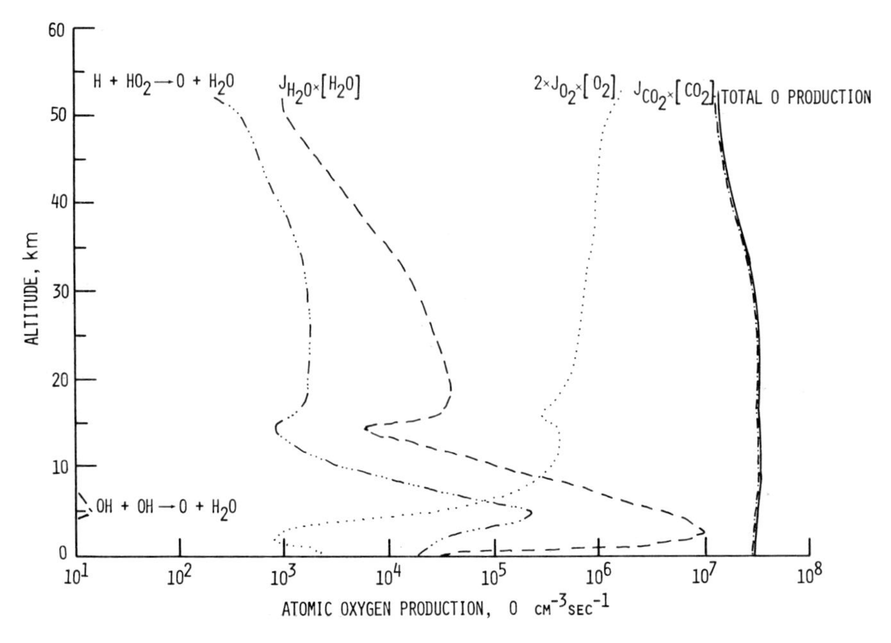

Fig. 13. Vertical distribution of the production rate of atomic oxygen (O) for CO_2 = 28,000 ppmv, H_2 = 17 ppmv, and solar UV = 100. Also shown is the photolysis rate of H_2O.

emphasized that it wasn't until the origin and evolution of photosynthetic organisms that O_2 accumulated in the atmosphere to become the second most abundant constituent. (See Table I for a possible timetable for the evolution of atmospheric O_2 resulting from photosynthetic activity.)

VI. Photochemistry and Chemical Evolution

Chemical evolution is the abiotic synthesis of organic molecules of increasing complexity, leading ultimately to the appearance of life. The chemical evolution process can be divided into two distinct phases: (1) the abiotic production of key precursor molecules, such as H_2CO and HCN, via atmospheric gas-phase reactions, and (2) the aqueous solution reactions of these precursor molecules in the early oceans, leading to the synthesis of more complex organic molecules, for example, amino acids, purines, pyrimidines, and carbohydrates (the components of nucleic acids and proteins), eventually leading to living systems (Schwartz, 1981). The classic laboratory chemical evolution experiments of Miller and Urey (Miller, 1953) clearly demonstrated that organic molecules could be abiotically synthesized in highly reduced mixtures of CH_4, NH_3, and H_2 exposed to an electric discharge. More recent laboratory studies indicate that organic molecular synthesis can also occur in more mildly reduced mixtures of N_2, CO_2, CO, H_2O, and H_2, with and without CH_4 and NH_3 (Schlesinger and Miller, 1983a,b). Two of the key precursor molecules required for the synthesis of more complex organic molecules in the early ocean were H_2CO and HCN (Schwartz, 1981). In solution, H_2CO leads to the formation of carbohydrates, and HCN forms amino acids, purines, and pyrimidines (Schwartz, 1981).

Photochemical calculations, as well as laboratory experiments, indicate that H_2CO is readily formed from CH_4. But what about the production of H_2CO in an atmosphere of N_2, CO_2, and H_2O, free of CH_4? A chemical mechanism for the photochemical production of H_2CO, initiated by the photolysis of CO_2 and H_2O has been proposed by Pinto *et al.* (1980). The photochemical scheme for the production of H_2CO is initiated by reactions (18) and (21),

$$H_2O + h\nu \longrightarrow OH + H \qquad \lambda \leq 240 \text{ nm} \tag{18}$$

$$CO_2 + h\nu \longrightarrow CO + O \qquad \lambda \leq 230 \text{ nm} \tag{21}$$

leading to the production of H and CO, which results in the formation of the formyl radical (CHO):

$$H + CO + M \longrightarrow CHO + M \tag{28}$$

Table VII

CALCULATED SURFACE CONCENTRATION AND RAINOUT RATE OF FORMALDEHYDE[a]

	UV = 1		UV = T Tauri	
H_2	$CO_2 = 1$	$CO_2 = 100$	$CO_2 = 1$	$CO_2 = 100$
H_2CO surface concentration (molec cm^{-3})				
17 ppmv	1.62(6)[b]	2.65(8)	3.38(4)	7.17(7)
10^{-3}	4.14(8)	4.12(9)	3.79(8)	2.43(10)
H_2CO rainout rate (molec cm^{-2} sec^{-1})				
17 ppmv	5.55(5)	7.34(7)	5.98(3)	5.45(7)
10^{-3}	8.6(8)	2.02(9)	3.7(8)	1.07(10)

[a] From Canuto *et al.* (1983). Reprinted by permission from *Nature*. Copyright 1983 Macmillan Journals Ltd.

[b] $1.62(6) \equiv 1.62 \times 10^6$.

Two formyl radicals combine to produce H_2CO:

$$CHO + CHO \longrightarrow H_2CO + CO \tag{29}$$

Formaldehyde is water soluble, and ~1% of the total atmospheric production of H_2CO may have rained out of the early atmosphere (Pinto *et al.*, 1980). The rain transported H_2CO to the oceans, where it underwent polymerization reactions leading to more complex organics. The important conclusion of this study is that significant quantities of H_2CO produced photochemically in the prebiological paleoatmosphere of CO_2 and H_2O, devoid of CH_4, could have been delivered to the early oceans.

The sensitivity of the photochemical production of H_2CO via reactions (18), (21), (28), and (29) to varying atmospheric levels of CO_2, H_2, and solar UV radiation was assessed by Canuto *et al.* (1983). These calculations are summarized in Table VII, which gives the calculated surface concentration (molec cm^{-3}) of H_2CO for values of CO_2 of 1 and 100 times the preindustrial atmospheric level (280 ppmv), for H_2 mixing ratios of 1.7×10^{-5} and 1×10^{-3}, and for solar UV = 1 and T Tauri emission spectra. The table also gives the H_2CO rain out rate (molec cm^{-2} sec^{-1}) to the early oceans for these levels of CO_2, H_2, and solar UV for the present-day rain out coefficient of 10^{-6} sec^{-1}, which corresponds to 11.6 days. The vertical distribution of H_2CO in the prebiological troposphere for the present solar UV flux and CO_2 = 1, 10, and 100 times the preindustrial level for H_2 = 17 ppmv and 10^{-3} are given in Figs. 14 and 15, respectively.

Laboratory studies indicate that once the H_2CO in the oceans reached a concentration of 10^{-3} *M*, aqueous polymerization reactions leading to the synthesis of more complex organics could have begun (Pinto *et al.*, 1980). The

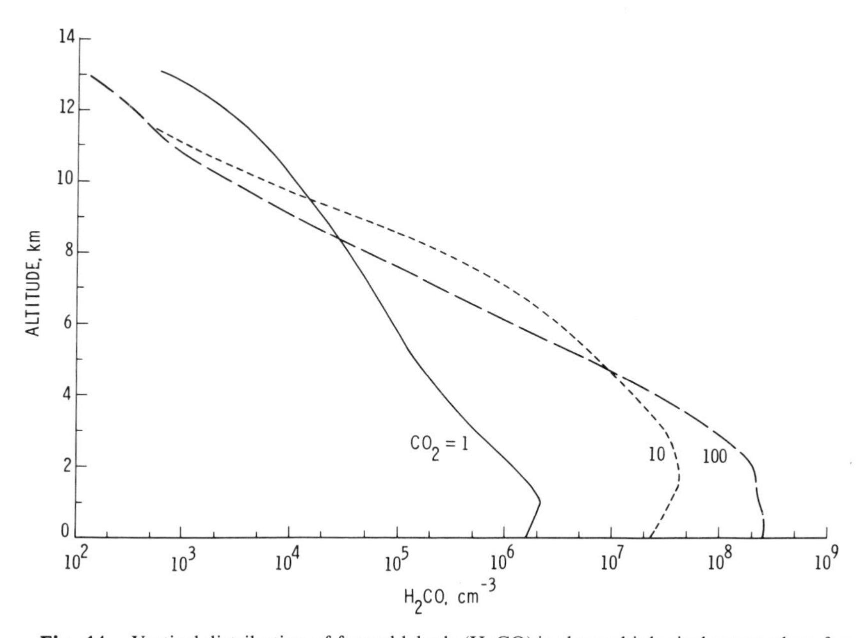

Fig. 14. Vertical distribution of formaldehyde (H_2CO) in the prebiological troposphere for $H_2 = 17$ ppmv, solar UV = 1, and CO_2 = 1, 10, and 100 times the preindustrial level (280 ppmv).

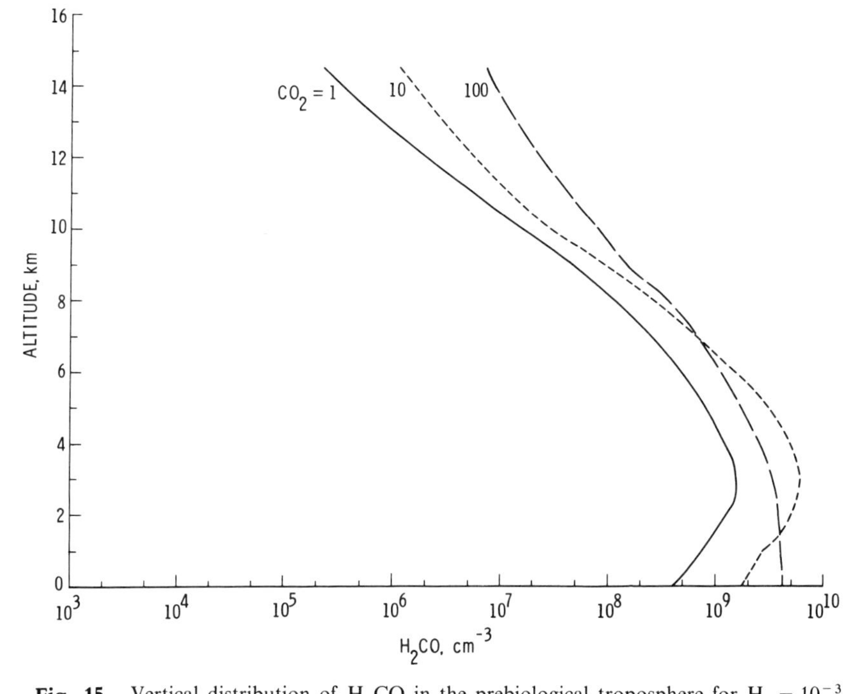

Fig. 15. Vertical distribution of H_2CO in the prebiological troposphere for $H_2 = 10^{-3}$, solar UV = 1, and CO_2 = 1, 10, and 100 times the preindustrial level (280 ppmv).

photochemical calculations of Pinto *et al.* (1980) and Canuto *et al.* (1983) indicate that H_2CO is photochemically produced in sufficient quantity to fill the oceans, at their present volume, to that concentration in a period of 10 million years or less!

It appears to be very difficult to produce HCN by atmospheric photochemical and chemical processes with solar radiation as the energy source. For example, producing nitrogen atoms needed for the formation of HCN from N_2 is very difficult to do via photochemical and chemical processes, since the N_2 molecular band is very stable and hard to break. Solar radiation less than 100 nm, which is only available in the upper regions of the atmosphere, can provide the needed dissociation energy. Hence, other sources of energy available in the lower atmosphere and/or other production mechanisms are required for the formation of HCN. The production of HCN in the early prebiological atmosphere by atmospheric lightning has been investigated theoretically by Chameides and Walker (1981), and experimentally by Schlesinger and Miller (1983b). These studies indicate that HCN is indeed produced in mixtures of N_2 with trace amounts of CO_2, CO, H_2O, and H_2. Hence, it appears that both key species involved in the chemical evolution process, H_2CO and HCN, can be produced in mildly reducing atmospheres of N_2, CO_2, CO, H_2O, and H_2.

VII. Photosynthesis, Oxygen, and Ozone

The origin and evolution of photosynthetic organisms and the accompanying production of O_2, as a by-product of the photosynthesis process, transformed the atmosphere from mildly reducing to strongly oxidizing. In photosynthesis, atmospheric H_2O and CO_2, in the presence of sunlight ($h\nu$) and chlorophyll, are biochemically transformed to a carbohydrate $[C_m(H_2O)_n]$, used by the organism for food, and O_2, which is released to the atmosphere:

$$n\,H_2O + m\,CO_2 + h\nu \xrightarrow{\text{chlorophyll}} C_m(H_2O)_n + m\,O_2 \tag{30}$$

As a result of photosynthetic production, O_2 became a major constituent of the atmosphere. A possible timetable for the evolution of O_2 in the atmosphere is given in Table I (Cloud, 1983). Accompanying and directly controlled by the buildup of O_2 was the evolution of O_3, which is formed photochemically from O_2. The evolution of O_3 in the atmosphere resulted in the shielding of the Earth's surface from lethal solar UV radiation between 200 and 300 nm. It has been suggested that the development of the O_3 layer and the accompanying shielding of the Earth's surface from lethal solar UV radiation permitted early life to leave the safety of the oceans and go ashore for the first time (Berkner and Marshall, 1965).

As already noted, the bulk of O_3 (~90%) in the present atmosphere is found in the stratosphere, with only ~10% in the troposphere. The photochemical production of O_3 in the early atmosphere was initiated by the photolysis of O_2 [reaction (24)], followed by the three-body recombination of O, O_2, and M [reaction (31)]. There are a number of photochemical and chemical

$$O + O_2 + M \longrightarrow O_3 + M \qquad (31)$$

processes that lead to the destruction of O_3, including

$$O_3 + h\nu \longrightarrow O + O_2 \qquad \lambda \leq 1100\ \text{nm} \qquad (32)$$

$$O_3 + O \longrightarrow 2\,O_2 \qquad (33)$$

In addition, O_3 is chemically destroyed through a series of catalytic cycles involving the oxides of hydrogen (OH and HO_2), nitrogen [nitric oxide (NO) and nitrogen dioxide (NO_2)], and chlorine [atomic chlorine and chlorine oxide (ClO)]. The oxides of hydrogen were produced photochemically and chemically from H_2O. Sources of nitrogen oxides in the early biological atmosphere included lightning, biogenic activity, and the oxidation of nitrous oxide (N_2O) by excited oxygen atoms. Volcanic emissions and sea salt spray were sources of chlorine in the early atmosphere. These catalytic cycles leading to the chemical destruction of O_3 are summarized here:

$$O_3 + OH \longrightarrow HO_2 + O_2 \qquad (34)$$

$$\underline{O_3 + HO_2 \longrightarrow OH + 2\,O_2} \qquad (35)$$

$$2\,O_3 \longrightarrow 3\,O_2 \qquad (36)$$

$$O_3 + NO \longrightarrow NO_2 + O_2 \qquad (37)$$

$$\underline{O + NO_2 \longrightarrow NO + O_2} \qquad (38)$$

$$O_3 + O \longrightarrow 2\,O_2 \qquad (39)$$

$$O_3 + Cl \qquad ClO + O_2 \qquad (40)$$

$$\underline{O + ClO \longrightarrow Cl + O_2} \qquad (41)$$

$$O_3 + O \longrightarrow 2\,O_2 \qquad (42)$$

We have investigated the origin and evolution of atmospheric O_3 as a function of the buildup of O_2 by considering reactions (24) and (31)–(42). The results of these calculations are presented in Fig. 16, where the vertical profiles of O_3 with and without the inclusion of chlorine-species chemistry are given in terms of present atmospheric level of O_2, ranging from 10^{-4} PAL to the present atmospheric level (1 PAL). The associated total atmospheric burden or column density of O_3 corresponding to the five profiles in Fig. 16 is given in Table VIII (Levine, 1982). It has been suggested that biological shielding of the Earth's surface was achieved when the O_3 column density reached $\sim 6 \times 10^{18}$ O_3 molec cm^{-2} (Berkner and Marshall, 1965), which is

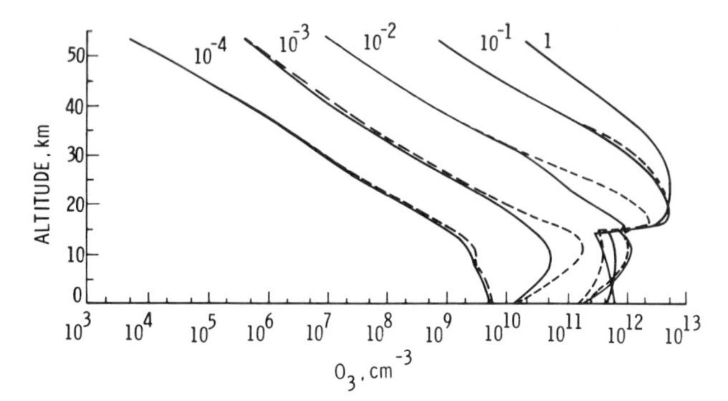

Fig. 16. Vertical distribution of ozone (O_3), with (——) and without (---) the inclusion of chlorine-species chemistry, as a function of O_2 level [given in terms of present atmospheric level (PAL = 1) of O_2]. From Levine (1982).

approximately half of the total O_3 burden in the present atmosphere. According to the calculations presented in Table VIII this burden of O_3 was reached when O_2 reached 10^{-1} PAL.

Once O_3 reached sufficient levels to absorb solar UV radiation, and hence, protect the Earth's surface from this lethal radiation, life "exploded" both in numbers and diversification (Berkner and Marshall, 1965). Through various metabolic and biochemical processes, different microorganisms produced a variety of trace, but environmentally important, atmospheric constituents,

Table VIII

EVOLUTION OF OZONE AS A FUNCTION OF INCREASING OXYGEN LEVELS[a]

O_2 Level (PAL)	O_3 Column density (cm^{-2})	Height of O_3 peak (km)	O_3 Density at peak (cm^{-3})
Without chlorine-species chemistry			
1	9.93(18)[b]	20.5	5.53(12)
10^{-1}	6.07(18)	19	4.57(12)
10^{-2}	2.47(18)	16	2.48(12)
10^{-3}	1.88(17)	11.5	1.92(11)
10^{-4}	5.58(15)	0	5.63(09)
With chlorine-species chemistry			
1	9.70(18)	20.5	5.40(12)
10^{-1}	5.94(18)	19	4.62(12)
10^{-2}	1.59(18)	10	1.16(12)
10^{-3}	6.98(16)	9	5.72(10)
10^{-4}	5.18(15)	0	5.42(09)

[a] From Levine (1982).
[b] $9.93(18) \equiv 9.93 \times 10^{18}$.

including CH_4, NH_3, N_2O, NO, hydrogen sulfide (H_2S), dimethyl sulfide [$(CH_3)_2S$], dimethyl disulfide [$(CH_3)_2S_2$], methyl chloride (CH_3Cl), methyl bromide (CH_3Br), and methyl iodide (CH_3I). Other microorganisms metabolically recycled N_2 and CO_2 between the atmosphere and biosphere. The atmosphere became a complex mixture of these trace gases, with equally complex reactions controlling its photochemistry.

VIII. Directions for Future Research

To a large extent, the study of the photochemistry of the early atmosphere is similar to studies of the photochemistry of other planetary atmospheres, prior to planetary exploration. The hard facts are uncertain or nonexistent, and there is considerable speculation. Most if not all studies are strongly model and assumption dependent. Future discovery and continued analysis of the early geological record will provide important information. The following list gives some first-order questions that may be amenable to future study with existing photochemical models:

1. How, when, and over what time period did the early atmosphere lose the vast amounts of H_2O and CO_2 released from the interior?
2. What were the atmospheric levels of H_2O and CO_2 in the early history of the Earth?
3. How did cloud cover and precipitation vary over geological time, and how did they affect photochemical processes?
4. How did the Earth's climate respond to changes in atmospheric composition, variations in solar radiation, and changes in volcanic emissions?
5. What was the precise timetable for the growth of O_2 in the atmosphere?
6. How did biogenic production of various trace gases vary over geological time and with increasing oxygen levels, and what were their effects on the photochemistry of the atmosphere?

The answers to these and many more related questions will provide a better understanding of the origin and evolution of the atmosphere, and of our cosmic roots.

References

Berkner, L. V., and Marshall, L. C. (1965). On the origin and rise of oxygen concentration in the Earth's atmosphere. *J. Atmos. Sci.* **22,** 225–261.

Cameron, A. G. W. (1968). A new table of abundances of the elements in the solar system. *In* "Origin and Distribution of the Elements" (L. H. Ahrens, ed.), pp. 125–143. Pergamon, New York.

Canuto, V. M., Levine, J. S., Augustsson, T. R., and Imhoff, C. L. (1982). UV radiation from the young Sun and oxygen and ozone levels in the prebiological paleoatmosphere. *Nature* (*London*) **305,** 281–286.

Canuto, V. M., Levine, J. S., Augustsson, T. R., Imhoff, C. L., and Giampapa, M. S. (1983). The young Sun and the atmosphere and photochemistry of the early Earth. *Nature* (*London*) **305,** 281–286.

Chameides, W. L., and Walker, J. C. G. (1981). Rates of fixation by lightning of carbon and nitrogen in possible primitive atmospheres. *Origins Life* **11,** 291–302.

Cloud, P. (1983). The biosphere. *Sci. Am.* **249,** 176–189.

Hart, M. H. (1978). The evolution of the atmosphere of the Earth. *Icarus* **33,** 23–39.

Hart, M. H. (1979). Was the prebiotic atmosphere of the Earth heavily reducing? *Origins Life* **9,** 261–266.

Henderson-Sellers, A., and Schwartz, A. W. (1980). Chemical evolution and ammonia in the early Earth's atmosphere. *Nature* (*London*) **287,** 526–528.

Holland, H. D. (1962). Model for the evolution of the Earth's atmosphere. *In* "Petrologic Studies: A Volume in Honor of A. F. Buddington" (A. E. J. Engle, H. L. James, and B. F. Leonard, eds.), pp. 447–477. Geol. Soc. Am., New York.

Kasting, J. F., and Walker, J. C. G. (1981). Limits on oxygen concentration in the prebiological atmosphere and the rate of abiotic fixation of nitrogen. *JGR, J. Geophys. Res.* **86,** 1147–1158.

Kuhn, W. R., and Atreya, S. K. (1979). Ammonia photolysis and the greenhouse effect in the primordial atmosphere of the Earth. *Icarus* **37,** 207–213.

Levine, J. S. (1982). The photochemistry of the paleoatmosphere. *J. Mol. Evol.* **18,** 161–172.

Levine, J. S., Augustsson, T. R., and Natarajan, M. (1982). The prebiological paleoatmosphere: stability and composition. *Origins Life* **12,** 245–259.

Levine, J. S., and Graedel, T. E. (1981). Photochemistry in planetary atmospheres. *EOS Trans. Am. Geophys. Union* **62,** 1177–1181.

Miller, S. L. (1953). A production of amino acids under possible primitive Earth conditions. *Science* **117,** 528–529.

Oró, J. (1961). Comets and the formation of biochemical compounds on the primitive Earth. *Nature* (*London*) **190,** 389–390.

Pinto, J. P., Gladstone, G. R., and Yung, Y. L. (1980). Photochemical production of formaldehyde in the Earth's primitive atmosphere. *Science* **210,** 183–185.

Schlesinger, G., and Miller, S. L. (1983a). Prebiotic synthesis in atmospheres containing CH_4, CO, and CO_2: I. Amino acids. *J. Mol. Evol.* **19,** 376–382.

Schlesinger, G., and Miller, S. L. (1983b). Prebiotic synthesis in atmospheres containing CH_4, CO, and CO_2: II. Hydrogen cyanide, formaldehyde, and ammonia. *J. Mol. Evol.* **19,** 383–390.

Schwartz, A. W. (1981). Chemical evolution—the genesis of the first organic compounds. *In* "Marine Organic Chemistry" (E. K. Duursma and R. Dawson, ed.), pp. 7–30. Elsevier, Amsterdam.

Turekian, K. K., and Clark, S. P. (1969). Inhomogeneous accumulation of the Earth from the primitive solar nebula. *Earth Planet. Sci. Lett.* **6,** 346–348.

Walker, J. C. G. (1976). Implications for atmospheric evolution of the inhomogeneous accretion model of the origin of the Earth. *In* "The Early History of the Earth" (B. F. Windley, ed.), pp. 537–546. Wiley, New York.

Walker, J. C. G. (1977). "Evolution of the Atmosphere." Macmillan, New York.

Zahnle, K. J., and Walker, J. C. G. (1982). The evolution of solar ultraviolet luminosity. *Rev. Geophys. Space Phys.* **20,** 280–292.

2

The Photochemistry of the Troposphere

T. E. GRAEDEL

AT&T Bell Laboratories
Murray Hill, New Jersey

I. Historical Perspective

The quality of the air in and near urban areas has been of concern for centuries. Brimblecombe (1976) discovered references to high levels of contaminants from as early as AD 1257 and correlated the air quality in London since that time with the amount of coal burned in the city. (Other emission

ISBN 0-12-444920-4

sources were, of course, present: early industries, open sewers, decaying refuse piles, etc.) The effects of coal burning were twofold: the coal dust inhibited visibility and blackened buildings, and the sulfurous gases produced distasteful smells and resulted in corrosion.

It was not until the scientific instrumentation developments of the 1930s and 1940s that specific trace gases in the air began to be determined in a reliable manner. Some of the studies used wet chemical techniques, others investigated nonsolar absorption features in the solar spectrum. By 1950 there was direct evidence of a number of trace gases in cities and their environs but no suggestions of chemical reactions among them. The photochemical link was provided by Haagen-Smit (1952), who demonstrated that ozone (O_3) and other oxidants are produced by the irradiation of low concentrations of nitrogen dioxide (NO_2) and a variety of organic compounds in air. It was then clear that the smog (a contraction formed from the words smoke and fog) of Los Angeles is totally different from that of London: its fuel sources are different (gasoline versus coal), its maximum concentrations occur at different times (early afternoon versus early morning), and its ambient environment is different (abundant sunshine and low humidity versus fog and high humidity).

Throughout the 1960s, much analytical effort was expended on studying the detailed composition of the urban atmosphere. Hundreds of compounds were detected, and the diurnal patterns of the more abundant ones were established. It was clear, however, that the reactions of ozone were insufficient to explain the observations of emittants and products. The missing reactant was suggested in 1971 by Levy to be the hydroxyl radical ($HO\cdot$). With this proposal, the gas-phase chemistry of the urban atmosphere could be reasonably well understood, and work since that time has centered on laboratory and theoretical studies relating field observations to the photochemistry of O_3 and $HO\cdot$.

With the development of increasingly sensitive analytical instrumentation in the 1970s, air-quality observations began to be made away from urban areas. It was promptly discovered that trace gases, including O_3, NO_2, and numerous hydrocarbons, are present at elevated concentrations as far downwind of urban regions as several hundred kilometers. An example of such an effect is given in Fig. 1, which shows O_3 concentrations measured by a gas detector within a research aircraft.

It is only since about the late 1960s and early 1970s that chemical studies of the atmosphere began to involve the liquid phase as well as the gas phase. These efforts originated in Scandinavia, where the transport and transformation of molecules generated in urban areas were shown to result in acidified rain. [Scandinavian monitoring activities during the period 1946–1970 were described by Granat (1972).] The chemical complexities of these liquid-phase systems are much greater than those in gas-phase systems and

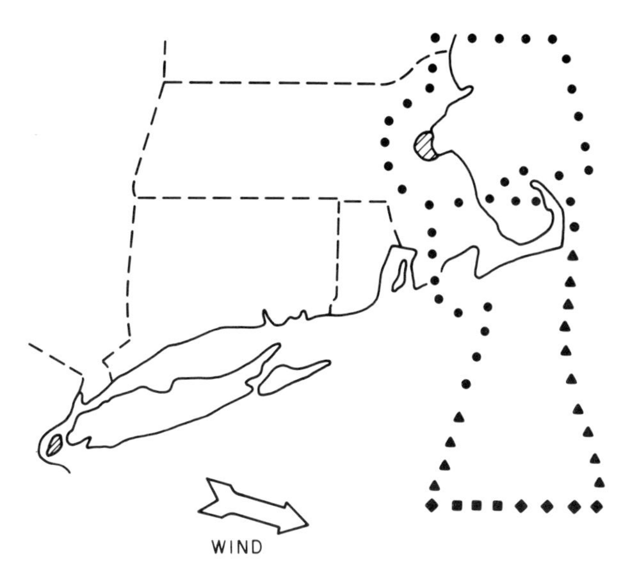

Fig. 1. Ozone concentrations (ppb) at 330 to 490 m altitude off the coast of the northeastern United States on the afternoon of 14 August 1975: ●, <100; ▲, 100–149; ◆, 150–199; ■, >200. The highest concentrations were seen about 250 km east of the New York City metropolitan complex, a region of high precursor emission fluxes. Trajectory analysis demonstrated that the high ozone air mass passed over the metropolitan area during the morning of the day on which measurements were made. From Levine and Graedel (1981); reproduced by permission of the American Geophysical Union.

include such new (to atmospheric chemists) topics as solute–solvent effects and trace-metal catalysis (Clarke, 1981; Graedel and Weschler, 1981). If the chemical processes are presently obscure, however, the effects are not, as seen in Fig. 2.

Even far from major population centers, the troposphere is a complex chemical mixture. Measureable amounts of largely anthropogenic emissions are seen throughout the world, as shown in the carbon monoxide (CO) Space Shuttle data of Fig. 3. Natural sources play important chemical roles as well. Decaying vegetation, forest fires, animal wastes, and a variety of soil and ocean processes have impacts on tropospheric chemistry. Linking all the systems together are the atmospheric motions, which mix and disperse the trace species of the atmosphere. A proper perspective on tropospheric chemistry is thus impossible without some understanding of tropospheric meteorology as well.

Contamination of the troposphere in populous areas has occurred since the advent of controlled fires, and has been chronicled for some 700 years. We now know that a variety of trace species are present in the troposphere throughout

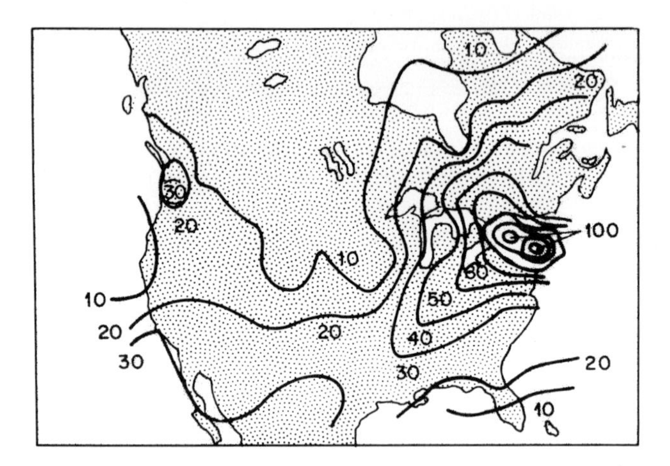

Fig. 2. Map of the average concentration of hydrogen ion (μequiv liter^{-1}) in precipitation in the continental United States and Canada. From Munger and Eisenreich (1983); reproduced by permission of the American Chemical Society.

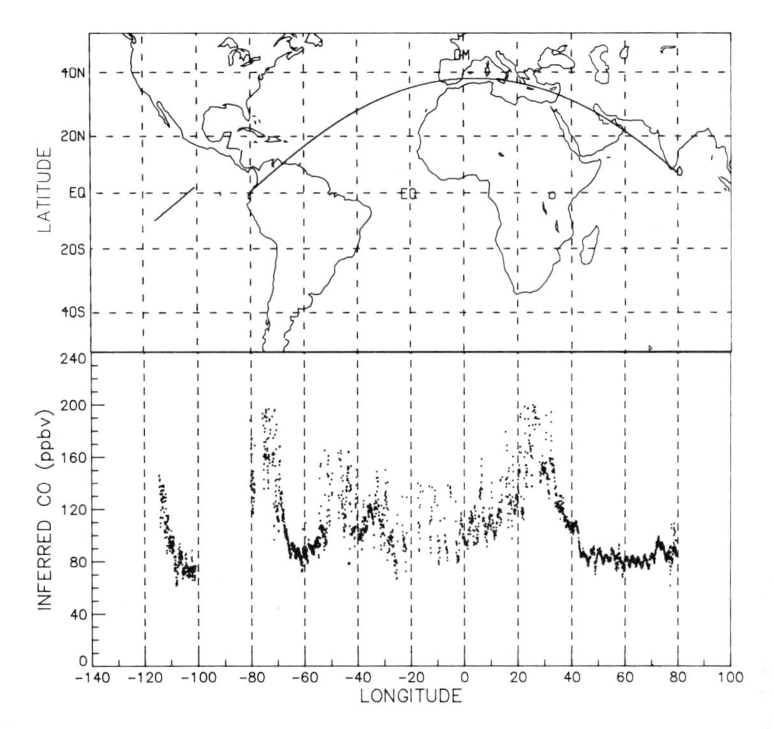

Fig. 3. Inferred middle tropospheric CO mixing ratio and corresponding ground track from the MAPS (measurement of air pollution from satellites) experiment on the flight of the NASA Space Shuttle, 13 November 1981. From Reichle *et al.* (1982); reproduced by permission of the American Association for the Advancement of Science.

the entire planet, not solely near emissions areas, and that knowledge of the chemical reactions of these species are crucial to an understanding of their effects. The relatively recent advent of extensive high-temperature combustion of fossil fuels has created a tropospheric photochemical system fueled by anthropogenic emissions. Its elucidation is partly accomplished but still fragmentary in scope. In this chapter we review what is known of the photochemistry of the Earth's troposphere.

II. Chemical Perspective

About 21% of the Earth's present atmosphere consists of oxygen (O_2), and molecules entering the atmosphere tend to be chemically driven toward increased oxidation. In contrast, molecules created or resident in regions removed from direct contact with the atmosphere are often in reduced chemical form. This situation obtains with the molecules produced by anaerobic bacteria in marshes, with petroleum deposits resulting from the decay of organic detritus, and with industrially important compounds used for a variety of purposes.

Combustion of fossil fuels is of great importance to humans, and the by-products of combustion are in many ways the most severe modifiers of the atmosphere. Complete combustion of hydrocarbon (C_xH_y) fuels produces carbon dioxide (CO_2) and water (H_2O):

$$C_xH_y + \left(\frac{x+y}{4}\right) O_2 \xrightarrow{\Delta} x\,CO_2 + \frac{y}{2}\,H_2O \tag{1}$$

and the high heat produced forms nitric oxide (NO) from the oxygen and nitrogen (N_2) in the air:

$$N_2 + O_2 \xrightarrow{\Delta} 2\,NO \tag{2}$$

In practice, combustion is never complete, and a variety of partially oxygenated products occur. These products, together with the incompletely oxidized nitrogen, are active participants in atmospheric chemistry.

Molecules emitted from natural processes, such as the hydrogen sulfide (H_2S) produced by bacterial action in sediments and the terpenes (isomers with the empirical formula $C_{10}H_{16}$) produced by trees, are often fully reduced compounds. They are also directly involved in the oxidation–reduction chemistry of the atmosphere.

Despite its large abundance, molecular oxygen is a poor oxidizer, and more active species are needed to initiate the chemical processes of the troposphere. These species arise from photochemical processes driven by solar radiation.

Table I

TYPICAL TROPOSPHERIC LIFETIMES FOR SELECTED ATMOSPHERIC TRACE GASES

Trace gas	Principal oxidizer	Rate constant $k_{25°C}$[a] (cm^3 $molec^{-1}$ sec^{-1})	Lifetime sec	Lifetime hr
H_2	HO·	6.7×10^{-15}	3.0×10^7	8.3×10^3
H_2O_2	HO·	1.7×10^{-12}	1.2×10^5	33
NH_3	HO·	1.6×10^{-13}	1.3×10^6	360
NO	O_3	1.8×10^{-14}	1.1×10^2	0.03
NO_2	HO·	1.6×10^{-11}	1.3×10^4	3.6
HNO_3	HO·	1.3×10^{-13}	1.5×10^6	420
CO	HO·	2.8×10^{-13}	7.1×10^5	200
CH_4	HO·	8.0×10^{-15}	2.5×10^7	6.9×10^3
C_2H_2	HO·	7.3×10^{-13}	2.7×10^5	75
C_2H_4	HO·	8.7×10^{-12}	2.3×10^4	6.4
α-Pinene	O_3	1.4×10^{-16}	1.4×10^2	0.04
C_2H_6	HO·	2.9×10^{-13}	7.0×10^5	190
Toluene	HO·	6.4×10^{-12}	3.1×10^4	8.6
H_2S	HO·	5.3×10^{-12}	3.8×10^4	11
OCS	HO· (?)	$\leqslant 9 \times 10^{-15}$	$\geqslant 2.2 \times 10^7$	$\geqslant 6.1 \times 10^3$
SO_2	HO·	2.5×10^{-12}	8.0×10^4	22
HCl	HO·	6.6×10^{-13}	3.0×10^5	83
CH_3Cl	HO·	4.2×10^{-14}	4.8×10^6	1.3×10^3

[a] Atkinson *et al.* (1979), except for NO + O_3 (Baulch *et al.*, 1982) and α-pinene + O_3 (Atkinson *et al.*, 1982).

Once formed, they inaugurate reactions of the type

$$[\text{Emittant}] \xrightarrow{\text{O}} [\text{Product}] \tag{3}$$

where the square brackets denote composition and O indicates "any oxidizer."

Reactions of type (3) produce (ultimately) CO_2 from hydrocarbons, nitric acid (HNO_3) from oxides of nitrogen, sulfate (SO_4^{2-}) from H_2S and sulfur dioxide (SO_2), etc. The products of the chemistry can be placed in perspective only when studied in conjunction with the rates at which production occurs. For a chemical reaction of the form

$$\text{A} + \text{O} \longrightarrow \text{B} + \text{O}' \tag{4}$$

where the rate constant for the reaction at a given temperature is k, the rate of change of concentration of species A as a result of reaction with oxidizer O is given by

$$d\text{A}/dt = -k[\text{A}][\text{O}].$$

The lifetime τ of species A in the troposphere may be estimated by dividing its concentration by the rates of the reactions in which it participates:

$$\tau(\mathrm{A}) = \frac{[\mathrm{A}]}{\sum_i k_i[\mathrm{A}][\mathrm{O}_i]}.$$

The usual case is that the lifetime of a given species is approximately determined by the rate of the most rapid of its removal reactions. In Table I, lifetimes are given for several of the more important atmospheric trace gases, using oxidizer concentrations appropriate for the mid-troposphere at mid-day in middle latitudes. With few exceptions (and none shown in the table), $HO\cdot$ and O_3 turn out to be the dominant oxidizers. The lifetimes in the table vary widely, from small fractions of an hour to many days. As a result, the perspective one adopts toward atmospheric chemistry is directly dependent on the time scales of concern in any particular situation.

III. The Role of Meteorology

The Earth's troposphere is a turbulent fluid, and the effects on chemical compositions and reactions that are produced by the turbulence are very great. At locations near sources of trace species, such as automobile exhausts and sewage treatment plants, turbulent diffusion reduces trace species concentrations rapidly with distance, as mixing into adjacent air parcels occurs. Thus emissions from very localized sources can lead to trace species dispersion over very wide regions and sometimes throughout the entire global atmosphere.

On a local scale of a few kilometers, the density of sources and the local topography and climatology tend to control the atmospheric concentrations of trace species. The well-known tendencies for concentrations to be relatively high in urban street "canyons" or in well-shielded valleys are examples of such control (Johnson *et al.*, 1973). On larger scales, personal experience cannot be of use, and extensive data analyses must be undertaken to demonstrate meteorological effects. An example of air quality on the synoptic scale (a few hundred to a few thousand kilometers) is shown in Fig. 4. The figure demonstrates the ability of atmospheric motions to affect air quality over a large geographical region on a time scale of a few days.

On the global scale (greater than a few thousand kilometers), the effects of individual weather patterns tend to merge into the general circulation. The mean surface-level winds embodied in this circulation have been studied for centuries, since they provided the motive power for seagoing trade. (Figure 5 shows an early schematic diagram of the circulation.) Transport of material

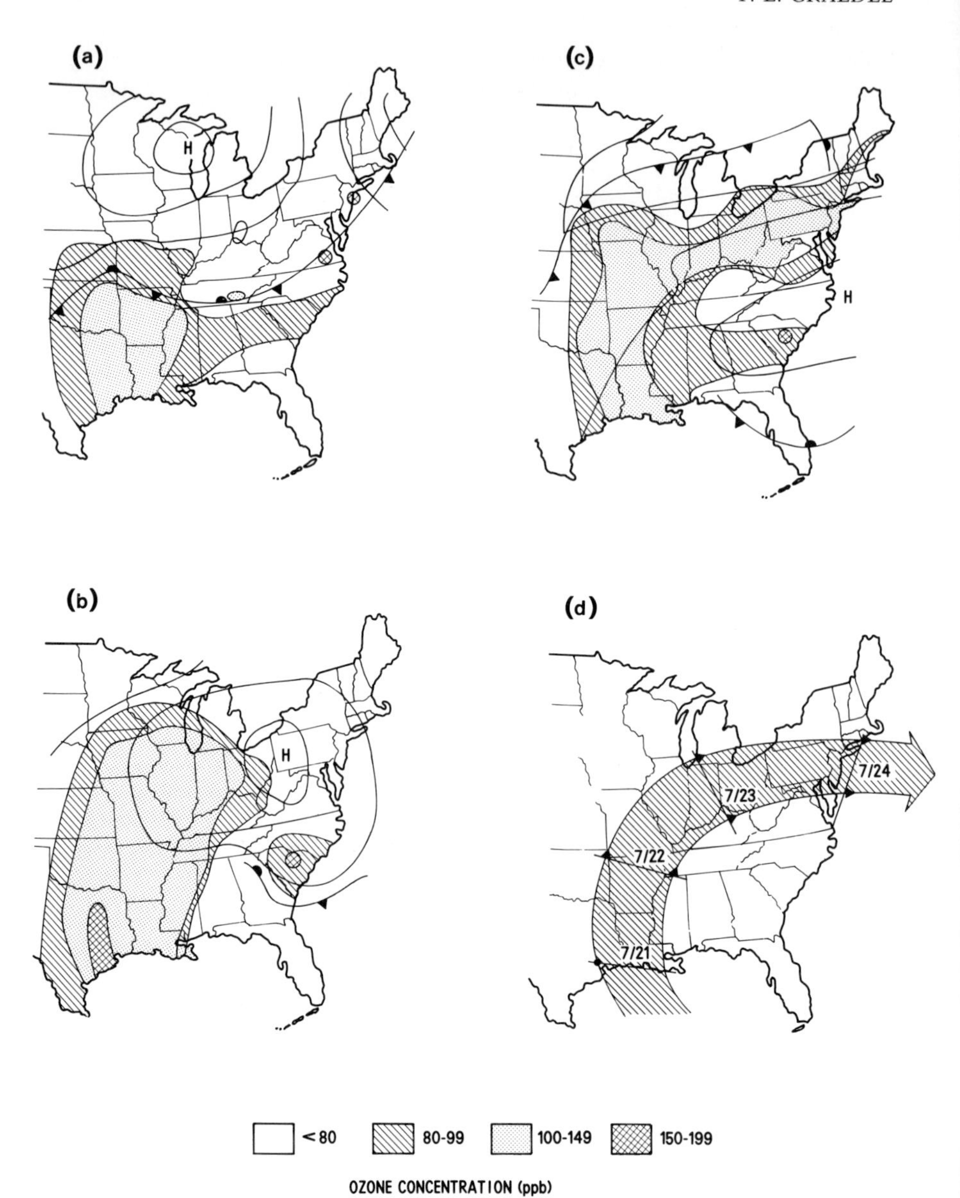

Fig. 4. An illustration of synoptic scale transport. In (a), high concentrations of ozone are seen over Texas and Louisiana. One day later (b), the clockwise circulation around the moving high-pressure region has drawn this ozone-laden air into the midwestern states. On the third day (c), the high ozone concentrations extend east to the Atlantic coast. The mean air flow throughout the boundary layer during this period (21–24 July 1977; d) is shown. Adapted from Wolff (1980).

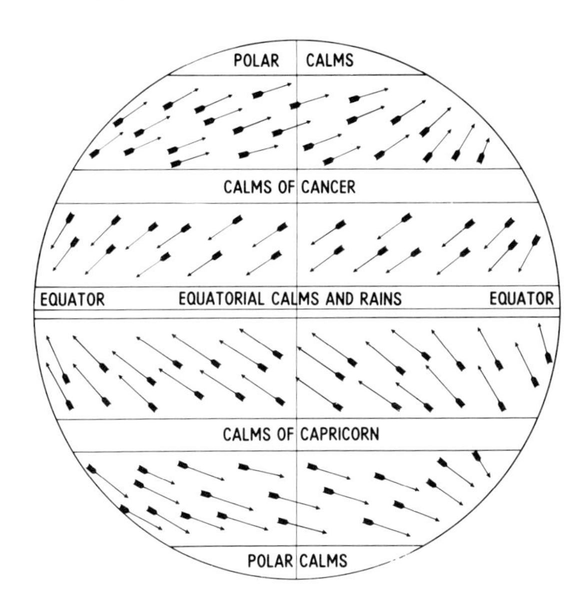

Fig. 5. Zonally averaged circulation in the atmospheric boundary layer, as drawn in 1857 by Thomson.

(and sailing ships) is quite efficient in the lower latitudes, the regions of the trade winds. At high latitudes the circulation is less vigorous but still important for our considerations. The presence of averaged circulations is reflected by recent observations in Florida of material from Sahara dust storms (Prospero and Nees, 1977) and in Hawaii of material from Chinese dust storms (Darzi and Winchester, 1982).

At the trade wind interface [Thomson's region of *equatorial calms and rains*, now termed the *intertropical convergence zone* (ITCZ)] mixing is strongly inhibited. As a result, the tropospheric air in the two hemispheres is distinctly different, particularly near the surface. An example is shown in Fig. 6, where the concentration of CO (emitted almost entirely in the northern hemisphere) is seen to drop by a factor of ~ 3 across the ITCZ. At higher altitudes these limitations are somewhat less severe, and height-integrated concentrations of CO show less dramatic variations (Fig. 3).

The reaction lifetimes of a number of trace gases were shown in Table I to vary widely. When those lifetimes are combined with typical meteorological transport velocities, the distance scales appropriate to the lifetimes can be specified, as seen in Table II. (This table includes the trace gases of Table I, as well as several others of interest for which no tropospheric chemical reaction is known.) The striking feature of the table is that few gases react rapidly enough for their effects to be confined to the local scale. Most are primarily global in

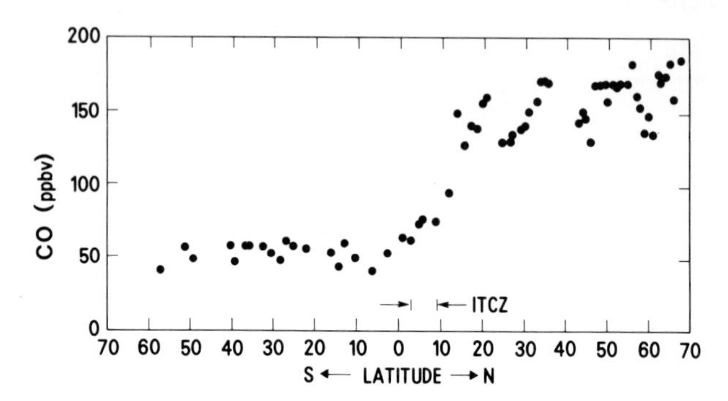

Fig. 6. Latitudinal distribution of carbon monoxide, as measured from aircraft at ~6 km altitude over the Pacific Ocean. The latitudinal extent of the intertropical convergence zone (ITCZ) is indicated. Adapted from Heidt *et al.* (1980).

Table II

DISTANCE SCALES APPROPRIATE TO THE GAS-PHASE REACTIONS OF SELECTED ATMOSPHERIC TRACE GASES

Trace gas	Local (≤20 km)	Regional[a] (20–1000 km)	Global[a] (>1000 km)
H_2			x
H_2O_2		x	
NH_3			x
NO	x		
NO_2		x	
N_2O			x
HNO_3			x
CO			x
CO_2			x
CH_4			x
C_2H_2			x
C_2H_4		x	
α-Pinene	x		
C_2H_6			x
Toluene			x
H_2S		x	
OCS			x
SO_2		x	
HCl			x
CH_3Cl			x
$CFCl_3$			x
CF_2Cl_2			x

[a] The synoptic scale of a few hundred to a few thousand kilometers is intermediate between the regional and global scales as defined here.

effect, while a few [notably the sulfur gases H_2S and SO_2 and the strong oxidizer hydrogen peroxide (H_2O_2)] have regional distance scales. On balance, therefore, the effects of local emissions are felt throughout the troposphere.

IV. Atmospheric Oxidizers and Their Sources

In the urban and regional scale troposphere, combustion of fossil fuels is an important source of the atmosphere's reactive species. As the fuel burns, N_2 and O_2 combine to form NO. A few percent of the NO is oxidized in the flame to gaseous NO_2. The latter compound plays a key role, since it absorbs ground-level solar radiation and photodissociates to produce atomic oxygen:

$$NO_2 + h\nu \longrightarrow NO + O \qquad \lambda < 420 \text{ nm} \tag{5}$$

This reaction is rapidly followed by

$$O + O_2 + M \longrightarrow O_3 + M \tag{6}$$

to form O_3. Ozone itself is a vigorous oxidizer, regenerating NO_2 through

$$O_3 + NO \longrightarrow O_2 + NO_2 \tag{7}$$

and attacking a variety of unsaturated hydrocarbons. Perhaps an even more important function is its generation of hydroxyl radicals (HO·) through photolysis of O_3, with the first excited state of the oxygen atom, $O(^1D)$, as an intermediate:

$$O_3 + h\nu \longrightarrow O_2 + O(^1D) \qquad \lambda \leqslant 320 \text{ nm} \tag{8}$$

$$O(^1D) + H_2O \longrightarrow HO\cdot + HO\cdot \tag{9}$$

The hydroxyl radical reacts with virtually all trace atmospheric species and is the primary link between emittants and products in all tropospheric chemical systems.

A second source of HO· derives from the incomplete combustion of the fossil fuel organic constitutents to produce aldehydes, ketones, and other oxygenated organic compounds. Many are photosensitive. Formaldehyde (HCHO), the simplest example, photolyzes to produce hydroperoxyl radicals ($HO_2\cdot$):

$$HCHO + h\nu \longrightarrow H\cdot + CHO\cdot \qquad \lambda < 335 \text{ nm} \tag{10}$$

$$H\cdot + O_2 + M \longrightarrow HO_2\cdot + M \tag{11}$$

$$CHO\cdot + O_2 \longrightarrow HO_2\cdot + CO \tag{12}$$

$HO_2\cdot$ is a source of $HO\cdot$ by disproportionation* and photodissociation of the resulting hydrogen peroxide:

$$HO_2\cdot + HO_2\cdot \longrightarrow H_2O_2 + O_2 \qquad (13)$$

$$H_2O_2 + h\nu \longrightarrow HO\cdot + HO\cdot \qquad \lambda < 350\ \text{nm} \qquad (14)$$

and by oxidation of nitric oxide:

$$NO + HO_2\cdot \longrightarrow NO_2 + HO\cdot \qquad (15)$$

The alkylperoxyl radicals ($RO_2\cdot$) produced by similar chemical chains from the higher aldehydes and from ketones also generate alkoxyl radicals ($RO\cdot$) and oxidize NO.

Thus the gas-phase atmospheric oxidizers O_3 and $HO\cdot$ are largely consequences of fossil fuel combustion followed by emission of the combustion products into the sunlit troposphere, although the injection of stratospheric air into the troposphere is a source as well. In aerosol particles, cloud droplets, and other atmospheric condensed water systems, dissolved O_3 and $HO\cdot$ continue to initiate oxidation reactions, as does H_2O_2.

V. Photochemistry in the Troposphere

A. Urban Scale

The sequence of chemical reactions in the troposphere begins when an oxidizer attacks an oxidizable molecule. As seen above, oxidizers are produced readily from several precursors; solar radiation is generally required as well. The molecule under attack is usually a hydrocarbon. If we use the symbol $R\cdot$ to represent any saturated hydrocarbon chain less its terminal hydrogen atom [i.e., the methyl radical ($CH_3\cdot$), the ethyl radical ($CH_3CH_2\cdot$), etc., also called alkyl radicals], we can write the reaction between the hydrocarbon molecule and a hydroxyl radical as

$$RH + HO\cdot \longrightarrow R\cdot + H_2O \qquad (16)$$

In the Earth's atmosphere, $R\cdot$ promptly adds molecular oxygen to form an alkylperoxyl radical ($RO_2\cdot$):

$$R\cdot + O_2 + M \longrightarrow RO_2\cdot + M \qquad (17)$$

* A reaction between two identical radicals in which one is chemically oxidized and the other chemically reduced.

Alkylperoxyl radicals are not particularly reactive, and the vigor of the chemistry depends on the rate at which they can be recycled to oxidizing species. Although this can be accomplished by disproportionation,

$$RO_2 + RO'_2\cdot \longrightarrow ROOR' + O_2 \tag{18}$$

followed by photolysis,

$$ROOR' + h\nu \longrightarrow RO\cdot + R'O\cdot \qquad \lambda < 350\ \text{nm} \tag{19}$$

a much more effective reaction is that with nitric oxide:

$$RO_2\cdot + NO \longrightarrow RO\cdot + NO_2 \tag{20}$$

The alkoxyl radical (RO·) is now available as an oxidizer, and, in addition, NO_2 can photolyze to produce O_3. The reaction thus produces two oxidizing molecules in a single process. The alkoxyl radical is rapidly converted to a carbonyl molecule (a molecule with an oxygen atom doubly bonded to a carbon atom) by reaction with O_2:

$$VRO + O_2 \longrightarrow R_{-1}CHO + HO_2\cdot \tag{21}$$

Thus, if the initial saturated hydrocarbon is ethane (CH_3CH_3), this reaction sequence will produce acetaldehyde (CH_3CHO), if propane ($CH_3CH_2CH_3$), the product will be acetone [$CH_3C(O)CH_3$] or propionaldehyde (CH_3CH_2-CHO), depending on the point of initial HO· attack. It is a general characteristic of oxygenated derivatives of hydrocarbon compounds that the derivative is more toxic than its precursor.

The degree of intensity of atmospheric chemistry in urban areas is due in large part to the abundant supply of nitric oxide, a species in short supply elsewhere. Adding to the effects of NO is a high concentration of hydrocarbons, also from anthropogenic sources. The chemistry feeds upon itself, as shown in Fig. 7. Once oxidizing molecules are present, chemical processing of a host of different chemicals can occur.

Hydrocarbon compounds in urban air are divided approximately equally (by total carbon concentration) into alkanes (compounds with saturated hydrocarbon chains), olefins (compounds with unsaturated hydrocarbon chains), and aromatics (compounds with aromatic ring structures). Unlike the relatively straightforward reaction paths of the alkanes, those of the olefins and aromatics are complex. For example, ethylene, the simplest of the olefins, follows the HO· reaction chain shown in Fig. 8. This sequence, in which conversion of an alkylperoxyl radical to an alkoxyl radical by NO is again a necessary step, results in the production of oxoalkane (aldehyde and ketone) products. The importance of the oxoalkanes in atmospheric chemistry is a consequence of their ability to absorb a solar photon and dissociate to oxidizing radicals (as illustrated for formaldehyde in the previous section). In

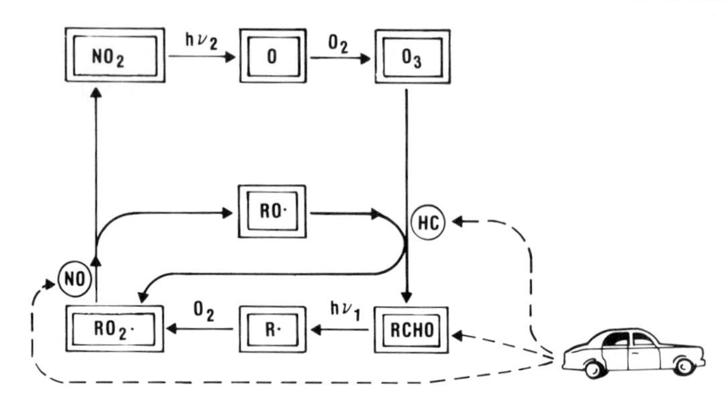

Fig. 7. Initiation and propagation of free radical chemical chains in the urban atmosphere. The sequence is initiated by photodissociation of aldehyde (RCHO) molecules, followed by oxidation of the nitric oxide and hydrocarbon emittants.

addition to reacting with HO·, olefins react with O_3. Such reactions proceed through a molozonide to give an oxoalkane and an excited diradical intermediate, as shown in Fig. 9 for propylene. The energetic intermediates undergo a variety of decomposition, isomerization, and reaction processes to form a host of small oxygenated molecules and radicals.

The atmospheric chemistry of the aromatic compounds is intricate and has not yet been fully described. It is clear that the principal oxidizer is the hydroxyl radical and that a variety of aromatic aldehydes, alcohols, and nitrates are produced by the reaction chains. These intermediate products subsequently suffer scission of the aromatic ring, a process that leads to oxygenated fragment molecules. A reaction mechanism for toluene that is

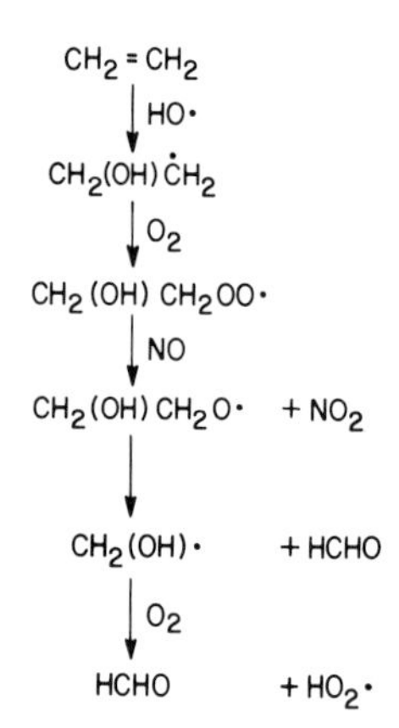

Fig. 8. Mechanistic sequence for the hydroxyl radical–initiated oxidation of ethylene in the troposphere.

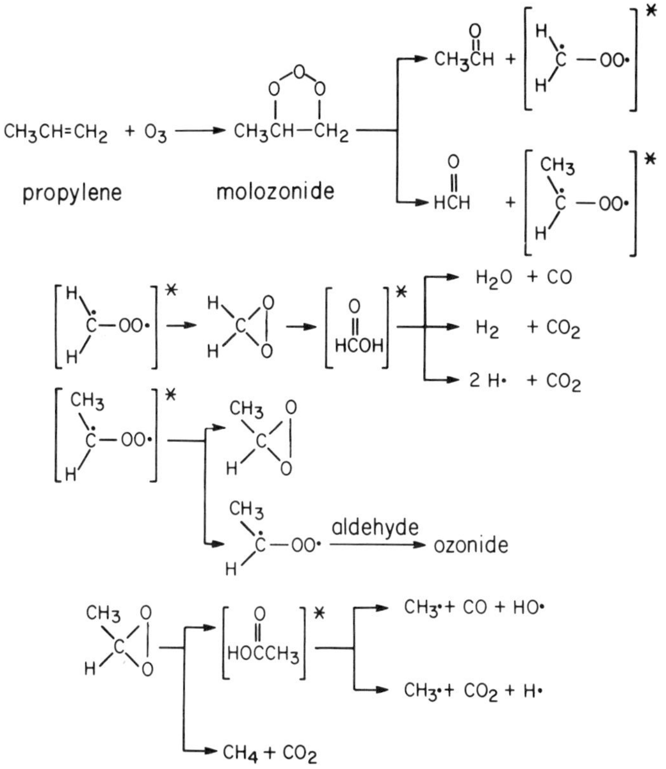

Fig. 9. Reaction pathways for the ozonation of propylene in the troposphere.

consistent with experiments and atmospheric observations is given in Fig. 10, where mechanism (a) is inaugurated by abstraction* of a hydrogen atom by $HO\cdot$ and mechanism (b) begins with $HO\cdot$ addition to the aromatic ring. Some of the many products identified in laboratory and field studies are shown. More detailed but chemically similar schemes apply to other common urban aromatic compounds such as benzene or the xylenes.

The other reactions of major interest in and near urban areas involve the conversion of sulfur and nitrogen oxides to the strong mineral acids. In each case $HO\cdot$ initiates the process:

$$NO_2 + HO\cdot + M \longrightarrow HNO_3 + M \quad (22)$$

$$SO_2 + HO\cdot + M \longrightarrow HSO_3\cdot + M \quad (23a)$$

$$HSO_3\cdot \xrightarrow{\text{intermediate steps}} H_2SO_4 \quad (23b)$$

* A chemical process in which an atom is withdrawn from a stable molecule.

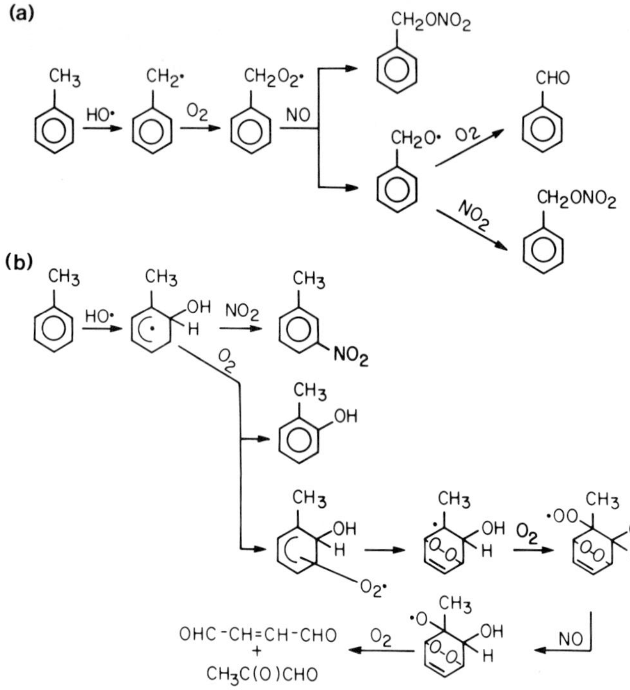

Fig. 10. Reaction pathways (a and b) for the hydroxyl radical–initiated oxidation of toluene in the troposphere.

The eventual fate of these acids and of the low vapor pressure organic compounds is deposition on atmospheric particulates (Junge, 1977), as shown schematically in Fig. 11.

B. Regional Scale

In many instances, chemical reactions on a regional scale occur with molecules that are emitted in urban areas by anthropogenic sources but react slowly enough that meteorological transport away from the urban region dominates chemical loss. Many of the reactions of Fig. 11 thus occur over a substantial range of distance (see Fig. 2). Natural emissions sources are present on the regional scale, however, and can have significant effects on some processes. A few of these are indicated schematically in Fig. 12. Most of these sources produce compounds in reduced oxidation states, so that the transition to the final chemical state is more complex and takes a longer time than for anthropogenic emittants. For hydrogen sulfide, the intermediate oxidation

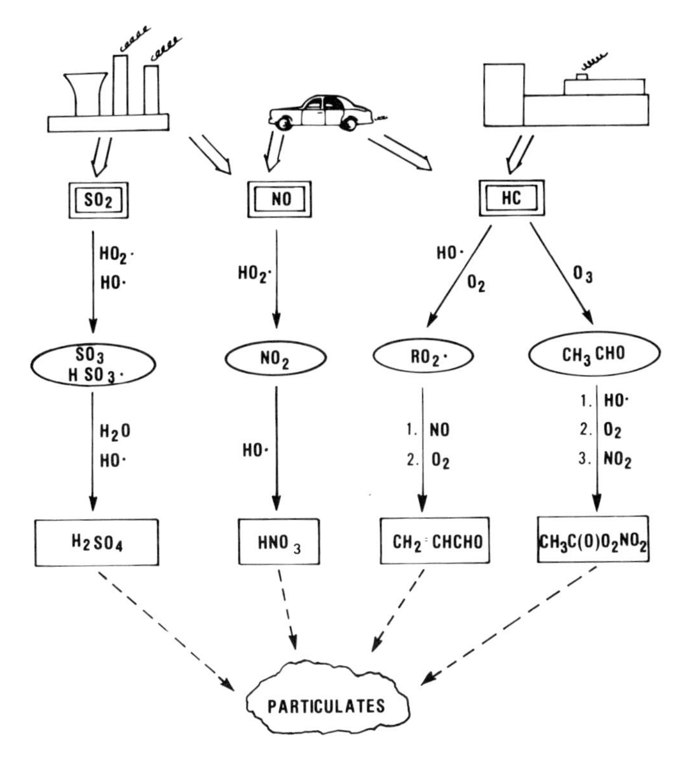

Fig. 11. Reaction sequences for formation of oxidized chemical products from urban precursors. Reactants are indicated by double-edged boxes, intermediates by ovals, and products by single-edged boxes. HC, Hydrocarbons.

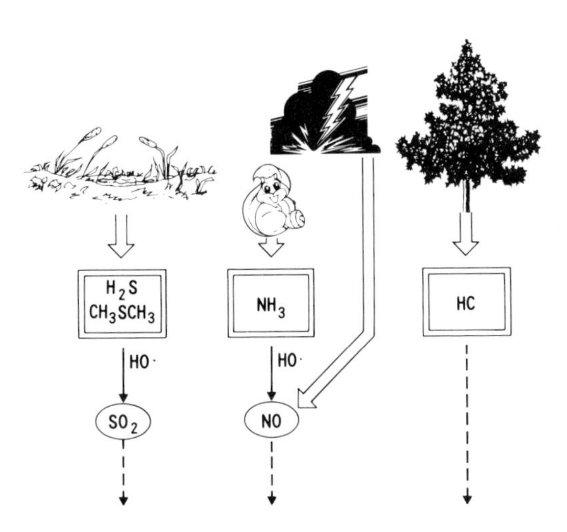

Fig. 12. Involvement of naturally generated molecules in atmospheric reaction sequences. (The sequences are continued by the processes shown in Fig. 11.)

product SO_2 is produced by

$$H_2S + HO\cdot \longrightarrow HS\cdot + H_2O \tag{24a}$$

$$HS\cdot + O_2 + M \longrightarrow HSO_2\cdot + M \tag{24b}$$

$$HSO_2\cdot + O_2 \longrightarrow SO_2 + HO_2\cdot \tag{24c}$$

The reaction chain for dimethyl sulfide (CH_3SCH_3), a major sulfur compound on the global scale, is more complex:

$$CH_3SCH_3 + HO\cdot \longrightarrow CH_3SCH_2\cdot + H_2O \tag{25a}$$

$$CH_3SCH_2\cdot + O_2 + M \longrightarrow CH_3SCH_2O_2\cdot + M \tag{25b}$$

$$CH_3SCH_2O_2\cdot + NO \longrightarrow CH_3SCH_2O\cdot + NO_2 \tag{25c}$$

$$CH_3SCH_2O\cdot \longrightarrow CH_3S\cdot + HCHO \tag{25d}$$

$$CH_3S\cdot + O_2 + M \longrightarrow CH_3SO_2\cdot + M \tag{25e}$$

$$CH_3SO_2\cdot \xrightarrow{\text{intermediate steps}} CH_3SO_3H \tag{25f}$$

The final product of Eqs. (25a) through (25f), methane sulfonic acid (CH_3SO_3H), has been detected in atmospheric particulate matter (Saltzman *et al.*, 1983).

In the case of ammonia (NH_3), it has been proposed that NO is generated by the reaction sequence

$$NH_3 + HO\cdot \longrightarrow \cdot NH_2 + H_2O \tag{26a}$$

$$NH_2 + O_2 + M \longrightarrow NH_2O_2\cdot + M \tag{26b}$$

$$NH_2O_2\cdot + NO \longrightarrow NH_2O\cdot + NO_2 \tag{26c}$$

$$NH_2O\cdot + O_2 \longrightarrow HNO + HO_2\cdot \tag{26d}$$

$$HNO + O_2 \longrightarrow NO + HO_2\cdot \tag{26e}$$

but there is no definitive field evidence of NO generation from an ammonia source. It appears more likely that the principal fate of the highly soluble ammonia molecules is absorption into atmospheric water droplets. Natural sources of NO include lightning and biogenic production, but the magnitudes of these sources have not yet been established.

Vegetation emits large fluxes of organic molecules into the atmosphere, the most abundant of which are the terpenes. Although the terpenes react with $HO\cdot$ and NO_3, reaction with O_3 is dominant under most circumstances (Graedel, 1979). The atmospheric reaction chains are thought to proceed similarly to those of the olefins and aromatics (Lloyd *et al.*, 1983). Laboratory studies of α-pinene, one of the most abundant of the terpenes, are consistent with the partial reaction mechanism shown in Fig. 13.

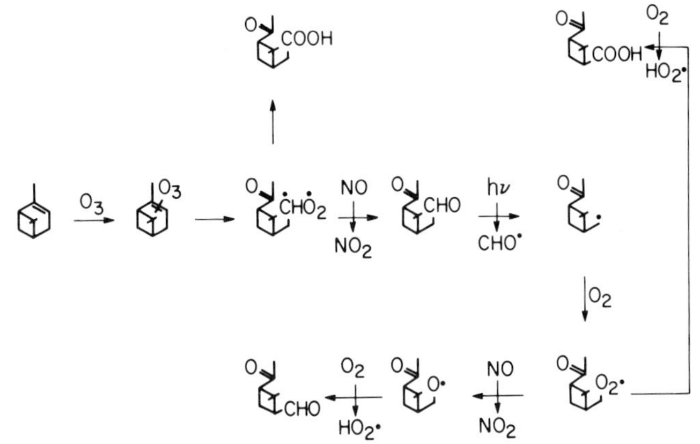

Fig. 13. Reaction pathways for the ozonation of α-pinene in the troposphere.

Some natural sources, such as forest fires, emit compounds in intermediate oxidation states. The chemistry of these compounds follows that described above for the urban scale, except that in many cases a shortage of NO inhibits the flow of material through the multistep reaction chains.

C. Global Scale

If by *global-scale photochemistry* one means chemistry that is independent of geographical location, there is little or none in the troposphere, since emittants are not generally fully mixed before they leave the troposphere by

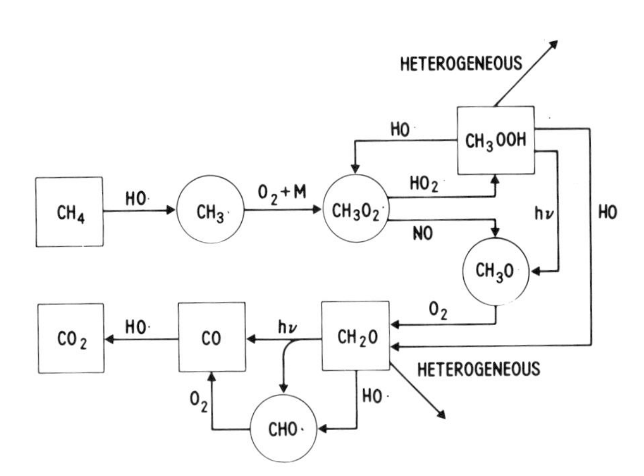

Fig. 14. Atmospheric oxidation cycle of methane. Adapted from Logan *et al.* (1981).

upward diffusion or surface loss. Neglecting the species that are virtually unreactive in the troposphere, such as nitrous oxide (N_2O), CO_2, and dichlorodifluoromethane (CF_2Cl_2), and neglecting interhemispheric differences that nearly always exist, the tropospheric chemistry that most nearly can be regarded as global is that of methane (CH_4) and carbon monoxide. As shown in Fig. 14, these species react (slowly) with HO· in a chain that ultimately produces CO_2. The rate of reaction of the steps in this chain is influenced by the concentrations of NO and by heterogeneous loss to droplets, both of which may depend on location and altitude. The principal rate control, however, is imposed by the rates of the initial reactions of CH_4 and CO with HO·. There is evidence that the atmospheric concentrations of CH_4 and CO are increasing due to increased rates of emission over the past few decades (Graedel and McRae, 1980; Rasmussen and Khalil, 1981).

VI. Chemistry in Atmospheric Water Droplets

The study of chemical reactions in cloud and fog droplets and in raindrops is in its infancy. As this is written, it appears that one can identify some of the major products of this chemistry, but the reaction paths that link these products to their precursors are known with considerably less confidence (Graedel and Weschler, 1981; Chameides and Davis, 1982).

A central concern in aqueous atmospheric chemistry is the formation of the inorganic acids sulfuric acid (H_2SO_4) and nitric acid (HNO_3). In solution, each acid is in equilibrium with its ions,

$$H_2SO_4 \rightleftharpoons H^+ + HSO_4^- \tag{27a}$$

$$HSO_4^- \rightleftharpoons H^+ + SO_4^{2-} \tag{27b}$$

$$HNO_3 \rightleftharpoons H^+ + NO_3^- \tag{28}$$

and the acids are referred to as *strong* because the equilibria favor the ionic forms.*

The generation of the acids in solution must involve oxidizing species. Molecular oxygen is among the primary oxidizing agents, with a typical solution concentration of $\sim 3 \times 10^{-4}$ *M*. Hydrogen peroxide, with an aqueous-phase concentration of $\sim 3 \times 10^{-6}$ *M*, has perhaps a greater chemical role: it serves both as an oxidant and as a source of solution hydroxyl radicals when it is photolyzed:

$$H_2O_2 + h\nu \longrightarrow 2\,HO\cdot \qquad \lambda < 380 \text{ nm} \tag{29}$$

---

* Note that all reactions and equilibria in this section occur in aqueous solution. For visual clarity, this is not indicated in the individual equations.

Ozone dissolved in water is a source of HO· as well:

$$O_3 + HO_2\cdot \longrightarrow HO\cdot + 2\,O_2 \tag{30}$$

$$O_3 + h\nu \longrightarrow O_2 + O(^1D) \qquad \lambda < 320\ \text{nm} \tag{31a}$$

$$O(^1D) + H_2O \longrightarrow 2\,OH \tag{31b}$$

and will probably be a significant radical source in situations where H_2O_2 solution concentrations are low.

The initial pH of atmospheric water droplets is established by the equilibrium between gaseous CO_2 and solution bicarbonate ion (HCO_3^-). Except for this function, the presence of HCO_3^- in solution does not appear to affect solution chemistry in any significant way. Other common but relatively uninteresting trace components are the chloride salts. The conversion of chloride ions to chlorine atoms leads to a number of radical processes, including the generation of HCl. However, chloride's greatest influence is likely to be as an electrolyte, that is, as a contributor to the overall ionic strength of the solution.

Ammonia [and the ammonium ion (NH_4^+)] and nitric acid [and the nitrate ion (NO_3^-)] are the most important inorganic nitrogen compounds in atmospheric water droplets. Ammonia is the principal species that reacts with strong acid anions, as evidenced by the large concentrations of ammonium salts found in aerosols. The NO_x compounds (NO and NO_2) are not highly soluble in aqueous solution, and they have yet to be detected in any atmospheric water system. Their most common fate is conversion to either HNO_3 or nitrous acid (HNO_2). Since the latter will subsequently be oxidized to nitrate ion, any NO_x present in water droplets will eventually add to the nitrate concentration.

Sulfur compounds are important constituents of atmospheric aqueous solutions. The most abundant are the sulfates, including sulfuric acid (H_2SO_4), acid ammonium sulfate (NH_4HSO_4), and ammonium sulfate [$(NH_4)_2SO_4$]. The simple inorganic sulfur gases H_2S, SO_2, carbonyl sulfide (OCS), and carbon disulfide (CS_2), however, are all appreciably soluble in water and can be expected to participate in the chemistry. The solution concentration of SO_2 is the highest of any of these species, high enough, in fact, to make it the most important reducing agent. Its principal reactions (in the customary bisulfite form) are thought to be with H_2O_2 and O_3:

$$HSO_3^- + H_2O_2 \xrightarrow{H^+} HSO_4^- \tag{32}$$

$$HSO_3^- + O_3 \longrightarrow HSO_4^- \tag{33}$$

although there is evidence for catalysis of the sulfite (SO_3^{2-}) to sulfate (SO_4^{2-}) oxidation by transition metal ions and by soot particles (Martin, 1983).

As a result of these considerations, the principal solution oxidation chains are deduced to be those of Fig. 15 (possible catalytic processes neglected). The third reaction path in the figure is that of the organic acids [RC(O)OH]. These acids ionize in solution as do the strong acids, that is

$$\mathrm{RC(O)OH} \rightleftharpoons \mathrm{H^+} + \mathrm{RC(O)O^-} \tag{34}$$

but the equilibria are such that the acids are only partly ionized and are thereby termed "weak." Nonetheless, organic acids can be the major acid constituents of atmospheric water droplets if the precursors of the inorganic acids are present only at very low concentrations. Such conditions exist in some of the more remote parts of the Earth (Keene *et al.*, 1983).

Organic hydroperoxides (ROOH) are readily formed in the gas phase and may be acid precursors. Precursors more likely to be important are the relatively more abundant aldehydes (RCHO), which are hydrolyzed in solution to the glycol form $RC(OH)_2$. Oxidation of these precursors will lead to the organic acids, but the precise mechanisms remain to be defined.

It is impossible to say at present whether incorporation or solution

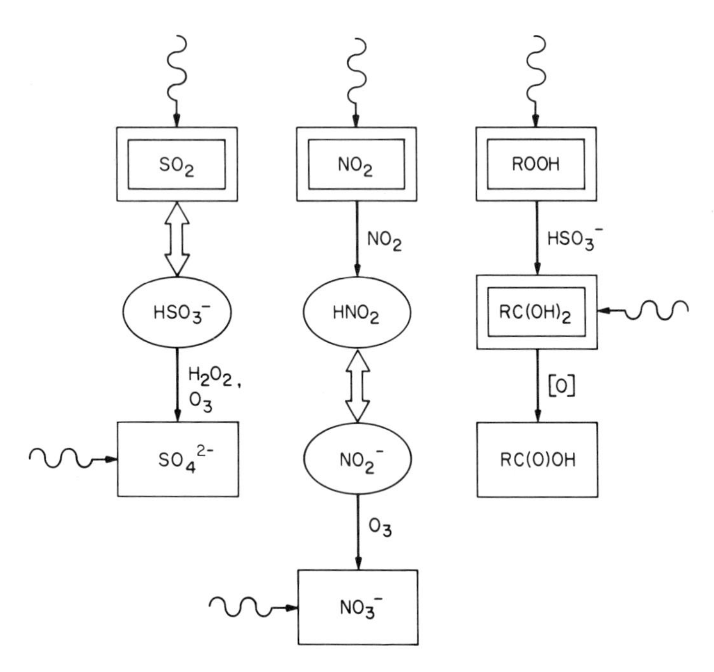

Fig. 15. Chemical reaction paths for formation of oxidized products in atmospheric water droplets. The wavy lines signify incorporation of gaseous molecules, the open double arrows indicate rapid chemical equilibria, and the closed single arrows indicate aqueous-phase reactions.

chemistry (i.e., the wavy lines or the solid lines of Fig. 15) is more important. It is likely that all processes operate in all situations, with different ones becoming dominant under different conditions. Current evidence suggests that solution reactions are often important for sulfuric acid (Martin, 1983) and often unimportant for nitric acid (Lee and Schwartz, 1981). The organic acids have received little study, and researchers have yet to determine the processes required to explain their presence in atmospheric water droplets.

VII. Aerosol Chemistry

Aerosol particles are present in the atmosphere in great numbers and in a wide variety of sizes. The principal sources of the small particles (diameter $<0.5\ \mu m$) are combustion processes such as forest fires or the burning of diesel fuel. Larger particles are produced in abundance by soil erosion and other "mechanical" processes.

Many aerosol particles are hydrophilic and attract layers or shells of water when the humidity is moderate or high. Subsequent contact with organic vapors may leave a surface film on the water. The resulting picture of the aerosol particle is shown in Fig. 16. The aqueous solution may include molecules or ions dissolved from the core or incorporated from the gas phase. The surface film may or may not be complete; if present, it is expected to impede the transfer of molecules into and out of the solution (Gill *et al.*, 1983).

Because of the very small size of aerosol particles, analytical chemical techniques have generally been applied to an ensemble of particles rather than to individual ones. Analyses have shown the presence of ammonium, nitrate,

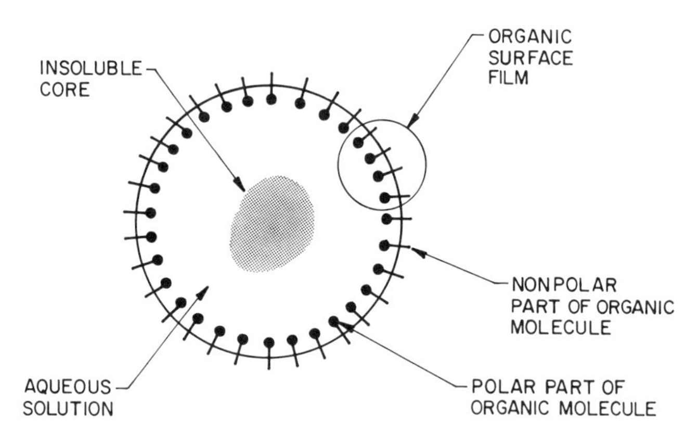

Fig. 16. Schematic diagram of the possible structure of aged atmospheric aerosol particles. From Graedel and Weschler (1981); reprinted by permission of the American Geophysical Union.

and sulfate ions, inorganic carbon, a wide variety of saturated and oxidized organic compounds, and a number of trace metals. Among the chemical processes thought likely to occur on aerosol particles are the oxidation of sulfite to sulfate (probably by trace-metal or soot catalysis), the partial neutralization of the acidic aerosol by dissolved ammonia, and the oxidation of a variety of saturated and unsaturated hydrocarbons.

Aerosol particle processes in atmospheric chemistry have not yet been fully assessed but are certain to be important. For example, lifetimes of aerosol particles are several hours to several days, during which time they can travel long distances. Thus, chemical species deposited on aerosols in a local area can be carried hundreds of miles away. In their new location, they may be incorporated into cloud droplets and raindrops to produce an impact far from their sources.

VIII. Concentration of Tropospheric Trace Species

It has been pointed out that the concentrations of trace species in the troposphere vary greatly because of source distributions, atmospheric mixing, and chemical lifetimes. In Fig. 17, information on concentrations has been assembled to demonstrate this variation. [These data are an updated version of those presented by Graedel (1980).] As an example, consider hydrogen sulfide (H_2S). The highest concentrations, >10 ppm, are found in selected urban areas, although urban concentrations of 1 to 10 ppb are more common and 0.1 to 1 ppb occur occasionally. Concentrations over marshland can be in the 1- to 10-ppb range but are usually lower. In other locations, 0.1- to 1-ppb levels are sometimes seen, but 10- to 100-ppt (parts per trillion) levels are more common, and 1- to 10-ppt levels occasionally occur. The measured distribution of concentrations demonstrates the following:

1. The principal H_2S source regions are urban areas and marshland.
2. H_2S is reasonably reactive, since its concentrations show wide variations in the troposphere and decrease away from the source regions.
3. No significant upper atmospheric sources exist, since the free troposphere (i.e., the troposphere above the surface boundary layer) concentrations are significantly lower than the concentrations near the ground.

In contrast to the concentration behavior for H_2S, that for OCS is everywhere uniform at a level of 0.1 to 1 ppb. This latter behavior indicates that the gas does not have any high flux sources and that its atmospheric lifetime is very long so that it has become well mixed. A third type of behavior is indicated by SO_2, which has a very strong urban source and no natural sources of any significance. Thus, the SO_2 concentrations in urban regions can be very high and are elsewhere quite low.

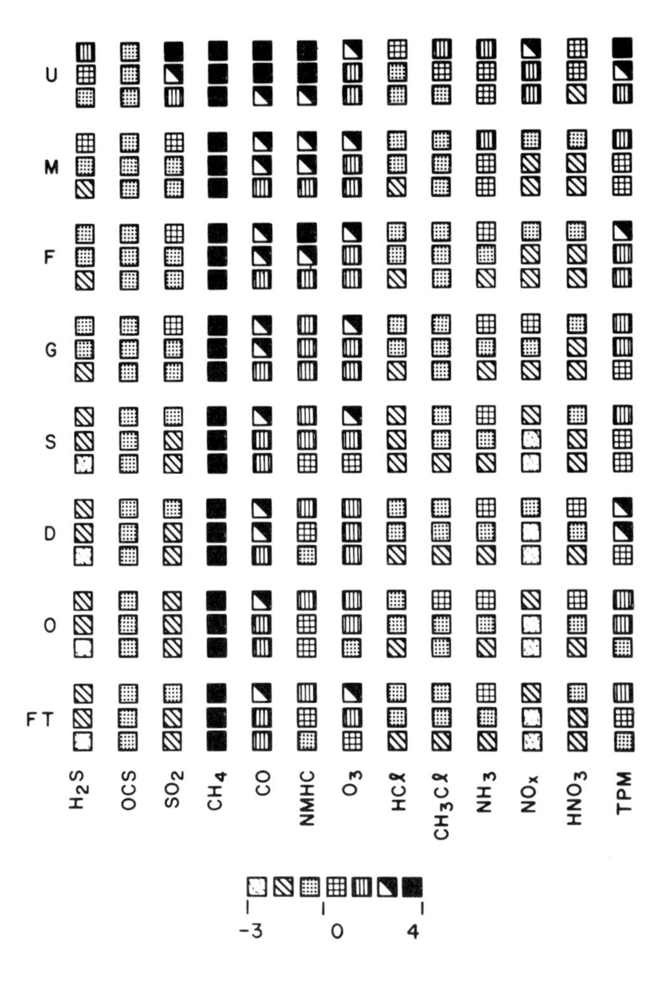

Fig. 17. Observed concentrations of trace species in different tropospheric regimes. Concentrations are in ppb except for total particulate matter (TPM; $\mu g\ m^{-3}$). The letters on the ordinate refer to the geographical location: U, urban; M, marshland; F, farmland; G, grassland; S, steppes and mountains; D, deserts; O, oceans; FT, free troposphere. In each triad of symbols, the upper symbol represents high concentrations, the center symbol mean concentrations, and the lower symbol low concentrations.

The fourth through sixth columns of Fig. 17 refer to carbon-budget gases (see below). The CH_4 concentrations are high (>1 ppm) and are relatively uniform—again, an indication of a long atmospheric lifetime. Carbon monoxide has a long lifetime as well, but its strong urban sources are shown in the measurements. Nonmethane hydrocarbons (NMHC) have strong sources in both urban and forest areas as well as moderate marshland sources.

Ozone has its highest concentrations in urban areas, but high concentrations sometimes occur over any land area. The upper tropospheric concentrations can also be high, generally as a result of intrusion from the ozone-rich stratosphere. Concentrations are somewhat lower over forests as a result of the rapid reaction of ozone with terpenes. Oceanic and desert ozone levels are low, presumably because of a shortage of precursors.

The eighth and ninth columns of the figure refer to the two principal atmospheric halogenated compounds. Hydrogen chloride (HCl) has a variety of sources, none of them large, and is found in urban areas at moderate concentrations. Methyl chloride (CH_3Cl) is produced almost entirely by the oceans; it has low concentrations elsewhere.

The nitrogen-containing molecules of the tenth through twelfth columns have very different chemical patterns. Ammonia is released by a variety of bacterial decay processes, such as sewage treatment, as well as in animal urine. Its concentrations are moderate and apparently fairly uniform, although measurements are not very widely available. Nitrogen oxides (NO_x, i.e., NO and NO_2) are produced almost solely by high temperature combustion; they are abundant in urban areas and rare in remote areas of the Earth. Nitric acid is a product of the oxidation of NO_x. As a result, HNO_3 concentrations are highest near the NO_x sources.

The last column in Fig. 17 is for total particulate matter (TPM). A variety of anthropogenic sources produce the highest TPM concentrations in urban areas. Next are the desert regions, as a result of dust storms, and the forest regions, as a result of vegetation-induced haze. Since the TPM value is based on dry weight, it is heavily influenced by the large particles, which fall out rapidly by gravitational settling. Regions without strong TPM sources, such as the oceanic areas, thus tend to have low TPM concentrations.

IX. Chemical Budgets in the Troposphere

The scientific understanding of why a particular trace species (or an element contained in several species) is present in the atmosphere at the measured spatial and temporal distribution of concentrations is best evaluated by constructing a chemical budget. A budget is an attempt to identify and quantify those processes that supply the species to the atmosphere or remove it from the atmosphere; it is thus an atmospheric "balance sheet" for the species. For a species demonstrably stable with time (on the time scale of interest here), such as O_2, the budget may be assumed to be in steady state, that is,

$$d[\mathrm{A}]/dt = \sum \text{sources} - \sum \text{sinks} = 0. \tag{35}$$

It is generally the case, however, that one is not sure whether the budget is in balance or knows because of a time series of observations that an imbalance is present (CO_2 is the best example). Thus

$$d[\mathrm{A}]/dt \neq 0, \tag{36}$$

and the magnitude of the concentration change is a measure of the budget imbalance.

Budgets may be global in scope, or they may pertain to specified geographical areas. In any case a budget will assume the form diagrammed schematically in Fig. 18. In this figure the sources of the budgeted species are on the left, the sinks on the right. Chemical transformations are often, but not always, involved. Individual budgets may omit some of the components, since they do not all apply in all cases. For example, methane is not generated chemically in the troposphere, carbon dioxide is not transported from the stratosphere, nitrous oxide is not chemically transformed in the troposphere, and loss to transport applies to budgets on local and synoptic scales but not on global scales.

The observed atmospheric concentrations of trace species provide the benchmark against which chemical budget assessments are evaluated. It is often difficult to perform a series of observations sufficiently detailed to serve as a guide to atmospheric processes involving a particular chemical compound. Once observations are acquired, however, many properties of the

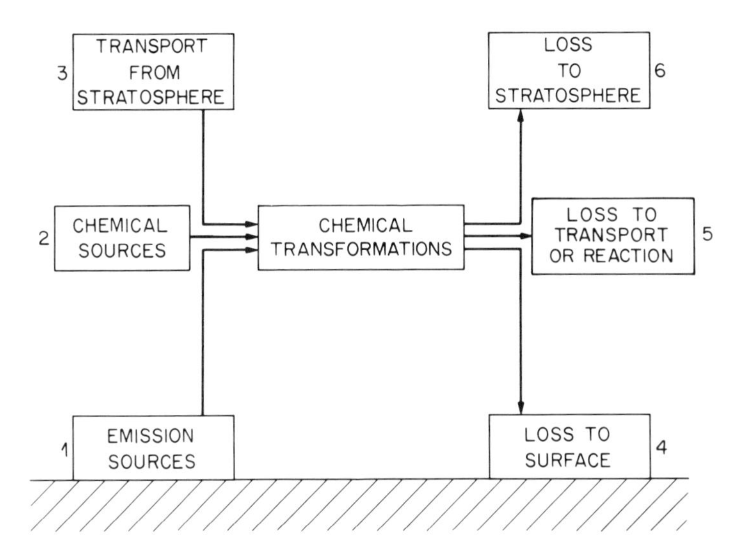

Fig. 18. Principal components of an atmospheric chemical budget.

budget for a given species can be deduced readily. Examples of this are:

1. Concentrations that are relatively uniform throughout the world indicate that chemical reactions of the species are slow and thus the species has time to become well mixed.
2. Concentrations that increase with altitude can indicate loss at the ground or generation by photochemistry in the atmosphere (since the solar flux increases with altitude).

Budgets can be constructed for an individual species, for an element contained in several species, or for an assemblage such as atmospheric particulate matter. In Table III, budget figures for several assessments are assembled. Such is the state of information that nearly all these numbers could be strongly contested; many are uncertain by a factor of two or more. The purpose in presenting them here is not to justify the numbers but to illustrate how budgets are assembled and how they differ. The first is for atmospheric sulfur, including H_2S, CH_3SCH_3, OCS, SO_2, and particulate forms. Some details are provided by Fig. 19. The natural and anthropogenic sources are seen to be similar in magnitude. Since all sulfur compounds are included in the budget, chemical production and loss do not apply. Dry deposition is estimated to be about one-third as important as removal by rain and snow. It is uncertain whether or not the budget is in balance. The assessment of Fig. 19 was done so that sources and sinks balance, but uncertainties in many of the estimates do not permit an unequivocal answer.

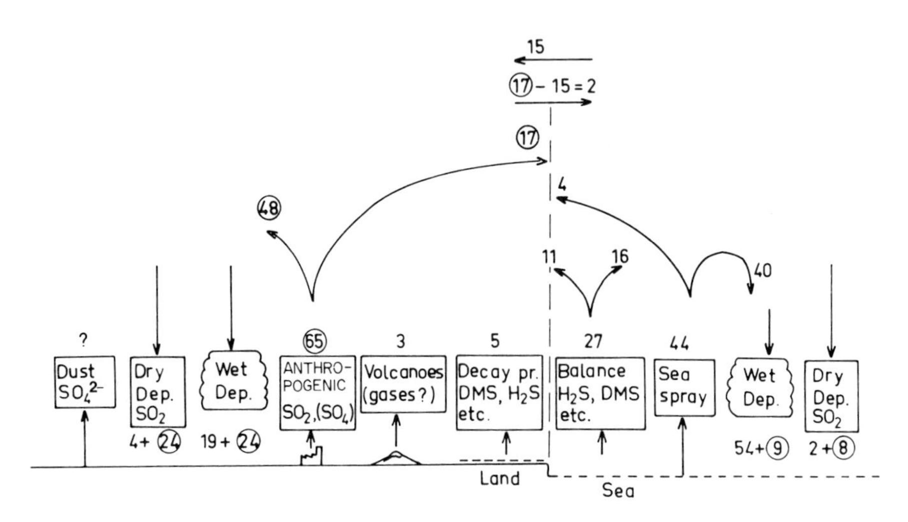

Fig. 19. A global sulfur budget for the atmosphere. The fluxes emanating from anthropogenic emissions are circled. From Granat *et al.* (1976).

Table III

SELECTED ATMOSPHERIC GLOBAL TROPOSPHERIC BUDGET ASSESSMENTS[a]

Budget	Atmospheric burden (Tg)[b]	Sources (Tg/year)[c]			Sinks (Tg/year)			Balanced budget
		1(a)	1(b)	2	4(a)	4(b)	5	
S	3800	79	65	0	38	106	0	?
CO	470	200	980	1460	250	0	2400	?
CO_2	7.0×10^6	0	1.9×10^4	3300	7300	0	0	No
CH_4	9600	553	50	0	0	0	580	No
$CFCl_3$	3.5	0	0.3	0	0	0	0	No
TPM[d]	76	800	50	820	300	1400	0	?

[a] The source and sink terms are drawn from the following references: S, Granat *et al.* (1976); CO, Logan *et al.* (1981); CO_2, Walsh *et al.* (1981); CH_4, Khalil and Rasmussen (1983); $CFCl_3$, Cunnold *et al.* (1983a); TPM, SMIC (1971).

[b] Tg ≡ Teragram (10^{12} g).

[c] Numbers refer to Fig. 18, except that 1(a) refers to natural emissions, 1(b) to anthropogenic emissions, 4(a) to dry deposition, and 4(b) to wet deposition. None of these budgets has a stratospheric transport component. Since they are global budgets, there is no transport loss term either; the eighth column reflects entirely loss by chemical reaction.

[d] TPM, Total particulate matter.

The second through fourth rows of Table III are for gaseous components of the atmospheric carbon cycle. (Nonmethane hydrocarbon vapors and inorganic and organic particulate carbon would be added to assemble a full atmospheric carbon budget.) The molecular budgets are quite different. For CO and CO_2, fossil fuel combustion is the major source. Carbon monoxide is generated in the atmosphere by methane and NMHC chemistry, and it is converted to CO_2 by atmospheric chemical reactions. Thus, both CO and CO_2 have secondary chemical sources. Methane, in contrast, is produced largely by natural processes and is removed almost totally by chemical reactions. It will be seen that the figures for CO chemical loss and CO_2 chemical gain are not identical, a common problem in budget construction.

The fifth row of the table contains data for fluorotrichloromethane ($CFCl_3$), one of the halogenated compounds capable of interacting destructively with stratospheric ozone. No tropospheric sinks are known for $CFCl_3$, since it is insoluble in water and does not react with atmospheric oxidizers; the only known atmospheric chemical process in which it participates is photodissociation in the stratosphere.

The final budget is for total particulate matter; TPM has very strong natural sources (dust storms, sea salt spray, etc.) and a large chemical source. The latter represents the conversion of gases to particles in the air. Most of the gases that undergo conversion are anthropogenic.

The consequences of unbalanced budgets are shown in Figs. 20 through 22, which are measurements of the concentrations of CO_2, CH_4, and $CFCl_3$, respectively, over extended periods. These species are becoming increasingly

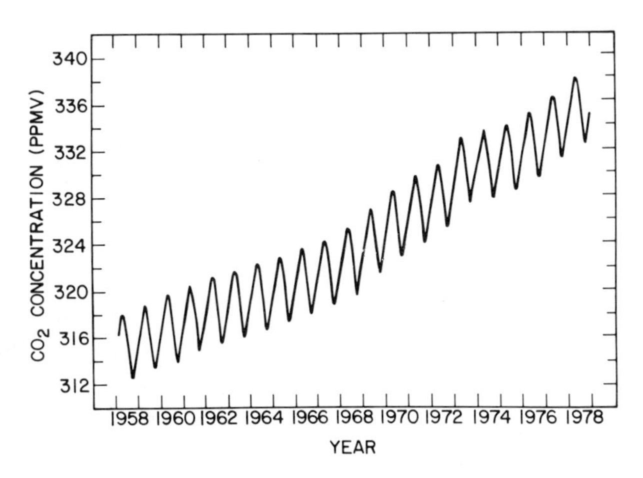

Fig. 20. Concentration of atmospheric CO_2 at Mauna Loa Observatory, Hawaii, 19.5°N, 155.6°W. From Bacastow and Keeling (1981).

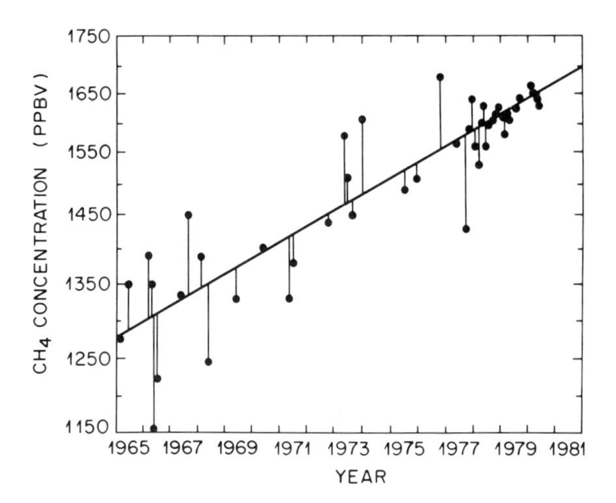

Fig. 21. Concentration of atmospheric CH_4 at different sites in the northern hemisphere. From Rasmussen and Khalil (1981).

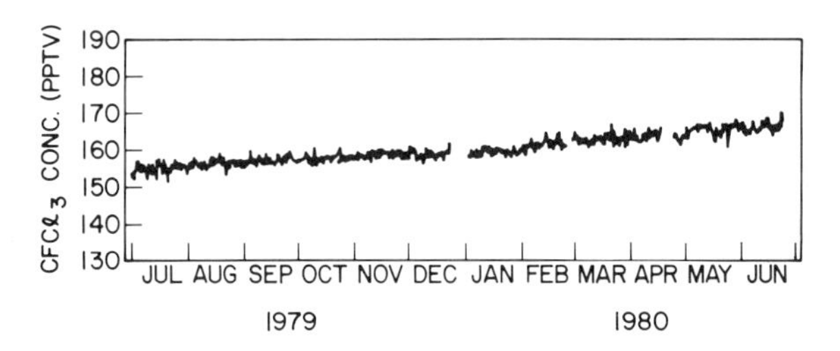

Fig. 22. Concentration of atmospheric $CFCl_3$ at Cape Grim, Tasmania, 41°S, 145°E. From Cunnold *et al.* (1983a).

abundant in the atmosphere, with probable climatic consequences as discussed in Chapter 4. In addition to those for these species, increasing concentration patterns have been detected for CF_2Cl_2 (Cunnold *et al.*, 1983b), trichloroethane (CH_3CCl_3; Prinn *et al.*, 1983), N_2O (Weiss, 1981), and carbon tetrachloride (CCl_4; Simmonds *et al.*, 1983), and perhaps for CO (Graedel and McRae, 1980).

X. Consequences of Tropospheric Photochemistry

A. Photochemical Smog

Smog refers to the mixture of oxidized compounds resulting from the emission of hydrocarbons and oxides of nitrogen into the sunlit atmosphere. Many of the compounds produced by the chemical reactions of these emittants cause eye irritation or decreased pulmonary function. Examples are acrolein ($H_2C{=}CHCHO$) and peroxyacetylnitrate [$H_3CC(O)O_2NO_2$], produced by

$$H_2C{=}CHCH{=}CH_2 + HO\cdot \longrightarrow H_2C{=}CH\dot{C}HCH_2OH \quad (37a)$$

$$H_2C{=}CH\dot{C}HCH_2OH + O_2 + M \longrightarrow H_2C{=}CHC(\dot{O}_2)HCH_2OH + M \quad (37b)$$

$$H_2C{=}CHC(\dot{O}_2)HCH_2OH + NO \longrightarrow H_2C{=}CHC(\dot{O})HCH_2OH + NO_2 \quad (37c)$$

$$H_2C{=}CHC(\dot{O})HCH_2OH \longrightarrow H_2C{=}CHCHO + CH_2OH\cdot \quad (37d)$$

$$CH_3CH_3 \xrightarrow{\text{several steps}} H_3CCHO \quad (38a)$$

$$H_3CCHO + HO\cdot \longrightarrow H_3CC(O)\cdot + H_2O \quad (38b)$$

$$H_3CC(O)\cdot + O_2 + M \longrightarrow H_3CC(O)O_2\cdot + M \quad (38c)$$

$$H_3CC(O)O_2\cdot + NO_2 + M \longrightarrow H_3CC(O)O_2NO_2 + M \quad (38d)$$

These and other photochemical oxidants also cause severe crop damage when present in sufficiently high concentrations.

Ozone concentrations are often used as indications of the severity of smog, since ozone is produced under the same conditions favoring the formation of other oxidants. As can be surmised from the reaction chains outlined above, NMHC and NO_x are concomitantly involved in photochemical oxidant production. Reduction of photochemical oxidants requires that the components be reduced jointly, maintaining the NMHC/NO_x concentration ratio near a value of 10. Such a controlled reduction has generally proved difficult to achieve, given the myriad of sources and the complexities and engineering difficulties of reducing their emissions in well-defined, designated amounts.

B. Acidic Rain

Although rain is often pictured as highly distilled, uncontaminated water, such a picture is quite inaccurate. Rainwater (and cloud water, its precursor) tends toward equilibrium with the gaseous molecules that surround it. In addition, chemicals leached from scavenged particles are present in the rain. The result is an aqueous solution of substantial complexity.

The acidity of rain is established largely by interaction of the water droplets with gaseous carbon dioxide, nitric acid, organic acids, compounds of sulfur, and ammonia. Carbon dioxide is a weak acid that sets the pH benchmark in droplets at ~5.6 (i.e., cloud water and rainwater are weakly acidic throughout the world even in the absence of the strong inorganic acids). In regions in or downwind from high emission sources of NO_x and sulfur gases, HNO_3 and H_2SO_4 raise the hydrogen ion concentration and lower the pH significantly, as seen in Fig. 2. In the northeastern United States, sulfur is more important than

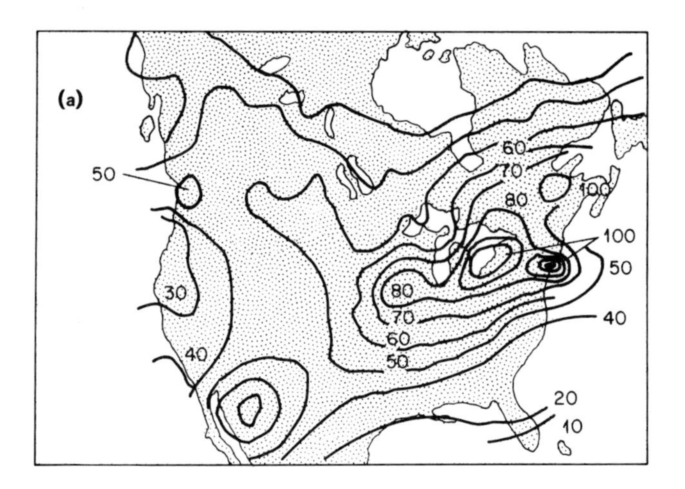

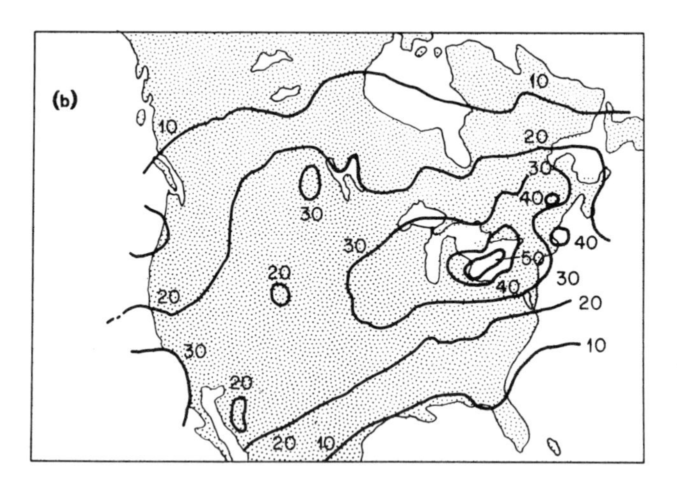

Fig. 23. Average concentration of inorganic acid anions, (a, SO_4^{2-}; b, NO_3^-) in precipitation in the continental United States and Canada. From Munger and Eisenreich (1983); reproduced by permission of the American Chemical Society.

nitrogen in the acidification of the droplets (Fig. 23). In remote regions of the world where anthropogenic emissions are low, naturally emitted sulfur species or organic acids influence the acidity. Ammonia acts as a neutralizing gas, generally reducing the acidity but not eliminating it.

Acidic rain is a major concern because of its detrimental influence on forests, crops, lakes, fish, and structures. Assessing the problem in a quantitative sense requires a rather complete understanding of the interplay among gas-phase chemistry, liquid-phase chemistry, and meteorology. Such a degree of understanding is not yet at hand, and much research remains to be done.

C. Visibility

One of the most noticeable consequences of chemistry in the atmosphere is a restriction in visibility (White *et al.*, 1981). Some visibility changes are perceived as desirable (e.g., the Smoky Mountains), but most are quite negative in effect. One example among many that have been displayed is shown by the contrast of Figs. 24 and 25.

Reductions in visibility are caused principally by the presence of airborne particles of sizes near that of the wavelengths of visible light (i.e., about 0.4–0.8 μm). At the lower end of this range, gaseous NO_2 can also be a significant absorber. These particles may be directly injected into the atmosphere, often from combustion sources, or they may be transformed from gas-phase molecules. Small particles are generally found to be rich in sulfate and organic and inorganic carbon. In the eastern United States, visibility degradation has been shown to be correlated with coal consumption (Lyons *et al.*, 1978).

Fig. 24. New York City on a day with good visibility. The buildings in the foreground are the towers of the World Trade Center during their construction.

Fig. 25. New York City on a day with visibility impaired by high levels of atmospheric particulate matter.

D. Corrosion

The damage caused by corrosion is very great. Although damages resulting from corrosion are difficult to estimate, studies suggest that losses may amount to 70 billion dollars annually (Bennett *et al.*, 1978). Of this total, a significant fraction is atmospheric corrosion. The results of such corrosion can be dramatic, as in an occasional bridge collapse, but are more often slow and steady. Among the consequences of the latter are the formation of nonconducting films on electrical contacts, the discoloration of structural and decorative metals on buildings, and the disintegration of marble statuary.

The principal agents of atmospheric corrosion are compounds of chlorine and sulfur, aided by high humidity, solar radiation, and the presence of atmospheric oxidants (Ailor, 1982). Near the oceans, where sea spray provides an abundant source of chlorine, substitution or protection of materials is the only defense against atmospheric corrosion. Elsewhere, a general decrease in the amount of airborne sulfur would be expected to reduce atmospheric corrosion effects.

XI. Future Study

Although a great deal has been learned about the photochemistry of the Earth's troposphere, a number of areas are still poorly understood. These include

1. A definitive assessment of the role of natural emittants in atmospheric budgets. Current, but preliminary, evidence suggests that natural sources play important roles in the budgets for sulfur, organic carbon, and the halogens. Natural contributions to the nitrogen cycle are probably small. These assessments need to be buttressed with more research, however, since natural processes tend to be of modest local importance but broad geographical impact.
2. An understanding of atmospheric liquid-phase chemistry. It is clear that the liquid phase is important to many atmospheric processes, but measurements are difficult and exacting and theoretical studies are incomplete.
3. Defining the atmospheric aerosol budget. The fluxes of aerosol particles from a variety of sources are still poorly known, as are the rates of aerosol loss.
4. Developing high-quality regional scale chemical–meteorological models. Only with such models can scientists forecast the effects that might be produced by changes in emission fluxes.

It has been clearly demonstrated that the Earth's troposphere is a complex system produced by a mixture of natural and anthropogenic sources. Its study has called upon chemistry, physics, meteorology, microbiology, and a host of allied fields. Explaining its operation and its changes offers perhaps an extreme test in synthesizing scientific thought and observation.

References

Ailor, W. H., ed. (1982). "Atmospheric Corrosion." Wiley, New York.

Atkinson, R., Darnall, K. R., Lloyd, A. C., Winer, A. M., and Pitts, J. N., Jr. (1979). Kinetics and mechanisms of the reaction of the hydroxyl radical with organic compounds in the gas phase. *Adv. Photochem.* **11,** 375–488.

Atkinson, R., Winer, A. M., and Pitts, J. N., Jr. (1982). Rate constants for the gas phase reactions of O_3 with the natural hydrocarbons isoprene and α- and β-pinene. *Atmos. Environ.* **16,** 1017–1020.

Bacastow, R. B., and Keeling, C. D. (1981). Atmospheric carbon dioxide concentration and the observed airborne fraction. *SCOPE* [*Rep.*] **16,** 103–112.

Baulch, D. L., Cox, R. A., Crutzen, P. J., Hampson, R. F., Jr., Kerr, J. A., Troe, J., and Watson, R. T. (1982). Evaluated kinetic and photochemical data for atmospheric chemistry. Supplement I. *J. Phys. Chem. Ref. Data* **11,** 327–496.

Bennett, L. H., Kruger, J., Parker, R. L., Passaglia, E., Reimann, C., Ruff, A. W., Laskowitz, H., and Berman, E. B. (1978). "Economic Effects of Metallic Corrosion in the United States," NBS Spec. Publ. 511. Natl. Bur. Stand., Washington, D.C.

Brimblecombe, P. (1976). Attitudes and responses towards air pollution in medieval England. *J. Air Pollut. Control Assoc.* **26,** 941–945.

Chameides, W. L., and Davis, D. D. (1982). The free radical chemistry of cloud droplets and its impact upon the composition of rain. *JGR, J. Geophys. Res.* **87,** 4863–4877.

Clarke, A. G. (1981). Electrolyte solution theory and the oxidation rate of sulphur dioxide in water. *Atmos. Environ.* **15,** 1591–1595.

Cunnold, D. M., Prinn, R. G., Rasmussen, R. A., Simmonds, P. G., Alyea, F. N., Cardelino, C. A., Crawford, A. J., Fraser, P. J., and Rosen, R. D. (1983a). The atmospheric lifetime experiment, 3, lifetime methodology and application to three years of $CFCl_3$ data. *JGR, J. Geophys. Res.* **88,** 8379–8400.

Cunnold, D. M., Prinn, R. G., Rasmussen, R. A., Simmonds, P. G., Alyea, F. N., Cardelino, C. A., and Crawford, A. J. (1983b). The atmospheric lifetime experiment. 4. Results for CF_2Cl_2 based on three years data. *JGR, J. Geophys. Res.* **88,** 8401–8414.

Darzi, M., and Winchester, J. W. (1982). Aerosol characteristics at Mauna Loa Observatory, Hawaii, after East Asian dust storm episodes. *JGR, J. Geophys. Res.* **87,** 1251–1258.

Gill, P. S., Graedel, T. E., and Weschler, C. J. (1983). Organic films on atmospheric aerosol particles, fog droplets, cloud droplets, raindrops, and snowflakes. *Rev. Geophys. Space Phys.* **21,** 903–920.

Graedel, T. E. (1979). Terpenoids in the atmosphere. *Rev. Geophys. Space Phys.* **17,** 937–947.

Graedel, T. E. (1980). Atmospheric photochemistry. *In* "Handbook of Environmental Chemistry" (O. Hutzinger, ed.), Vol. 2A, pp. 107–143. Springer-Verlag, Berlin and New York.

Graedel, T. E., and McRae, J. E. (1980). On the possible increase of the atmospheric methane and carbon monoxide concentrations during the last decade. *Geophys. Res. Lett.* **7,** 977–979.

Graedel, T. E., and Weschler, C. J. (1981). Chemistry in aqueous atmospheric aerosols and raindrops. *Rev. Geophys. Space Phys.* **19,** 505–539.

Granat, L. (1972). "Deposition of Sulfate and Acid with Precipitation over Northern Europe," Rep. AC-20. University of Stockholm, Dept. of Meteorology/International Meteorological Institute, Stockholm.

Granat, L., Rodhe, H., and Hallberg, R. O. (1976). The global sulphur cycle. *SCOPE* [*Rep.*] **7,** 89–134.

Haagen-Smit, A. J. (1952). Chemistry and physiology of Los Angeles smog. *Ind. Eng. Chem.* **44,** 1342–1351.

Heidt, L. E., Krasnec, J. P., Lueb, R. A., Pollack, W. H., Henry, B. E., and Crutzen, P. J. (1980). Latitudinal distributions of CO and CH_4 over the Pacific. *JGR, J. Geophys. Res.* **85,** 7329–7336.

Johnson, W. B., Ludwig, F. L., Dabberdt, W. F., and Allen, R. J. (1973). An urban diffusion simulation model for carbon monoxide. *J. Air Pollut. Control Assoc.* **23,** 490–498.

Junge, C. E. (1977). Basic considerations about trace constituents in the atmosphere as related to the fate of global pollutants. *In* "Fate of Pollutants in the Air and Water Environment" (I. H. Suffet, ed.), Part 1, pp. 7–25. Wiley, New York.

Keene, W. C., Galloway, J. N., and Holden, J. D., Jr. (1983). Measurement of weak organic acidity in precipitation from remote areas of the world. *JGR, J. Geophys. Res.* **88,** 5122–5130.

Khalil, M. A. K., and Rasmussen, R. A. (1983). Sources, sinks, and seasonal cycles of atmospheric methane. *JGR, J. Geophys. Res.* **88,** 5131–5144.

Lee, Y.-N., and Schwartz, S. E. (1981). Evaluation of the rate of uptake of nitrogen dioxide by atmospheric and surface liquid water. *JGR, J. Geophys. Res.* **86,** 11971–11983.

Levine, J. S., and Graedel, T. E. (1981). Photochemistry in planetary atmospheres. *EOS, Trans. Am. Geophys. Union* **62,** 1177–1181.

Levy, H., II (1971). Normal atmosphere: large radical and formaldehyde concentrations predicted. *Science* **173,** 141–143.

Lloyd, A. C., Atkinson, R., Lurmann, F. W., and Nitta, B. (1983). Modeling potential ozone impacts from natural hydrocarbons—I. Development and testing of a chemical mechanism for the NO_x–air photooxidation of isoprene and α-pinene under ambient conditions. *Atmos. Environ.* **17,** 1931–1950.

Logan, J. A., Prather, M. J., Wofsy, S. C., and McElroy, M. B. (1981). Tropospheric chemistry: a global perspective. *JGR, J. Geophys. Res.* **86,** 7210–7254.

Lyons, W. A., Dooley, J. C., Jr., and Whitby, K. T. (1978). Satellite detection of long-range pollution transport and sulfate aerosol hazes. *Atmos. Environ.* **12,** 621–631.

Martin, L. R. (1983). Kinetic studies of sulfite oxidation in aqueous solution. *In* "Acid Precipitation: SO_2, NO and NO_2 Oxidation Mechanisms: Atmospheric Considerations" (J. G. Calvert, ed.), pp. 63–100. Ann Arbor Sci. Publ., Ann Arbor, Michigan.

Munger, J. W., and Eisenreich, S. J. (1983). Continental-scale variations in precipitation chemistry. *Environ. Sci. Technol.* **17,** 32A–42A.

Prinn, R. G., Rasmussen, R. A., Simmonds, P. G., Alyea, F. N., Cunnold, D. M., Lane, B. C., Cardelino, C. A., and Crawford, A. J. (1983). The atmospheric lifetime experiment. 5. Results for CH_3CCl_3 based on three years of data. *JGR, J. Geophys. Res.* **88,** 8415–8426.

Prospero, J. M., and Nees, R. T. (1977). Dust concentrations in the atmosphere of the equatorial North Atlantic: possible relationship to the Sahelian drought. *Science* **196,** 1196–1198.

Rasmussen, R. A., and Khalil, M. A. K. (1981). Atmospheric methane (CH_4): trends and seasonal cycles. *JGR, J. Geophys. Res.* **86,** 9826–9832.

Reichle, H. G., Jr., Beck, S. M., Haynes, R. E., Hesketh, W. D., Holland, J. A., Hypes, W. D., Orr, H. D., III, Sherill, R. T., Wallio, H. A., Casas, J. C., Saylor, M. S., and Gormsen, B. B. (1982). Carbon monoxide measurements in the troposphere. *Science* **218,** 1024–1026.

Saltzman, E. S., Savoie, D. L., Zika, R. G., and Prospero, J. M. (1983). Methane sulfonic acid in the marine atmosphere. *JGR, J. Geophys. Res.* **88,** 10897–10902.

Simmonds, P. G., Alyea, F. N., Cardelino, C. A., Crawford, A. J., Cunnold, D. M., Lane, B. C., Lovelock, J. E., Prinn, R. G., and Rasmussen, R. A. (1983). The atmospheric lifetime experiment. 6. Results for carbon tetrachloride based on three years data. *JGR, J. Geophys. Res.* **88,** 8427–8441.

SMIC (1971). "Inadvertant Climate Modification: Report of a Study of Man's Impact on Climate," pp. 188–192. MIT Press, Cambridge, Massachusetts.

Thomson, J. (1857). "Grand Currents of Atmospheric Circulation," pp. 65, 115, 136. Br. Assoc. Meet., Dublin.

Walsh, J. J., Rowe, G. T., Iverson, R. L., and McRoy, C. P. (1981). Biological export of shelf carbon is a sink of the global CO_2 cycle. *Nature (London)* **291,** 196–201.

Weiss, R. F. (1981). The temporal and spatial distribution of tropospheric nitrous oxide. *JGR, J. Geophys. Res.* **86,** 7185–7195.

White, W. H., Moore, D. J., and Lodge, J. P., Jr., eds. (1981). Plumes and visibility: measurements and model components. *Atmos. Environ.* **15**(10/11), 1785–2406.

Wolff, G. T. (1980). Mesoscale and synoptic scale transport of aerosols. *Ann. N. Y. Acad. Sci.* **338,** 379–388.

3

The Photochemistry of the Stratosphere

RICHARD P. TURCO

R & D Associates
Marina del Rey, California

I. Introduction

Interest in the photochemistry of the Earth's stratosphere is primarily a result of interest in stratospheric ozone (O_3). The presence of large quantities of O_3 in the atmosphere was first recognized by Hartley (1881b), based on the observed attenuation of near ultraviolet (UV) solar radiation (Cornu, 1879) in the O_3 "Hartley" absorption bands (Hartley, 1881a). It was nearly half a

ISBN 0-12-444920-4

century before Sidney Chapman (1930) devised a photochemical reaction scheme that could explain the formation of O_3 at stratospheric altitudes (about 10–50 km). It was another four decades before further major advances in stratospheric photochemistry occurred. Then, between 1970 and 1980, an explosive growth in stratospheric science took place, and highly sophisticated schemes to describe the photochemical, radiative, and dynamic processes of the region appeared. In addition to O_3, interrelated chemical cycles involving water vapor (H_2O), nitrous oxide (N_2O), and hydrogen chloride (HCl) were discovered. Theories based on laboratory data, atmospheric measurements, and learned speculations were proposed and refined, culminating in the development of detailed photochemical/dynamic models of stratospheric composition (World Meteorological Organization, 1982).

The dynamic meteorology of the stratosphere is a specialized and complex subject (Holton, 1979), which is not treated here. It is important, however, to note that stratospheric air is buoyantly stable, which limits its mixing rate with tropospheric air. As a result, gaseous compounds typically have long "residence times" within the stratospheric photochemical "reaction chamber," perhaps a year or more. Stratospheric composition therefore tends to be fairly homogeneous on a global scale, unlike tropospheric composition. Likewise, the same photochemistry applies generally to the entire stratosphere, which is convenient for studying this region (notwithstanding, of course, substantial variations in composition with season, altitude, and other geophysical indices).

II. Composition and Chemical Families

The Earth's stratosphere is composed of dozens of trace gases. Many are listed, along with typical relative abundances, in Table I. Figure 1 gives the concentration profiles of a number of important stratospheric species. It is apparent that the stratosphere is composed of several long-lived "source" compounds from which families, or groups of closely related reactive species, are derived. For example, the "odd-oxygen family" derived from molecular oxygen (O_2) consists of O_3 and atomic oxygen (O) in two distinct electronic states [$O(^3P)$, or ground state O, and $O(^1D)$]. By the early 1970s, it was appreciated that families of related compounds are strongly coupled through photochemical reaction cycles (Nicolet, 1972). The family concept has since been developed into a formal mathematical technique for numerical modeling (e.g., Turco and Whitten, 1977). Within a family, individual species are rapidly cycled one into another. The overall production and loss of species in the family, however, occurs at a much slower rate. In other words, there are (at least) two fundamental time scales that distinguish a family: a short time scale,

Table I
STRATOSPHERIC GASEOUS CONSTITUENTS[a]

Source species[b]	Derivative species[c]
O_2 [21%]	O [0.1–100 b]$_u$, O(^{1}D) [0.1–10 q]$_u$, O_3 [2–10 m]
N_2 [78%]	—
Ar [0.9%]	—
CO_2 [350 m]	CO [2–40 b]
H_2O [3–5 m]	H [≲20 t]$_u$, OH [0.01–0.8 b]$_u$,
H_2 [0.5 m]	HO_2 [0.1–0.4 b]$_u$, H_2O_2 [≲0.3 b]
N_2O [0.15–0.3 m]$_l$	N [1–100 q]$_u$, NO [≲15 b], NO_2 [≲5 b], NO_3 [≲0.3 t], N_2O_5 [≲0.6 b], HNO_2 [≲10 t], HNO_3 [≲5 b], HO_2NO_2 [≲0.5 b]
CH_4 [1–1.5 m]$_l$	CH_3O_2 [≲20 t], CH_3OOH [≲30 t], CH_3O [≲10 q], CH_2O [≲0.1 b], CHO [≪1 q], $CH_3O_2NO_2$ [≪1 b]
C_2H_2 [<0.1 b]$_l$; C_2H_4 [≲0.1 b]$_l$; C_2H_6 [0.1–1 b]$_l$	
CH_3Cl [0.5 b]$_l$	Cl [≲20 t]$_u$, ClO [≲0.5 b], ClO_2 [≪1 t],
HCl [≲2 b]	HOCl [≲0.1 b], $ClONO_2$ [≲0.5 b]
$CFCl_3$ [≲0.1 b]$_l$	
CF_2Cl_2 [≲0.2 b]$_l$	HF [0.1–1 b]
CCl_4 [≲0.1 b]$_l$	
CF_4 [75 t]	
OCS [0.5 b]	SO [≪1 t], HSO_3 [≲1 t], H_2SO_4 [≲10 t]
SO_2 [0.05 b]	
CH_3Br [≲20 t]	Br [≲4 t], BrO [≲10 t], HBr [≲3 t], $BrONO_2$ [≲3 t].
HCN [160 t]; CH_3CN [≲10 t]	

[a] Typical species concentrations, expressed as mixing fractions relative to the total concentration of air molecules (and averaged over 24 hr), are given in brackets: m ≡ parts per million by number, b ≡ parts per billion, t ≡ parts per trillion, q ≡ parts per quadrillion. Thus 3 m indicates 3 ppm, or 3×10^{-6}. Subscripts on the mixing-fraction brackets: l ≡ lower stratosphere (≲30 km), u ≡ upper stratosphere (≳30 km). When so subscript is present, a rough average mixing ratio for the entire stratosphere is indicated.

[b] Isotopic differentiation is neglected. Potentially important atmospheric isotopes are ^{18}O [0.204%], ^{15}N [0.37%], ^{2}H [0.015%], ^{34}S [4.22%], and ^{37}Cl [24.47%], where the natural fractional abundances are indicated in brackets. In addition, ^{3}H and ^{14}C, formed by cosmic rays and nuclear explosions, may be used to trace stratospheric air motions.

[c] Atoms and molecules in vibrational and metastable electronic states are not included as separate species, except for O(^{1}D). Potentially important excited species are $O_2(^1\Delta_g)$, $O_2(^1\Sigma_u^+)$, OH (v = 1–9), and N(^{2}D).

which represents the rapid equilibration among the family members, and a long time scale, which represents the lifetime for production or loss of the entire family.

These simple concepts may be illustrated using the odd-oxygen family, $O_x \equiv O_3 + O + O(^1D)$. The basic photochemical reactions controlling O_3

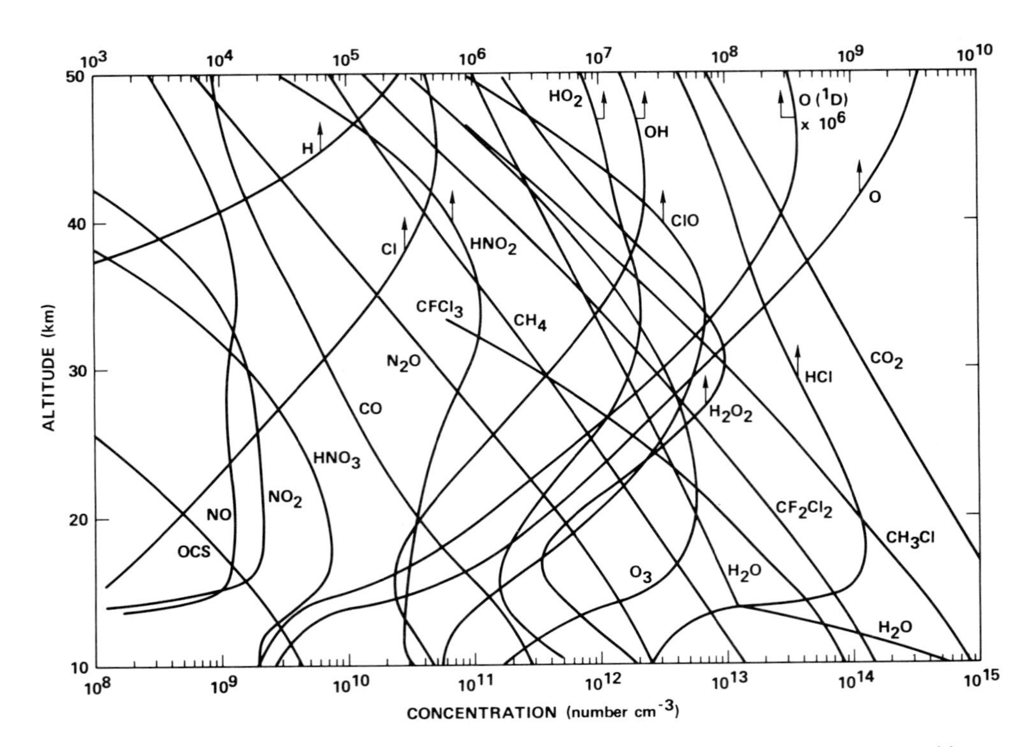

Fig. 1. Vertical concentration profiles of stratospheric gaseous constituents. Note the upper and lower scales.

and O were outlined by Chapman (1930):

$$O_2 + h\nu \longrightarrow O + O \tag{1}$$

$$O + O_2 + M \longrightarrow O_3 + M \tag{2}$$

$$O_3 + h\nu \longrightarrow O + O_2 \tag{3}$$

$$O + O_3 \longrightarrow O_2 + O_2 \tag{4}$$

$$O + O + M \longrightarrow O_2 + M \tag{5}$$

where $h\nu$ indicates a photon of sunlight that is sufficiently energetic to cause molecular dissociation, and M represents the total concentration of air molecules (i.e., $O_2 + N_2$) that can act as "third bodies" in association reactions. Assuming that $[O_2]$ is fixed (where square brackets are used to denote a species' number concentration per cubic centimeter of air) and ignoring the dynamic transport of material, the chemical rate equations for the O_3 and O concentrations are

$$[\dot{O}_3] = R_2 - R_3 - R_4, \tag{6}$$

$$[\dot{O}] = 2R_1 + R_3 - R_2 - R_4 - 2R_5, \tag{7}$$

where the chemical rates (molec cm^{-3} sec^{-1}) are indicated by appropriately subscripted reaction R terms. The rate of change of $[O_x]$ is simply the sum of Eqs. (6) and (7):

$$[\dot{O}_x] = 2R_1 - 2R_4 - 2R_5 \doteq 2R_1 - 2R_4. \tag{8}$$

The rate of reaction (5) is much slower than that of reaction (4) in the stratosphere.

The time constants for odd-oxygen recycling processes (2) and (3) are of the order of milliseconds to minutes. For example, the lifetime of O against association with O_2 is only 1 msec at 20 km. In contrast, the lifetime of the O_x family, as determined by reaction (4) through Eq. (8), is several months at 20 km.

The immense range of photochemical time constants characteristic of stratospheric chemical systems poses a classic "stiffness" problem in obtaining numerical solutions. The family concept allows, to some degree, a division of the problem into time-constant regimes. Family analysis also allows one to gain insights into the important processes and interactions that affect stratospheric photochemistry. For example, it seems apparent that relatively slow reaction (4) controls the loss of O_x in the Chapman reaction scheme. The other processes determine the concentration of O relative to O_3. One can show that

$$[O] \doteq r_3[O_3]/r_2[O_2][M], \tag{9}$$

where the r's represent the chemical kinetic or solar photolysis constants for the appropriate photochemical processes. Because of the close coupling between O and O_3 reflected by Eq. (9), the chemical loss rate R_4 can be expressed as

$$R_4 = r_4[O][O_3] \doteq r_4 r_3 [O_3]^2 / r_2 [O_2][M]. \tag{10}$$

Thus it is revealed that processes (2) and (3) are as important as reaction (4) in determining O_x (and so, O_3) concentrations in isolated air parcels (for this simplified reaction scheme).

Equations (8)–(10) can be solved for the "equilibrium" Chapman O_3 concentration by making the reasonable assumption that $[\dot{O}_x] = 0$. Then,

$$[O_3] \doteq \sqrt{r_1 r_2 [O_2]^2 [M] / r_3 r_4} \doteq [O_2]^{3/2} (5 r_1 r_2 / r_3 r_4)^{1/2}. \tag{11}$$

This relationship is only approximate for two reasons: important chemical processes involving hydrogen, nitrogen, and chlorine species are ignored (see below), and transport of odd-oxygen is neglected. Nevertheless, Eq. (11) provides a reasonable first-order estimate of stratospheric O_x concentrations above $\sim$30 km.

A number of stratospheric chemical families are described in Table II, and the primary stratospheric budgets are summarized in Table III. The consideration of families can provide insights into the workings of stratospheric photochemistry. The definition of a family depends on the time scales of interest. For example, on a time scale of hours, the species H, OH, and HO_2 are closely coupled and together comprise the HO_x family. On a longer time scale (days, approximately), H_2O_2 may be included as an HO_x species. Families may also overlap. For example, NO_2 is a member of both the odd-nitrogen (NO_x) and odd-oxygen families. The connection between the NO_x and O_x families is made through the rapid reaction cycle:

$$NO_2 + h\nu \longrightarrow NO + O$$

$$O + O_2 + M \longrightarrow O_3 + M$$

$$NO + O_3 \longrightarrow NO_2 + O_2$$

which leaves the composition of both families unchanged. The so-called *odd* families may have *even* members (e.g., N_2O_5 in the NO_x family). In such cases, a key chemical bond (e.g., N—N in N_2O_5) is weak and subject to (relatively) rapid dissociation, in contrast to the stronger bonds characteristic of more stable compounds (e.g., N—N in N_2 and N_2O).

In subsequent sections, the photochemistry of the principal stratospheric chemical families, as well as the kinetic coupling between the families, are discussed in more detail. In reality, the photochemistry of the stratosphere involves more than 60 key species and 200 photochemical processes. An accurate numerical analysis requires the use of a sophisticated computer

Table II

FAMILIES OF STRATOSPHERIC CONSTITUENTS

Family	Species[a] and weights[b]
Odd-oxygen O_x	O [1], O_3 [1], $O(^1D)$ [1], NO_2 [1], NO_3 [2], N_2O_5 [3], ClO [1], $ClONO_2$ [2], $BrONO_2$ [2]
Odd-hydrogen HO_x	H [1], OH [1], HO_2 [1], H_2O_2 [2], HNO_2 [1], HNO_3 [1], HOCl [1], CH_3OOH [1]
Odd-nitrogen NO_x	N [1], NO [1], NO_2 [1], NO_3 [1], N_2O_5 [2], HNO_2 [1], HNO_3 [1], $ClONO_2$ [1], $BrONO_2$ [1]
Active chlorine Cl_x	Cl [1], ClO [1], ClO_2 [1], ClOO [1], HOCl [1], $ClONO_2$ [1], HCl [1]
Methane derivatives CH_xO_y	CH_3O_2 [1], CH_3O [1], CH_3OOH [1], CH_2O [1], CHO [1]
Sulfur oxides SO_x	SO [1], SO_2 [1], SO_3 [1], HSO_3 [1], H_2SO_4 [1]
Bromine Br_x	Br [1], BrO [1], $BrONO_2$ [1], HBr [1]

[a] Several of the species are members of two or more families. The definition of a family is flexible, depending on the time scales under consideration.

[b] The weights (in brackets) give the number of times each species is to be counted in calculating the total family concentration.

model based on a set of coupled species continuity equations (Chapter 1). Even so, a thorough understanding of the fundamental photochemical processes is necessary to construct such a model and to interpret properly its numerical predictions.

III. Photochemical Data Base

To study the photochemistry of the stratosphere, a suitable chemical kinetic and photolytic data base is needed. Temperature- and pressure-dependent chemical reaction rate coefficients, deduced from extensive laboratory measurements, are summarized in Appendix II [detailed tabulations are also provided by Baulch *et al.* (1982) and by DeMore *et al.* (1983)]. To calculate photolysis rates at a given location in the stratosphere, the following expression must be evaluated:

$$p_{ij}(\bar{x}, t) = \int d\lambda \{I(\lambda, \bar{x}, t)\sigma_i(\lambda, T[\bar{x}])\phi_{ij}(\lambda, T[\bar{x}])\}, \tag{12}$$

where p_{ij} is the photolysis rate ($molec^{-1}\ sec^{-1}$) of species i by dissociation process j, λ the wavelength of the impinging light (nm), I the total solar flux

Table III

CHEMICAL BUDGETS OF THE STRATOSPHERE[a]

Chemical family or species	Photochemical production in stratosphere	Photochemical destruction in stratosphere	Flux from troposphere	Flux from mesosphere
O_x (O)	35,000	35,000	−300	~0
HO_x (H)	14	14	−0.1	~0
H_2O (H_2O)	55	$\lesssim 1$	−55	−0.3
NO_x (N)	0.9	0.3	−0.5	−0.1
N_2O (N_2O)	0[b]	20	20	<1
CH_4 (CH_4)	0	30	30	~0
Cl_x (Cl)	0.5	0	−0.5	~0
SO_x (S)	0.1	0.2[c]	0.1[d]	~0

[a] Rough estimates of global annual photochemical production and loss rates, and dynamic transfer rates, are given. Units of 10^6 metric tons (1 metric ton = 10^6 g) are used throughout, with the tabulated masses keyed to the specific atom or species indicated in the parentheses in the first column.

[b] Zipf (1984) and Zipf and Prasad (1984) proposed a substantial stratospheric source of N_2O from the reactions

$$N_2(A\,^3\Sigma_u^+) + O_2 \longrightarrow N_2O + O$$
$$OH(A\,^2\Sigma^+) + N_2 \longrightarrow N_2O + H$$

[c] Conversion to sulfate aerosol.

[d] The flux can be enhanced 10-fold in years when major volcanic eruptions occur, such as Agung (1963) and El Chichón (1982).

(photons cm^{-2} sec^{-1} nm^{-1}, both direct and diffuse) at point $\bar{x}$ at time t, σ_i the molecular absorption cross section (cm^2 $molec^{-1}$), which may be sensitive to temperature T and pressure, and ϕ_{ij} the quantum yield for the particular dissociation process of interest (see Chapter 1 for a more complete discussion of solar photodissociation mechanisms).

While σ and ϕ are fundamental molecular parameters, which may be determined experimentally, the solar intensity depends on several variable factors such as atmospheric composition, altitude, and solar zenith angle. Figure 2 illustrates the spectrum of sunlight in the Earth's stratosphere and mesosphere. The most prominent features are the altitude-independent solar flux at wavelengths above ~330 nm, the strong attenuation with altitude of the flux in the UV wavelength region below 300 nm, and the relatively large solar intensities in the stratospheric spectral "window" between 200 and 230 nm. The calculation of stratospheric photodissociation rates involves a convolution of absorption cross sections, quantum yields, and light fluxes in the near-UV and visible-wavelength regions.

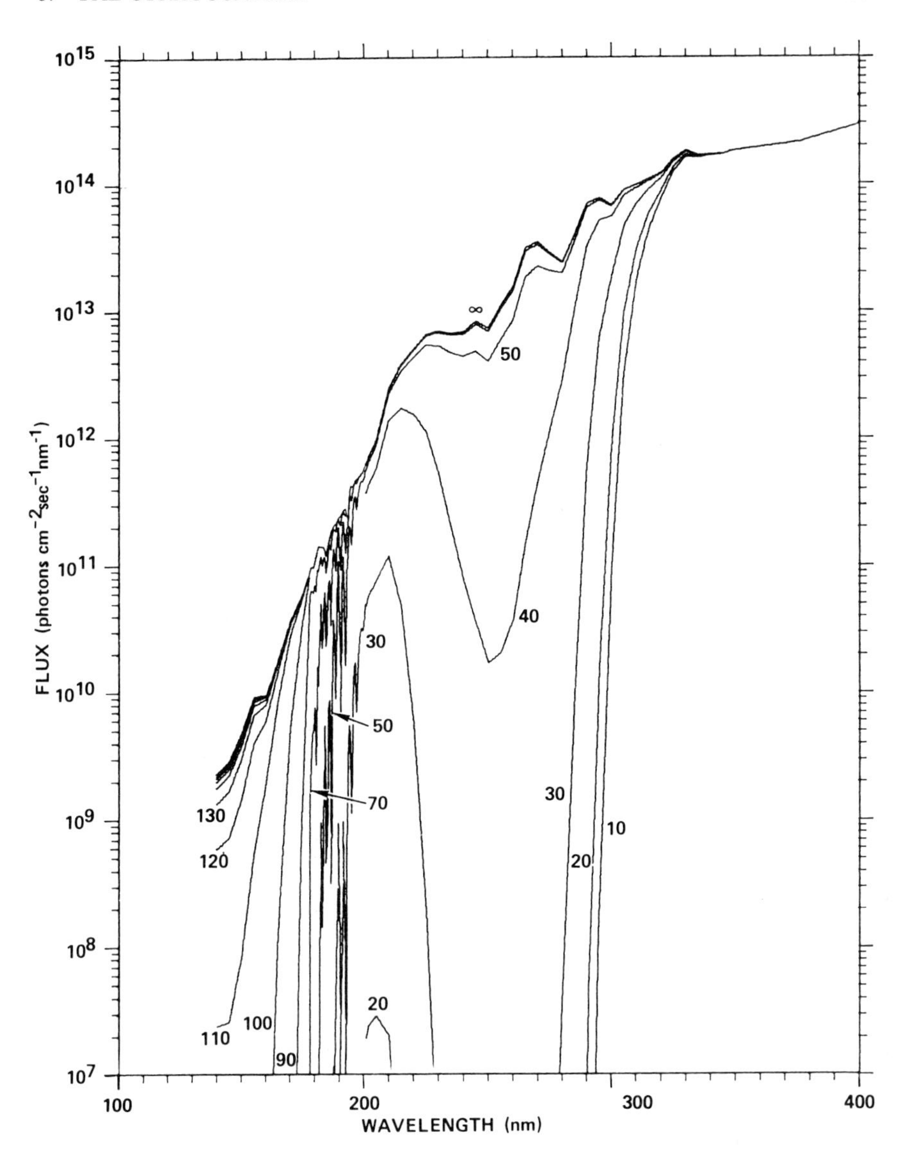

Fig. 2. Solar fluxes in the atmosphere as a function of wavelength and altitude at a solar zenith angle of 60°. The flux spectra are labeled by altitude (km). In the O_2 Schumann–Runge band region (178–198 nm), the attenuated solar fluxes are shown only at altitudes of 30, 50, 70, and 90 km for clarity.

IV. The Chapman Ozone Cycle

The basic Chapman reactions for O_3 and atomic oxygen are given in Section II [reactions (1)–(5)]. The complete photochemical cycle of the oxygen species is illustrated in Fig. 3. The photodissociation of molecular oxygen is the source of atmospheric O_3 and the fundamental driving force of

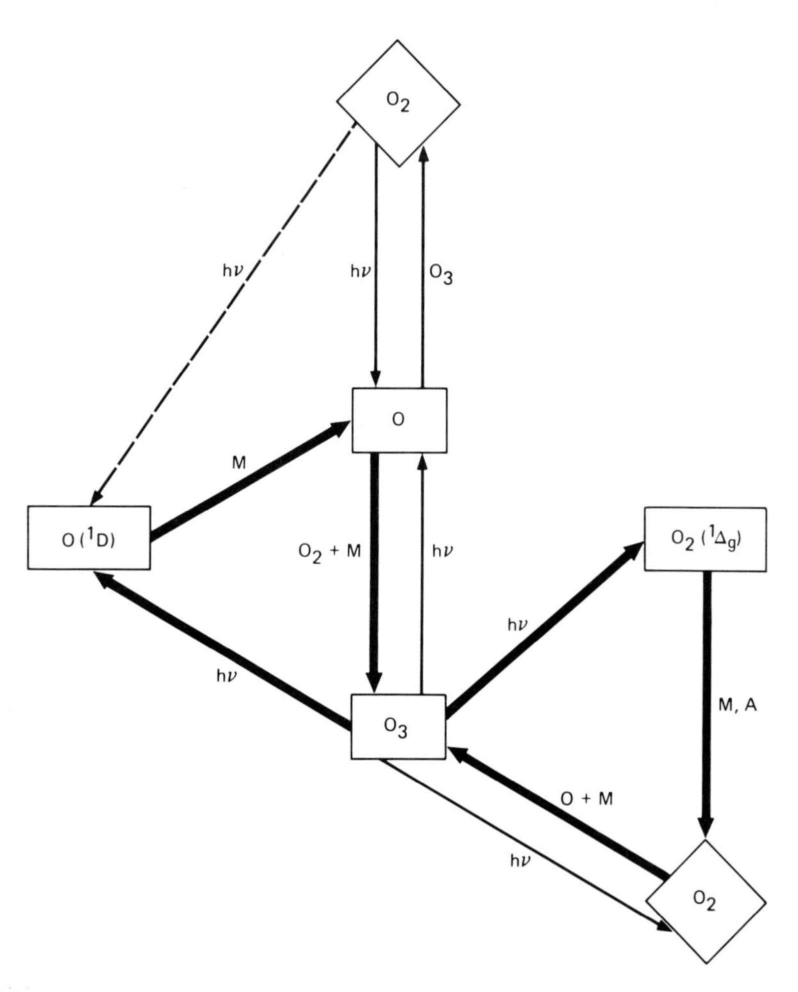

Fig. 3. The pure oxygen photochemical cycle of the stratosphere. Source gases are shown within diamonds, reactive derivative species within boxes. The major reaction pathways are indicated by the heaviest arrows, minor pathways by dashed arrows (see following figures). Reactant species are shown alongside the arrows indicating reaction pathways. *hν*, Solar photolysis; M, a third body (any air molecule); A, spontaneous light emission.

stratospheric photochemistry:

$$O_2 + h\nu \xrightarrow{129 < \lambda < 176 \text{ nm}} O(^3P) + O(^1D) \quad (13)$$

$$O_2 + h\nu \xrightarrow{176 < \lambda < 195 \text{ nm}} O(^3P) + O(^3P) \quad (14)$$

$$O_2 + h\nu \xrightarrow{185 < \lambda < 242 \text{ nm}} O(^3P) + O(^3P) \quad (15)$$

[Note that the symbols O and $O(^3P)$ are used interchangeably to indicate ground-state oxygen atoms.] The first process—photodissociation of O_2 in the Schumann–Runge (S–R) continuum—occurs only in the upper mesosphere and thermosphere. Oxygen predissociation in the S–R bands [process (14)] dominates O_2 photolysis throughout the mesosphere. The attenuating effect of the O_2 S–R band absorption on the solar radiation field also controls the photolysis in the upper stratosphere of molecules with dissociation thresholds below 200 nm (e.g., H_2O and CF_2Cl_2). Photodissociation in the Herzberg continuum [process (15)] is the major source of odd-oxygen in the stratosphere and thus is responsible for most of the atmospheric ozone burden.

Absorption by O_2 in the Herzberg continuum region from about 200 to 230 nm is particularly important because it lies within the stratospheric spectral "window" (Fig. 2) between the O_2 S–R bands at shorter wavelengths and the O_3 Hartley bands at longer wavelengths. The Herzberg continuum absorption cross sections have always been somewhat controversial (e.g., Turco, 1975). Direct stratospheric measurements of solar light intensities in this window, however, seem to have established cross sections *lower*, by as much as 40%, than previously accepted (Frederick and Mentall, 1982; Herman and Mentall, 1982; Anderson and Hall, 1983). Reduced Herzberg continuum absorption implies increased fluxes of dissociating UV radiation leaking through the stratospheric spectral window and attendant changes in photodissociation rates.

Ozone photodissociation can occur by several processes:

$$O_3 + h\nu \xrightarrow{450 < \lambda < 750 \text{ nm}} O(^3P) + O_2(^3\Sigma_g^-) \quad (16)$$

$$O_3 + h\nu \xrightarrow{310 < \lambda < 350 \text{ nm}} O(^3P) + O_2(^1\Delta_g) \quad (17)$$

$$O_3 + h\nu \xrightarrow{\lambda < 310 \text{ nm}} O(^1D) + O_2(^1\Delta_g) \quad (18)$$

$$O_3 + h\nu \xrightarrow{\lambda < 267 \text{ nm}} O(^1D) + O_2(^1\Sigma_g^+) \quad (19)$$

where $O_2(^3\Sigma_g^-)$ is the ground electronic state of molecular oxygen. The first process [(16)] occurs in the ozone Chappuis bands. In the lower stratosphere (below ~30 km), this process dominates ozone photolysis. Photodissociation in the weak (spin-forbidden) Huggins bands [process (17)] is never the dominant mode of decomposition.

Photolysis in the Hartley bands [processes (18) and (19)] is the primary O_3 decomposition mechanism above approximately 30–35 km but is less important at lower altitudes because the solar flux in the Hartley spectral region is strongly attenuated by absorption in the overlying O_3 layer. Nevertheless, pathway (18), which leads to $O(^1D)$ production, is critical to the chemistry of the atmosphere: $O(^1D)$ reacts with H_2O and N_2O, initiating the odd-hydrogen and odd-nitrogen chemical cycles, respectively. These cycles in turn drive the hydrocarbon, chlorine, sulfur, and other chemical cycles (see below).

The absorption of sunlight by O_3 leads to significant heating of the stratosphere. Ozone heating is balanced primarily by CO_2 emission of infrared, or thermal, radiation to space. Detailed aspects of solar absorption, atmospheric heating, and radiative transfer, although closely related to the process of photodissociation, are not discussed here.

The production of electronically excited oxygen atoms [$O(^1D)$] by O_3 photolysis suggests that the basic Chapman reactions should be modified as follows:

$$O_3 + h\nu \longrightarrow O + O_2 \qquad \text{for reactions (16) and (17)}$$

$$O_3 + h\nu \longrightarrow O(^1D) + O_2 \qquad \text{for reactions (18) and (19)}$$

$$O(^1D) + M \longrightarrow O + M \tag{20}$$

The chemical rate equations for O and $O(^1D)$ are then

$$[\dot{O}] = 2R_1 + R_{16+17} + R_{20} - R_2 - R_4 - 2R_5, \tag{21}$$

$$[\dot{O}(^1D)] = R_{18+19} - R_{20}. \tag{22}$$

Equations (21) and (22) may be combined (added) to yield the original Chapman equation [Eq. (7)] for atomic oxygen.

The concentration of $O(^1D)$ is given, to a high degree of accuracy, by the relationship

$$[O(^1D)] \doteq r_{18+19}[O_3]/r_{20}[M]. \tag{23}$$

In the stratosphere, the odd-oxygen species concentrations fall into a distinct hierarchy: $[O(^1D)] \ll [O] \ll [O_3]$. The lifetime of $O(^1D)$ against quenching [reaction (20)] is $\lesssim 10\ \mu$sec throughout the stratosphere. Despite its low abundance, $O(^1D)$ is a key stratospheric constituent that is ultimately responsible for profound modifications of the Chapman photochemical scheme.

The excited oxygen molecules [$O_2(^1\Delta_g)$] produced by O_3 photolysis are largely responsible for the prominent "infrared atmospheric" oxygen airglow emission,

$$O_2(^1\Delta_g) \xrightarrow{1.27\ \mu m} O_2 + h\nu \tag{24}$$

Most of the $^1\Delta_g$ is quenched, however,

$$O_2(^1\Delta_g) + M \longrightarrow O_2 + M \tag{25}$$

The corresponding photochemical equilibrium equation for $O_2(^1\Delta_g)$ can be solved for the volumetric photoemission rate (i.e., the number of photons emitted per cubic centimeter of air per second),

$$\epsilon_{1.27} \doteq \frac{r_{24}r_{17+18}}{r_{24} + r_{25}[M]}[O_3]. \tag{26}$$

The form of Eq. (26) suggests that O_3 concentrations may be deduced from measurements of the 1.27 μm airglow intensity. In fact, this has been accomplished (Noxon, 1982).

Circerone and McCrumb (1980) noted that "heavy" isotopic molecular oxygen ($^{18}O^{16}O$), which has an abundance of 0.4% of the total atmospheric molecular oxygen, may undergo photodissociation more efficiently than $^{16}O_2$, providing an important secondary source of odd-oxygen for the stratosphere. Enhanced photolysis would occur because the individual rotational lines of the O_2 S–R band system would be slightly offset in $^{18}O^{16}O$ relative to $^{16}O_2$. Self-absorption normally limits the rate of photodissociation in the S–R bands. Heavy oxygen, on the other hand, could be dissociated by light leaking through the $^{16}O_2$ bands. The overall increase in ozone formation might be 10% at 30 to 50 km. Preliminary atmospheric measurements have detected "heavy" O_3, presumably formed by this process (Mauersberger, 1981) although with a different altitudinal distribution than expected. On the other hand, Kaye and Strobel (1983) have presented theoretical arguments suggesting that heavy odd-oxygen would be rapidly lost by isotopic mixing with O_2. These new developments demonstrate that the field of stratospheric photochemistry, inaugurated by Sidney Chapman more than 50 years ago, is still growing and evolving along unexpected and interesting pathways.

V. Water Vapor and the Odd-Hydrogen Cycle

Bates and Nicolet (1950) first discussed, in a thorough fashion, the photochemistry of H_2O in the upper atmosphere. The major initiating process above 50 km was assumed to be photolysis,

$$H_2O + h\nu \xrightarrow{\lambda < 242\ \text{nm}} H + OH \tag{27}$$

which happens to be quite slow in the stratosphere (Nicolet, 1972). Hunt (1966) introduced alternative decomposition reactions, which were discovered in experiments on the photolytic decomposition of O_3/H_2O mixtures:

$$O(^1D) + H_2O \longrightarrow 2\,OH \tag{28}$$

$$O(^1D) + H_2 \longrightarrow H + OH \tag{29}$$

Reactions (28) and (29) both have rate coefficients of the order of $\sim 10^{-10}$ cm^3 sec^{-1} and dominate odd-hydrogen (HO_x) production in the stratosphere (other sources of HO_x from the decomposition of methane, and nonmethane hydrocarbons, are described below).

Odd-hydrogen interacts strongly with the oxygen allotropes through a series of cycling reactions:

$$H + O_2 + M \longrightarrow HO_2 + M \tag{30}$$

$$HO_2 + O \longrightarrow OH + O_2 \tag{31}$$

$$OH + O \longrightarrow H + O_2 \tag{32}$$

$$H + O_3 \longrightarrow OH + O_2 \tag{33}$$

$$OH + O_3 \longrightarrow HO_2 + O_2 \tag{34}$$

$$HO_2 + O_3 \longrightarrow OH + 2\,O_2 \tag{35}$$

The HO_x species are closely coupled by these reactions, and their relative abundances may be roughly estimated using simple chemical equilibrium relationships. For example, it has been deduced (Leovy, 1969) that

$$[H]/[OH] \simeq r_{32}[O]/r_{30}[O_2][M], \tag{36}$$

$$[HO_2]/[OH] \simeq (r_{32}[O] + r_{34}[O_3])/(r_{31}[O] + r_{35}[O_3]), \tag{37}$$

where reaction (33) has been ignored. From Eqs. (36) and (37) it is readily determined (e.g., Crutzen, 1969) that $[H] \ll [OH]$ and $[OH] \lesssim [HO_2]$ in the middle and upper stratosphere [the lower stratosphere is more complicated, and Eqs. (36) and (37) do not apply there].

The basic odd-hydrogen reactions may be rearranged to define an important set of reaction cycles that consume odd-oxygen (e.g., Johnston and Podolske, 1978):

$$\begin{array}{c} OH + O \longrightarrow H + O_2 \\ H + O_2 + M \longrightarrow HO_2 + M \\ \underline{HO_2 + O \longrightarrow OH + O_2} \\ O + O \xrightarrow{M} O_2 \end{array} \tag{38}$$

$$\begin{array}{c} OH + O_3 \longrightarrow HO_2 + O_2 \\ \underline{HO_2 + O \longrightarrow OH + O_2} \\ O + O_3 \longrightarrow 2\,O_2 \end{array} \tag{39}$$

$$\begin{array}{c} OH + O_3 \longrightarrow HO_2 + O_2 \\ \underline{HO_2 + O_3 \longrightarrow OH + 2\,O_2} \\ 2\,O_3 \longrightarrow 3\,O_2 \end{array} \tag{40}$$

Note that in each of these reaction cycles the HO_x molecules are conserved, while the O_x species are converted to O_2. Thus the HO_x radicals are continuously recycled as O_x is consumed, prompting the definition of these reaction sets as "catalytic" cycles. It is worth remembering, however, that the HO_x concentrations are not *independent* of these cycles. Indeed, the partitioning of the HO_x species, and thus the rate of chemical production and loss in the HO_x family is essentially determined by the catalytic reaction steps.

Catalytic reaction cycle (38) operates primarily in the upper stratosphere, (39) in the middle stratosphere, and (40) in the lower stratosphere. These cycles describe the fundamental coupling between O_x and HO_x in the upper atmosphere.

A series of reactions terminates the HO_x cycle by regenerating chemically stable H_2O ad H_2 molecules:

$$2\,OH \longrightarrow H_2O + O \tag{41}$$

$$OH + HO_2 \longrightarrow H_2O + O_2 \tag{42}$$

$$H + HO_2 \longrightarrow H_2 + O_2 \tag{43}$$

$$H + HO_2 \longrightarrow H_2O + O \tag{44}$$

Of these, reaction (42) is by far the most important in the stratosphere. Certain hydrogen radical chemical processes, such as reaction (42), exhibit pressure and humidity dependences (DeMore *et al.*, 1983), which in the past have caused problems in interpreting laboratory kinetic data (see Appendix II). Additional chemical processes that terminate the HO_x cycle but involve nitrogen oxides and other species are discussed in subsequent sections.

Hydrogen peroxide (H_2O_2) is formed exclusively by the reaction

$$2\,HO_2 \longrightarrow H_2O_2 + O_2 \tag{45}$$

and it is depleted by the following processes:

$$H_2O_2 + h\nu \xrightarrow{<578\,\text{nm}} 2\,OH \tag{46}$$

$$H_2O_2 + OH \longrightarrow HO_2 + H_2O \tag{47}$$

$$H + HO_2 \longrightarrow OH + HO_2 \tag{48}$$

On the time scale of stratosphere–troposphere mixing, H_2O_2 may by treated as an HO_x family member (this is not true when diurnal variations are considered, however, as OH and HO_2 have lifetimes of $\ll 1$ day, while H_2O_2 has a lifetime of > 1 day). With H_2O_2 in the HO_x family, reaction (45) is not an HO_x termination reaction, and (46) and (48) simply recycle HO_x. Reaction (47), on the other hand, provides a net sink for HO_x. It appears that H_2O_2 has only a minor role in stratospheric photochemistry, even though it is the most abundant HO_x species that has been discussed up to this point.

The (pure) hydrogen photochemical cycle is summarized in Fig. 4. The HO_x termination reactions regenerate H_2O and H_2, which are only slightly affected

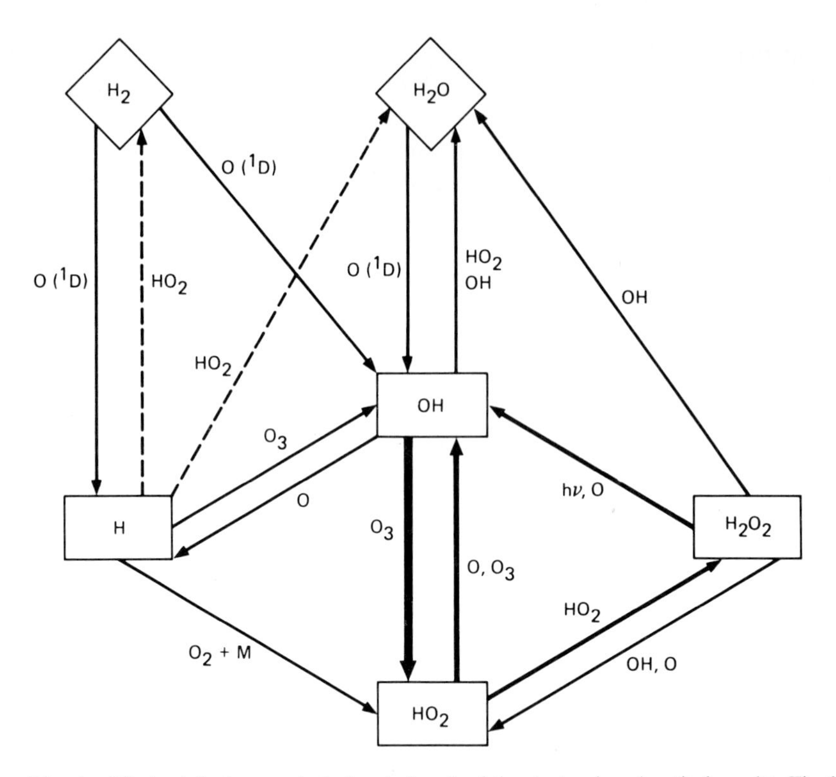

Fig. 4. The basic hydrogen photochemical cycle of the stratosphere (see the legend to Fig. 3 for an explanation of the symbols used).

by chemical transformation in the stratosphere. Many photochemical processes have been neglected in the present survey, because they have only a minor stratospheric role (e.g., $HO_2 + h\nu \rightarrow OH + O$, $2\,OH + M \rightarrow H_2O_2 + M$, $O + H_2 \rightarrow OH + H$). The processes outlined in this section describe the essential behavior of a pure O_x/HO_x reaction system.

When night falls in the stratosphere, many species disappear rapidly—e.g., O [by reaction (2)] and H [by reaction (30)]—other species remain relatively constant, e.g., O_3 and H_2O_2, and a few species increase substantially, e.g., NO_3 and N_2O_5. In the nighttime environment, the source of HO_x is shut off, while the loss of HO_x continues through reactions such as (42) and (47). Accordingly, OH and HO_2 concentrations decrease steadily during the night. At sunrise, recovery of HO_x concentrations through processes (28) and (46) is rapid. Because O, OH, and HO_2 all are depleted at night, the catalytic ozone reaction cycles (38), (39), and (40) are suppressed in darkness.

The OH "airglow" is an aspect of odd-hydrogen chemistry that has fascinated aeronomers ever since its discovery by Meinel in 1950 (1950a,b). The emission is caused by the relaxation of vibrationally excited OH molecules that are formed in the reactions

$$H + O_3 \longrightarrow OH(v) + O_2 \qquad v \leqslant 9 \tag{49}$$

$$HO_2 + O \longrightarrow OH(v) + O_2 \qquad v \leqslant 6 \tag{50}$$

A series of spontaneous transitions and quenching reactions follow:

$$OH(v) \longrightarrow OH(v') + h\nu \tag{51}$$

$$OH(v) + M \longrightarrow OH(v') + M \tag{52}$$

$$OH(v) + O \longrightarrow H + O_2 \tag{53}$$

Although most of the OH excitation occurs in the mesosphere and the emission is most prominent in the nightglow originating above 70 km, substantial OH excitation occurs throughout the stratosphere (Nagy *et al.*, 1976; Streit *et al.*, 1976). Students of aeronomy are well advised to remember that excited states and airglow emissions are important diagnostics of photochemical processes.

VI. Nitrous Oxide and the Odd-Nitrogen Cycle

Bates and Hayes (1967) introduced nitrous oxide (N_2O) into stratospheric photochemistry. They deduced most of the important aspects of the N_2O problem: N_2O is produced in soils by microbial action; N_2O reaches the stratosphere by atmospheric transport and diffusion; photodissociation at near-UV wavelengths is the major sink for atmospheric N_2O; and N_2O decomposition leads to odd-nitrogen (NO_x) production.*

To date, only two important modifications of the Bates and Hayes theory have been introduced. First, odd-nitrogen is not produced through the photodissociation of N_2O but rather through the reaction of N_2O with $O(^1D)$ (Greenberg and Heicklen, 1970; Nicolet, 1970; Crutzen, 1971). Second, the sources of nitrous oxide are more diverse than originally thought and include anthropogenic emissions; for example, N_2O is produced by lightning, by natural and manmade combustion processes, in fertilizer and sewage de-nitrification, and by auroral activity. Modern observations suggest that background N_2O concentrations are increasing at a rate of $\sim$0.2% per year

* Here the odd-nitrogen family is referred to as NO_x. Some authors use NO_x to indicate the subfamily of $NO + NO_2$, and NO_y to indicate the entire odd-nitrogen family. Such a distinction is unnecessary in the present discussion.

(Weiss, 1981). While all of the principal sources of atmospheric N_2O reside outside of the stratosphere, a secondary local source may be attributable to the reaction (Zipf, 1984)

$$OH(A\,^2\Sigma^+) + N_2 \longrightarrow N_2O + H \tag{54}$$

The photodissociation of nitrous oxide proceeds as

$$N_2O + h\nu \xrightarrow{\lambda < 341\ \text{nm}} N_2 + O(^1D) \tag{55}$$

The only other critical photochemical process for N_2O is decomposition by $O(^1D)$,

$$N_2O + O(^1D) \longrightarrow NO + NO \tag{56}$$

$$N_2O + O(^1D) \longrightarrow N_2 + O_2 \tag{57}$$

Reaction (56), which generates NO_x from N_2O, accounts for $\sim 60\%$ of the overall reaction with $O(^1D)$.

Although the primary source of stratospheric NO_x is associated with N_2O decomposition, other sources include downward transport of NO_x produced in the mesosphere by auroral activity; *in situ* production by solar proton events (SPEs), galactic cosmic rays, and large meteors; injection by nuclear explosions; and emission from high-altitude aircraft engines. Major cosmic events such as nearby supernovae may also have profound effects on stratospheric NO_x and its photochemical impact.

The significance of odd-nitrogen to stratospheric photochemistry was proposed independently by Crutzen (1970) and Johnston (1971), although earlier, Nicolet (1965) had investigated the aeronomic implications of nitrogen oxides in the mesosphere. The fundamental oxygen–nitrogen reaction cycle involves the processes

$$NO + O_3 \longrightarrow NO_2 + O_2 \tag{58}$$

$$NO_2 + h\nu \xrightarrow{\lambda < 398\ \text{nm}} NO + O \tag{59}$$

$$NO_2 + O \longrightarrow NO + O_2 \tag{60}$$

Note that processes (58) and (59), together with reaction (2), which regenerates O_3 from O and O_2, constitute a "do-nothing" cycle that leaves the original reacting species unmodified. On the other hand, reactions (58) and (60) form one of the most powerful odd-oxygen catalytic cycles of the stratosphere, leading to $O + O_3$ recombination:

$$\begin{array}{c} NO + O_3 \longrightarrow NO_2 + O_2 \\ \underline{NO_2 + O \longrightarrow NO + O_2} \\ O + O_3 \longrightarrow 2\,O_2 \end{array} \tag{61}$$

The photolysis of NO_2 is very rapid in full sunlight. The average lifetime of an NO_2 molecule against photodissociation is ~2 min. Thus NO_2 is often considered as an odd-oxygen species; the NO—O bond is very fragile, and the labile oxygen atom is extremely mobile within the stratospheric photochemical system. Processes (58)–(60) determine the relative abundances of NO and NO_2 in the stratosphere during the day,

$$[NO]/[NO_2] \doteq (r_{59} + r_{60}[O])/r_{58}[O_3]. \quad (62)$$

Because the relationship between O and O_3 is also established by photochemical processes [Eq. (9)], a straightforward dependence of the NO:NO_2 ratio on the O_3 concentration can be deduced.

Nitrogen dioxide reacts to form more complex compounds:

$$NO_2 + O_3 \longrightarrow NO_3 + O_2 \quad (63)$$

$$NO_3 + h\nu \xrightarrow{\lambda < 580\text{ nm}} NO_2 + O \quad (64)$$

$$NO_3 + h\nu \xrightarrow{\lambda < 750\text{ nm}} NO + O_2 \quad (65)$$

$$NO_2 + NO_3 + M \longrightarrow N_2O_5 + M \quad (66)$$

$$N_2O_5 + M \longrightarrow NO_2 + NO_3 + M \quad (67)$$

$$N_2O_5 + h\nu \xrightarrow{\lambda < 495\text{ nm}} 2\,NO_2 + O \quad (68)$$

The formation of additional NO_x compounds such as N_2O_3 and N_2O_4 is not favored under normal stratospheric conditions. The NO_3 molecule strongly absorbs radiation in the visible spectrum and has a lifetime against photolysis of only a few seconds in full sunlight. At night, however, its concentration may build up significantly. Likewise, N_2O_5 is formed efficiently at night through the reaction sequence

$$NO \xrightarrow{O_3} NO_2 \xrightarrow{O_3} NO_3 \xrightarrow{NO_2} N_2O_5 \quad (69)$$

Reaction sequence (69) is reversed in daylight through photodissociation processes (68) and (64) + (65). Generally speaking, NO_3 and N_2O_5 have only secondary roles in stratospheric photochemistry, except in cases of extended darkness, such as occurs during the polar winter, when their abundances may build up considerably.

The major sink for stratospheric NO_x is transport to other regions of the atmosphere (e.g., to the troposphere, where NO_x can be scavenged by rainfall). However, some NO_x destruction occurs in the stratosphere through the photolytic mechanism:

$$NO + h\nu \xrightarrow{\lambda \simeq 191,\, 198\text{ nm}} N + O \quad (70)$$

$$N + NO \longrightarrow N_2 + O \quad (71)$$

$$N + O_2 \longrightarrow NO + O \quad (72)$$

which is driven by NO photodissociation in the δ bands (Nicolet and Cieslik, 1980; Frederick and Hudson, 1979). Most of the nitrogen atoms produced quickly react with O_2 to form NO via reaction (72), but some of the nitrogen atoms react with and destroy NO by reaction (71). This odd-nitrogen destruction mechanism is most effective in the mesosphere but also occurs in the upper stratosphere.

A fundamental difference between the HO_x and NO_x families is that the former species are rapidly cycled with their source molecules (e.g., H_2O), while the latter species are not. The average lifetime of NO_x in the stratosphere is several years, while the HO_x lifetime is only hours to days.

The HO_x and NO_x families interact in a number of ways. The overall hydrogen–nitrogen photochemical system is depicted in Fig. 5. An important reaction is

$$NO + HO_2 \longrightarrow NO_2 + OH \tag{73}$$

which recycles odd-oxygen atoms from HO_2 to NO_2, and thus to ozone. For example, recalling the basic O_3-catalytic HO_x cycles (39) and (40), reaction (73) affects these cycles as follows:

$$OH + O_3 \longrightarrow HO_2 + O_2 \tag{34}$$

$$HO_2 + NO \longrightarrow OH + NO_2 \tag{73}$$

$$NO_2 + h\nu \longrightarrow NO + O \tag{59}$$

$$\underline{O + O_2 + M \longrightarrow O_3 + M} \tag{2}$$

$$\text{(zero net change)}$$

Hence, reaction (73) has a fundamental influence on stratospheric O_3 photochemistry. Reaction (73) is also rapid enough to affect the $OH:HO_2$ concentration ratio in the stratosphere.

A more important set of reactions involves the formation of mixed hydrogen–nitrogen compounds, HNO_x—principally HNO_2, HNO_3, and HO_2NO_2. Each compound undergoes a similar sequence of photochemical reactions, for example,

$$OH + NO + M \longrightarrow HNO_2 + M \tag{74}$$

$$HNO_2 + h\nu \xrightarrow{\lambda < 591\ \text{nm}} OH + NO \tag{75}$$

$$OH + HNO_2 \longrightarrow H_2O + NO_2 \tag{76}$$

$$O + HNO_2 \longrightarrow OH + NO_2 \tag{77}$$

The photodissociation of HNO_2, which occurs in the visible spectrum, is so rapid that HNO_2 concentrations and chemical reactions can be safely ignored.

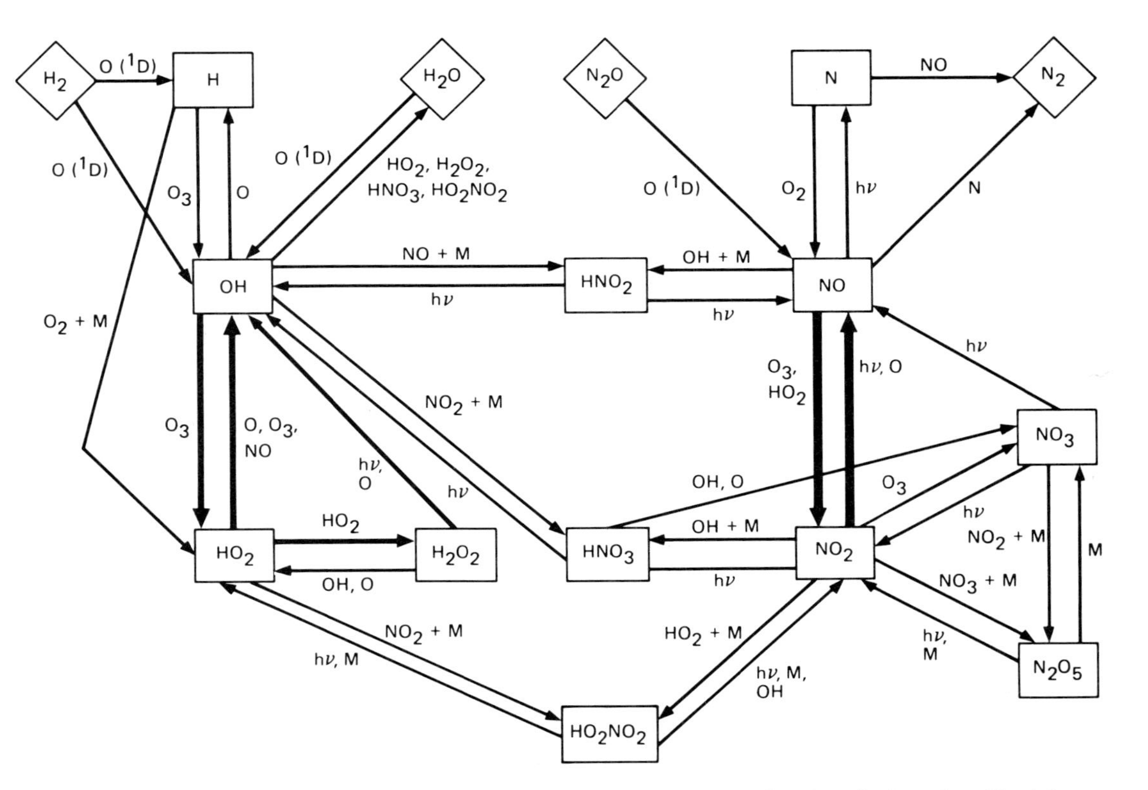

Fig. 5. The coupled hydrogen–nitrogen photochemical cycle of the stratosphere (see the legend to Fig. 3 for an explanation of the symbols used).

Nitric acid (HNO_3) is the most important HNO_x species. Its reactions are

$$OH + NO_2 + M \longrightarrow HNO_3 + M \tag{78}$$

$$HNO_3 + h\nu \xrightarrow{\lambda < 340\ nm} OH + NO_2 \tag{79}$$

$$OH + HNO_3 \longrightarrow H_2O + NO_3 \tag{80}$$

$$O + HNO_3 \longrightarrow OH + NO_3 \tag{81}$$

Because the photochemical lifetime of HNO_3 is long (weeks to months) in the lower stratosphere, it acts as the primary reservoir of odd-nitrogen below ~ 30 km.

As a rough approximation, the $NO_2:HNO_3$ concentration ratio may be estimated as

$$[NO_2]/[HNO_3] \simeq (\bar{r}_{79} + r_{80}[\overline{OH}])/r_{78}[\overline{OH}][M], \tag{82}$$

where the overbars indicate average values for a 24-hr period. Equation (82) suggests that the $NO_2:HNO_3$ ratio is sensitive to the stratospheric OH abundance and may be utilized to gauge the consistency between measured NO_x and HO_x species concentrations.

The formation and destruction of HO_2NO_2 occur as follows:

$$HO_2 + NO_2 + M \longrightarrow HO_2NO_2 + M \tag{83}$$

$$HO_2NO_2 + h\nu \xrightarrow{\lambda < 350\ nm} HO_2 + NO_2 \tag{84}$$

$$HO_2NO_2 + M \longrightarrow HO_2 + NO_2 + M \tag{85}$$

$$OH + HO_2NO_2 \longrightarrow H_2O + NO_2 + O_2 \tag{86}$$

$$O + HO_2NO_2 \longrightarrow OH + NO_2 + O_2 \tag{87}$$

HO_2NO_2 is not as important as HNO_3 in the photochemistry of stratospheric NO_x. However, HO_2NO_2 has only recently been introduced into photochemical schemes (Simonaitis and Heicklen, 1975), and future revisions in the rates of the basic processes can be expected.

HNO_3 and HO_2NO_2 have an important role in the loss of odd-hydrogen through reactions (78) and (80), and (83) and (86), respectively. For example,

$$OH + NO_2 + M \longrightarrow HNO_3 + M \tag{78}$$

$$\underline{OH + HNO_3 \longrightarrow H_2O + NO_3} \tag{80}$$

$$2\,OH + NO_2 \longrightarrow H_2O + NO_3$$

The rate-limiting step in this NO_x-mediated HO_x recombination mechanism is reaction (80) [note that most of the odd-hydrogen participating in reaction (78) is recycled by photolysis, (79), or reaction, (81)]. The same analysis applies to HO_2NO_2.

The catalytic effect of NO_x on stratospheric O_3 occurs principally through reactions (58) and (60)—or cycle (61)—as discussed earlier. Secondary cycles also exist, however; for example,

$$\begin{aligned} NO + O_3 &\longrightarrow NO_2 + O_2 \\ NO_2 + O_3 &\longrightarrow NO_3 + O_2 \\ NO_3 + h\nu &\longrightarrow NO + O_2 \\ \hline 2\,O_3 &\longrightarrow 3\,O_2 \end{aligned} \tag{88}$$

$$\begin{aligned} HO_2 + NO &\longrightarrow OH + NO_2 \\ NO_2 + h\nu &\longrightarrow NO + O \\ O + O_2 + M &\longrightarrow O_3 + M \\ \hline HO_2 + O_2 &\longrightarrow OH + O_3 \end{aligned} \tag{89}$$

The reaction sequence (89) generates O_3 from peroxy radicals (HO_2) in the presence of NO_x and sunlight. This process is similar to the mechanism of photochemical smog formation in the troposphere. In the next section, it is shown that the decomposition of methane and other hydrocarbons in the lower stratosphere also leads to ozone production through this process.

VII. Methane–Hydrocarbon Photochemistry

The photochemistry of stratospheric methane (CH_4) was introduced by Crutzen (1971) and Wofsy *et al.* (1972). Methane is emitted at the Earth's surface, and some is eventually transported into the stratosphere. Natural biogenic sources appear to dominate the CH_4 cycle (Sheppard *et al.*, 1982), although anthropogenic emissions are large and may be causing an increase in background concentrations. The primary mode of decomposition of CH_4 in the stratosphere is reaction with OH,

$$CH_4 + OH \longrightarrow CH_3 + H_2O \tag{90}$$

Photodissociation of CH_4 may be ignored, but reactions with $O(^1D)$ and chlorine atoms must be considered:

$$O(^1D) + CH_4 \longrightarrow OH + CH_3 \tag{91}$$

$$O(^1D) + CH_4 \longrightarrow H_2 + CH_2O \tag{92}$$

$$Cl + CH_4 \longrightarrow HCl + CH_3 \tag{93}$$

The rate of reaction (92) is less than 10% that of reaction (91). Methane is not regenerated in the stratosphere.

The methyl radical (CH_3) formed when CH_4 is decomposed initiates a chain of reactions, the most important of which are (Nicolet, 1972; Wofsy *et al.*,

1972; Whitten *et al.*, 1973)

$$CH_3 + O_2 + M \longrightarrow CH_3O_2 + M \quad (94)$$

$$CH_3O_2 + NO \longrightarrow CH_3O + NO_2 \quad (95)$$

$$CH_3O_2 + HO_2 \longrightarrow CH_3OOH + O_2 \quad (96)$$

$$CH_3OOH + OH \longrightarrow CH_3O_2 + H_2O \quad (97)$$

$$CH_3OOH + h\nu \xrightarrow{\lambda < 360\ \mathrm{nm}} CH_3O + OH \quad (98)$$

$$CH_3O + O_2 \longrightarrow CH_2O + HO_2 \quad (99)$$

$$CH_2O + h\nu \xrightarrow{\lambda < 334\ \mathrm{nm}} H + CHO \quad (100)$$

$$CH_2O + h\nu \xrightarrow{\lambda < 370\ \mathrm{nm}} H_2 + CO \quad (101)$$

$$CHO + O_2 \longrightarrow HO_2 + CO \quad (102)$$

$$CO + OH \longrightarrow H + CO_2 \quad (103)$$

The CH_4 oxidation chain is depicted in Fig. 6. In the stratosphere, the branching ratios for the two formaldehyde (CH_2O) photolysis pathways [(100) and (101)] are roughly comparable. The overall decomposition and oxidation of CH_4 leads to the production of CO_2, H_2O, and H_2. In fact, CH_4 is a major source of stratospheric H_2O above ~20 km. Globally, about 6×10^7 metric tons of H_2O are formed in the stratosphere each year from CH_4.

Odd-hydrogen is generated as an intermediate species during CH_4 oxidation. In general, the decomposition of one CH_4 molecule leads to the net production of at least one HO_x molecule. Figure 6 shows the HO_x inventory for CH_4 oxidation initiated by reaction (90). The CH_4 source of HO_x is important in the lower stratosphere.

Methane oxidation in the presence of NO also leads to O_3 (odd-oxygen) formation (Fig. 6). The key steps involve the reactions of NO with the peroxy compounds CH_3O_2 and HO_2. Similar reactions produce O_3 in urban photochemical smog. Several O_3 molecules may be generated during the oxidation of one CH_4 molecule.

Carbon monoxide (CO) is a well-known stratospheric species, which has only a minor role in stratospheric photochemistry. Its main source is CH_4 decomposition [reactions (101) and (102)]. Some additional CO is transported into the stratosphere from the troposphere and the mesosphere (Hayes and Olivero, 1970) or is generated *in situ* by the oxidation of hydrocarbons other than CH_4 (see below). At stratospheric concentrations, CO does not significantly affect the other chemical cycles.

Methane oxidation produces a number of intermediate hydrocarbon species that react with other stratospheric constituents. One reaction is (Nicolet, 1975)

$$CH_3O_2 + NO_2 + M \longrightarrow CH_3O_2NO_2 + M \quad (104)$$

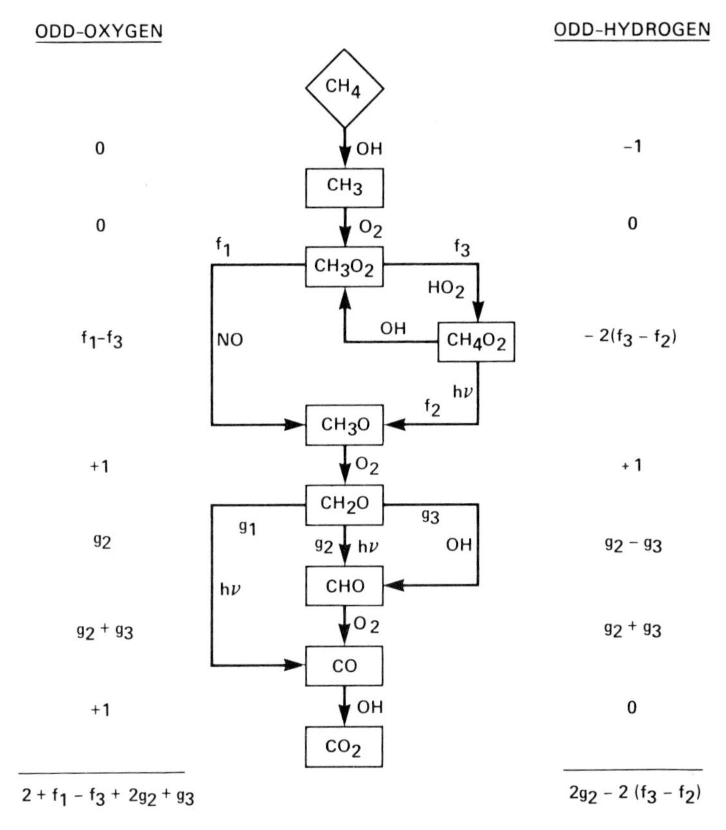

Fig. 6. The stratospheric methane oxidation mechanism. The changes in odd-oxygen and odd-hydrogen at each reaction step are given in terms of the fractional branching ratios for that reaction (the f's and g's, with $f_1 + f_2 = 1$ and f_3 normalized relative to f_1). To obtain the odd-oxygen fractional yields, unit efficiency is assumed for the chemical transformation sequence

$$H \xrightarrow{O_2} HO_2 \xrightarrow{NO} NO_2 \xrightarrow{h\nu} O$$

which is followed by

$$CH_3O_2NO_2 + M \longrightarrow CH_3O_2 + NO_2 + M \qquad (105)$$

These more complex hydrocarbon–nitrogen hybrid compounds appear to have only a secondary role in stratospheric photochemistry.

A number of nonmethane hydrocarbons (NMHCs) of tropospheric origin have been detected in the stratosphere (Chameides and Cicerone, 1978; Aikin *et al.*, 1982). These include acetylene (C_2H_2), ethylene (C_2H_4), and ethane (C_2H_6). Although the decomposition of complex NMHCs with carbon–carbon bonds proceeds in a different manner than the decomposition of CH_4, the oxidation chain involves similar reactions and end products. The overall oxidation processes may be expressed in terms of stoichiometric formulas (Aikin *et al.*, 1982; Brewer *et al.*, 1983):

$$C_2H_4 + 2\,O_2 + 2\,NO \longrightarrow 2\,CH_2O + 2\,NO_2 \qquad (106)$$

$$C_2H_6 + 5\,O_2 + 5\,NO \longrightarrow CH_2O + 2\,H_2O + 5\,NO_2 + CO_2 \qquad (107)$$

These are not the only possible oxidation mechanisms, of course. The decomposition of the formaldehyde (CH_2O) product also has several alternative pathways. Note that odd-oxygen (and thus ozone) is generated by NMHC oxidation (i.e., $NO \rightarrow NO_2$ in the presence of O_2). Existing stratospheric concentrations of NMHCs are too low to modify significantly the major photochemical reaction cycles.

VIII. The Chlorine Cycle

The chlorine (Cl_x) photochemical cycle is coupled to all of the other chemical cycles of the stratosphere. The suggestion that chlorine would be found in the stratosphere and could affect ozone concentrations, was first made by Stolarski and Cicerone (1974), Crutzen (1974a), and Wofsy and McElroy (1974). Initially, research focused on hydrogen chloride (HCl) emitted directly into the stratosphere by Space Shuttle rocket motors (Stolarski and Cicerone, 1974). Attention quickly shifted, however, to the industrial chlorofluorocarbons F-11 ($CFCl_3$) and F-12 (CF_2Cl_2) when Molina and Rowland (1974a) forwarded their now-famous theory of chlorocarbon-induced O_3 depletions (also see Rowland and Molina, 1975). A series of photochemical modeling studies was subsequently carried out (Cicerone *et al.*, 1974 1975; Crutzen, 1974b; Wofsy *et al.*, 1975a; Turco and Whitten, 1975). In these studies the chlorine photochemical cycle was developed and tested against observational data.

The sources of stratospheric chlorine are diverse (Ryan and Mukherjee, 1975; Cicerone, 1981). Natural sources include biogenic emissions of methyl chloride (CH_3Cl), which diffuses into the stratosphere, and direct volcanic injection of HCl (Stolarski and Cicerone, 1974). Anthropogenic chlorine sources include Space Shuttle emissions of HCl and industrial emissions of a broad suite of chlorocarbon compounds including the fluorocarbons F-11, F-12, and F-22 ($CHClF_2$), carbon tetrachloride (CCl_4), and methyl chloroform (CH_3CCl_3) (Crutzen *et al.*, 1978; McConnell and Schiff, 1978). The total chlorine budget of the present-day stratosphere is roughly 5×10^5 metric tons of chlorine per year. Chlorine is, for the most part, transported into the stratosphere in the form of stable source compounds and is removed as HCl, eventually to be scavenged by precipitation below ~ 5 km.

The overall chlorine chemical cycle of the stratosphere is depicted in Fig. 7. The major chlorine source gases are photodissociated by UV light at altitudes above approximately 20 to 30 km, releasing free chlorine atoms (Cl):

$$CH_3Cl + h\nu \xrightarrow{\lambda < 220\text{ nm}} CH_3 + Cl \tag{108}$$

$$CF_2Cl_2 + h\nu \xrightarrow{\lambda < 200\text{ nm}} CF_2Cl + Cl \tag{109}$$

$$CFCl_3 + h\nu \xrightarrow{\lambda < 265\text{ nm}} CFCl_2 + Cl \tag{110}$$

$$CCl_4 + h\nu \xrightarrow{\lambda < 280\text{ nm}} CCl_3 + Cl \tag{111}$$

The chlorine-containing dissociation products react rapidly with oxygen, releasing additional chlorine (as ClO). The stratospheric fate of the stable end products, CF_2O, CFClO and CCl_2O, is not completely known. However, it appears that photochemical decomposition continues (DeMore *et al.*, 1983) until all of the chlorine atoms are released.

Chlorine source molecules with C—H bonds are subject to chemical attack by the hydroxyl radical:

$$CH_3Cl + OH \longrightarrow CH_2Cl + H_2O \tag{112}$$

$$CHClF_2 + OH \longrightarrow CClF_2 + H_2O \tag{113}$$

The more stable molecules can also react with electronically excited oxygen atoms:

$$CF_2Cl_2 + O(^1D) \longrightarrow CF_2Cl + ClO \tag{114}$$

$$CFCl_3 + O(^1D) \longrightarrow CFCl_2 + ClO \tag{115}$$

Again, the product species are quickly converted to Cl_x and stable fluorine compounds in the stratosphere.

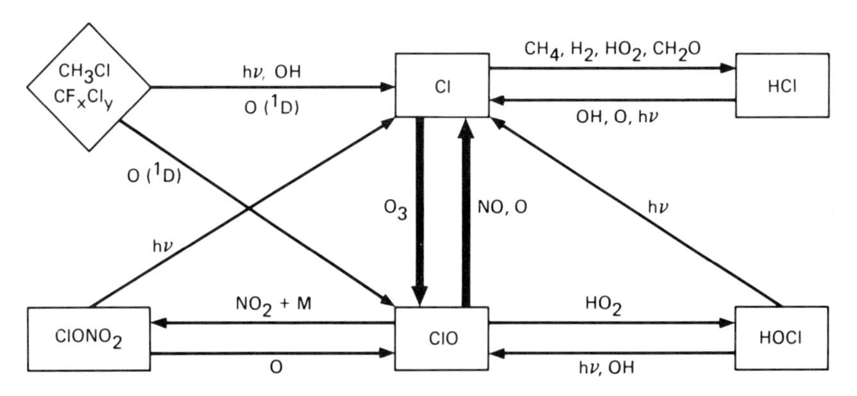

Fig. 7. The chlorine photochemical cycle of the stratosphere (see the legend to Fig. 3 for an explanation of the symbols used).

Hydrogen chloride may act as both a member of chlorine family Cl_x—produced by the decomposition of source gases such as CH_3Cl—and as a reservoir for the O_3-active, or "free," chlorine species Cl and ClO. Hence, the definition of the Cl_x family can be flexible in that HCl need not be included. Hydrogen chloride has a relatively long photochemical lifetime (~days) in the stratosphere, and is slowly decomposed by the following processes:

$$HCl + OH \longrightarrow Cl + H_2O \tag{116}$$

$$HCl + O \longrightarrow Cl + OH \tag{117}$$

$$HCl + O(^1D) \longrightarrow Cl + OH \tag{118}$$

$$HCl + h\nu \xrightarrow{\lambda < 225\ \text{nm}} Cl + H \tag{119}$$

The reaction with OH [(116)] is responsible for most of the chlorine atoms released from HCl. The chlorine atoms, in turn, abstract hydrogen atoms from a number of species; the most important reactions are

$$Cl + CH_4 \longrightarrow HCl + CH_3 \tag{120}$$

$$Cl + H_2 \longrightarrow HCl + H \tag{121}$$

$$Cl + HO_2 \longrightarrow HCl + O_2 \tag{122}$$

Reaction (122) has a secondary branch leading to OH + ClO, which can be neglected in the present discussion. Some less-important Cl-to-HCl transformation reactions involve (as reactants) CH_2O, H_2O_2, C_2H_4, and C_2H_6. A reaction between ClO and OH, which produces mainly Cl and HO_2, may also yield some HCl, although this latter reaction path has not been verified experimentally.

The rates of the photochemical processes that continously cycle HCl and Cl [reactions (116)–(122)] determine the quantity of "free" chlorine that is available to interact with ozone. Obviously, the effectiveness of chlorine in modifying stratospheric photochemistry is strongly dependent on the HO_x, methane, and other family cycles.

The basic O_3-destructive chlorine cycle comprises two reactions:

$$(125)\left\{\begin{array}{ll} Cl + O_3 \longrightarrow ClO + O_2 & (123) \\ \underline{ClO + O \longrightarrow Cl + O_2} & (124) \\ O + O_3 \longrightarrow 2\,O_2 & \end{array}\right.$$

Reaction cycle (125) may occur hundreds of times before the Cl atom is recycled into (inert) HCl. With regard to ozone depletion, ClO is much more effective per molecule than OH or NO.

In addition to reaction (124), ClO is recycled into Cl by reaction with NO:

$$ClO + NO \longrightarrow Cl + NO_2 \tag{126}$$

It is significant that reaction (126) preserves odd-oxygen. That is, the efficient photochemical cycle,

$$Cl + O_3 \longrightarrow ClO + O_2 \quad (123)$$

$$ClO + NO \longrightarrow Cl + NO_2 \quad (126)$$

$$NO_2 + h\nu \longrightarrow NO + O \quad (59)$$

$$O + O_2 + M \longrightarrow O_3 + M \quad (2)$$

conserves all of the species involved and "short-circuits" chlorine-catalyzed ozone destruction. The presence of NO_x in the stratosphere therefore moderates the effect of chlorine on O_3.

Additional chlorine oxides are formed through ClO reactions. Two of the most important species are chlorine nitrate ($ClONO_2$) and hypochlorous acid (HOCl). Chlorine nitrate has the following photochemistry:

$$ClO + NO_2 + M \longrightarrow ClONO_2 + M \quad (127)$$

$$ClONO_2 + h\nu \xrightarrow{\lambda < 450\ \text{nm}} Cl + NO_3 \quad (128)$$

$$ClONO_2 + h\nu \xrightarrow{\lambda < 391\ \text{nm}} ClONO + O \quad (129)$$

For some time, it was uncertain which isomer(s) of chlorine nitrate are produced in reaction (127) (e.g., $ClONO_2$, $OClNO_2$, or OClONO). It now appears certain that the "nitrate" form, $ClONO_2$, nearly always results. The ClONO generated by process (129) is rapidly photolyzed, yielding chlorine atoms. Data suggest that the first photolysis branch [(128)] dominates $ClONO_2$ photodecomposition. Chlorine nitrate acts as a secondary reservoir of Cl_x and NO_x, reducing the O_3-catalytic impact of both cycles [reaction (126), in comparison, reduces only the O_3-catalytic effect of Cl_x].

The principal reactions of HOCl are

$$ClO + HO_2 \longrightarrow HOCl + O_2 \quad (130)$$

$$HOCl + h\nu \xrightarrow{\lambda < 310\ \text{nm}} Cl + OH \quad (131)$$

$$HOCl + OH \longrightarrow ClO + H_2O \quad (132)$$

$$HOCl + O \longrightarrow ClO + OH \quad (133)$$

The importance of HOCl was first discussed by Prasad *et al.* (1978). Hypochlorous acid is a secondary Cl_x reservoir (which is not as effective as HCl or $ClONO_2$); it also participates in photochemical cycles that influence ozone. For example, consider the reaction sequence

$$OH + O_3 \longrightarrow HO_2 + O_2 \quad (34)$$

$$Cl + O_3 \longrightarrow ClO + O_2 \quad (123)$$

$$ClO + HO_2 \longrightarrow HOCl + O_2 \quad (130)$$

$$HOCl + h\nu \longrightarrow Cl + OH \quad (131)$$

$$\overline{\qquad 2\,O_3 \longrightarrow 3\,O_2 \qquad}$$

A number of other chlorine species are formed in small quantities in the stratosphere: OClO, ClOO, Cl_2O, Cl_2, and ClNO, for example. These interesting, but less important, species are discussed in the references cited in this section. Anthropogenic chlorine emissions and attendant O_3 perturbations are discussed in Section XII.

IX. The Sulfur Cycle

The sulfur (SO_x) cycle of the stratosphere is unique in two respects. First, under ambient conditions it is decoupled from all the other chemical cycles; that is, its influence on other cycles is negligible. (Volcanic eruptions, however, occasionally enhance the sulfur cycle to such a degree that interactions become important.) Second, the sulfur cycle is the only one in which the condensed (aerosol) phase plays a major role (Whitten, 1982).

Although the presence of dust in the stratosphere was known since the 1920s from twilight studies (Gruner and Kleinert, 1927), the first samples and analyses revealing sulfate aerosols were not reported until the early 1960s by C. E. Junge and co-workers (1961). Jaeschke *et al.* (1976) first detected sulfur dioxide (SO_2), the precursor of sulfate aerosols, in the lower stratosphere, and Inn *et al.* (1979) later measured the vertical distribution of OCS. Crutzen (1976) had developed a photochemical theory for OCS that matched Inn's observations, and Turco *et al.* (1979, 1981b) extended Crutzen's theory to include HSO_x, H_2SO_4, and sulfate aerosols.

Under nonvolcanic conditions, the source of stratospheric sulfur is dominated by the upwelling of tropospheric OCS and SO_2, with small injections of meteoric sulfur and SO_2 from aircraft exhaust. The OCS is decomposed as follows:

$$OCS + h\nu \xrightarrow{\lambda < 288\ \text{nm}} S(^1D) + CO \tag{134}$$

$$S(^1D) + O_2 \longrightarrow SO + O \tag{135}$$

$$OCS + O \longrightarrow SO + CO \tag{136}$$

$$SO + O_2 \longrightarrow SO_2 + O \tag{137}$$

This reaction set is not complete for S and SO, both of which have very short chemical lifetimes and low concentrations in the stratosphere.

Sulfur dioxide, injected directly or generated from OCS, reacts with OH and O:

$$SO_2 + OH + M \longrightarrow HSO_3 + M \tag{138}$$

$$SO_2 + O + M \longrightarrow SO_3 + M \tag{139}$$

Reaction (138) is by far the most important reaction of SO_2 in the stratosphere.

The photochemistry of the sulfur radical (HSO_3) is only poorly known, although H_2SO_4 is almost certainly the end product of HSO_3 oxidation. Reactions of HSO_3 that may occur in the stratosphere are

$$HSO_3 + OH \longrightarrow SO_3 + H_2O \tag{140}$$

$$HSO_3 + O \longrightarrow SO_3 + OH \tag{141}$$

McKeen *et al.* (1984) adopted for the stratosphere a mechanism proposed by Stockwell and Calvert (1983) to explain the oxidation of SO_2 in the troposphere:

$$HSO_3 + O_2 \longrightarrow HO_2 + SO_3 \tag{142}$$

Reaction (142) might involve HSO_3 molecules clustered with H_2O molecules. The rate coefficients of such HSO_3 reactions have not been measured, and the UV photolysis of HSO_3 has never been studied.

Under stratospheric conditions, the fate of SO_3 produced by reactions (139) through (142) is rapid hydration to sulfuric acid (Castleman *et al.*, 1975),

$$SO_3 + H_2O \longrightarrow H_2SO_4^* \longrightarrow H_2SO_4 \tag{143}$$

where the asterisk indicates an excited isomeric state.

In volcanic eruption clouds, the concentrations of sulfur gases are greatly enhanced. There are also high concentrations of H_2O and other compounds that can affect SO_x photochemistry, and of ash and sulfuric acid droplets that can provide sites for "heterogeneous" chemistry (see Section X). While reaction (138) remains a key oxidation step for SO_2 (Turco *et al.*, 1982), reactions (141) and (142) may have a critical role in determining the overall rate of SO_2 oxidation in volcanic eruption clouds. Because both of these reactions *recycle* odd-hydrogen, HO_x can oxidize SO_2 *catalytically*. On the other hand, if reactions such as (140) dominate SO_2 oxidation, HO_x is depleted as SO_2 is converted to sulfate and the SO_2 lifetime increases in the volcanic cloud.

Volcanoes also emit large quantities of hydrogen sulfide (H_2S), which is not present in the ambient stratosphere. A major decomposition pathway for H_2S in volcanic eruption clouds is

$$H_2S + OH \longrightarrow HS + H_2O \tag{144}$$

$$HS + O \longrightarrow SO + H \tag{145}$$

$$HS + O_2 \longrightarrow SO + OH \tag{146}$$

This is by no means a complete reaction set for H_2S, and, in particular, the exact fate of the HS radical is uncertain. For example, the reaction between HS and O_2 [(146)] may not even occur.

In the stratosphere, the photochemical lifetime of injected H_2S is shorter than the lifetime of injected SO_2. Moreover, according to presently accepted chemical schemes, H_2S oxidation does not cause a net depletion of HO_x. Thus, H_2S represents a major potential source of SO_x and sulfate in fresh volcanic clouds.

Sulfur may be recycled into less oxidized forms in the upper stratosphere. For example, H_2SO_4 and HSO_3 might be photolyzed by solar UV radiation, although molecular absorption spectra are not available.

Sulfur dioxide (SO_2) is predissociated in the near-UV region:

$$SO_2 + h\nu \xrightarrow{\lambda < 220\ \text{nm}} SO + O \tag{147}$$

Efficient SO_2 photolysis suggests the possibility of catalytic ozone production at the top of volcanic eruption clouds, where SO_2 concentrations and solar UV intensities are great (Crutzen and Schmailzl, 1983):

$$\begin{array}{l} SO_2 + h\nu \longrightarrow SO + O \\ \underline{SO + O_2 \longrightarrow SO_2 + O} \\ O_2 + h\nu \longrightarrow O + O \end{array} \tag{148}$$

In the lower stratosphere, H_2SO_4 molecules generated by SO_2 oxidation condense with water vapor to form a mist of fine aqueous sulfuric acid droplets (Junge *et al.*, 1961). Figure 8 illustrates the basic processes involved in the sulfur oxidation and aerosol formation cycle of the stratosphere. A detailed description of the chemical and physical mechanisms is given by Turco *et al.*

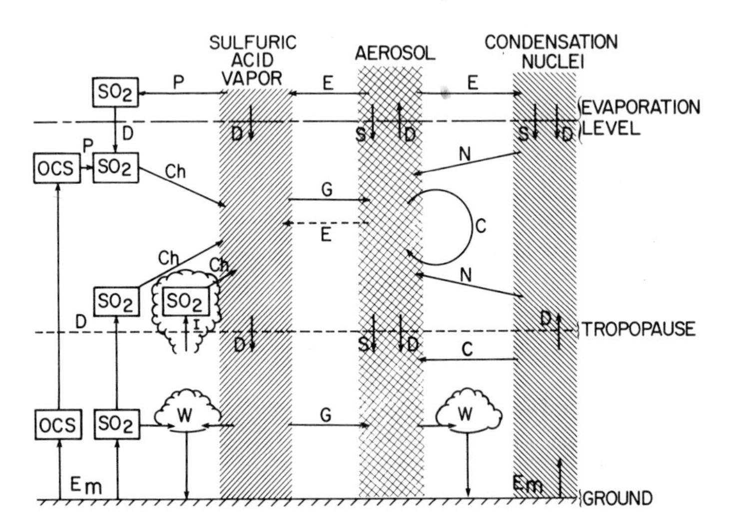

Fig. 8. A schematic description of the chemical and physical processes that control the stratospheric sulfur cycle. C, Coagulation; Ch, chemical reaction; D, diffusion (vertical); E, aerosol evaporation; Em, emission; G, aerosol condensation and growth; I, injection; N, particle nucleation; P, photolysis; S, particle sedimentation; W, wash out and rain out.

(1982). The aerosol (or "Junge") layer is concentrated below an altitude of 25 km, is quite uniform on a global scale, and consists of just a few particles with radii $\lesssim 0.2\ \mu m$ per cubic centimeter of air. The aerosols are the sink for stratospheric sulfur in the form of sulfates (primarily condensed H_2SO_4, but also some ammonium, nitrosyl, and metallic sulfates).

The stratospheric sulfate aerosols are not just an interesting chemical oddity. The particles are effective in scattering incident sunlight back to space, thereby reducing the solar insolation. This "albedo" effect provides a direct link between stratospheric composition and terrestrial climate. The climatic effects of aerosols are most enhanced following major volcanic eruptions, when the particle optical thickness (or turbidity) may be enhanced by a factor of 100 or more (Pollack *et al.*, 1976). It has been suggested that in past geological epochs volcanoes exercised a substantial influence on global climate variations. Some human activities (e.g., industrial emissions of OCS, and aircraft and rocket emissions of SO_2 and solid exhaust particles) also affect the aerosol layer and might conceivably become agents of climatic change.

X. Heterogeneous Chemistry

The presence of aerosols in the stratosphere implies that chemical reactions can occur on particle surfaces and in aqueous solution. In addition to sulfate particles, the stratosphere contains meteoric debris, aluminum oxide dust, and, occasionally, nacreous ice clouds. Figure 9 provides an estimate of the frequency at which gas molecules encounter aerosols in the ambient stratosphere. The time between collisions varies from ~ 2 hr in the lower stratosphere to ~ 1 day in the upper stratosphere (in volcanic clouds the collision frequencies can be several orders of magnitude larger).

The overall rate of reaction of a gaseous species i with stratospheric particles may be expressed as the product of the collision frequency, k_c (Fig. 9), and the reaction efficiency (per collision) for a specific process j, γ_{ij}:

$$k_{ij} = \gamma_{ij} k_c. \tag{149}$$

The reaction efficiencies of several important stratospheric species on H_2SO_4/H_2O surfaces are summarized in Table IV (note that product species are not distinguished). For the most part, the efficiencies are very small. Accordingly, surface decomposition reactions such as

$$N_2O_5 \xrightarrow{H_2O} HNO_3 + HNO_3 \tag{150}$$

$$ClONO_2 \xrightarrow{H_2O} HOCl + HNO_3 \tag{151}$$

can generally be ignored. It has also been shown by direct measurements that

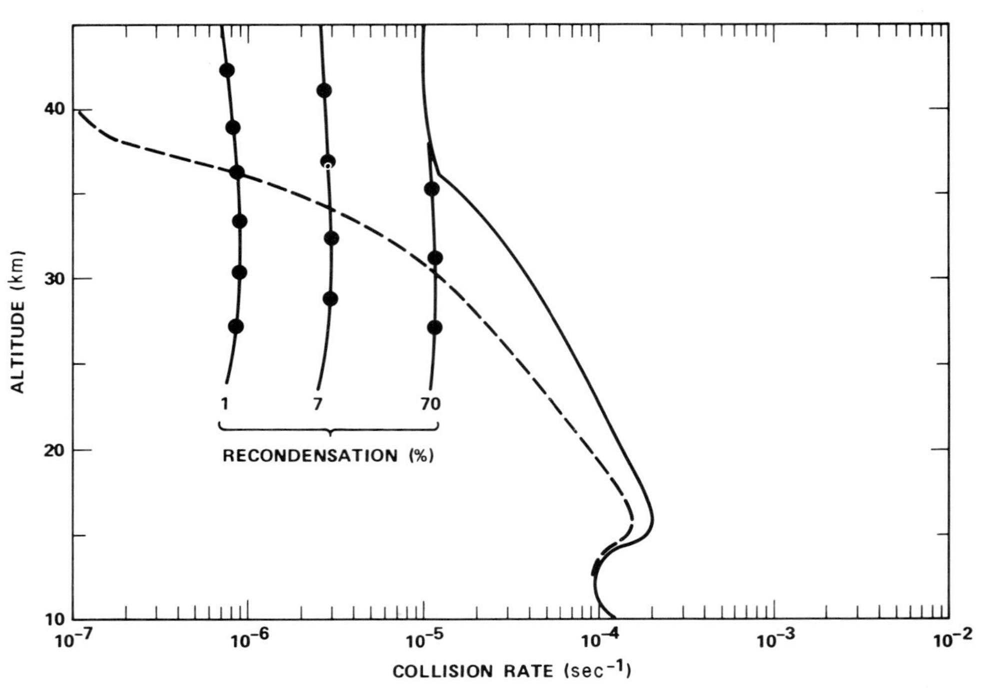

Fig. 9. Collision rate of an air molecule with stratospheric particles. A mean molecular mass of 30 amu and a U.S. Standard Atmosphere (1976) temperature profile are assumed. Collision frequencies are given for ambient aerosol concentrations with and without meteoric dust (the dust profiles correspond to various percentages of recondensation of meteor ablation vapors into smoke particles). ——, Aerosols plus meteoric dust; – – –, aerosols only; —●—, meteoric dust only.

Table IV

REACTION EFFICIENCIES ON SULFURIC ACID/WATER SURFACES[a]

Species	Efficiency	Comments[b]
O_3	$1.7 \times 10^{-6} e^{-1081/T}$	100% H_2SO_4, 217–263 K
	$3.3 \times 10^{-5} e^{-2360/T}$	20% H_2SO_4, 217–263 K
	$>1 \times 10^{-11}$	75% H_2SO_4, 223 K
	$>5 \times 10^{-11}$	75% H_2SO_4, 223 K, with 0.1% Cu, Ni, Al, Cr cations
	$>4 \times 10^{-9}$	75% H_2SO_4, 223 K, with 0.02% Fe cations
	$<10^{-6}$	95% H_2SO_4, 300 K
O	$<10^{-6}$	95% H_2SO_4, 300 K
OH	4.9×10^{-4}	95% H_2SO_4, 300 K
H_2O_2	7.8×10^{-4}	95% H_2SO_4, 300 K
H_2O	$\sim 2 \times 10^{-3}$	Accommodation coefficient, 95% H_2SO_4, 300 K
N	$<10^{-6}$	95% H_2SO_4, 300 K
NO	$<10^{-6}$	95% H_2SO_4, 300 K
NO_2	$<10^{-6}$	95% H_2SO_4, 300 K
N_2O_5	$\geqq 3.8 \times 10^{-5}$	95% H_2SO_4, 300 K
HNO_3	$\geqq 2.4 \times 10^{-4}$	95% H_2SO_4, 300 K
HO_2NO_2	2.7×10^{-5}	95% H_2SO_4, 300 K
NH_3	$>1 \times 10^{-3}$	95% H_2SO_4, 300 K
	0.2–0.3	Aerosols, radius $\sim$0.1–0.7 μm, 296 K
Cl	$3.8 \times 10^{-8} e^{2120/T}$	75% H_2SO_4, 230 K
ClO	$3.7 \times 10^{-10} e^{3220/T}$	75% H_2SO_4, 230 K
$ClONO_2$	1.0×10^{-5}	95% H_2SO_4, 300 K
SO_2	$<10^{-6}$	95% H_2SO_4, 300 K

[a] Detailed references to sources of data are given in Turco *et al.* (1982).
[b] Reactions occur on flat surfaces, except as noted.

stratospheric aerosols are not a significant reservoir or sink for NO_x (either as NO_3^- or nitrosyl compounds) or Cl_x (as Cl^-).

Reactions on the surfaces of nonsulfate particles are probably even less important. An exception may be the reaction of acidic vapors (H_2SO_4, HNO_3, and HCl) on the metallic surfaces of meteoric debris particles. Such reactions could lead to the formation of inert metal sulfate coatings on the particle surfaces (Turco *et al.*, 1981a).

Reactions between species dissolved within aqueous stratospheric aerosol droplets are also possible. Several reaction mechanisms, which are believed to occur in raindrops, have been suggested for the stratosphere. One potentially important mechanism involves H_2O_2 and SO_2,

$$H_2O_2 \xrightarrow{\text{sol.}} H_2O_2 \tag{152}$$

$$SO_2 \xrightarrow{\text{sol.}} H_2SO_3 \tag{153}$$

$$H_2SO_3 \overset{\text{aq.}}{\rightleftarrows} H^+ + HSO_3^- \tag{154}$$

$$HSO_3^- + H_2O_2 \longrightarrow HSO_4^- + H_2O \tag{155}$$

where (152) and (153) are (reversible) gas-to-droplet dissolution processes with appropriate Henry constants, (154) the equilibrium process for the first dissociation of dissolved $SO_2 \cdot H_2O$, and (155) the overall reaction of sulfite with peroxide to form sulfate, which may proceed in several steps. A serious problem arises in evaluating the potential importance of solution processes because kinetic data are not available that apply to the prevailing stratospheric conditions of very low temperature (220 K) and high solution acidity (pH ~ -1). A number of aqueous reaction mechanisms have been assessed for the stratosphere and tentatively found to be unimportant (Turco *et al.*, 1982).

XI. Additional Photochemistry

A variety of other stratospheric species and their photochemical reactions can be considered. Fluorine is released by the decomposition of fluorocarbons in the middle stratosphere (Section VIII). For example, free fluorine is generated by the photolysis of the F-12 by-product CF_2O:

$$CF_2O + h\nu \xrightarrow{\lambda < 226 \text{ nm}} CFO + F \tag{156}$$

This may be followed by the reaction sequence (Stolarski and Rundel, 1975)

$$F + O_3 \longrightarrow FO + O_2 \tag{157}$$

$$FO + O \longrightarrow F + O_2 \tag{158}$$

$$FO + NO \longrightarrow F + NO_2 \tag{159}$$

Fluorine, like chlorine, can consume O_3 (odd-oxygen) in a catalytic reaction cycle. Its importance in this regard is negligible, however, because the cycle is efficiently terminated by the formation of hydrogen fluoride (HF):

$$F + H_2O \longrightarrow HF + OH \tag{160}$$

$$F + CH_4 \longrightarrow HF + CH_3 \tag{161}$$

Hydrogen fluoride is immune to photochemical attack by O, OH, and UV radiation at wavelengths longer than ~ 160 nm [but it may be decomposed by $O(^1D)$]. Removal of HF from the stratosphere occurs by transport to the troposphere followed by rain out. It should be mentioned that measured and calculated stratospheric HF concentration profiles agree only to within a factor of 2 to 3.

As with the chlorine family, higher oxides of fluorine (i.e., FO_2, FONO, $FONO_2$) can evolve in the stratosphere, but they have little photochemical significance.

A particularly intriguing fluorine compound is CF_4 (Cicerone, 1979). This gas is apparently produced only by man and is so stable in the environment that it may have an atmospheric lifetime exceeding 10,000 years. While only traces of CF_4 presently exist in the atmosphere, it will continue to accumulate steadily in future centuries. The actual decomposition pathways for CF_4 are not well defined. They may include vacuum UV photolysis (at wavelengths below 103 nm in the upper mesosphere) and reactions with electronically excited oxygen atoms.

Bromine has been detected in the stratosphere, and its photochemistry has been widely discussed (e.g., Wofsy *et al.*, 1975b; Yung *et al.*, 1980). The major source of stratospheric bromine appears to be CH_3Br of biogenic (and partly anthropogenic) origin. As with the analogous decomposition of methyl chloride, CH_3Br is attacked by the hydroxyl radical:

$$CH_3Br + OH \longrightarrow CH_2Br + H_2O \tag{162}$$

CH_2Br is rapidly oxidized, releasing "free" bromine (Br, BrO, etc). The fundamental bromine chemical cycle consists of the reactions

$$Br + O_3 \longrightarrow BrO + O_2 \tag{163}$$

$$BrO + O \longrightarrow Br + O_2 \tag{164}$$

$$BrO + NO \longrightarrow Br + NO_2 \tag{165}$$

Parallels between chlorine and bromine photochemistry are very strong, as indicated by comparing reactions (163) through (165) with the equivalent chlorine reaction system. For example, reactions (163) and (164) form an ozone catalytic cycle that is analogous to the chlorine cycle (125). However, there are significant differences between the Br_x and Cl_x photochemical cycles. Formation rates of HBr are slower, and HBr is much less stable than HCl. Thus HBr is not as effective a bromine reservoir as HCl is a chlorine reservoir (likewise, HOBr and $BrONO_2$ are photolytically less stable than their chlorine counterparts). Also, BrO undergoes rapid self-disproportionation, unlike ClO:

$$2\,BrO \longrightarrow 2\,Br + O_2 \tag{166}$$

Because of reaction (166), bromine is capable of odd-oxygen destruction through the mechanism

$$\begin{array}{c} 2(Br + O_3 \longrightarrow BrO + O_2) \\ 2\,BrO \longrightarrow 2\,Br + O_2 \\ \hline 2\,O_3 \longrightarrow 3\,O_2 \end{array} \tag{167}$$

Catalytic cycle (167) is unique because it is as effective at night as during the day (all of the other O_3-catalytic cycles discussed so far require sunlight to operate).

The bromine and chlorine families are also coupled through the reactions

$$BrO + ClO \longrightarrow Br + Cl + O_2 \quad (168)$$

$$BrO + ClO \longrightarrow Br + OClO \quad (169)$$

which is the basis for perhaps the most important bromine catalytic cycle:

$$(170)\left\{\begin{array}{ll} Br + O_3 \longrightarrow BrO + O_2 & (163) \\ Cl + O_3 \longrightarrow ClO + O_2 & (123) \\ \underline{BrO + ClO \longrightarrow Br + Cl + O_2} & (168) \\ 2O_3 \longrightarrow 3O_2 & \end{array}\right.$$

This coupled cycle is referred to as "synergistic" because it enhances the capabilities of both Cl_x and (particularly) Br_x to destroy O_x.

Iodine can undergo a series of photochemical transformations similar to that of bromine (Chameides and Davis, 1980). Iodine has never been detected in the stratosphere, however, even in trace quantities, and it will not be considered further.

Metals released by meteor ablation can generate a variety of trace compounds. The most important metals in this regard are sodium, calcium, and magnesium, and their oxides, hydroxides, chlorides, and sulfates. Only preliminary studies of stratospheric metal photochemistry have been carried out. For example, metals can act as a chemical sink for chlorine and as a photoionization source for electrically charged species. Neither role has been confirmed. In fact, calculations suggest that the metals exist primarily in the condensed (aerosol) phase in the stratosphere. If true, the metals would have only a minor role in stratospheric photochemistry.

Cyanogenic compounds (HCN and CH_3CN) have been detected in trace quantities in the stratosphere (Table I). Both compounds are thought to have industrial origins. Although these species are relatively stable, they may be decomposed by reactions with OH and $O(^1D)$, and by UV photolysis. The cyanogen radicals formed in this manner are readily oxidized in air, eventually producing CO_2 and NO_x. At their present concentrations, however, the intriguing cyanogenic compounds do not significantly affect the major photochemical cycles of the stratosphere.

The stratosphere is filled with electrically charged ions. There are equal concentrations of positive and negative ions, produced by galactic cosmic rays that stream in from space and collide with and ionize air molecules. While the total concentration of ions is small ($\sim 10^4$ ions cm^{-3}), the electrical charge causes some of their reactions with neutral molecules to be very rapid. A detailed treatment of the ion chemistry of the stratosphere would involve dozens of species and hundreds of reactions. Despite the scope and diversity of ion-related processes, the impact of ion–molecule reactions on stratospheric composition is relatively small. A few potentially important and interesting roles for ions are

1. Use in the detection of trace neutral molecules that readily cluster to ions (e.g., HNO_3, H_2SO_4, and CH_3CN) (the ions act as sensitive and selective microscopic samplers whose molecular constituency can be deduced through mass spectrometry)
2. Sites for the transformation of chemical species, particularly when ionization levels are enhanced by solar proton events (SPEs) (the generation of HO_x from water cluster ions is an example)
3. Action as nucleation centers for stratospheric sulfate aerosol formation

XII. Ozone Perturbations

Natural and human influences on the photochemistry of the stratosphere can lead to perturbations of O_3—with serious environmental consequences. Therefore, a substantial effort has been expended in the last decade to understand the O_3 layer. A number of authoritative assessments are available that deal with O_3 perturbations arising from a variety of causes (Grobecker, 1975; National Academy of Sciences, 1975, 1976, 1982, 1984; World Meteorological Organization, 1982). Tables V and VI list some of the important chemical and physical agents that can perturb stratospheric O_3. Only a brief overview of the subject is possible here.

Ozone perturbations are estimated using mathematical models of the atmosphere that incorporate the photochemistry described above as well as the effects of atmospheric transport. One-dimensional (1-D, vertical column) and two-dimensional (2-D, meridional cross section) models are available that treat stratospheric photochemistry in sufficient detail. Unfortunately, the representations of atmospheric transport in these models have serious deficiencies. Most ozone perturbation calculations have been carried out with 1-D models, which are only adequate to describe average global conditions.

The effects of different pollutants on O_3 can be understood in part by considering the rates of the key chemical reaction steps that deplete odd-oxygen. Figure 10 shows these rates for the ambient stratosphere. It should be noted that NO_x dominates O_3 loss in the middle stratosphere, and HO_x in the upper stratosphere, while Cl_x contributes to O_3 loss at altitudes of between 25 and 50 km. Thus, as an example, NO_x injected at 20 km would have a potent effect in reducing O_3. However, Fig. 10 does not define the complex couplings between the chemical families that modify the overall impact of pollutant injections on O_3. NO_x injected below $\sim$15 km, for example, would interrupt the HO_x catalytic cycle and cause a small O_3 increase. The change in the total column amount of stratospheric O_3 is the sum of the changes at each altitude

Table V

ANTHROPOGENIC CHEMICAL AGENTS LEADING TO STRATOSPHERIC OZONE PERTURBATIONS

Agent[a]	Source	Effect[b]
N_2O	By-product of combustion and fertilizer decomposition	N_2O doubling leads to a 10 to 16% O_3 depletion; N_2O is currently increasing at a rate of ~0.2% per year.
NO_x	Injection below 12 km by commercial jet aircraft	With present air traffic, O_3 change could range from 0 to +2%
	Injection by supersonic transport aircraft at 17 to 20 km	For a fleet of several hundred aircraft injecting 1×10^9 kg NO_2 per year at 20 km, the O_3 reduction could be 4–8%
	Production and injection of NO_x in fireballs during a nuclear war	For a 5000-megaton exchange, peak hemispherical average O_3 depletions of 30 to 40% are possible the first year, recovering by half in 2 years
$CFCl_3$, CF_2Cl_2	Industrial production and release of chlorofluorocarbons	Chlorofluorocarbon release to the atmosphere at current production rates could, in 50 to 100 years, reduce global ozone by 2 to 8%
Other chlorocarbons	Industrial release of chlorinated compounds such as CCl_4, $CHCl_3$, CH_3CCl_3, and $CHClF_2$	Current levels of industrial emission might eventually cause a 1 to 3% decrease in O_3
CO_2	Biomass burning and fossil fuel combustion generate CO_2, which affects the infrared radiation balance and temperature structure of the troposphere and stratosphere	Doubling of CO_2 concentrations could increase total O_3 by 3 to 6%; an increase in CO_2 causes the stratosphere to cool; the temperature dependences of O_3 reactions are such that cooling leads to an O_3 enhancement

[a] Other chemical agents not included in the table, but which have been studied and are potentially important, are HCl emitted by solid-fueled rocket engines, H_2O injected by high-altitude aircraft (SSTs) and liquid-fueled H_2/O_2 rocket motors, and CH_4 modified by industrial and agricultural activity.

[b] Unless otherwise noted, the effect is given as the percentage change in the average vertical O_3 column concentration (molec cm^{-2}) through the stratosphere for a steady-state perturbation (e.g., for continuous pollutant emission at a fixed rate over a period exceeding the atmospheric transport and photochemical equilibration times).

Table VI

NATURAL PHYSICAL AGENTS LEADING TO STRATOSPHERIC OZONE PERTURBATIONS

Agent	Description	Effect
Solar UV variations	Natural cyclic variations (e.g., with periods of 37 days and 11 years) in solar UV intensities modulate O_2 photodissociation	O_3 concentrations co-vary with the UV flux, by as much as 10% at some altitudes
Solar eclipses	Short-term interruption of sunlight	No discernible effect on stratospheric O_3 has yet been detected
Solar proton events (SPEs)	Protons emitted in solar storms penetrate the Earth's upper atmosphere, producing NO_x	A large SPE (e.g., August 1972) might cause a maximum O_3 depletion of $\sim 20\%$ at 40 km, but $\leq 2\%$ total O_3 column change
Meteors	High-velocity meteoric debris heats the upper atmosphere by friction, generating NO_x	Rare large meteors such as Tunguska (Russia, 1908) may cause $\sim 10\%$ O_3 depletions; smaller, more common meteoroids have little effect on O_3
Supernovae	Energetic particles and radiation emitted by terminal stellar explosions ionize the atmosphere and generate NO_x	For supernovae within ~ 100 light-years of Earth, the cosmic rays might deplete O_3 by 30 to 80% for ~ 100 years; such events are rare (~ 1 every 100 million years)
Volcanoes	Volcanic emissions of H_2O and HCl and heating by aerosols cause O_3 perturbations	No unambiguous O_3 changes have been detected; trend analysis shows that Agung (1963) could have caused only $\lesssim 2\%$ change in total O_3

where a pollutant is injected or is transported. Accordingly, the interpretation of O_3 perturbation calculations is a complex scientific problem requiring considerable practice.

The first stratospheric pollutant to receive wide attention was NO_x (Johnston, 1971). Concern centered around the proposed development of a commercial supersonic transport (SST) aircraft, which would inject NO_x directly into the stratosphere at altitudes up to 20 km. Figure 11 illustrates the history of calculated O_3 perturbations caused by SST emissions of NO_x and

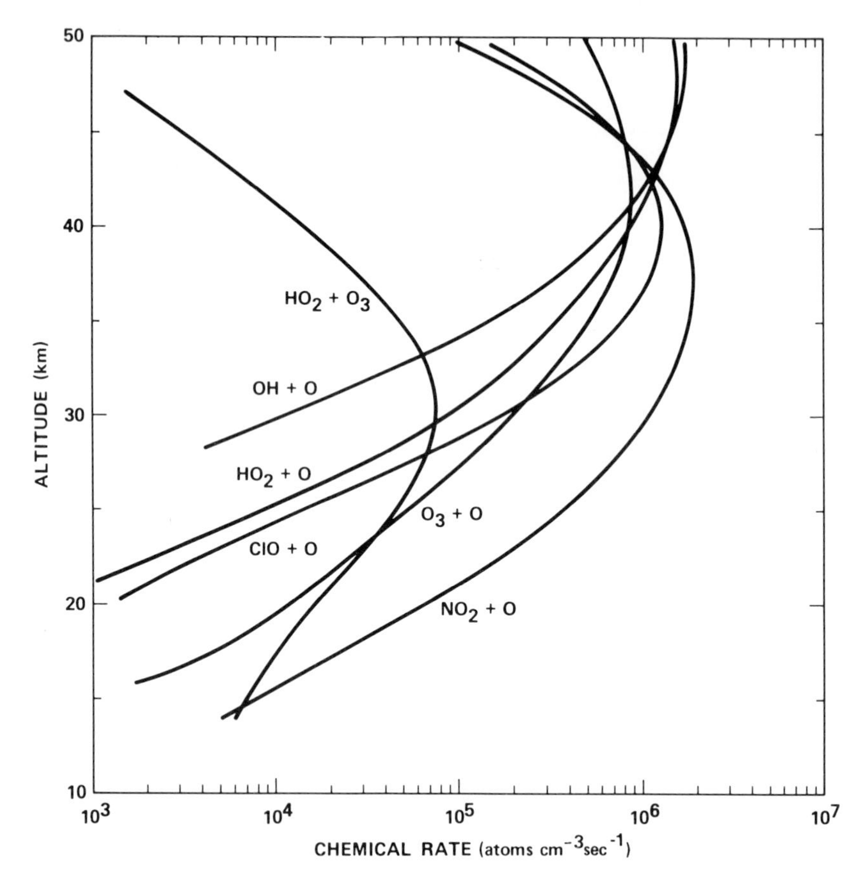

Fig. 10. Rates of key chemical reactions that deplete odd-oxygen in the stratosphere. The rates are averaged over 24 hr and reflect the total loss of odd-oxygen (e.g., the NO_2 + O reaction involves the loss of two odd-oxygen species).

H_2O. It is particularly noteworthy that the NO_x-induced O_3 perturbations have fluctuated from large decreases to small increases and back again to large decreases!

Such changes in predictions are mainly the result of improvements in our understanding of stratospheric photochemistry. There are two mechanisms by which these improvements have occurred. First, basic photochemical kinetic data, obtained through laboratory studies, have become increasingly reliable and precise. Second, new chemical species and families that affect the O_3 balance have been discovered. Most atmospheric scientists and aeronomers believe that although there are still significant uncertainties in

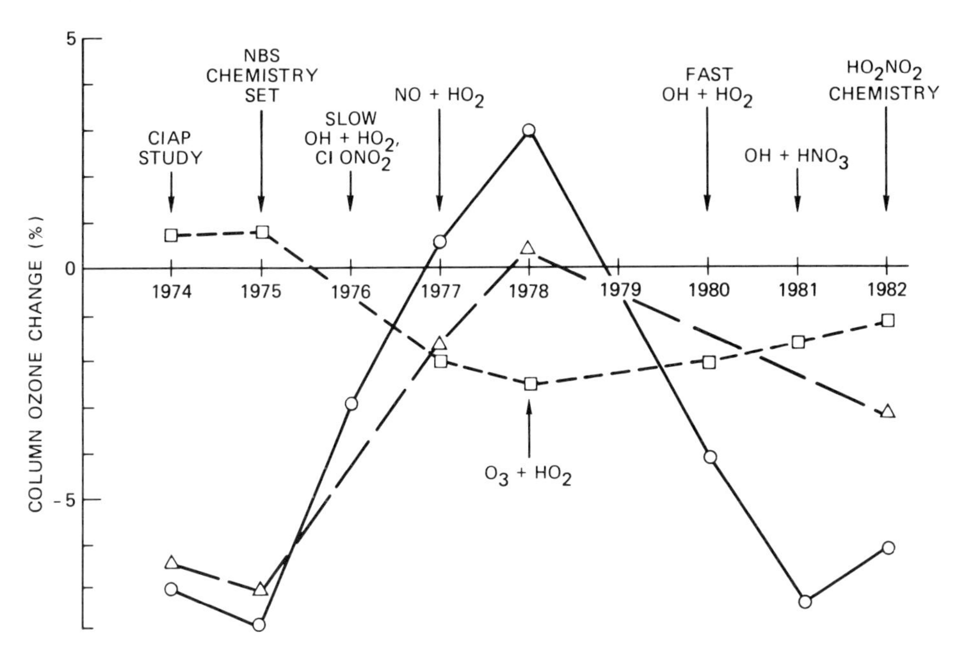

Fig. 11. Historical trends in the prediction of ozone perturbations caused by SST emissions. Steady-state globally averaged column O_3 changes are given for fixed global emission rates of NO_2 and H_2O. Not all of the significant improvements in basic knowledge of stratospheric photochemistry that occurred between 1971 and 1983 are shown. The O_3 perturbations are representative of calculations from each period, corresponding to the following annual worldwide injections at 20 km: ○, 1×10^9 kg NO_2; □, 2.1×10^{11} kg H_2O; △, simultaneous NO_x and H_2O emissions. From Turco (1984).

stratospheric photochemical and dynamic models, current estimates of O_3 change are probably fairly reliable.

The SST has not proved to be an economic success, and its current utilization remains limited. However, commercial subsonic long-haul air traffic has grown steadily. These flights deposit NO_x in the upper troposphere (near the tropopause), and may presently be responsible for a small columnar O_3 increase of up to 2%. This O_3 change is not of great concern to most scientists.

Increases in background N_2O concentrations are potentially serious, because N_2O generates NO_x in the middle stratosphere where the O_3/NO_x catalytic cycle is most powerful. Tables V and VI indicate that large O_3 depletions are possible when N_2O abundances increase by 20% or more. Observations suggest that background N_2O concentrations are increasing by ~0.2% per year. The reason for the increase is unknown, but is thought to be related to the combustion of fossil fuels. Such an increase would have to continue unabated for ~100 years to cause a significant O_3 perturbation.

The theory of stratospheric O_3 depletion by chlorofluorocarbons (CFCs) F-11 and F-12 was first proposed by Molina and Rowland in 1974 (1974a). Chlorine generated during the photochemical decomposition of CFCs and other chlorocarbon compounds attack O_3 catalytically as described in Section VIII. Figure 12 illustrates the history of CFC O_3-depletion cal-

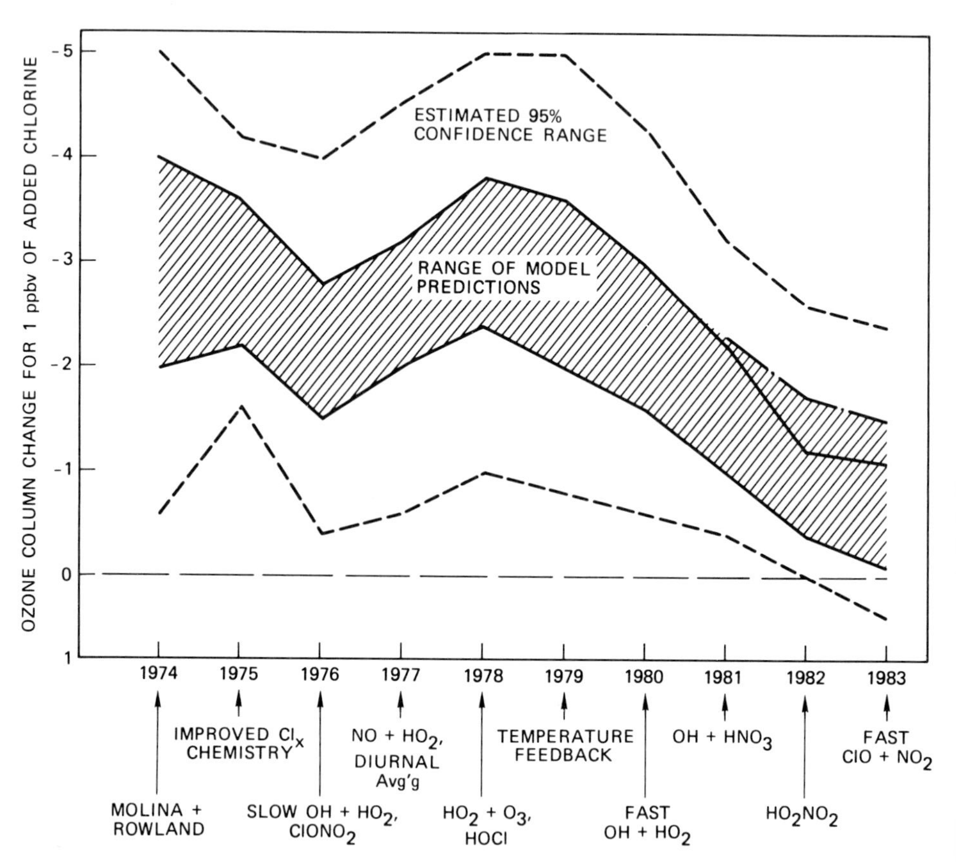

Fig. 12. Historical trends in model predictions of steady-state column O_3 depletions caused by 1 ppbv (parts per billion by volume) of chlorine added to the stratosphere as chlorofluorocarbons. The range of O_3 perturbations calculated by different modelers is roughly bounded by the solid lines. A crude appraisal of the uncertainty range for 95% confidence in the predictions, held at the time of the calculations, is indicated by the dashed lines. Key improvements in modeling techniques and photochemical data are shown in chronological order. The dash-dot curve defines the average O_3 response (i.e., $\%\Delta O_3$ per ppbv Cl_x) for larger chlorine injections in cases where the O_3 response is significantly nonlinear with respect to the chlorine increase. From Turco (1984).

culations. As with NO_x-induced perturbations, the model predictions have varied over wide limits. Utilizing the most recent photochemical schemes, model forecasts of chlorine-induced O_3 depletions have fallen to quite low values (a columnar depletion of $\sim 1\%$ for an increase of 1 ppbv of Cl_x). If the current emission rates of F-11 and F-12 remain constant indefinitely, the eventual increase in background Cl_x concentrations could be ~ 6 ppbv, and the decrease in stratospheric O_3, approximately 3 to 6%.

The height abundance of the O_3 reduction caused by fluorocarbons is illustrated in Fig. 13 for two distinct sets of photochemical rate coefficients. With both chemistry sets, large O_3 depletions are calculated near 40 km. This feature has remained essentially unaltered through nearly a decade of theoretical analysis. Moreover, satellite UV backscatter and Umkehr (ground-based) measurements of the O_3 profile indicate that a decrease at 40 km may already have occured due to chlorocarbons now present in the atmosphere.

The major differences between the old and new predictions in Fig. 13 appear in the lower stratosphere. The older chemistry yields substantial local O_3 decreases below 25 km, while the newer chemistry yields moderate local O_3 *increases*. Thus more recent forecasts of small O_3 depletions by CFCs are the net result of large decreases at high altitudes offset by significant increases at low altitudes. Because the photochemistry of the lower stratosphere is still quite uncertain, the predictions of Cl_x-induced O_3 increases in this region must be regarded with caution.

The photochemical explanation for the chemical behavior of the lower stratosphere is complex. Importantly, the most recent chemical kinetic schemes predict smaller OH concentrations below 30 km than previous schemes (there are also no measurements of OH in this region to check the models!). Reduced OH causes the NO_x cycle to be emphasized, and the Cl_x cycle to be deemphasized, with respect to O_3 catalysis. As chlorine is added to the stratosphere, NO_x is removed from catalytic activity as $ClONO_2$. Thus even though Cl_x catalytic activity increases, the *net* photochemical loss rate of O_x decreases because NO_x is diverted into chlorine nitrate.

A major goal of atmospheric science has been to observe and predict temporal trends in stratospheric O_3 (e.g., World Meteorological Organization, 1982). In addition to natural cyclic variations and synoptic-scale fluctuations, a secular trend in O_3 is expected due to the growing influence of humans on the composition of the atmosphere (Logan *et al.*, 1978). Human impact is the result of several pollutants acting simultaneously (Table V). Certain of the pollutants tend to decrease ozone (CFCs, N_2O, and SST NO_x), and others tend to increase O_3 (subsonic aircraft NO_x, CO_2, and CH_4). Accordinly, in future decades, only small net changes in total stratospheric O_3 are forecast (National Academy of Sciences, 1984). This will make it very difficult to detect anthropogenic signals in the O_3 burden using trend analysis.

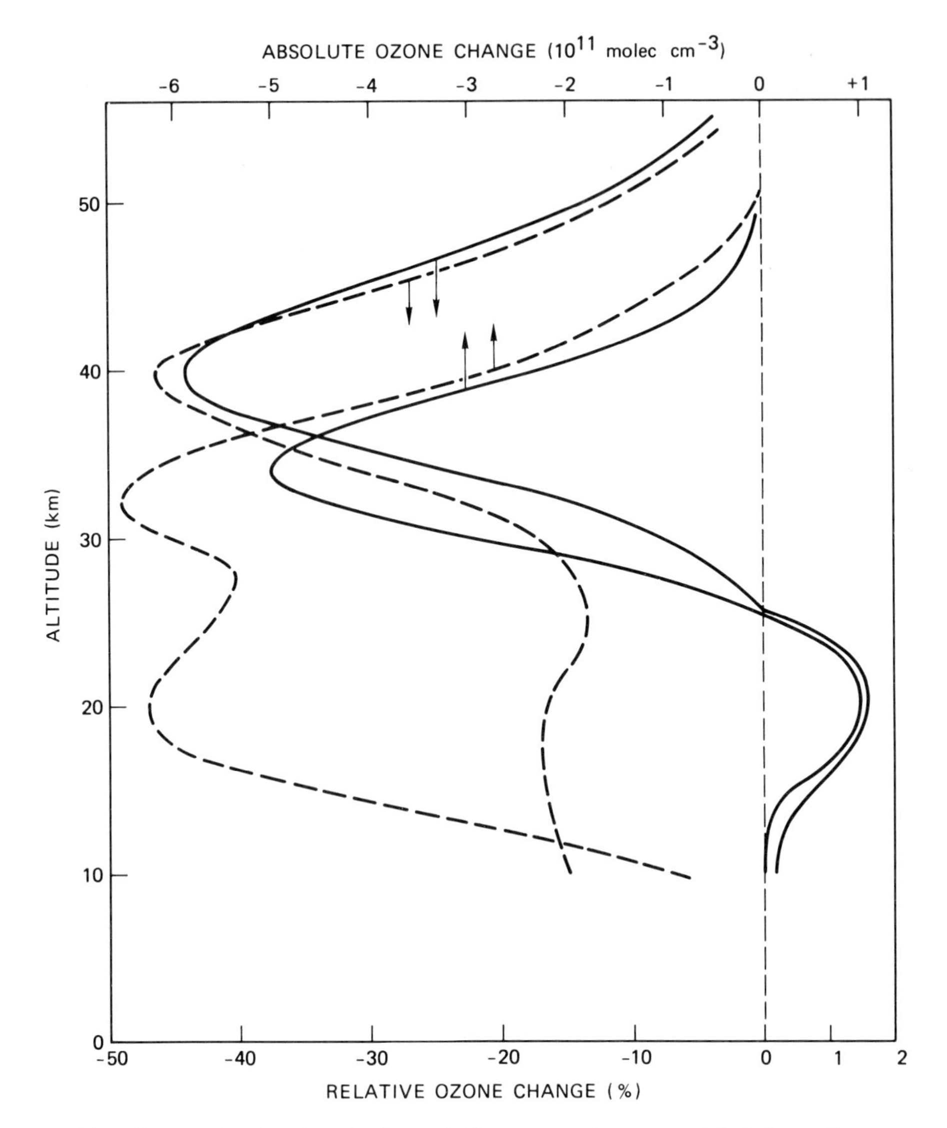

Fig. 13. Predicted steady-state changes in O_3 concentrations versus altitude for continuous chlorofluorocarbon ($CFCl_3$ and CF_2Cl_2, or F-11 and F-12) emissions at 1980 production rates. Both the relative and absolute O_3 changes are given. Results are shown for 1978 (dashed line) and 1982 (solid line) photochemistry schemes. At steady state, ~6 ppbv of chlorine (Cl_x) are produced as a result of chlorofluorocarbon decomposition. Note the change in the lower scale at the origin. From Turco (1984).

The average variation in the total abundance of stratospheric O_3 is expected to remain small for many years, which mitigates the danger of increased exposure to UV-B radiation. Nevertheless, there may be large adjustments in the *vertical distribution* of O_3, which can affect stratospheric dynamics through the mechanism of solar heating by O_3 absorption. Many of the pollutant gases that cause O_3 to change (e.g., N_2O, CF_2Cl_2, $CFCl_3$, and CO_2) also create a "greenhouse" effect; that is, each has strong infrared bands that trap thermal infrared radiation near the Earth's surface. Hence, in the future, climatic disturbances associated with these pollutants and the O_3 perturbations connected with them may prove to be a greater issue than the threat of UV-B exposure.

XIII. Summary and Research Requirements

It should be obvious by this point that the photochemistry of the stratosphere is exceedingly complex. Without attempting to be exhaustive, we have discussed nearly 130 key photochemical processes that control stratospheric composition. Literally hundreds of additional reactions could be considered. In many instances, even among the 130 key processes, chemical kinetic and photolytic data are inadequate. In other cases, fundamental chemical steps within complex reaction mechanisms are not well defined. Intricate computer models are required to solve all but the most modest photochemical reaction schemes. Nevertheless, analytical solutions of chemical rate equations, which apply to restricted sets of reactions or within isolated families of species (e.g., see Section IV), often provide useful insights.

Research is required in three major areas: laboratory photochemistry, atmospheric observations, and theoretical modeling.

A. Laboratory Photochemistry

Improved data are needed for key photochemical processes. It is particularly urgent to resolve uncertainties in the HO_x reaction system (e.g., $OH + HO_2$, $HO_2 + HO_2$, $HO_2 + O_3$, $OH + HO_2NO_2$, $HO_2 + Cl$). The critical O_2 Herzberg continuum absorption cross sections must also be precisely determined. The rate coefficients of the important chlorine reactions $ClO + O$ and $HCl + OH$ are still somewhat controversial, and contrary measurements should be reconciled. The formation reactions and photolysis of HOCl and $ClONO_2$ in the chlorine system also require attention. The photochemical processes of the sulfur-hydrogen radicals (HSO_x) are essentially unknown. Many important reactions in the fluorine and bromine systems must also be studied with greater precision.

B. Atmospheric Observations

Detailed measurements of atmospheric composition provide the ultimate key to understanding the photochemistry of the stratosphere. In addition to their intrinsic value, atmospheric observations often reveal new chemical species, establish relationships between species, and define temporal and spatial scales of variability. Measurements are critical for calibrating and validating models of stratospheric composition that utilize sophisticated photochemical schemes.

Detection and analysis of trends in stratospheric O_3 are essential if anthropogenic influences on O_3 are to be understood. The sparse measurements of critical species such as OH and ClO must be reinforced by additional observations using independent techniques. Simultaneous measurements of several related species are generally more useful than several individual measurements of one species. Firm quantitative observations are needed for H_2O_2, N_2O_5, HO_2NO_2, HOCl, and $ClONO_2$, as well as a number of secondary species in the hydrocarbon, sulfur, fluorine, and bromine families. Instrumental techniques presently in use must be cross-calibrated and observations intercompared. Wherever practical, alternative detection strategies should be attempted. Finally, continuous global-scale measurements from space-based platforms—if instruments can be constructed to observe a number of pertinent species with sufficient accuracy—should be developed.

C. Theoretical Modeling

While we have not discussed stratospheric photochemical/dynamic models in any detail, a few relevant points can be made. First, it is clear that three-dimensional stratospheric models with complete photochemical schemes will eventually be needed to illuminate outstanding problems, particularly those of the lower stratosphere. In the meantime, the one-dimensional model remains the most useful tool for studying stratospheric photochemistry and global chemical budgets. Future models of the stratosphere will be called on to treat accurately the physical processes that couple temperature, photochemistry, and dynamics. Analysis of the relationships between these processes may reveal subtle and important physical effects. Models that describe the climatic impacts associated with changes in stratospheric and tropospheric composition should continue to be refined, including the effects of "greenhouse" gases and of alterations in atmospheric dynamics. The transport mechanisms that couple the troposphere and the stratosphere also require further investigation, as these mechanisms determine the ultimate sources and sinks of most stratospheric constituents.

References

Aikin, A. C., Herman, J. R., Maier, E. J., and McQuillan, C. J. (1982). Atmospheric chemistry of ethane and ethylene. *JGR, J. Geophys. Res.* **87,** 3105–3118.

Anderson, G. P., and Hall, L. A. (1983). Attenuation of solar irradiance in the stratosphere: spectrometer measurements between 191 and 207 nm. *J. Geophys. Res.* **88,** 6801–6806.

Bates, D. R., and Hayes, P. B. (1967). Atmospheric nitric oxide. *Planet. Space Sci.* **15,** 189–197.

Bates, D. R., and Nicolet, M. (1950). The photochemistry of atmospheric water vapor. *J. Geophys. Res.* **55,** 301–327.

Baulch, D. L., Cox, R. A., Crutzen, P. J., Hampson, R. F., Jr., Kerr, J. A., Troe, J., and Watson, R. T. (1982). Evaluated kinetic and photochemical data for atmospheric chemistry: supplement I. *J. Phys. Chem. Ref. Data* **11,** 327–496.

Brewer, D. A., Augustsson, T. R., and Levine, J. S. (1983). The photochemistry of anthropogenic nonmethane hydrocarbons in the troposphere. *JGR, J. Geophys. Res.* **88,** 6683–6695.

Castleman, A. W., Jr., Davis, R. E., Munkelwitz, H. R., Tang, I. N., and Wood, W. P. (1975). Kinetics of association reactions pertaining to H_2SO_4 aerosol formation. *Int. J. Chem. Kinet. Symp.* **1,** 629–640.

Chameides, W. L., and Cicerone, R. J. (1978). Effects of nonmethane hydrocarbons in the atmosphere. *JGR, J. Geophys. Res.* **83,** 947–954.

Chameides, W. L., and Davis, D. D. (1980). Iodine: its possible role in tropospheric photochemistry. *JGR, J. Geophys. Res.* **85,** 7383–7398.

Chapman, S. (1930). A theory of upper atmospheric ozone. *Mem. R. Meteorol. Soc.* **3,** 103.

Cicerone, R. J. (1979). Atmospheric carbon tetrafluoride: a nearly inert gas. *Science* **206,** 59–61.

Cicerone, R. J. (1981). Halogens in the atmosphere. *Rev. Geophys. Space Phys.* **19,** 123–139.

Cicerone, R. J., and McCrumb, J. L. (1980). Photodissociation of isotopically heavy O_2 as a source of atmospheric O_3. *Geophys. Res. Lett.* **7,** 251–254.

Cicerone, R. J., Stolarski, R. S., and Walters, S. (1974). Stratospheric ozone destruction by man-made chlorofluoromethanes. *Science* **185,** 1165–1167.

Cicerone, R. J., Walters, S., and Stolarski, R. S. (1975). Chlorine compounds and stratospheric ozone. *Science* **188,** 378–379.

Cornu, A. (1879). Sur la limite ultra-violette du spectre solaire. *C. R. Hebd. Seances Acad. Sci.* **88,** 1101.

Crutzen, P. J. (1969). Determination of parameters appearing in the "dry" and in the "wet" photochemical theories for ozone in the stratosphere. *Tellus* **21,** 368–388.

Crutzen, P. J. (1970). The influence of nitrogen oxides on the atmospheric ozone content. *Q. J. R. Meteorol. Soc.* **96,** 320–325.

Crutzen, P. J. (1971). Ozone production rates in an oxygen–hydrogen–nitrogen oxide atmosphere. *J. Geophys. Res.* **76,** 7311–7327.

Crutzen, P. J. (1974a). Review of upper atmospheric photochemistry. *Can. J. Chem.* **52,** 1569–1581.

Crutzen, P. J. (1974b). Estimates of possible future ozone reductions from continued use of fluorochloromethanes (CF_2Cl_2,$CFCl_3$). *Geophys. Res. Lett.* **1,** 205–208.

Crutzen, P. J. (1976). The possible importance of CSO for the sulfate layer of the stratosphere. *Geophys. Res. Lett.* **3,** 73–76.

Crutzen, P. J., and Schmailzl, U. (1983). Chemical budgets of the stratosphere. *Planet. Space Sci.* **31,** 1009–1032.

Crutzen, P. J., Isaksen, I. S. A., and McAfee, J. R. (1978). Impact of the chlorocarbon industry on the ozone layer, *JGR, J. Geophys. Res.* **83,** 345–363.

DeMore, W. B., Watson, R. T., Golden, D. M., Hampson, R. F., Kurylo, M., Howard, C. J., Molina, M. J., and Ravishankara, A. R. (1983). "Chemical Kinetics and Photochemical Data for Use in Stratospheric Modeling," Publ. 83–62. Jet Propulsion Lab., Pasadena, California.

Frederick, J. E., and Hudson, R. D. (1979). Predissociation of nitric oxide in the mesosphere and stratosphere. *J. Atmos. Sci.* **36,** 737–745.

Frederick, J. E., and Mentall, J. E. (1982). Solar irradiance in the stratosphere: implications for the Herzberg continuum absorption of O_2. *Geophys. Res. Lett.* **9,** 461–464.

Greenberg, R. I., and Heicklen, J. (1970). Reaction of $O(^1D)$ with N_2O. *Int. J. Chem. Kinet.* **2,** 185–192.

Grobecker, A. J., ed. (1975). "The Natural Stratosphere of 1974," Climatic Impact Assessment Program Monograph 1, DOT-TST-75-51. U. S. Dept. of Transportation, Washington, D.C.

Graner, P., and Kleinert, H. (1927). Die Dämmerungerscheinen. *Probl. Kosm. Phys.* **10,** 1–113.

Hartley, W. N. (1881a). On the absorption spectrum of ozone. *J. Chem. Soc.* **39,** 57.

Hartley, W. N. (1881b). On the absorption of solar rays by atmospheric ozone. *J. Chem. Soc.* **39,** 111.

Hayes, P. B., and Olivero, J. J. (1970). Carbon dioxide and monoxide above the tropopause. *Planet. Space Sci.* **18,** 1729–1733.

Herman, J. R., and Mentall, J. E. (1982). O_2 absorption cross sections (187–225 nm) from stratospheric solar flux measurements. *JGR, J. Geophys. Res.* **87,** 8967–8975.

Holton, J. R. (1979). "An Introduction to Dynamic Meteorology." Academic Press, London.

Hunt, B. G. (1966). Photochemistry of ozone in a moist atmosphere. *J. Geophys. Res.* **71,** 1385–1398.

Inn, E. C. Y., Vedder, J. F., Tyson, B. J., and O'Hara, D. (1979). COS in the stratosphere. *Geophys. Res. Lett.* **6,** 191–193.

Jaeschke, W., Schmitt, R., and Georgii, H.-W. (1976). Preliminary results of stratospheric SO_2 measurements. *Geophys. Res. Lett.* **3,** 517–519.

Johnston, H. S. (1971). Reduction of stratospheric ozone by nitrogen oxide catalysts from supersonic transport exhaust. *Science* **173,** 517–522.

Johnston, H. S., and Podolske, J. (1978). Interpretations of stratospheric photochemistry. *Rev. Geophys. Space Phys.* **16,** 491–519.

Junge, C. E., Chagnon, C. W., and Manson, J. E. (1961). Stratospheric aerosols. *J. Meteorol.* **18,** 81–108.

Kaye, J. A., and Strobel, D. F. (1983). Enhancement of heavy ozone in the Earth's atmosphere? *J. Geophys. Res.* **88,** 8447–8452.

Leovy, C. B. (1969). Atmospheric ozone: an analytic model for photochemistry in the presence of water vapor. *J. Geophys. Res.* **74,** 417–426.

Logan, J. A., Prather, M. J., Wofsy, S. C., and EcElroy, M. B. (1978). Atmospheric chemistry: response to human influence. *Philos. Trans. R. Soc. London* **290,** 187–234.

McConnell, J. C., and Schiff, H. I. (1978). Methyl chloroform: impact on stratospheric ozone. *Science* **199,** 174–177.

McKeen, S. A., Liu, S. C., and Kiang, C. S. (1984). On the chemistry of stratospheric SO_2 from volcanic eruptions. *J. Geophys. Res.* (in press).

Mauersberger, K. (1981). Measurement of heavy ozone in the stratosphere. *Geophys. Res. Lett.* **8,** 935–937.

Meinel, A. R. (1950a). OH emission bands in the spectrum of the night sky. I. *Astrophys. J.* **111,** 555–564.

Meinel, A. R. (1950b). OH emission bands in the spectrum of the night sky. II. *Astrophys. J.* **112,** 120–130.

Molina, M. J., and Rowland, F. S. (1974a). Stratospheric sink for chlorofluoromethanes: chlorine atom–catalyzed destruction of ozone. *Nature (London)* **249,** 810–812.

Molina, M. J., and Rowland, F. S. (1974b). Predicted present stratospheric abundances of chlorine species from photodissociation of carbon tetrachloride. *Geophys. Res. Lett.* **1,** 309–312.

Nagy, A. F., Liu, S. C., and Baker, D. J. (1976). Vibrationally-excited hydroxyl molecules in the lower atmosphere. *Geophys. Res. Lett.* **3,** 731–734.
National Academy of Sciences (NAS) (1975). "Environmental Impact of Stratospheric Flight." Natl. Acad. Sci., Washington, D.C.
National Academy of Sciences (NAS) (1976). "Halocarbons: Effects on Stratospheric Ozone." Natl. Acad. Sci., Washington, D.C.
National Academy of Sciences (NAS) (1982). "Causes and Effects of Stratospheric Ozone Reduction: An Update." Natl. Acad. Sci., Washington, D.C.
National Academy of Sciences (NAS) (1984). "Causes and Effects of Changes in Stratospheric Ozone: Update 1983." Natl. Acad. Sci., Washington, D.C.
Nicolet, M. (1965). Nitrogen oxides in the chemosphere. *J. Geophys. Res.* **70,** 679–689.
Nicolet, M. (1970). The origin of nitric oxide in the terrestrial atmosphere. *Planet. Space Sci.* **18,** 1111–1118.
Nicolet, M. (1972). Aeronomic chemistry of the stratosphere. *Planet. Space Sci.* **20,** 1671–1702.
Nicolet, M. (1975). Stratospheric ozone: an introduction to its study. *Rev. Geophys. Space Phys.* **13,** 593–636.
Nicolet, M., and Cieslik, S. (1980). The photodissociation of nitric oxide in the mesosphere and stratosphere. *Planet. Space Sci.* **28,** 105–115.
Noxon, J. F. (1982). A global study of $O_2(^1\Delta_g)$ airglow: day and twilight. *Planet. Space Sci.* **30,** 545–557.
Pollack, J. B., Toon, O. B., Sagan, C., Summers, A., Baldwin, B., and Van Camp, W. (1976). Volcanic explosions and climatic change: a theoretical assessment. *JGR, J. Geophys. Res.* **81,** 1071–1083.
Prasad, S. S., Jaffe, R. L., Whitten, R. C., and Turco, R. P. (1978). Reservoirs of atmospheric chlorine: prospects for HOCl revisited. *Planet. Space Sci.* **26,** 1017–1026.
Rowland, F. S., and Molina, M. J. (1975). Chlorofluoromethanes in the environment. *Rev. Geophys. Space Phys.* **13,** 1–35.
Ryan, J. A., and Mukherjee, N. R. (1975). Sources of stratospheric gaseous chlorine. *Rev. Geophys. Space Phys.* **13,** 650–658.
Sheppard, J. C., Westberg, H., Hopper, J. F., Ganesan, K., and Zimmerman, P. (1982). Inventory of global methane sources and their production rate. *JGR, J. Geophys. Res.* **87,** 1305–1312.
Simonaitis, R., and Heicklen, J. (1975). The reaction of HO_2 with NO and NO_2 and of OH with NO. *J. Phys. Chem.* **79,** 298–302.
Stockwell, W. R., and Calvert, J. G. (1983). The mechanism of the $HO–SO_2$ reaction. *Atmos. Environ.* **17,** 2231–2235.
Stolarski, R. S., and Cicerone, R. J. (1974). Stratospheric chlorine: a possible sink for ozone. *Can. J. Chem.* **52,** 1610–1615.
Stolarski, R. S., and Rundel, R. D. (1975). Fluorine chemistry in the stratosphere. *Geophys. Res. Lett.* **2,** 443–444.
Streit, G. E., Whitten, G. Z., and Johnston, H. S. (1976). The fate of vibrationally excited hydroxyl radicals HO ($v \leqslant 9$), in the stratosphere. *Geophys. Res. Lett.* **3,** 521–523.
Turco, R. P. (1975). Photodissociation rates in the atmosphere below 100 km. *Geophys. Surv.* **2,** 153–192.
Turco, R. P. (1984). Stratospheric ozone perturbations. *In* "Stratospheric Ozone" (R. C. Whitten and S. S. Prasad, eds.), Chapter 6. Van Nostrand-Reinhold, Princeton, New Jersey (in press).
Turco, R. P., and Whitten, R. C. (1975). Chlorofluoromethanes in the stratosphere and some possible consequences for ozone. *Atmos. Environ.* **9,** 1045–1061.
Turco, R. P., and Whitten, R. C. (1977). "The NASA Ames Research Center One Dimensional Stratospheric Model," NASA Tech. Pap. 1002. Natl. Aeron. Space Admin., Washington, D.C.
Turco, R. P., Hamill, P., Toon, O. B., Whitten, R. C., and Kiang, C. S. (1979). A one-dimensional

model describing aerosol formation and evolution in the stratosphere. I. Physical processes and mathematical analogs. *J. Atmos. Sci.* **36,** 699–717.

Turco, R. P., Toon, O. B., Hamill, P., and Whitten, R. C. (1981a). Effects of meteoric debris on stratospheric aerosols and gases. *JGR, J. Geophys. Res.* **86,** 1113–1128.

Turco, R. P., Whitten, R. C., Toon, O. B., Inn, E. C. Y., and Hamill, P. (1981b). Stratospheric hydroxyl radical concentrations: new limitations suggested by observations of gaseous and particulate sulfur. *JGR, J. Geophys. Res.* **86,** 1129–1139.

Turco, R. P., Whitten, R. C., and Toon, O. B. (1982). Stratospheric aerosols: observation and theory. *Rev. Geophys. Space Phys.* **20,** 233–279.

U.S. Standard Atmosphere (1976). U.S. Govt. Printing Office, Washington, D.C.

Weiss, R. W. (1981). The temporal and spatial distribution of tropospheric nitrous oxide. *JGR, J. Geophys. Res.* **86,** 7185–7195.

Whitten, R. C., ed. (1982). "The Stratospheric Aerosol Layer," Top. Curr. Phys. Vol. 28. Springer-Verlag, Berlin and New York.

Whitten, R. C., Sims, J. S., and Turco, R. P. (1973). A model of carbon compounds in the stratosphere and mesosphere. *J. Geophys. Res.* **78,** 5362–5374.

Wofsy, S. C., and McElroy, M. B. (1974). HO_x, NO_x and ClO_x: their role in atmospheric photochemistry. *Can. J. Chem.* **52,** 1582–1591.

Wofsy, S. C., McConnell, J. C., and McElroy, M. B. (1972). Atmospheric CH_4, CO and CO_2. *J. Geophys. Res.* **77,** 4477–4493.

Wofsy, S. C., McElroy, M. B., and Sze, N. D. (1975a). Freon consumption: implications for atmospheric ozone. *Science* **187,** 535–537.

Wofsy, S. C., McElroy, M. B., and Yung, Y. L. (1975b). The chemistry of atmospheric bromine. *Geophys. Res. Lett.* **2,** 215–218.

World Meteorological Organization "The Stratosphere 1981: Theory and Measurements. Global Ozone Res. Proj., Rep. No. 11. WMO, Geneva.

Yung, Y. L., Pinto, J. P., Watson, R. T., and Sander, S. P. (1980). Atmospheric bromine and ozone perturbations in the lower stratosphere. *J. Atmos. Sci.* **37,** 339–353.

Zipf, E. C. (1984). On the formation of nitrous oxide by reactions between N_2 and excited OH radicals. 1. OH ($A\,^2\Sigma^+$). *Nature (London)*, in press.

Zipf, E. C. and Prasad, S. S. (1984). Nitrous oxide formation by metastable $N_2(A\,^3\Sigma_u^+)$ chemistry: a new perspective on its prospects. *J. Geophys. Res.*, submitted.

4

Photochemistry, Composition, and Climate

WILLIAM R. KUHN

Department of Atmospheric and Oceanic Science
University of Michigan
Ann Arbor, Michigan

I. Introduction

One might well wonder why a chapter on climate should be included in a book devoted to photochemistry. The primary gas responsible for maintaining the Earth's surface and air temperatures—water vapor (H_2O)—is not controlled by photochemistry but by evaporation from the ocean surface. Another important gas, carbon dioxide (CO_2), is primarily controlled by the

ISBN 0-12-444920-4

oceans and by the biosphere—photosynthesis, respiration, decay, and the burning of fossil fuels are the primary production and loss processes. These two gases, in addition to H_2O in the form of clouds, are responsible for the radiative contribution to the mean temperature of the Earth to within several degrees.

Yet this is not to say that photochemistry is unimportant to climate studies. Global surface temperature variations of the order of a degree or two are thought to be quite important to the overall climate, and the combined effect of several minor constituents in the atmosphere that are photochemically controlled may well influence the surface temperature by this amount. Even on a geological time scale, it appears that the mean temperature of the Earth has varied by perhaps only 10°C, yet major glacial advances have at times covered 30% of the surface. An increase of only a degree or two at mid latitudes could produce a heating in polar regions several times as large; because of the melting of ice and exposure of the land or ocean, which have lower reflectivities than ice or snow, a greater absorption of solar radiation and an amplification of the temperature increase would occur. Current estimates indicate that an increase in temperature of the west Antarctic ice pack to 5°C would cause it to become unstable, break apart, and drift to lower latitudes and melt, increasing mean sea level by ~6 m. Such an increase would lead to the inundation of parts of the eastern and southeastern coasts of the United States, for example, and would represent a major climatic change. One might well imagine that temperature changes of even a few tenths of a degree would shift present climate zones, leading to social and political turmoil.

The radiative processes that provide energy to the surface are the absorption of solar radiation and the long-wave emission from the atmosphere and clouds, commonly known as the greenhouse effect. The long-wave radiation that is returned to the surface is equivalent to ~92% of the average solar energy incident on the atmosphere (345 W m^{-2}). If outgoing infrared (IR) radiation was the only process for removing energy from the surface, then the surface temperature would be some 340 K rather than 288 K. In addition to radiation, evaporation of water from the surface (latent heat flux) and the transfer of sensible heat are important in cooling the surface. The average temperature of the surface is somewhat warmer than the air layer a few millimeters thick adjacent to the surface. Conduction of heat occurs across this surface–air interface, and the heat is then transported by convection into the troposphere, a process known as sensible heat flux. The latent heat and sensible heat fluxes are about 20 and 6%, respectively, of the long-wave radiation emitted by the surface.

Among the gases that provide a greenhouse effect are some that are photochemically controlled. One that is of obvious importance in ozone (O_3).

Present estimates indicate that the mean surface temperature is about a degree or two higher than it would be if O_3 was not present in the atmosphere; there would also be no temperature maximum in the stratosphere, the temperature being some 50 K lower. Other gases of importance include methane (CH_4), nitrous oxide (N_2O), ammonia (NH_3), and the Freons.

One should also note that the gases that control our climate are not those that make up the bulk of the atmosphere, namely, nitrogen (N_2) and oxygen (O_2), but rather the minor or trace constituents. Even CO_2 and H_2O occupy less than a few percent of the volume of the atmosphere—CO_2 $\sim 0.034\%$ and H_2O less than 3% near the surface, decreasing to one–ten-thousandth that amount near the tropopause. In the troposphere, O_3 has a concentration of 30 ppb, but the concentration increases to several parts per million in the mid stratosphere. Even small amounts of photochemically active gases should be considered as potentially important to climatic variations, through their influence not only on the chemical reactions of radiatively active species but also on the temperature and climate directly.

Climate is generally defined as weather averaged over some specified period of time but including some measure of the frequency of extremes of the weather variables. Such variation is important since two regions might have the same average temperature, yet the temperature range and therefore climates could be decidedly different. The weather variables used are determined by the particular application of the climate classification as well as by the availability of data. Early classifications were quite descriptive and emphasized the weather elements of air temperature and precipitation, the two variables most important to human activity. A more detailed climatology would include pressure, humidity, winds, and clouds. Other climatologies can also be developed. For example, O_3 amount, turbidity, and cloudiness would be the primary climate variables if one were concerned with erythemal doses.

One must also define a spacial extent for a particular climate. The microclimate for a leaf, measured in centimeters, would include leaf temperature, transpiration rate, diffusion resistance, wind speed, and incident solar radiation. At the other extreme would be global climate, measured in thousands of kilometers, where the usual climate variable is a global average surface temperature. The time over which a climate is defined can vary greatly, extending from several months representing seasonal climates to billions of years if one is interested in the temperature history of the Earth.

In this chapter we shall be primarily concerned with the models that have been developed for large-scale climate variations and the influence on temperature of changes in minor atmospheric constituents including those that are photochemically controlled.

II. Radiative Climate Modeling of the Earth

A. Zero-Dimensional Models

Climate models can be conveniently classified according to their dimensionality. A zero-dimensional model, for example, would give a climate parameter(s) that represents an average for the entire system. The usual parameter is temperature, representative of the mean temperature of the Earth's surface. Since the primary energy input to the Earth–atmosphere system is solar radiation (cosmic rays and the solar wind provide less than 10^{-5} of the total), an energy balance equation for the surface temperature T_s can be written

$$\frac{S(1 - A)}{f} = \epsilon \sum_i B_{\lambda_i}(T_s)\,\Delta\lambda_i + \sum_j B_{\lambda_j}(T_i)\,\Delta\lambda_j, \tag{1}$$

where the left-hand side represents the solar energy absorbed by the Earth–atmosphere system and the right-hand side the energy emitted to space in the form of IR radiation; S is the solar energy at the mean distance of the Earth from the Sun and is known as the solar constant (1370 W m^{-2}). Not all this energy is absorbed by the atmosphere and Earth; the albedo A is the fraction scattered by the clouds, atmosphere, and ground back to space; the global albedo for the Earth is $\sim$0.30. The factor f is known as the flux factor and is the ratio of the area of the planet emitting IR radiation to the area receiving solar radiation. If the planet rotates rapidly and has an appreciable atmosphere, such as the Earth, then the heat energy from the Sun is distributed in a short time over the planet so that the area for emission of IR radiation is the entire surface area of the planet. If a planet rotates slowly and in addition has a tenuous atmosphere, then the effective area for emission is only one-half the planetary surface. Earth satisfies the former conditions, so the flux factor is essentially 4.

The two terms on the right hand side of the equation represent emission to space from the Earth's surface and the atmosphere, respectively. The first term is summed over only those wavelength intervals $\Delta\lambda$ for which there is little or no atmospheric absorption of the surface radiation, while the second term includes only those wavelength regions where there is absorption. Atmospheric gases absorb IR radiation from the surface in selected wavelength regions as shown in Fig 1. The vibration–rotation bands of CO_2, for example, absorb radiation primarily in the spectral regions of 4.3 and 15 μm, while H_2O absorbs at 6.3 μm and at wavelengths beyond 15 μm, where the pure rotational band is important. Radiation in these spectral regions is transferred up through the atmosphere, with eventual emission to space occurring from an assumed isothermal layer of temperature T_i, which can be determined from the Eddington approximation (Kourganoff, 1963),

$$T_i = 2^{-1/4} T_e, \tag{2}$$

where T_e is the mean global temperature of the surface in the absence of an atmosphere. For the radiation from both the surface and the atmosphere the spectral distribution is given by the Planck function B_λ. The emissivity ϵ takes into account that the surface does not quite emit as a blackbody but rather some fraction of the Planck function; the emissivity in general ranges from about 0.9 to 1 (deserts, e.g., have an emissivity of about 0.9, ice 0.96, while snow ranges from 0.82 to 0.995).

If there were no H_2O, CO_2, and O_3 in the atmosphere, then essentially all the long-wave radiation from the surface would escape to space. In that case, the last term in Eq. (1) would be zero, and the remaining term integrated over all wavelengths would be the Stefan–Boltzmann equation,

$$\frac{S(1-A)}{f} = \epsilon \sigma T_s^4, \tag{3}$$

where σ is the Stefan–Boltzmann constant (5.7×10^{-8} W m^{-2} deg^{-4}). If

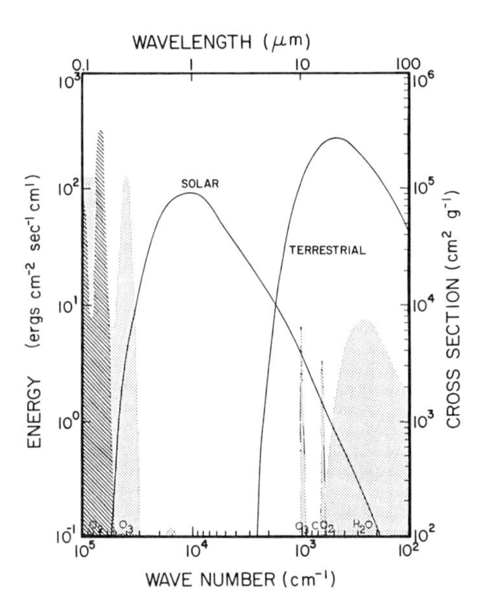

Fig. 1. Spectral distribution of the incoming solar and outgoing terrestrial radiation. Also shown (shaded) are the absorption cross sections of the major radiatively active gases in the atmosphere.

15% is used for the albedo, which is representative of the surface, a temperature of 260 K is obtained, whereas the globally averaged value is some 28 K higher. This increase, the greenhouse effect, is primarily due to water vapor absorbing the surface radiation and then reradiating a fraction back down to the surface, thus providing an additional heat source. When account is taken of H_2O and CO_2 absorption and clouds, Eq. (1) does indeed give a temperature of ~ 290 K. While one cannot draw definitive conclusions from such a simple model, the model can be used to estimate the relative importance of an absorbing constituent.

B. One-Dimensional Horizontal Models

One-dimensional models are used to study climate in either a horizontal or vertical direction. The former are known as models of the Budyko–Sellers type, named for the two scientists who pioneered their use. In these models, a latitude-dependent surface temperature is the climate parameter of major concern. Those models that yield a vertical temperature distribution, known as radiative–convective models, have been extensively used to study the effects on the surface temperature of changes in concentrations of minor constituents.

A simple latitudinal one-dimensional model can be generated from Eq. (1) (see, e.g., North, 1975). The solar radiation S' and the absorbed solar radiation $a = 1 - A$ can be approximated as two-term series expansions in the Legendre functions,

$$S' = S[1 + s_2 P_2(x)],$$

$$a = a_0 + a_2 P_2(x),$$

if there is no ice, and $a = \text{const}$ if there is a polar ice cap; P_2 is the second Legendre function, x the sine of the latitude, and a_0, a_2, and s_2 are coefficients in the expansions determined from solar radiation and albedo data. These Legendre functions are orthogonal functions, and have the important property that the integral of any two different functions in the set over some specified interval is zero while the integral of the square of the function has some constant value. The advantage of using orthogonal functions is that the coefficients of the series expansions a_0, a_2, and s_2 are easy to compute. The ice–albedo feedback has been included since the albedo for ice may reach 0.80 while an ice- or snow-free surface rarely reaches 0.20 except for large solar zenith angles over water. The right-hand side of Eq. (1), representing the outgoing radiation I, is usually given by an empirical relation, first developed by Budyko (1969),

$$I = A + BT_s, \tag{4}$$

where A and B are coefficients determined from observations. The resulting equation can then be written as

$$S[1 + s_2 P_2(x)][a_0 + a_2 P_2(x)]/4 = A + BT(x) \tag{5}$$

and solved for the temperature. A typical solution appears in Fig. 2. Note that this particular model does not allow for any transport of energy between pole and equator, thus the equatorial regions are some 30 K warmer than what is observed while the poles are colder. A zero-dimensional model, also shown, represents one for which there is an infinite transport such that the temperature is uniform between pole and equator.

Meridional heat transport occurs in both the oceans and atmosphere; heat is transported poleward by not only large-scale winds and currents, but also by eddy motions in the air and water. In addition, there is also the transport of latent heat by the atmosphere. The total of this heat transport is sometimes approximated by assuming a diffusive thermal heat transport dependent on the latitudinal surface temperature gradient. Equation (5) then can be written as

$$-d/dx[C(1 - x^2)\,dT(x)/dx] + A + BT(x) = S'a(x)/4, \tag{6}$$

where C is a free parameter. If we assume, in addition to the expansion of S' and a in a series of Legendre functions, that the temperature can be likewise represented, then Eq. (6) becomes an algebraic equation since the Legendre functions are eigenfunctions of the diffusion term; furthermore, if there is no transport of heat across the pole and equator, the temperature will be symmetric with respect to the equator and pole, and there will be no odd-numbered Legendre functions appearing in the expansion. The resulting solution for temperature can be made to approximate the present-day zonally averaged temperature distribution as shown in Fig. 2. The model can then be

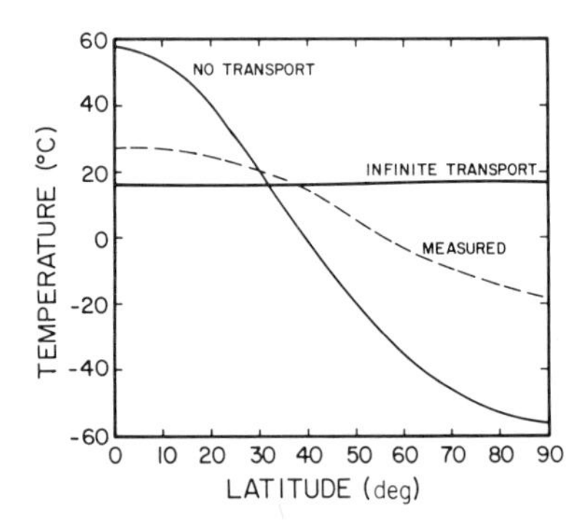

Fig. 2. Surface temperature from a one-dimensional (horizontal) climate model for infinite transport and no transport. Also shown is the actual latitudinal temperature distribution.

used to carry out sensitivity studies for changes in various parameters such as solar radiation.

C. Three-Dimensional Models

At the opposite extreme from these very simple climate models are the general circulation models (GCMs). A complete description of climate would include results for both directions in the horizontal (i.e., latitude and longitude) as well as for the vertical. Six scalar equations are solved for the three components of the wind velocity and the pressure, temperature, and air density. For dry air, these are

$$\frac{d\mathbf{V}}{dt} = -\frac{1}{\rho}\nabla p - 2\boldsymbol{\omega} \times \mathbf{V} + \mathbf{g} + \mathbf{F}, \tag{7}$$

$$\frac{\partial \rho}{\partial t} + \nabla \cdot \rho \mathbf{V} = 0, \tag{8}$$

$$p = \rho R T, \tag{9}$$

$$Q = c_p \frac{dT}{dt} - \frac{1}{\rho}\frac{dp}{dt}, \tag{10}$$

The first of these is the equation of motion for a uniformly rotating coordinate system. This equation states that the acceleration a fluid parcel undergoes is due to the pressure gradient and to Coriolis, gravitational, and frictional forces. The $\mathbf{V}$ is the velocity, t time, ρ density, p pressure, $\boldsymbol{\omega}$ the angular velocity of rotation of the Earth, $\mathbf{g}$ the acceleration of gravity, and $\mathbf{F}$ the frictional force. The second equation represents the conservation of mass, and the third the equation of state for air, which we can assume is a perfect gas and represent by the ideal gas equation; R is the gas constant. The final equation is the thermodynamic energy equation; c_p is the specific heat of air at constant pressure and Q the rate of heat addition. It is this term that is most influenced when one considers climate change brought about by a modification in a radiatively active constituent such as O_3 or N_2O.

Equations (7) through (10) are applicable for dry air; if water vapor is considered, then a conservation equation for that constituent must be added which would have the form of Eq. (8) but include an additional term representing the sources and sinks for H_2O.

Equations (7) through (10) represent a complete set for the atmosphere and can be numerically solved given appropriate initial conditions and boundary values. Even with such a detailed model, many processes must still be parameterized, such as sensible heat flux from the surface, surface evaporation, and cloud formation. Also there is an exchange of momentum, water, and heat across the air–ocean interface, and a similar set of equations for the ocean should be solved since sea surface temperature and ice thickness are boundary conditions needed for the atmospheric model. On the other hand, runoff, ice formation, evaporation rates, precipitation, sensible heat flux, radiation, and surface wind stress can be obtained by the atmospheric model and are needed for the ocean model. Coupling of these models poses a serious problem since the ocean does not immediately come to a new thermal equilibrium when there is a change in the energy input at the ocean surface as occurs, for example, if there is an increase in atmospheric CO_2 which will cause an increase in the downward long-wave radiation. The time to reach equilibrium depends on both the heat capacity of the ocean and the depth to which the increased heating at the surface is mixed. If this mixing was to occur over the full extent of the ocean, then several hundred years would be required to reach thermal equilibrium. However, mixing is known to occur much more rapidly in the upper region of the ocean because of wind-generated waves. This "mixed layer" may extend from the surface down to tens or hundreds of meters. If the mixed layer is 50 m thick, then the time to reach thermal equilibrium for a doubling of atmospheric CO_2 would be a few years, while if the layer is 500 m thick, then 30 years would be required (National Research Council, 1982).

D. Radiative–Convective Models

Models that are one-dimensional in the vertical direction are frequently used to study the effects of composition changes on the temperature structure. We shall examine this model in some detail to illustrate the procedures used in radiation calculations.

The formulation of these models can be traced to the early 1960s (Manabe and Möller, 1961; Manabe and Strickler, 1964). They were developed primarily to incorporate radiation into general circulation models. Since that time, these models have frequently been used to represent a globally averaged vertical temperature profile. As such, solar energy absorbed by the Earth–atmosphere system must equal the energy lost by the system to space which is in the form of long-wave or IR emission. Furthermore, if radiation is the only process responsible for producing the temperature profile, then in an equilibrium state, the net flux (the difference between the outward- and downward-directed radiation) at every level in the atmosphere must be zero, a condition known as radiative equilibrium.

A typical radiative equilibrium profile is shown in Fig. 3. Note that in the stratosphere, the actual temperature is near that of the radiation calculation, indicating that radiation is the primary process responsible for the stratospheric temperature. In the troposphere, on the other hand, the calculated temperature is much too low. Processes other than radiation such as latent

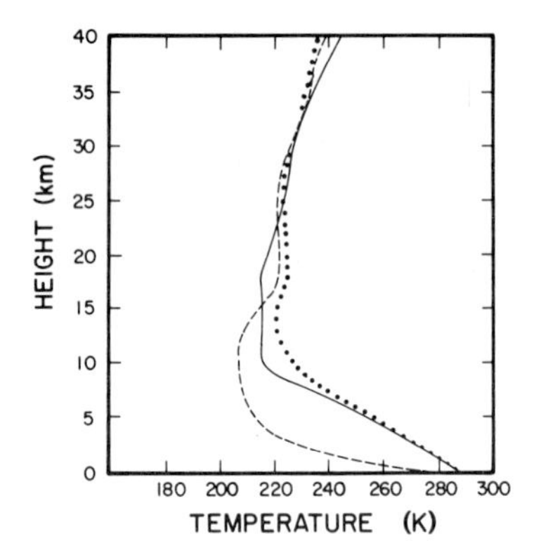

Fig. 3. Radiative equilibrium (– – –) and radiative–convective (· · ·) temperature profiles and a comparison with the observed mean global temperature (——).

heat release and convection are important in this part of the atmosphere. These nonradiative processes are crudely taken into account by adjusting the calculated lapse rate (the change in temperature with height) to a mean tropospheric lapse rate whenever the calculated lapse rate becomes smaller than the mean lapse rate. This procedure, known as convective adjustment, is used in virtually all one-dimensional vertical models. The adjusted lapse rate is usually assumed to be 6.5 K km^{-1}, taken from the work of Manabe and Strickler (1964), although a lapse rate of 5.2 K km^{-1} has been shown to better approximate an average tropospheric value (Stone and Carlson, 1979). Use of a convective adjustment will not allow one to determine how the shape of the tropospheric temperature profile changes; however, with only minor variations in concentrations of the radiatively active constituents, the departures in calculated temperature from an unperturbed or standard state should approximately represent what would occur.

The procedure frequently used to generate a radiative equilibrium temperature profile is to calculate the asymptotic solution to an initial value problem. A "guessed" solution to the problem is used to determine net radiative fluxes at selected levels in the atmosphere from which can be determined the vertical flux divergences. If the guessed temperature represented radiative equilibrium, then the flux divergences would all be zero. In an atmospheric region where the divergences are positive, the temperature is too high, and radiation is attempting to cool the atmosphere; the converse is true when the divergence is negative. These flux divergences are related to the time rate of change of temperature by

$$\frac{\partial T}{\partial t} = -\frac{1}{\rho c_p}\frac{\partial F}{\partial z}, \tag{11}$$

where F is the net flux, z the height, ρ the air density, c_p the specific heat of air at constant pressure, T the temperature, and t the time. These temperature changes can then be multiplied by a specified time interval and the initial temperature adjusted accordingly at each level. The calculations are repeated until the heating rates are within a specified limit so that the temperature profile approximates the radiative equilibrium case at those heights where convective adjustment was not necessary.

The radiative flux calculations are time consuming since they involve integrations over wavelength, height, and angle. For the planetary or long-wave radiation, the integral form of the transfer equation can be written:

$$I_\lambda^\uparrow(u_0,\theta) = \epsilon B_\lambda(0)\Pi_\lambda(u_0,\theta) + \int_0^{u_0} S_\lambda(u)\kappa_\lambda \sec\theta\, \Pi_\lambda(|u_0 - u|,\theta)\, du,$$

$$I_\lambda^\downarrow(u_0,\theta) = \int_{u_0}^{u_\infty} S_\lambda(u)\kappa_\lambda \sec\theta\, \Pi_\lambda(|u - u_0|,\theta)\, du, \tag{12}$$

where I_λ is the specific intensity, the energy passing through unit area in unit solid angle about the direction θ per unit wavelength per unit time. The arrows refer to radiation passing in an upward and downward direction through the level corresponding to the mass depth of the absorbing gas u_0. This mass depth is the amount of absorbing gas in a vertical column of unit cross-sectional area extending from the surface to the level at which the intensity is to be determined. The θ is the zenith angle referenced from the vertical direction; azimuthal symmetry is assumed. The $B_\lambda(0)$ is the blackbody function at the temperature of the surface, and ϵ the surface emissivity. The S_λ is the atmospheric source function, which can be approximated as a blackbody as long as the emission is from elevations less than ~80 km. The κ_λ is a mass absorption coefficient and represents the total cross-sectional area presented to the radiation field for each gram of absorbing gas.

Infrared radiation, also known as long-wave or planetary radiation, is transferred through the atmosphere primarily by absorption and emission by H_2O, CO_2, and O_3. Not all of the radiation interacts with these molecules; only those wavelengths that correspond to the energies associated with the allowable vibrational transitions in the molecule do. Many rotational states are possible for each vibrational frequency, so that many vibrational rotational transitions are allowed. These make up the vibration–rotation band, an example of which is shown in Fig. 4. In the far IR, the energy in the radiation field is small, and so only the rotation of the molecule can be increased with absorption of energy. The pure rotational spectrum of H_2O is an example.

The wavelength dependence of this absorption, known as line broadening (see, e.g., Goody, 1964), is due to both collisions of the absorbing molecule

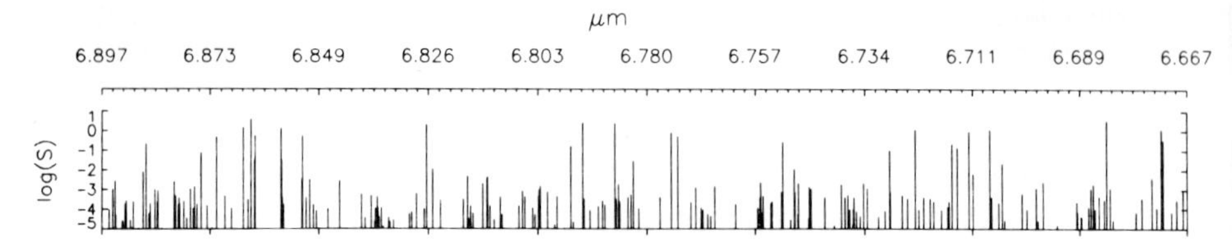

Fig. 4. Line positions and strengths (cm^{-2} atm^{-1}) in the 6.3-μm water vapor vibration–rotation band. Only a small section of the spectrum is shown. From Park *et al.* (1981).

with the other gas molecules (pressure broadening) and the apparent shift of the frequency of the radiation due to the velocity of the absorbing molecule (Doppler broadening). Pressure broadening dominates in the troposphere and decreases rapidly with height, while Doppler broadening is of major importance in the mesosphere. Both processes are important in the mid atmosphere, and produce a line broadening known as a Voigt profile. Examples are shown in Fig. 5.

The monochromatic transmission function Π_λ is the fractional amount of radiation that is transmitted through a layer of absorbing gas in direction θ

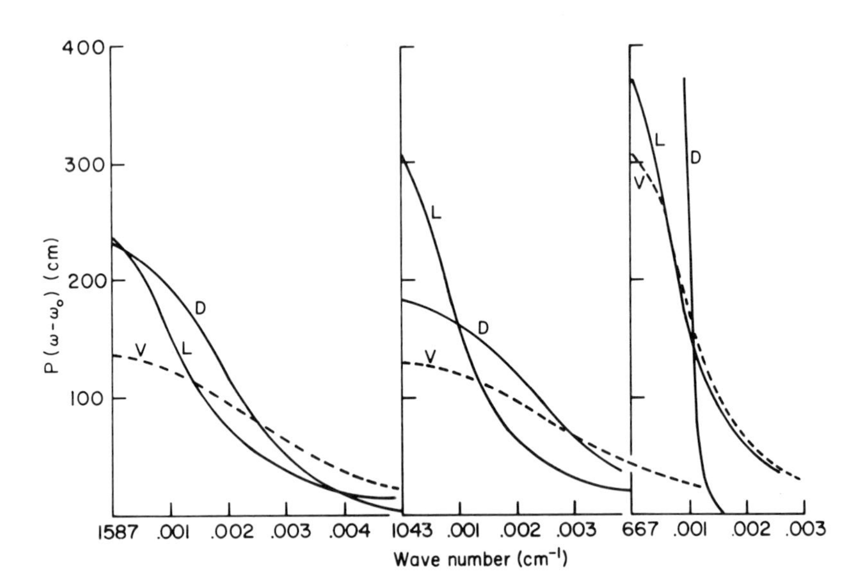

Fig. 5. Comparison of pressure (Lorentz, L) and Doppler (D) broadening at 30 km (V, Voigt). The profile function P is related to the absorption coefficient $\kappa_\omega = P(\omega - \omega_0)\mathscr{S}$, where the line strength $\mathscr{S} = \int \kappa_\omega \cdot d\omega$; $\omega - \omega_0$ is the distance from line center in wave numbers. The graph on the left refers to a line in the 6.3-μm (1587 cm^{-1}) water vapor band, in the center to an O_3 line in the 9.6-μm band, and on the right to a line in the 15-μm CO_2 band.

(Fig. 6). The transmission can be represented as

$$\Pi_\lambda = \exp[-\kappa_\lambda \sec\theta |u - u_0|]. \quad (13)$$

The upward component of the specific intensity [Eq. (12)] consists of two terms. The first represents radiation from the ground which is attenuated by the intervening atmosphere containing mass depth u_0. The last term gives the contributions to the radiation from the atmosphere; from Kirchoff's law, $S_\lambda \kappa_\lambda$ is the emission per unit mass of absorbing gas. Not all this radiation will reach the level for which the outward radiation is desired; some will be be absorbed, and the fractional amount transmitted is given by the transmission function.

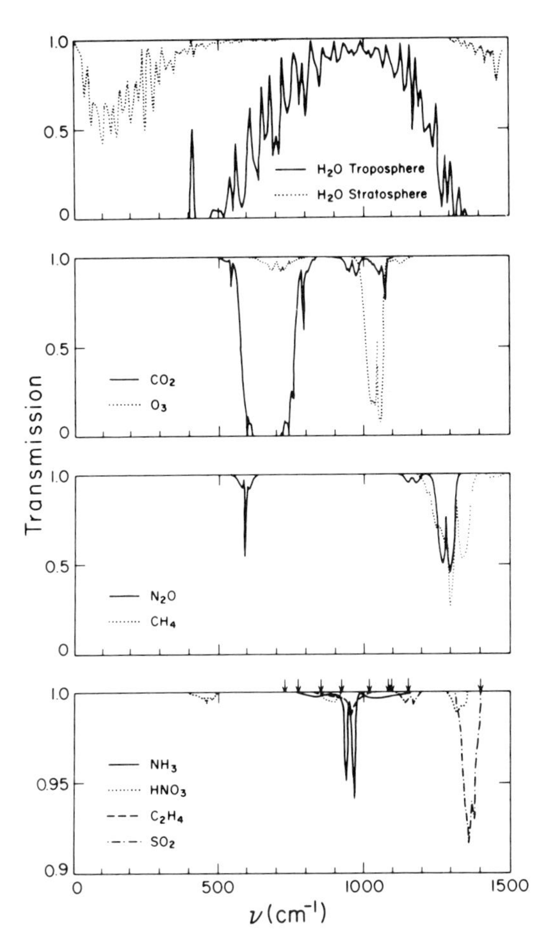

Fig. 6. Transmission profiles of selected radiatively active gasses in the Earth's atmosphere. Arrows indicate the Freon and chlorocarbon bands. From Wang *et al.* (1976).

Integration over the vertical extent of the atmosphere gives the total atmospheric contribution. The expression for the downward radiation is similar, except there is no IR radiation incident on the atmosphere and thus no boundary term.

Equation (12) must eventually be integrated over wavelength (or wave number) and angle since the radiative flux is required. Each broadened spectral line in a vibration–rotation band has a width of less than one-tenth of a wave number, however, and calculations for the transmission would need to be carried out for many wave numbers on each of the thousands of lines that make up each band. Since these line-by-line calculations are very time consuming, band models (see, e.g., Goody, 1964) have been developed that approximate the actual line spacing and line strengths over broad wave-number intervals. A single or average transmission function is then calculated for the interval. The two basic band models from which most of the others have made use are the Elsasser model and the Mayer–Goody model. In the Elsasser model, the pressure-broadened lines are equally spaced and have the same intensities. At the other extreme would be randomly spaced lines with some arbitrarily defined probability distribution for line strengths; variants of this Mayer–Goody model are used extensively in radiative–convective calculations.

The profile for the absorption coefficient must be approximated when a band model is used. The absorption coefficient explicitly depends on line strength and width of the absorption profile. Both depend on temperature, while the width also depends on pressure. Since temperature and pressure are inhomogeneous in the atmosphere, it becomes necessary to define an effective line strength and width that yield a line broadening that gives approximately the radiation actually transmitted through the inhomogeneous path. This mass-weighted line strength and width were suggested independently by A. R. Curtis and W. L. Godson and are known as the Curtis–Godson approximations. These expressions are sometimes simplified by defining a mass-weighted temperature and pressure directly, which are then used to calculate a mean line strength and width.

If Eq. (12) is averaged over wave number, if the source function and Planck function are considered constant over each spectral interval, and if the results summmed over the IR spectrum, then the energy in unit area in unit time in unit solid angle about a direction θ is

$$\begin{aligned} \mathscr{I}^{\uparrow}(u_0,\theta) &= \sum_i \delta_i \left[B_{\lambda_i}(0)\Pi_{\lambda_i} + \int_0^{u_0} S_\lambda(u)\frac{d\Pi_{\lambda_i}}{du}\,du \right], \\ \mathscr{I}^{\downarrow}(u_0,\theta) &= \sum_i \delta_i \int_{u_0}^{u_\infty} S_\lambda(u)\frac{d\Pi_{\lambda_i}}{du}\,du. \end{aligned} \tag{14}$$

Since we are interested in the total upward and downward energy passing through selected levels in the atmosphere, an integration must also be carried out over angle. Assuming azimuthal symmetry, and defining a flux transmission by

$$\tau_{\mathrm{f}} = 2\int_{0}^{\pi/2} \Pi_{\lambda_i} \cos\theta \sin\theta \, d\theta, \tag{15}$$

equations for the flux are yielded,

$$\begin{aligned} F^{\uparrow}(u_0) &= \sum_i \delta_i \left[\pi B_{\lambda_i}(0)\tau_{\mathrm{f}} + \int_{0}^{u_0} \pi S_{\lambda_i}(u) \frac{d\tau_f}{du}\, du \right], \\ F^{\downarrow}(u_0) &= \sum_i \delta_i \pi \int_{u_0}^{u_\infty} S_{\lambda_i}(u) \frac{d\tau_{\mathrm{f}}}{du}\, du. \end{aligned} \tag{16}$$

Rather than carrying out the integration shown in Eq. (15), one generally defines a diffusivity factor r that is the secant of an angle that represents an average zenith angle used in the calculation of the transmission, that is,

$$\tau_f \sim \Pi_{\lambda_i}(r|u - u_0|). \tag{17}$$

The diffusivity factor is generally chosen to be ~ 1.67.

Long-wave radiation is transferred through the atmosphere not only by the various gases but also by particulate matter in the form of clouds and aerosols, which also produce a greenhouse effect. The larger the temperature difference between the particulate layer and ground, the stronger will be the greenhouse warming. Water clouds absorb most of the long-wave radiation incident on them, so that scattering can generally be neglected. Furthermore, for water clouds with thicknesses of ~ 0.5 km or more, the emissivity is approximately greater than ~ 0.9 so that they can be treated as blackbodies (Paltridge and Platt, 1976). Ice clouds on the other hand, have a lower emissivity; a cirrus cloud with a typical thickness of 1.5 km has an emissivity of only ~ 0.6. A frequently used approach to include clouds in the IR calculations is to calculate the net flux at each atmospheric level for a clear and completely cloudy sky and then weight the fluxes by the fractional clear and cloudy sky coverages.

While the absorption and emission of long-wave radiation can either heat or cool the atmosphere, solar radiation can only provide a heating. However, as opposed to the long-wave radiation, the shorter-wavelength solar radiation can also be scattered by not only the air molecules but aerosols and clouds as well. Whether or not the particulate matter heats or cools the atmosphere depends on the amount of solar energy absorbed in comparison to that scattered.

Most of the absorption of solar radiation occurs in the ultraviolet (UV), visible, and near-IR regions of the spectrum, where the atmospheric emission is small. Furthermore, in the UV and visible, the transitions leading to the transfer of this radiation are electronic and consist of a multitude of vibrational bands. These spectral lines are so close together that the absorption coefficients overlap strongly, and the absorption appears as a slowly varying continuum. Thus, taking an average absorption coefficient over small spectral intervals and summing over these intervals gives for the flux at level z_0,

$$F(z_0) = \sum Q_i \exp\left\{-\sigma_i \int_{z_b}^{z_0} n dz \, \overline{\cos \zeta}\right\} \overline{\cos \zeta} \, \Delta\lambda_i, \tag{18}$$

where Q is the solar energy incident on and normal to the atmosphere per unit area per unit time per unit wavelength interval, σ the absorption cross section, n the number density of the absorbing constituent, ζ the solar zenith angle, and z_b the top of the atmospheric.

For the transfer of solar radiation by vibrational bands, the flux can be written as

$$F^{\downarrow}(u_0) = \sum_i \delta_i Q_i \overline{\cos \zeta} \, \Pi_{\lambda_i} \Delta\lambda_i. \tag{19}$$

In this case, the transmission does not include an angular integration, or the diffusivity approximation, since the solar beam is essentially collimated, and the transmission is evaluated for an average solar zenith angle ζ.

Air molecules scatter solar radiation, some of which is returned to space so that they effectively reduce the solar radiation available for absorption. This type of scattering was first investigated by Lord Rayleigh in the late 1800s and bears his name. In a clear atmosphere, molecules scatter $\sim 6.3\%$ of the incident solar radiation back to space. Unfortunately, even in the cleanest atmosphere, there is still enough particulate matter so that the scattering radiaticn is appreciably greater than the scattering from molecules alone. While the scattering pattern can be calculated if the particle is assumed spherical and the size and composition are known, the calculations are laborious (see, e.g., Hansen and Travis, 1974). Furthermore, multiple scattering will occur, and while the transfer of this radiation through the atmosphere can be computed, the calculations are very time consuming. The size, vertical distribution, and composition of particulate matter are also highly variable, and in one-dimensional radiation calculations, scattering is usually treated in a very approximate way.

Solar radiation scattered back to space can be either increased or decreased by the presence of an aerosol layer. If the absorption of visible radiation by an aerosol layer is small, then the planetary albedo will be increased and a cooling will occur. On the other hand, absorption of visible radiation by the aerosol layer and scattering of radiation from the underlying surface can combine to reduce the planetary albedo and cause a surface heating. As a simple example, consider the solar radiation $\mathscr{S}^{\uparrow}$ scattered back to space by an aerosol layer that reflects (r) and absorbs (a) the incoming solar radiation $\mathscr{S}^{\downarrow}$. Given that A is the albedo of the underlying surface, $\mathscr{S}^{\uparrow}$ can be written as

$$\mathscr{S}^{\uparrow} = r\mathscr{S}^{\downarrow} + \mathscr{S}^{\downarrow}(1 - a - r)^2 A + \mathscr{S}^{\downarrow}(1 - a - r)^2 A^2 r, \tag{20}$$

where the first term on the right-hand side is the incoming energy reflected back to space by the aerosol layer, the second the incoming energy transmitted through the layer, reflected from the surface and then transmitted back through the layer to space, and the third term the incoming energy transmitted through the layer, reflected from the surface back to the layer but then reflected again back to the surface and finally reflected from the surface back up through the layer. If one assumes that a and r are small, then dropping all terms in Eq. (20) of second degree or higher in a and r gives

$$\mathscr{S}^{\uparrow}/\mathscr{S}^{\downarrow} = A + r - 2Ar + A^2 r - 2Aa. \tag{21}$$

Note that the effective albedo can indeed be smaller than the surface albedo so that a heating can occur. The above is given only as an example of the complexities that can arise when scattering is considered. In any realistic calculation, the absorption of solar radiation by atmospheric gases would also need to be included.

Clouds not only scatter appreciable solar radiation back to space but also scater in a downward direction. This multiple scattering increases the H_2O mass path so that appreciable absorption takes place in the region of the cloud, effectively reducing the solar radiation available for absorption below the cloud. The absorption by liquid water drops is generally small. In radiative–convective models, cloud transmission and absorption are generally specified or calculated for a cloud model, and the downard fluxes weighted by a clear and cloudy sky calculation as in the long-wave case.

III. Chemical Constituents

A. Water Vapor

Water vapor is a nonlinear triatomic molecule that has three fundamental vibrational bands, ν_1 and ν_3 with band centers at 2.73 and 2.66 μm, respectively, and the strong bending mode ν_2 located at 6.27 μm. There is also the strong pure rotational band centered at $\sim$80 μm. In addition to the fundamentals, important H_2O bands occur throughout the visible and IR spectrum, and because of the quite different moments of inertia, the rotational structure of these bands is very complex. The most abundant isotope (99.7%) is $H_2{}^{16}O$.

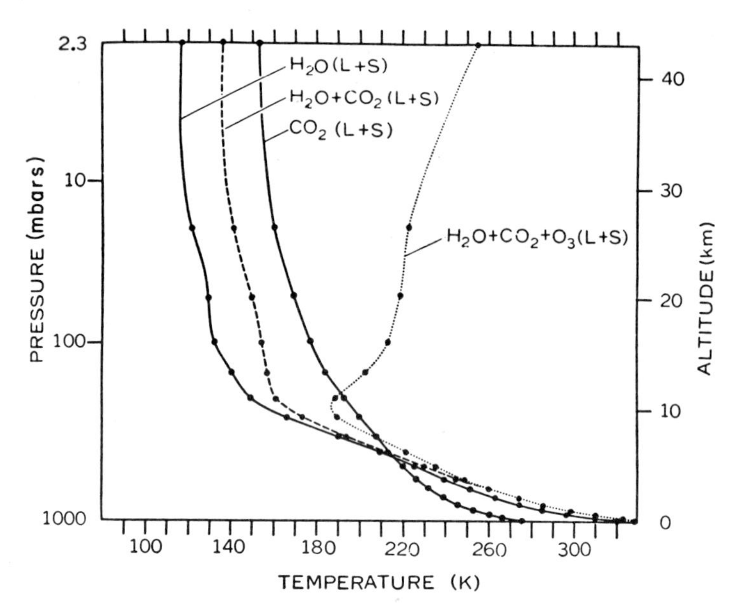

Fig. 7. Radiative equilibrium profiles for H_2O, CO_2, H_2O and CO_2, and H_2O, CO_2, and O_3. Calculations are for 35°N in April for a cloudless sky. From Manabe and Strickler (1964).

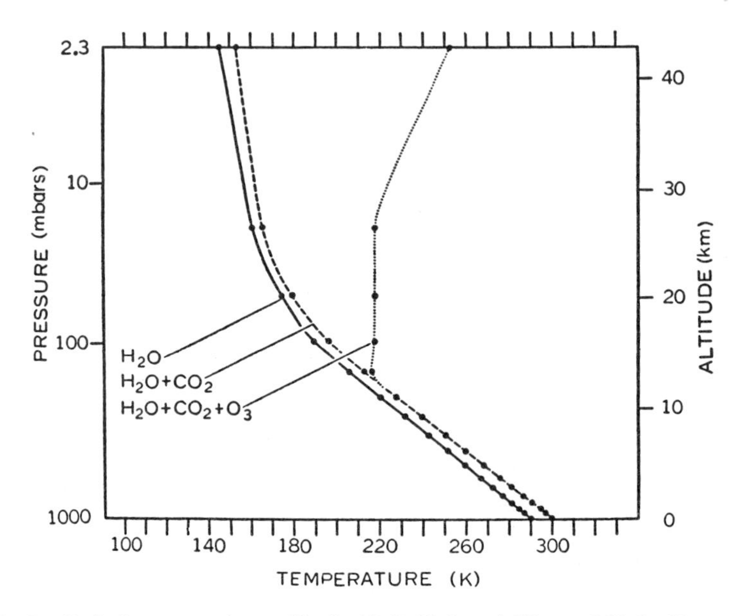

Fig. 8. Radiative–convective profiles for H_2O, H_2O and CO_2, and H_2O, CO_2, and O_3. Calculations are for 35°N in April for a cloudless sky. From Manabe and Strickler (1964).

The dominance of H_2O over the other radiatively active gases is clearly seen in Fig. 7. While the result is artificial in that clouds and a convective adjustment are not included, nevertheless one can qualitatively compare the importance of H_2O and CO_2 to the temperature. In the troposphere, there is a substantial H_2O greenhouse effect as a result of the large H_2O concentrations, while in the stratosphere the concentration of H_2O is small, leading to small absorption and emission and thus a low temperature.

A more realistic temperature profile results if one includes a convective adjustment, in this case 6.5 K km^{-1}, as shown in Fig. 8. Note that H_2O controls the overall temperature structure in the troposphere; CO_2 increases the temperature ~10 K.

The IR bands of H_2O produce a radiative cooling throughout the atmosphere (Fig. 9). This radiative heating rate represents the initial rate at

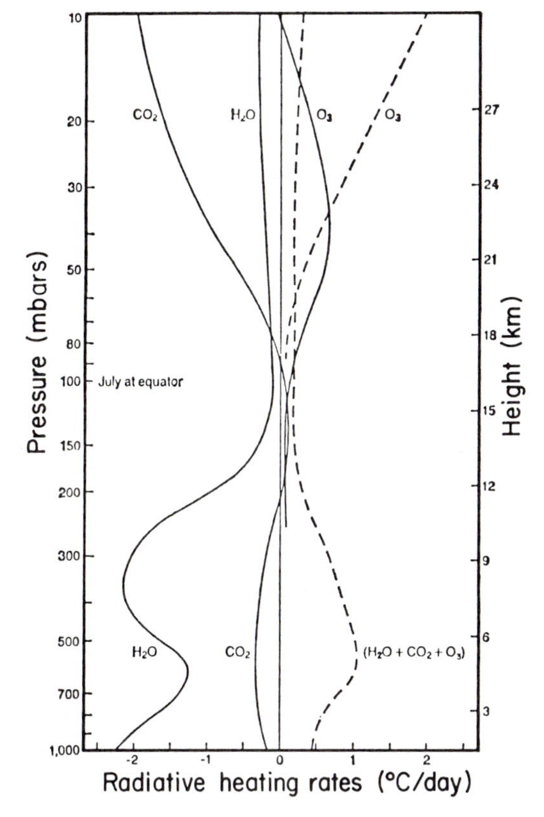

Fig. 9. Typical infrared (——) and solar (---) heating rates for CO_2, O_3, and H_2O. Results are for the equator in July and include clouds. From Climatic Impact Assessment Program (1975).

which the atmosphere would heat or cool if all processes affecting the temperature other than radiation ceased. Water vapor is dominant in the troposphere, but in regions above the tropopause where the concentrations are small, the cooling is less than 10% of the total. The heating provided by the absorption of solar radiation by the near IR bands of H_2O is slightly less than the cooling by the IR bands so that the net effect is a slight cooling in the troposphere.

B. Carbon Dioxide

Carbon dioxide is a linear triatomic molecule. The most abundant isotopic species (98.4%) is $^{12}C^{16}O_2$, with $^{13}C^{16}O_2$ and $^{12}C^{16}O^{18}O$ having concentrations of 1.1 and 0.4%, respectively. The $^{12}C^{16}O_2$ and $^{12}C^{16}O^{18}O$ molecules are symmetric with the IR spectrum, consisting of ν_2 (15 μm) and ν_3 (4.3 μm) fundamentals. In the 15- and 4.3-μm regions there are, beside the fundamentals, numerous overtone and combination bands as well; other weaker bands are located at 1.2, 1.4, 1.6, 2.0, 2.5, 5.0, and 10 μm. The 15-μm band is the most important CO_2 band for the transfer of IR radiation emitted by the Earth's surface and atmosphere. Although the 4.3-μm band is $\sim$10 times stronger, there is much less planetary radiation at that wavelength than in the vicinity of 15 μm. Carbon dioxide also absorbs a small amount of solar radiation in the near IR bands.

If the atmosphere contained no H_2O, clouds, or O_3, then the only major radiatively active gas remaining would be CO_2. If the atmosphere was in radiative equilibrium, the mean temperature of the surface and the atmosphere would be as appears in Fig. 7. The mean surface temperature would be near the freezing point of water, some 15 K lower than the temperature today yet some 20 K higher than the surface temperature would be if CO_2 were not present. The temperature of the atmosphere would decrease approximately exponentially with height, reaching a nearly isothermal value of 150 K at 40 km. Although CO_2 is not the primary greenhouse gas in the atmosphere, its concentration is increasing and is a cause for much concern, as discussed in Section IV. The radiative contribution made by CO_2 is shown in Fig. 8. Note that CO_2, in absorbing and emitting long-wave radiation, is attempting to cool the troposphere and the stratosphere; near the tropopause, there is a slight region of heating. This is to be expected since the radiation field is attempting to smooth the temperature distribution; the tropopause is emitting at a lower temperature than the temperatures of the regions above and below from which it is absorbing, the net effect being to increase the tropopause temperature. The influence of H_2O dominates CO_2 in the troposphere, while in the stratosphere and mesosphere the concentration of H_2O is small so that CO_2 is most important in transferring long-wave radiation.

C. Ozone

Ozone is a triatomic molecule and has three fundamental bands in the IR. Two of these, ν_1 and ν_3, located at 9.1 and 9.6 μm respectively, occur in a part of the spectrum that is relatively free from other spectral lines. The ν_2 band occurs at 14.3 μm and overlaps the 15-μm band of CO_2. The ν_3 band is at least an order of magnitude stronger than the other O_3 bands and is of most importance. Ozone also absorbs radiation in the UV part of the spectrum. These electronic bands, which consist of vibrational and associated rotational transitions, are the Hartley bands centered at 0.255 μm, the rather weak Huggins bands extending from 0.310 to 0.340 μm and Chappuis bands located between 0.450 and 0.740 μm (see Fig. 1).

The importance of O_3 to the stratospheric temperature is clearly seen in Fig. 8. The temperature maximum that occurs at the stratopause is due to the absorption of solar UV radiation by the Hartley bands. Although the O_3 concentrations are small at this elevation, nevertheless the small column abundance of ozone is enough to absorb most of the incoming UV radiation so that very little reaches the lower stratosphere, where O_3 concentrations are the highest. The vibration–rotation bands located in the IR, primarily in the vicinity of 9.6 μm, act to cool the stratosphere; if these IR bands are excluded from radiation calculations, then the stratopause temperature is $\sim$20 K higher. In the lower stratosphere, the temperature would be $\sim$4 K less, since near the tropopause the IR bands tend to cool the present atmosphere.

The contribution to the radiative heating and cooling by O_3 in the troposphere is minor (see Fig. 9). Maximum cooling and heating occur near the stratopause, with the cooling $\sim$4 K per day and a mid-latitude heating of 15 K per day; the heating of course strongly depends on the season and time of day (solar zenith angle).

D. Trace Gases

Although H_2O, CO_2, and O_3 are the most important gases that influence the atmospheric temperature structure, there are certain trace constituents that are of some importance because of the strength and spectral location of their bands. Among these are methane (CH_4) and nitrous oxide (N_2O).

Methane at present has a concentration of $\sim$1.65 ppmv in the troposphere. The major source of the gas is organic matter that is chemically changed by organisms in an oxygen-free environment, by chemical reactions caused by organisms in the intestines of mammals, as well as by the activity of termites. The dominant sink is reaction with the hydroxyl radical (OH):

$$CH_4 + OH \longrightarrow CH_3 + H_2O \tag{22}$$

Only two of the fundamental bands of CH_4 are IR active. One of these, the ν_4 band, is located at 7.66 μm and overlaps the H_2O band at 6.3 μm. The other is located at 3.3 μm, and although it is twice as strong as the ν_4 band, because there is little planetary radiation at 3 μm this band has not been considered in radiation calculations. Radiative equilibrium calculations indicate that if CH_4 were not present in the atmosphere, the surface temperature would be $\sim$1 K less. In the stratosphere, the temperature would increase by up to 0.25 K.

Nitrous oxide has a concentration of only $\sim$300 ppbv, but two of its three active fundamentals occur in the region of the spectrum where the Earth and atmosphere emit long-wave radiation. These two bands, located at 7.78 and 17.0 μm, contribute $\sim$1 K to the mean surface temperature of the Earth. The lower stratosphere would cool $\sim$0.1 K if there was no N_2O in the atmosphere, while there would be a heating of up to 0.2 K in the mid stratosphere. The other fundamental at 4.5 μm, although the strongest of the three, is located in a spectral region with little available energy.

There are numerous other trace gases that contribute to the greenhouse effect. Most of these influence the surface temperature by less than 0.1 K. Among these are sulfur dioxide (SO_2), the hydrocarbons, the chlorocarbons, NH_3 and nitric acid (HNO_3). Individually, each of these gases probably has a negligible effect on the climate. When taken together, however, their influence could rival that of a doubling of CO_2. There is also evidence that certain of these trace gases are increasing, which could pose a climatic concern in the years ahead.

IV. Impact of Anthropogenic Activities on Climate

The possible impact of human activities on climate has been of major concern since the early 1970s. The development of high-altitude commercial aircraft raised the possibility that nitrogen oxides, an exhaust product, would catalytically destroy stratospheric ozone, leading to an increase in the amount of UV radiation reaching the surface and to possible biospheric effects. Aircraft fuel also contains sulfur that when oxidized produces sulfuric acid (H_2SO_4) particles that may affect the Earth's energy balance. A few years following this "SST [supersonic transport] controversy" it was discovered that Freon, a commonly used refrigerant and gas in aerosol spray cans, can also destroy O_3. Although relatively inert in the troposphere, Freon is readily dissociated in the stratosphere by UV radiation, and in the ensuing reactions O_3 is catalytically destroyed. Throughout this period, and dating back to the late nineteenth century, CO_2 has been increasing through the burning of fossil fuels. Many studies indicate that if the trend continues, a climate signal will soon be seen. Most of our interest and concern about possible anthropogenic climate modification results from these studies.

A. Fossil Fuel Burning

1. Carbon Dioxide

The burning of coal, oil and natural gas increased markedly during the industrial revolution and has continued at an ever-increasing rate. Carbon dioxide, an oxidation product of this burning, has been found in increasing concentrations in the atmosphere. Around 1900, the atmospheric concentration was ~290 ppm and presently exceeds 340 ppm (National Research Council, 1983). This increase has averaged ~4.3% per year (Kellogg and Schware, 1981). From 1958 to 1981 the concentration increased by ~8% (World Meteorological Organization, 1982). Figure 10 summarizes CO_2 observations dating back to 1958 from several diverse locations extending from the pole to the equator, and all show a uniform increase. There can be no doubt that the concentration of CO_2 is indeed increasing. Although human influence on the total amount of CO_2 is most likely much less than the variations that have naturally occurred, these natural variations have taken place on a geological time scale while we are creating change on a time scale of decades. If these inadvertent modifications continue for even hundreds of years they would most certainly surpass the natural variations.

There are other sources of atmospheric CO_2 beside the burning of fossil fuels. Deforestation, for example, may release up to 40% as much CO_2 to the

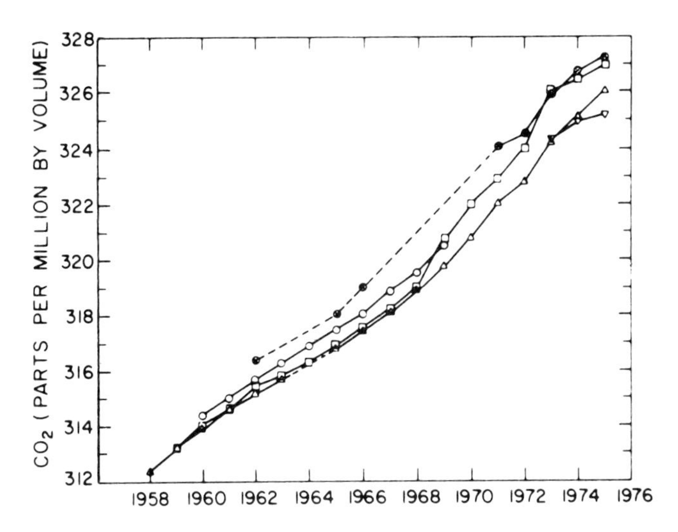

Fig. 10. Observed increase in atmospheric CO_2 from selected sites dating to 1958. ⊗, Point Barrow; ○, Swedish flights; □, Mauna Loa; ▽, American Samoa; △, South Pole. From Kellogg and Schware (1981).

atmosphere as the burning of fossil fuels (Kellogg and Schware, 1981). When exposed to the air, organic material in the soil will also oxidize and release CO_2; thus agricultural activities must also be considered when estimating the annual increase in CO_2. The primary sink for CO_2 is the ocean, although photosynthesis of plants is important. If the oceans were well mixed, then any increase in atmospheric CO_2 would be quickly absorbed. As discussed previously, however, only the first few hundred meters is well mixed and is separated from the very stable waters below by a narrow region known as the thermocline, where the temperature decreases rapidly with depth. There may be a delay of the order of a decade or so before CO_2 will be uniformly mixed in this upper layer, while a time scale of hundreds of years would be necessary for mixing throughout the ocean as a whole.

Prediction of the amount of CO_2 in the atmosphere in the years ahead is difficult. Indeed, even comparisons of predicted energy growth with actual data for the last two decades differ by 2%. The amount of oil consumption is particularly uncertain. Little reliance can be placed on extrapolation several decades into the future, although at some point our fossil fuel reserves will be expended, and we shall have to rely on alternate energy sources. Figure 11

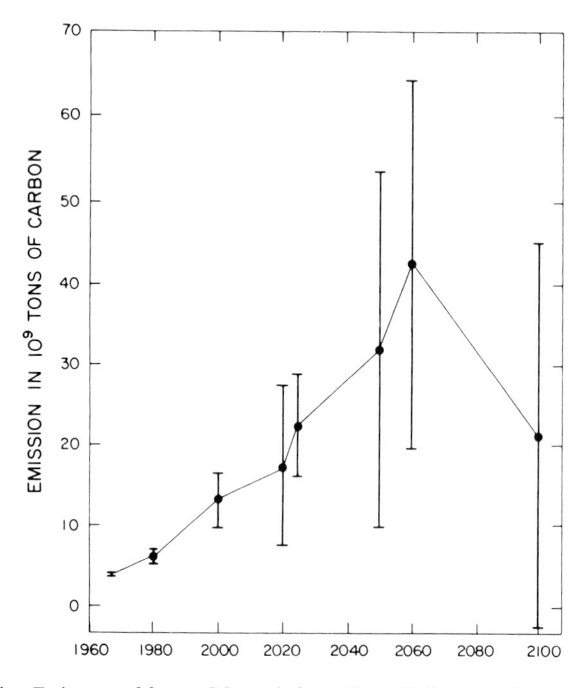

Fig. 11. Estimates of future CO_2 emissions. From Kellogg and Schware (1981).

gives an estimate of CO_2 emissions from fossil fuel extending to the twenty-second century; note the rather large uncertainty by the year 2000. Our best estimate to date is that by the year 2025, atmospheric CO_2 levels will be between 410 and 490 ppm, with a most likely value of 450 ppm (National Research Council, 1982).

One would expect that with continued increases in atmospheric CO_2, the warming predicted from the model studies would eventually be great enough so that it could be distinguished from the natural climate variability. A global temperature history has been determined by Hansen *et al.* (1981) for the period from 1888 to the present (Fig. 12). Quantitative information prior to this time

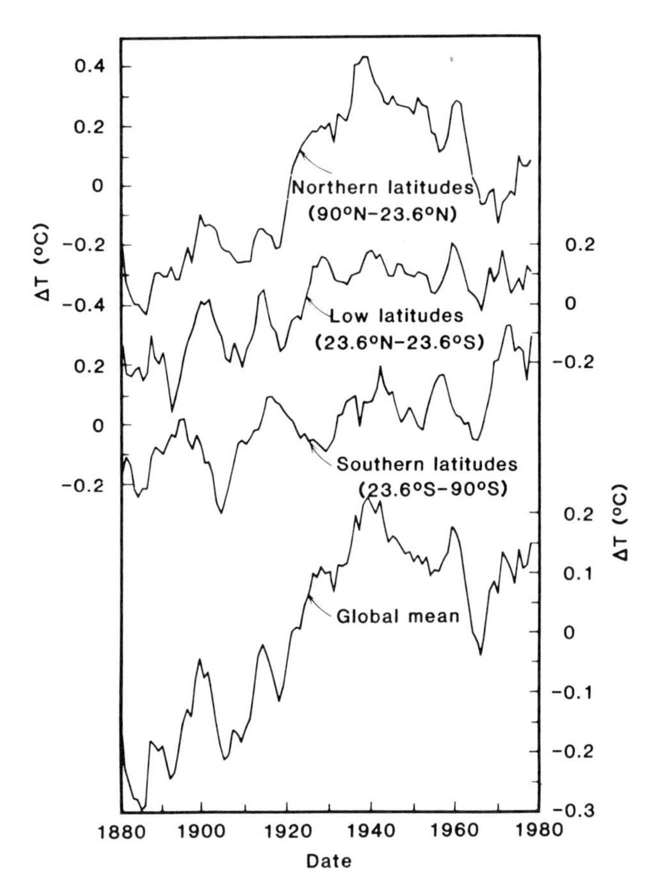

Fig. 12. Observed surface air temperature trends (5-year running mean) for three latitude bands and the entire globe. Temperature scales for low latitudes and global mean are on the right (note scale change). From Hansen *et al.* (1981).

is probably not possible to obtain since there were few stations. Note that there is significant latitudinal variation in the amount of warming; for example, the northern latitudes warmed ~0.8 K between the 1880s and 1940, while the low latitudes increased by only ~0.3 K. A global warming of ~0.4 K has occurred over the last century. Some calculations indicate that the "CO_2 signal" will be detected before the end of this century (Lacis *et al.*, 1981; Thompson and Schneider, 1982).

What are the temperature increases predicted for an increase of CO_2? For comparison, one generally calculates the temperature change for a doubling of CO_2. This assumed doubling is arbitrary, although it does represent the increase that will most likely occur sometime in the next century; furthermore, the temperature rise that most of the models generate is about what one would expect if there were a major climatic perturbation. This predicted temperature increase due to a doubling of CO_2 is generally quoted to be ~3 K, although estimates range from as low as a few tenths of a kelvin to as large as 4.5 K. Even increases of a few tenths of a kelvin, while not producing major climate variations, could cause changes in the boundaries of climate zones and cause political and economic problems.

The prediction of climate change is made difficult by the various feedback mechanisms. For example, a change in CO_2 will change surface and air temperatures, which will change the amount of water vapor in the air as well as cloud cover, height, and extent and the lapse rate, that is, the rate of change of temperature with height. Each of these will cause a positive or negative feedback, depending on whether or not there is an amplification or a decrease in the temperature change produced by CO_2 alone.

Most model calculations predict a temperature increase of 1 to 2 K if CO_2 is doubled without a change in any other atmospheric variables. An increase in temperature however, increases the atmosphere's capacity to retain water vapor, and one would expect the absolute humidity to increase rather than to remain constant. This condition is generally included in the models by assuming a constant distribution of relative humidity. Water vapor then contributes about one-half (1.5 K) of the total temperature increase for a doubling of CO_2.

The value chosen for the convective adjustment also influences the calculated surface temperature. In lower latitudes, the actual temperature decrease with height approximates the moist adiabatic rate. Convection transports H_2O to higher elevations where condensation occurs, releasing latent heat to the atmosphere; this lapse rate, although variable, has an average annual value of ~5.7 K km^{-1} in the troposphere. In mid and high latitudes, the actual lapse rates are more stable; the vertical temperature profile is controlled by eddies that are driven by horizontal temperature gradients and by topography. These so-called baroclinic processes produce an average lapse rate of 5.2 K km^{-1}. It is interesting to note that most radiative convective models have used a lapse rate of 6.5 K km^{-1}, which was based on date sets extending back to 1933. We know now that a better hemispherical annual lapse rate is closer to 5.2 K km^{-1}, although there may be significant seasonal variations. One can gain an appreciation of the importance of the lapse rate to climate modeling by comparing the surface temperature increase for a doubling of CO_2 with a fixed lapse rate of 6.5 K km^{-1} versus one with a moist adiabatic rate. The moist rate gives a temperature increase only 75% as large as that for a fixed 6.5 K km^{-1} lapse rate, ~1.4 rather than 1.9 K (Hansen *et al.*, 1981).

The changes in clouds that would occur with an increase in CO_2 are as yet unknown. We do know, however, that our present global cloud cover of ~50% keeps the surface temperature of the Earth some 20 K cooler than without clouds. Cloud cover, thickness, height, and optical properties all influence the surface temperature. An increase in cloud cover can cause either an increase or decrease in temperature depending on whether or not the increased reflection of solar radiation to space, which would lead to a cooling, dominates the increased absorption of IR radiation by the clouds from the surface, providing a greenhouse effect. An increase in cloud height will lower the effective temperature at which the cloud radiates to space, thus the surface temperature must increase to provide the same outgoing radiation. In addition, cloud thickness, liquid water content, and cloud droplet distribution all influence the transmissivity, reflectivity, and absorptivity of the cloud.

There are some studies that indicate clouds may not be very sensitive to small changes in climate. In one such study carried out with a general circulation model (Manabe and Wetherald, 1980), the cloud amount and height both decreased in low and mid latitudes for an increase in CO_2. The smaller cloud amount allowed more solar radiation to be absorbed, while the lower cloud height increased the outgoing radiation to space. Thus the net effect on the surface temperature was small. In the high latitudes, the cloud amount increased while there was no change in the cloud height. Again, the net effect on the surface temperature was small, since the larger amount of solar radiation reflected to space was compensated by the decrease in the surface IR radiation to space. These results must be considered speculative, however, since cloud prediction models are primitive and we have much to learn about the optical properties of clouds; at present, there are no models that simulate cloud cover on a seasonal basis.

In addition to CO_2, there are several other gases released in the burning of fossil fuels that could influence climate (see Table I). Among these are CH_4, carbon monoxide (CO), the nitrogen oxides (NO_x), and SO_2.

Table 1

CHANGES IN SURFACE TEMPERATURE THAT WOULD OCCUR FOR INCREASES IN THE CONCENTRATIONS OF RADIATIVELY ACTIVE GASES[a]

Trace gas	Band center (cm^{-1})	Band strength ($cm^{-2}\ atm^{-1}$)	Reference mixing ratio (ppb)	Perturbed mixing ratio (ppb)	Surface temperature change (K)
CO_2	667	220	330,000	660,000	2
H_2O	0–2,000	1,563	3,000 (stratosphere)	6,000 (stratosphere)	0.6
O_3	1,041			Doubled	0.9
	1,103	325	Troposphere	Troposphere	
N_2O	589	24			
	1,285	218	300	600	0.3
CH_4	1,306	134	1,500	3,000	0.3
SO_2	518	97			
	1,151	87	2	4	0.02
	1,361	763			
NH_3	950	534	6	12	0.09
$CFCl_3$	846	1,670	0	1	0.15
	1,083	781			
CF_2Cl_2	915	1,370	0	1	0.13
	1,100	1,330			
	1,150	893			

[a] Modified from Ramanathan *et al.* (1982).

2. Methane

Methane appears to have been increasing at least since 1965; the present rate of increase is a few percent per year. Concentrations today may be more than twice as great as they were prior to the sixteenth century (National Research Council, 1983a). Although it is difficult to estimate future concentrations, a doubling of methane from 1.5 to 3 ppm gives a temperature increase of ~ 0.3 K (World Meteorological Organization, 1982).

3. Nitrous Oxide

Nitrous oxide is the most important of the nitrogen oxides in producing a greenhouse warming. It is presently increasing at a rate of 0.2% per year (Lacis *et al.*, 1981); most of this increase is due to fossil fuel combustion. A doubling of N_2O from 300 to 600 ppb would cause a surface temperature increase of about 0.3 to 0.4 K (World Meteorological Organization, 1982).

4. Carbon Monoxide and the Nitrogen Oxides

Carbon monoxide and the other nitrogen oxides are not themselves significant in producing a greenhouse effect, but they do influence the IR-active gases O_3 and CH_4. Carbon monoxide can lead to an increase in CH_4 since it reacts with the hydroxyl radical (OH):

$$CO + OH \longrightarrow CO_2 + H \tag{23}$$

which is part of the reaction representing the major sink for CH_4 (Wang *et al.*, 1976), that is, reaction (22).

An increase in nitrogen oxides will cause an increase in O_3 through a series of reactions in which atomic oxygen is formed and, by a three-body reaction with O_2, produces O_3:

$$HO_2 + NO \longrightarrow NO_2 + OH \tag{24}$$

$$NO_2 + h\nu \longrightarrow NO + O \tag{25}$$

$$O + O_2 + M \longrightarrow O_3 + M \tag{26}$$

This increase in tropospheric O_3 will tend to decrease CH_4 and CO concentrations since photolysis of O_3 produces $O(^1D)$ which reacts with H_2O to give the hydroxyl radical:

$$O_3 + h\nu \longrightarrow O_2 + O(^1D) \tag{27}$$

$$O(^1D) + H_2O \longrightarrow 2\,OH \tag{28}$$

which, as seen above, destroys both CO and CH_4. On the other hand, since O_3 and CH_4 are "greenhouse" gases, any increase in their concentrations will affect the temperature of the Earth's atmosphere and surface. An increase in temperature should increase the amount of tropospheric H_2O, which would lead to enhanced levels of hydroxyl radical and therefore a decrease in CH_4 and CO_2 and thus O_3. This negative feedback is but one example of the complexities that can arise in coupled photochemical climate studies (see, e.g., Hameed *et al.*, 1980).

5. Sulfur Dioxide

Another trace gas that has the potential for influencing climate and whose primary source is the burning of fossil fuels is sulfur dioxide. It is as yet unknown whether or not the concentration of this gas, which at present is ~1 ppb, is increasing. Calculations indicate that if there were a doubling of SO_2 from 2 to 4 ppb, the mean surface temperature would increase ~0.02 K (World Meteorological Organization, 1982). In addition to providing a greenhouse effect, this gas can also influence climate by conversion to an aerosol particle.

In addition to gases, the atmosphere contains particulate matter, consisting of small liquid and/or solid particles. The composition is generally complex, consisting of water, water-soluble and -insoluble inorganic material, as well as some organics. Sizes in general range from about 0.01 to 1 μm, with concentrations of only ~100 cm^{-3} over the oceans, increasing to 10^5 cm^{-3} in rural areas. Since the source of these particles is at the surface, there one finds the largest concentrations; 80% of the aerosol mass resides in the first kilometer of the atmosphere.

These aerosols affect climate in two ways. Some are cloud condensation nuclei (CCNs), that is, they are the embryos upon which H_2O condenses to form cloud particles, which as we have seen strongly influence climate. These CCNs are necessary precursors to cloud formation. In order for water vapor to condense and form a pure water cloud droplet, supersaturations of several hundred percent relative humidity would be necessary; typically, relative humidities at most only a few percent above saturation are found in the atmosphere. As we have shown, aerosols also absorb and reflect solar radiation as well as produce a greenhouse effect similar to the gases we have been considering. Which process dominates, that is, whether or not aerosols cause the surface to heat or cool, will depend on their location, composition, size, and abundance. The surface temperature of the Earth would be 1 or 2 K warmer if there were no atmospheric aerosols. Results from a radiative–convective model (Charlock and Sellers, 1980) indicate that a doubling of CCNs would cause the reflectivity of low clouds to increase, with a lowering of surface temperature ~0.9 K. Thus any major changes in the atmospheric aerosols could have significant climatic effects.

How can SO_2 affect atmospheric aerosols? By a gas-to-particle conversion SO_2 can be converted to an H_2SO_4 droplet or an ammonium sulfate $[(NH_4)_2SO_4]$ particle. Among the several reactions in the former case, SO_2 can react with an odd-oxygen or a hydroxyl radical as follows:

$$SO_2 + O + M \longrightarrow SO_3 + M \tag{29}$$

$$SO_2 + OH + M \longrightarrow HOSO_2 + M \tag{30}$$

$HOSO_2$ is a free radical and ultimately forms an H_2SO_4 aerosol, although the elementary reaction pathways are unknown (National Research Council, 1983b).

In addition to being a combustion product of the burning of fossil fuels, SO_2 is a primary component of some volcanic emissions, for example, El Chichón in Mexico in April 1982. It has been estimated that 3 million tons of SO_2 were released into the atmosphere, some of which was transported into the stratosphere and eventually formed concentrated H_2SO_4 droplets having diameters of a few micrometers and smaller.

El Chichón has already perturbed the climate of the stratosphere; absorption of solar radiation has caused the temperature to increase by some 3 K and is the highest since measurements began in 1958. Several studies indicate that some of the visible radiation from the Sun should be reflected back to space by the H_2SO_4 droplets and that there should be a decrease in surface temperature by ~ 0.5 K, although as of fall 1983 this change was not detected.

The sulfate ion (SO_4^{2-}) is a major component of the atmospheric aerosols, and SO_2 contributes to its formation. The sulfite ion (SO_3^{2-}) is produced when SO_2 dissolves in H_2O (a cloud droplet), and in the presence of dissolved O_2 and certain metallic ions forms the sulfate radical. When NH_3 dissolves in H_2O, and the cloud droplet subsequently evaporates, an $(NH_4)_2SO_4$ aerosol particle can be formed and can further participate in cloud formation and perhaps influence climate. These are but two of the many ways in which gases, some of them released as a result of agricultural and industrial activities, can be converted to atmospheric aerosols.

B. Agricultural Activities

Agricultural activities also have the potential for influencing the climate. In the early 1970s, it was speculated that industrially produced nitrogen fertilizer could increase the flux of N_2O into the atmosphere, which would not only decrease the stratospheric O_3 abundance but also increase the temperature because of the greenhouse effect. Use of artificial fertilizers are increasing at the rate of 6% per year, and by the year 2000 their production will rival the natural fixation rate (Liu *et al*, 1977); most of the nitrogen fixed at present is due to microorganisms.

While most of the N_2O increase is presently due to fossil fuel combustion, the use of fertilizer may be a significant source in the future. While we do not know what this increase might be, a doubling of N_2O would be significant, with a temperature increase of several tenths of a kelvin. Some NH_3 would also be released from fertilizer, although again there are no good estimates for future concentrations; a twofold increase would cause a temperature increase of close to a tenth of a kelvin (World Meteorological Organization, 1982).

Any change in climate would have some influence on agricultural productivity. One would expect that an increase in temperature would cause the positions of agricultural regions to shift to higher latitudes if there is adequate soil moisture and nutrients. We can gain some insight into what a temperature increase might do to climate by examining the climate of the Altithermal Period, some 4500 to 8000 years ago (e.g., see Kellogg and Schware, 1981); during this time the temperature has been estimated to have been several degrees higher. Because of low summer rainfall, the U.S. Midwest was a dry prairie; this region now produces half the world's maize.

One should not infer from this that the total agricultural productivity would necessarily decrease with an increase in temperature. Indeed, one study (Council on Environmental Quality, 1981) indicates that a warming of 1.4 K could increase total agricultural productivity, although there would be a decrease in mid-latitude regions. A 1-K temperature increase and a 10% precipitation decrease could decrease crop yields in the United States and Russia by 20%. Different crops also respond differently to a temperature increase. While higher mean temperatures would cause a decrease in the maize crop, rice production would increase if there were little change in precipitation. Laboratory studies indicate that plants also respond differently to an increase in CO_2. Some do not experience an overall increase in photosynthetic rate. For others, the photosynthetic rate seems to peak at a CO_2 concentration about three times the present level. Clearly, little is known about the agricultural changes that could occur with a large-scale climate change. Field programs and laboratory studies are needed. Present models offer little help in that they lack the sophistication to predict the shifts in precipitation patterns and soil moisture that are so important to estimates of agricultural productivity.

V. Projections for Future Research

Although climate research has been actively supported for a decade and much progress has been made, there is a need for further study in all areas. The problem that has been given the most attention in recent years is that of the buildup of CO_2 since its effect is global and most modeling studies show some temperature increase in the near future. The verification of a CO_2 signal, that is, an increase in temperature, is made difficult by not only the natural variability but also changes in concentrations of trace constituents, aerosols, and solar radiation that will also change temperature. Thus there is the continued need to measure the concentrations of the minor gases, not only those that directly influence the temperature via the greenhouse effect, but also those that can exert an indirect effect such as CO, as discussed above. Concentration changes as low as a few tenths of a percent per year will need to be detected. It will be particularly important to monitor changes in tropospheric and stratospheric O_3 amounts and vertical distributions since many of the trace gases either directly or indirectly affect O_3, which influences not only the temperature but the amount of lethal UV radiation that reaches the surface.

Ocean studies that relate to the mixed layer must also be carried out; these will help us determine the delay in any temperature increase that might occur while the upper layer of the ocean and the atmosphere come to equilibrium. Measurements that will give us information on the rate of mixing include density distributions, currents, and tracers (release of inert substances in the ocean that can be followed as it diffuses).

In addition to field measurements, laboratory studies are needed. The absorption bands of some of the minor IR-active gases are not well known, especially for the appropriate atmospheric temperatures and pressures. In some cases, not even the total band intensity is known, which is needed for even a very approximate calculation of the greenhouse effect. Detailed spectroscopic information is necessary, including line positions, intensities, and half-widths in order to develop band models; while these parameters can be approximated from theory, comparison with measurements is necessary to validate the band model. Transmission functions determined from band models should also be compared with measurements with spectral resolution as small as five wavenumbers.

Significant improvement in the climate models is also needed. At present, for example, models cannot determine the response time for a climate change, nor are they capable of predicting temperature and precipitation on a regional scale. Although large-scale predictions may soon become possible, they will only tell us that we have incorporated the appropriate physics for such a scale. For such models to be practically useful, however, they must be able to be applied to specific geographical areas in order to assess social and political stress that might come about from inadvertent climate modification.

References*

Budyko, M. I., (1969). The effect of solar radiation variations on the climate of the Earth. *Tellus* **5,** 611–619.

Charlock, T. P., and Sellers, W. D. (1980). Aerosol, cloud reflectivity and climate. *J. Atmos. Sci.* **37,** 1136–1137.

Climatic Impact Assessment Program (1975). "The Natural and Radiatively Perturbed Troposphere, CIAP Monogr. 4. U.S. Dept. of Transportation, Washington, D. C.

Council on Environmental Quality (1981). "Global Energy Futures and the Carbon Dioxide Problem." U.S. Govt. Printing Office, Washington, D. C.

*Goody, R. M. (1964). "Atmospheric Radiation," Vol. I. Oxford Univ. Press, London and New York.

Hameed, S., Cess, R. D., and Hogan, J. S. (1980). Response of the global climate to changes in atmospheric chemical composition due to fossil fuel burning. *JGR, J. Geophys. Res.* **85,** 7537–7545.

Hansen, J., Johnson, D., Lacis, A., Lebedeff, S., Lee, P., Rind, D., and Russell, G. (1981). Climate impact of increasing atmospheric carbon dioxide. *Science* **213,** 957–966.

Hansen, J. E., and Travis, L. D. (1974). Light scattering in planetary atmospheres. *Space Sci. Rev.* **16,** 527–610.

*Kellogg, W. W., and Schware, R. (1981). "Climatic Change and Society." Westview Press, Boulder, Colorado.

Kourganoff, V. (1963). "Basic Methods in Transfer Problems." Dover, New York.

Lacis, A., Hansen, J., Lee, P., Mitchell, T., and Lebedeff, S. (1981). Greenhouse effect of trace gases 1970–1980. *Geophys. Res. Lett.* **8,** 1035–1038.

Liou, K.-N. (1980). "An Introduction to Atmospheric Radiation." Academic Press, New York.

Liu, S. C., Cicerone, R. J., and Donahue, T. M. (1977). Sources and sinks of atmospheric N_2O and the possible ozone reduction due to industrial fixed nitrogen fertilizers. *Tellus* **29,** 251–263.

*Manabe, S., and Möller, F. (1961). On the radiative equilibrium and heat balance of the atmosphere. *Mon. Weather Rev.* **89,** 503–532.

*Manabe, S., and Strickler, R. (1964). Thermal equilibrium of the atmosphere with a convective adjustment. *J. Atmos. Sci.* **21,** 361–385.

*Manabe, S., and Wetherald, R. T. (1980). On the distribution of climate change resulting from an increase in CO_2 content of the atmosphere. *J. Atmos. Sci.* **33,** 99–118.

National Research Council (1982). "Carbon Dioxide and Climate: A Second Assessment," Report of the CO_2/Climate Review Panel. Nat. Acad. Press, Washington, D. C.

*National Research Council (1983a). "Changing Climate, Report of the Carbon Dioxide Assessment Committee." Nat. Acad. Press, Washington, D. C.

National Research Council (1983b). "Acid Deposition: Atmospheric Processes in Eastern North America." Nat. Acad. Press, Washington, D. C.

North, G. R. (1975). Theory of energy balance climate models. *J. Atmos. Sci.* **32,** 2033–2043.

*Paltridge, G. W., and Platt, C. M. R. (1976). "Radiative Processes in Meteorology and Climatology." Am. Elsevier, New York.

Park, J. H., Rothman, L. S., Linsland, C. P., Smith, M. A. H., Richardson, D. J., and Larsen, J. C. (1981). Atlas of Absorption Lines from 0 to 17900 cm^{-1}," NASA Ref. Publ. 1084. Natl. Aeron. Space Admin., Washington, D. C.

---

* The references preceded by an asterisk provide a more detailed discussion of the material in this chapter.

Stone, P. H., and Carlson, J. H. (1979). Atmospheric lapse rate regimes and their parameterization. *J. Atmos. Sci.* **36,** 415–423.

Thompson, S. L., and Schneider, S. H. (1982). Carbon dioxide and climate: Has a signal been observed yet? *Nature* (*London*) **295,** 645–646.

Wang, W. C., Yung, Y. L., Lacis, A. A., Mo, T., and Hansen, J. E. (1976). Greenhouse effects due to man-made perturbations of trace gases. *Science* **194,** 685–690.

* World Meteorological Organization (1982). "Report of the Meeting of Experts on Potential Climatic Effects of Ozone and other Minor Trace Gases," Global Ozone Res. Monit. Proj., Rep. No. 14. Natl. Cent. Atmos. Res., Boulder, Colorado.

5

The Photochemistry of the Upper Atmosphere

DOUGLAS G. TORR

Center for Atmospheric and Space Sciences
Utah State University
Logan, Utah

I. Composition

The most common division of the Earth's atmosphere has been by its temperature profile (Chapter 1, Fig. 1). The previous chapters have already dealt with some important characteristics pertaining to the troposphere and stratosphere. We see that the first maximum in temperature occurs at ~ 50 km and constitutes the boundary between the stratosphere and mesosphere (the stratopause). Further increases in altitude are then accompanied by a falling temperature through the mesosphere until a minimum is reached again

THE PHOTOCHEMISTRY OF ATMOSPHERES
Earth, the Other Planets, and Comets

ISBN 0-12-444920-4

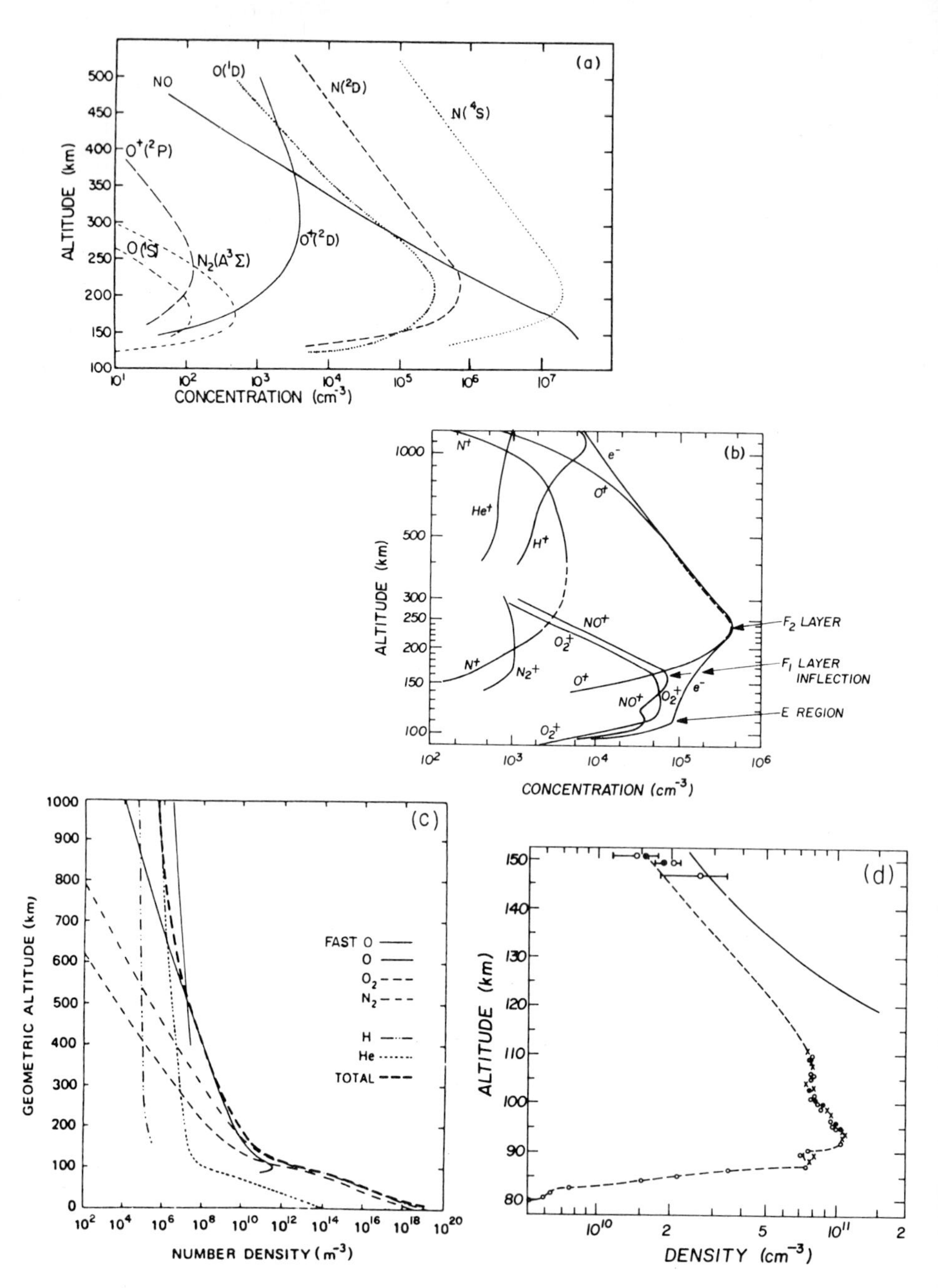

Fig. 1. Vertical profiles of thermospheric constituents illustrating the composition of the thermosphere. In (d), an example (2 November 1978, 1821 MST) of *in situ* measurements of atomic oxygen at a solar zenith angle of 104° in the thermosphere is shown. ——, MSIS model values. The measurements were made by a Michigan rocket payload which used resonance fluorescence techniques. The absolute values of the atomic oxygen densities below 120 km are controversial, and the values may be larger than those shown. From *Geophysical Research Letters.*

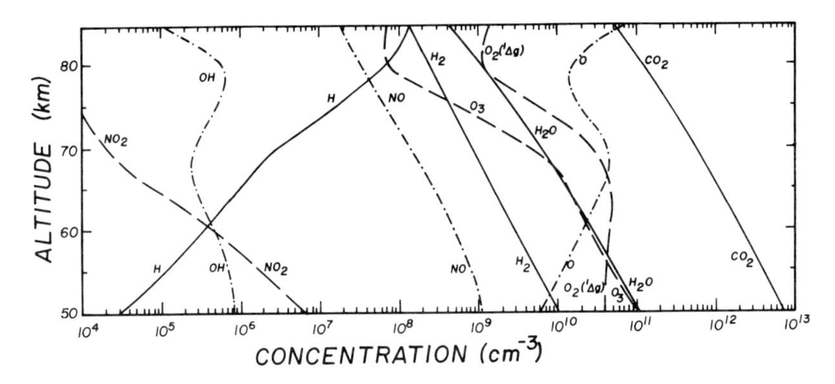

Fig. 2. Vertical profiles of neutral mesospheric constituents illustrating the composition of the mesosphere for noontime conditions at 45°N. From Wisemberg and Kockarts (1980).

at ~85 km (the mesopause). The minimum value of temperature at the mesopause is about 190–200 K at high latitudes. Further increases in altitude lead to a rapid rise in temperature (~5 K per kilometer at 150 km) through the thermosphere. The thermosphere constitutes the last region connecting the neutral atmosphere and interplanetary space. This chapter deals with the photochemistry of the mesosphere, thermosphere–exosphere, and ionosphere–magnetosphere.

The temperature profile owes its interesting characteristics largely to the chemical composition of the upper atmosphere, illustrated in Figs. 1 through 4. The Sun's radiation in the far-ultraviolet (far-UV) region of the spectrum (1000–2000 Å) is absorbed by several atmospheric constituents. The shorter extreme-UV wavelengths (200–1000 Å) are absorbed in the thermosphere, creating ionization, and the longer wavelengths in the mesosphere, resulting in photodissociation of molecular oxygen (O_2), ozone (O_3), and water vapor (H_2O). The absorption of the UV radiation results in the deposition of heat. The temperature minimum at the base of the thermosphere is caused by a decrease in energy deposition at these heights due to absorption at higher altitudes as well as to cooling by radiation at infrared (IR) wavelengths by carbon dioxide (CO_2), O_3, atomic oxygen, and nitric oxide (NO). This means that there must be a strong flow of heat from the hot thermosphere into the cooler mesosphere, where it is lost by radiation. Furthermore, since the collision frequency decreases with increasing altitude, the mean free path is long in the upper thermosphere, resulting in efficient heat conduction downward. The temperature profile in the thermosphere rapidly becomes isothermal above ~200 km. On the other hand, below 120 km the conduction time constant is of the order of 1 day, so a large temperature gradient is maintained.

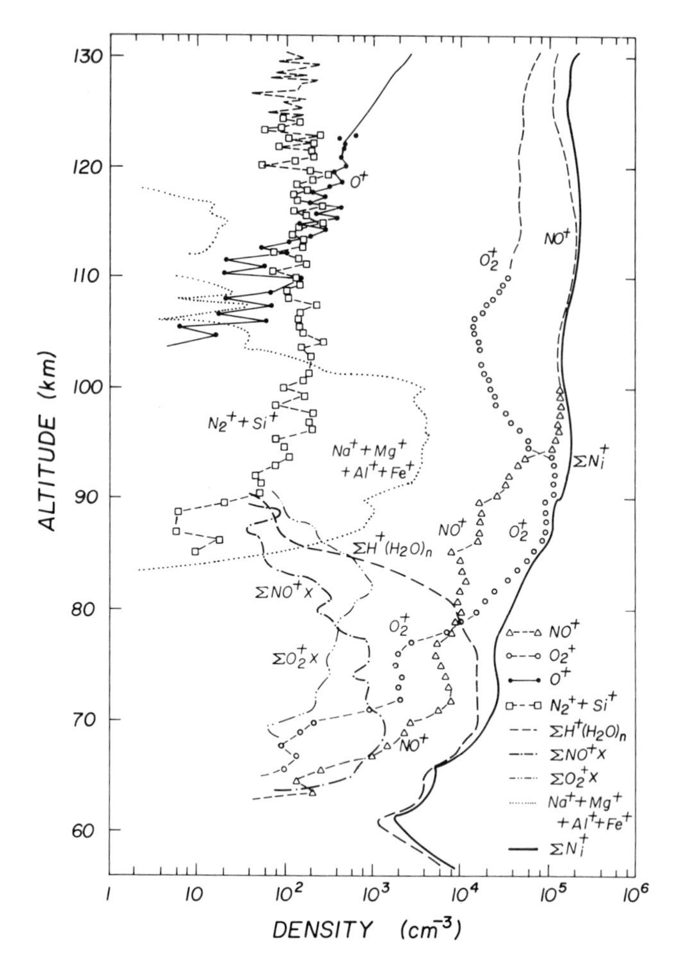

Fig. 3. Density profiles of the main ions and ion classes of the lower ionosphere; measurement above Red Lake, Ontario (50.9°N), 24 February 1979. From Kopp and Herrmann (1982).

Another way, therefore, of structuring the atmosphere is through its diffusion characteristics. In fact, the distribution of the neutral particles with height cannot be determined without knowing whether the atmosphere is completely mixed or not, and whether the time constants are dominated by transport or photochemical processes.

Figure 4a illustrates altitude profiles of the mean molecular mass and the characteristic times for transport by molecular diffusion and vertical eddy

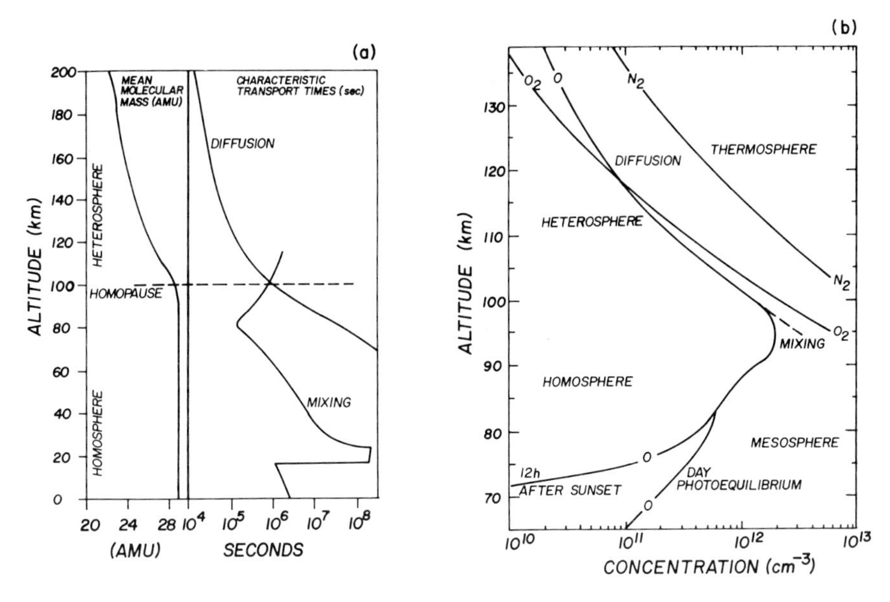

Fig. 4. (a) Altitude profiles of the mean molecular mass and of the characteristic times for transport by molecular diffusion and by vertical eddy mixing. Below the homopause, at 101 km, mixing is more rapid than diffusion, and the mean molecular mass is constant. Above the homopause, diffusion is more rapid than mixing, and the mean molecular mass decreases with increasing altitude. (b) Transition from the homosphere to the heterosphere. Dissociation of molecular oxygen and the transition from mixing to diffusion conditions are also shown. From Walker (1977).

mixing.* A level is shown at ~100 km, below which the mean mass is constant and above which molecular diffusion is more rapid than mixing. In this region each major atmospheric gas is distributed according to its own mass, and the mean molecular mass decreases with height.

As in the cases of the troposphere and stratosphere, the principal molecular species in the mesosphere and thermosphere are O_2 and N_2. They can absorb the Sun's radiation and photodissociate or ionize. The products of the latter processes may be formed in excited states. Some of these excited species may be sufficiently long-lived as to be capable of further interaction with other species. The concentrations of many of these species are photochemically

---

* Diffusion can arise from two sources. One of these is molecular diffusion (i.e., a molecule moves through the gas because of its individual velocity). This type of diffusion is readily computed from gas kinetic theory and varies inversely with pressure. At lower altitudes, turbulent mixing is the dominant diffusion term; this is referred to as eddy mixing.

controlled in certain altitude regions. In this way photoabsorption of solar UV energy gives rise to a complex subset of minor species that play an important role in controlling the basic thermal structure, composition, and energetics of the upper atmosphere. The ionized component resides mainly above the stratosphere, forming the region termed the ionosphere. The ionospheric ions tend to flow along the Earth's magnetic flux tubes, which form the magnetosphere, thereby constituting a significant source of ionization. Flow of ionization in the magnetosphere is relatively collision free, whereas it is collision dominated in the ionosphere. The transition region from collision-dominated to collision-free flows could be regarded as a boundary demarcating the magnetosphere and ionosphere. At mid latitudes the magnetospheric flux tubes tend to remain closed, forming the plasmasphere. The outflow of ionospheric material into the plasmasphere constitutes the main source of ionization for this region. At higher latitudes solar wind

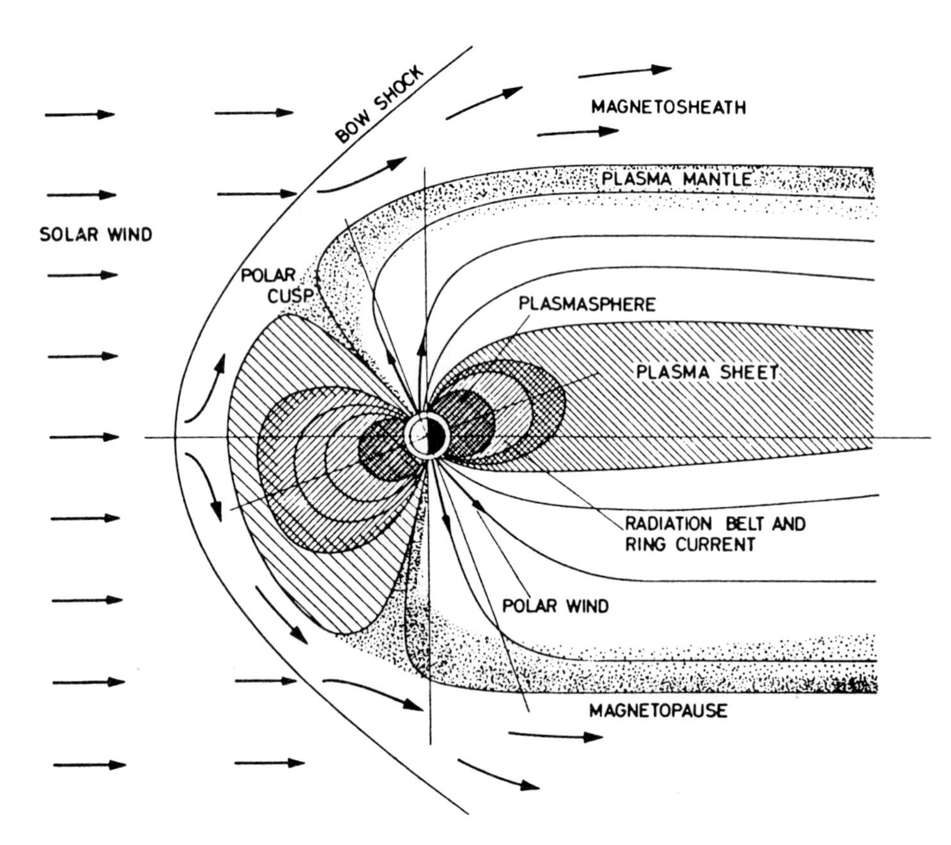

Fig. 5. Schematic illustration of the regions that constitute the magnetosphere.

pressure causes the field lines to be swept away in the anti-Sun-ward direction, leading to significant plasma depletions. Figure 5 schematically illustrates the magnetic field-line configuration of the magnetosphere.

In the next section we discuss the sources of these constituents. As we move higher in altitude the gas concentration decreases to a point where the collision frequency is so low that the neutral atoms and molecules undergo ballistic-type trajectories, with the Maxwellian tail of the lighter gas component energetic enough to achieve escape velocity. The theory of the loss of atmospheric gases to space is well developed, and the escape occurs in the region defined as the exosphere. There are a number of mechanisms that can cause an atmosphere to lose matter to space. Simple hydrodynamic considerations imply that an atmosphere must expand into the surrounding vacuum. When derived from kinetic theory as opposed to motion as a continuous fluid, this is called Jeans escape. In the terrestrial atmosphere only hydrogen (H) and helium (He) are lost in significant quantities in this way. Other factors that lead to energization of light atoms and even heavier constituents like atomic oxygen can lead to significant enhancements of the escape rate. For example, photochemical reactions and the precipitation of ions and neutral atoms give rise to hot hydrogen, helium, and oxygen atoms, which then may form a hydrogen–helium–oxygen geocorona around the Earth. These factors significantly affect the composition of the exosphere, as illustrated in Fig. 1, in which altitude profiles of oxygen are shown with and without the energization.

Figure 1 also shows profiles of ionic constituents observed up to altitudes in the topside ionosphere. In the plasmasphere, the primary constituents are singly and doubly ionized atomic oxygen, hydrogen, and helium, although atomic nitrogen ions were observed by the *Dynamics Explorer* satellite. The H^+ is generated via an energetically resonant charge-exchange reaction,

$$O^+ + H \rightleftharpoons H^+ + O$$

which regulates the flow rate between the ionosphere and the plamasphere.

II. Photochemistry

A. Photoabsorption of Solar Radiation

The photochemistry of the upper atmosphere is driven by the photoabsorption of solar radiation. Atmospheric gases absorb radiation strongly at some wavelengths and weakly at others. The wavelengths strongly absorbed will therefore be absorbed at lower gas concentrations and hence greater altitudes than the weakly absorbed wavelengths. Figure 6 shows the penetration of solar radiation and the principal absorbing species. Figure 7 shows the

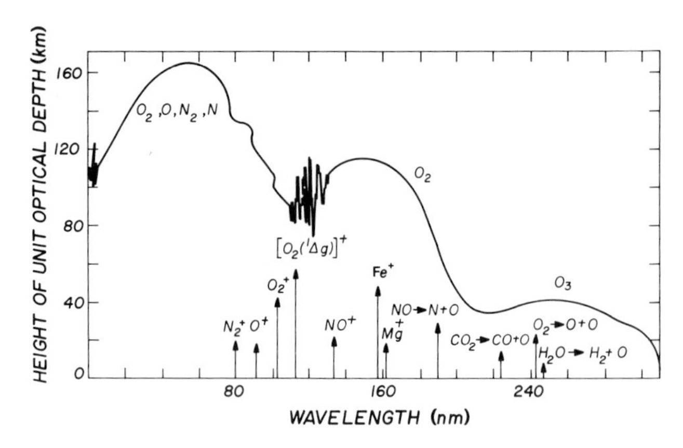

Fig. 6. Variation with wavelength of the height of unit optical depth [see Eq. (2c)] for solar radiation incident vertically. The wavelengths corresponding to the ionization and dissociation thresholds of certain constituents are indicated: ionization processes are represented by +. From Thomas (1980), with permission of the Royal Society.

absorption cross section as a function of wavelength for O_3 and O_2. Cross sections for O_2 and N_2 are given in Table A-5.* The rate of change of the photon flux at a height h is given by

$$\frac{dF(\lambda,h)}{ds} = \sum \sigma_i(\lambda) n_i(h) F(\lambda,h), \tag{1}$$

where ds is an element of distance along the slant path traversed by the photons in the atmosphere, $F(\lambda,h)$ the flux of solar photons at wavelength λ and height h, and $\sigma_i(\lambda)$ the cross section for absorption of these photons by constituent i at wavelength λ. Integration of Eq. (1) yields

$$F(\lambda,h) = F(\lambda,\infty) \exp\left[-\sum_i \sigma_i(\lambda) \int n_i \, ds \right], \tag{2a}$$

where $F(\lambda,\infty)$ is the unattenuated flux. For solar zenith angles χ less than $\pi/2$, $ds = dh \sec \chi$. Figures 8 and 9 show the solar flux as a function of wavelength from 20 to 3000 Å. Equation (2a) is usually written in the form

$$F(\lambda,h) = F(\lambda,\infty) e^{-\tau(\lambda)}, \tag{2b}$$

where τ is called the optical depth given by

$$\tau(\lambda) = \sum_i \sigma_i(\lambda) \int n_i \, ds = \sum_i \sigma_i(\lambda) \int n_i \sec \chi \, dh. \tag{2c}$$

---

* Tables prefixed with A are in the Appendix to this chapter.

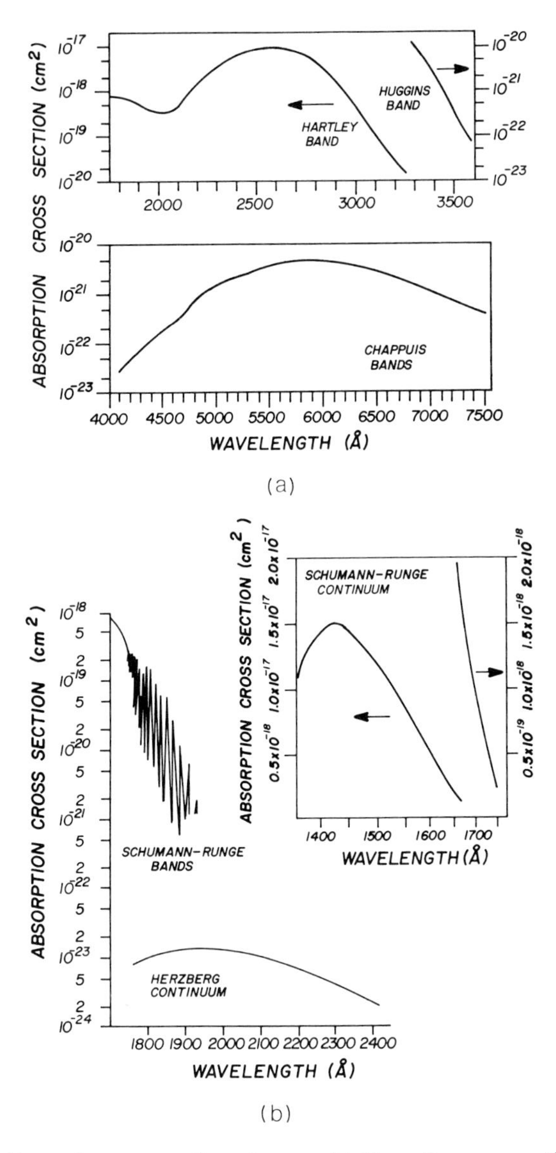

Fig. 7. (a) Absorption cross section of ozone. (b) Absorption cross section of molecular oxygen. The names designate different wavelength regions of the absorption spectrum. From Walker (1977), with permission of Macmillan Publishing Co., Inc., and the original authors.

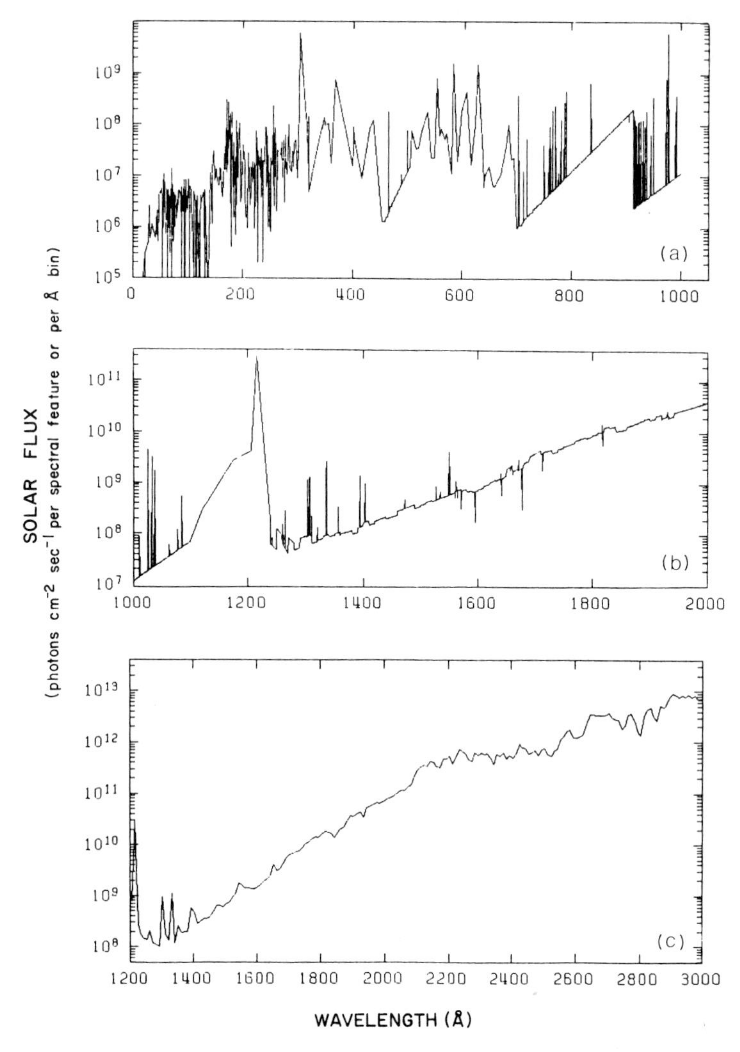

Fig. 8. (a and b) A reference spectrum of extreme ultraviolet fluxes in the wavelength range 20–1850 Å based on *Atmosphere Explorer* measurements. These data were supplied by H. E. Hinteregger. The values, which are illustrative of conditions at the July 1976 solar cycle minimum, are available in tabular form (reference identification: SC # 21REF). The main features are listed in Table A-6. (c) Solar ultraviolet spectrum from 1200 to 3000 Å. The spectrum is tabulated in Tables A-3 and A-4. The values below 1800 Å are Rottman's (1981) solar minimum reference spectrum, and the values above 1800 are measurements made on 12 January 1983, reproduced from Mount and Rottman (1983a). Although the latter were taken near solar maximum, the solar cycle variation at these wavelengths is not large. The data sets shown in these figures show significant unresolved differences in the wavelength overlap region (Lyα to $\sim$1800 Å). Note the units in (c) are photons cm^{-2} sec^{-1} nm^{-1}, and (a) and (b), photons cm^{-2} sec^{-1} Å^{-1} bin or per line feature.

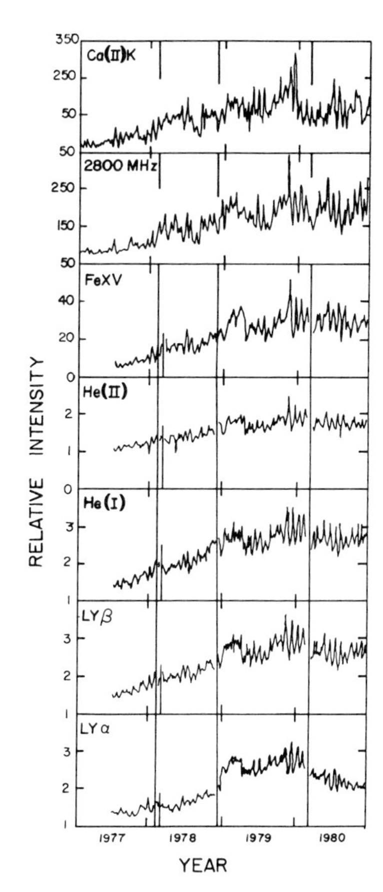

Fig. 9. Illustration of the relative variation of the main solar extreme-ultraviolet emission features as a function of solar cycle. Also shown are similar values for the Ca(II) plage and 10.7-cm radio indices. From the *Journal of Geophysical Research.*

Table A-1 gives the atmospheric transmission functions in the spectral region of the O_2 Schumann–Runge bands..

B. Mesospheric Photochemistry

An overview of the photochemistry of the mesosphere has been given by Nicolet (1974), who cited several historically important papers.

1. Odd-Oxygen Sources

The odd-oxygen family comprises O and O_3 and is referred to as O_x. The production of O_3 is dependent on the existence of sources of atomic oxygen. The photodissociation of O_2 is the primary source of O. This can take place either through the intense Schumann–Runge continuum, commencing at

1760 Å in the lower thermosphere,

$$O_2 + h\nu \xrightarrow{J_1} O(^3P) + O(^1D) \quad (3)$$

through the weak Herzberg continuum, commencing at 2424 Å in the stratosphere, and the Schumann–Runge bands,

$$O_2 + h\nu \xrightarrow{J_2} O(^3P) + O(^3P) \quad (4)$$

as illustrated in Fig. 10. Table A-2 tabulates dissociation rates of O_2 in the Schumann–Runge bands as a function of wavelength. Table A-3 shows several different measurements of solar fluxes in the wavelength range 1200–1900 Å for solar minimum. Table A-4 shows the same for the range 1800–3180 Å. The oxygen atoms formed by photodissociation of O_2 may recombine in the presence of a third body,

$$O + O + M \xrightarrow{k_1} O_2 + M \quad (5)$$

or they may unite with O_2,

$$O + O_2 + M \xrightarrow{k_2} O_3 + M \quad (6)$$

forming O_3, which may be destroyed by the two-body process

$$O_3 + O \xrightarrow{k_3} 2\,O_2 \quad (7)$$

and by photodissociation in the strong Hartley continuum ($\lambda < 3075$ Å),

$$O_3 + h\nu \xrightarrow{J_3} O_2(a^1\Delta_g) + O(^1D) \quad (8)$$

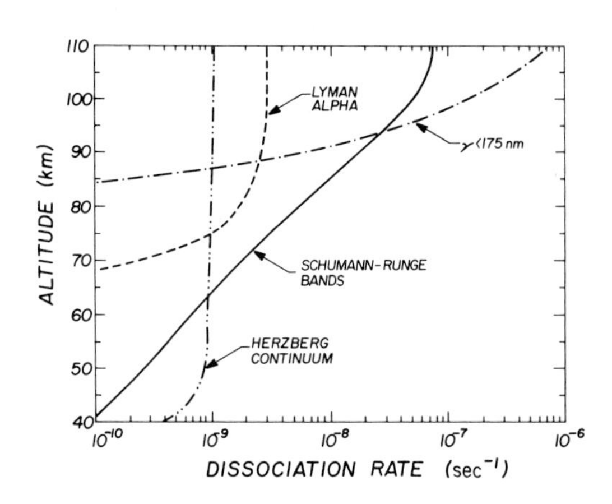

Fig. 10. Dissociation rates of molecular oxygen due to various spectral regions for an overhead Sun, at 40 to 110 km. From Frederick and Hudson (1980b), with permission of the American Meteorological Society.

or in the weak Chappius bands in the visible ($\lambda < 8000$ Å),

$$O_3 + h\nu \xrightarrow{J_4} O_2 + O(^3P) \tag{9}$$

The O_3 photodissociation rate due to reaction (8) is shown as a function of height in Fig. 11. Rate coefficients are given in Appendix II at the end of the volume.

The above chemical scheme was essentially proposed by Sydney Chapman in 1930. Over the years many additional processes have been added. It is nevertheless instructive to use this simple scheme to illustrate some fundamental characteristics of mesospheric chemistry.

The time-dependent variation of O is given by

$$d[O]/dt = (J_1 + 2J_2)[O_2] + (J_3 + J_4)[O_3] - \{2k_1[M][O] + k_2[M][O_2] + k_3[O_3]\}[O], \tag{10}$$

and of O_3 by

$$d[O_3]/dt = k_2[M][O_2][O] - \{(J_3 + J_4) + k_3[O]\}[O_3], \tag{11}$$

which leads to the following time-dependent equation for odd-oxygen:

$$d[O_x]/dt = 2J_O[O_2] - \{2k_1[M][O] + 2k_3[O_3]\}[O], \tag{12}$$

where $J_O = \frac{1}{2}J_1 + J_2$.

A calculation of the photochemical lifetimes of O, O_3, and O_x as a function of height yields an interesting result. Typical values are shown in Table I. (Note: These results were obtained using a more comprehensive chemical scheme.) We see that the individual photochemical lifetimes of O and O_3 are significantly shorter than that of the odd-oxygen family. If these lifetimes are compared with transport lifetimes, which may range between several hours at

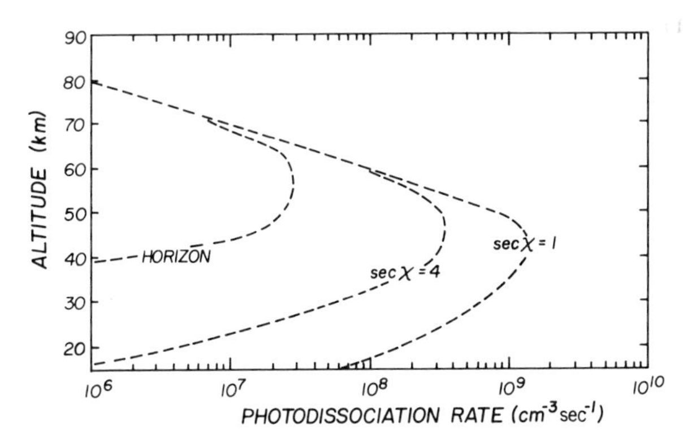

Fig. 11. Altitude of ozone photodissociation ($\lambda < 3075$ Å) for the Sun on the horizon and overhead ($\sec \chi = 1$).

Table I
PHOTOCHEMICAL LIFETIMES OF ODD-OXYGEN SPECIES

Altitude (km)	Lifetimes (sec)		
	O	O_3	O_x
50	10	10^2	10^4
70	5×10^3	10^2	5×10^3
90	$\sim 10^5$	10^2	$\sim 10^5$

60 km to $\sim$1 day above $\sim$80 km, we see that below $\sim$80 km, transport affects only the total odd-oxygen concentration, and that the partitioning between O and O_3 is determined entirely by photochemical considerations. Although the photochemical lifetime of O_3 may be relatively short (due to photodissociation), this does not mean that there is a net loss, because the atomic oxygen formed rapidly recombines to balance the O_3 loss rate. Hence we see that the processes that truly result in a net destruction of O and O_3 are those that do not subsequently result in a replenishment of O_x, such as reactions (5) and (7).

Equations (10) and (11) can be simplified considerably for altitude regimes in which photochemical equilibrium prevails. Ozone is in photochemical equilibrium throughout the mesosphere, and O below $\sim$70 km. As mentioned above, the net removal of O_x via reaction (7) is slow compared to photodissociation of O_3, hence Eq. (11) can be written

$$[O_3] \simeq k_2[M][O_2][O]/J_{O_3}, \tag{13}$$

where $J_{O_3} = J_3 + J_4$. Substituting in Eq. (10) we obtain

$$\frac{d[O]}{dt} = 2J_O[O_2] - \left\{2k_1 + \frac{k_3k_2}{J_{O_3}}[O_2]\right\}[M][O]^2. \tag{14}$$

The same result can be derived from Eq. (12) for the upper mesosphere, where $[O_x] \simeq [O]$.

2. Oxygen–Hydrogen Photochemistry

In the stratosphere and mesosphere, atomic hydrogen reacts with O_2 and O_3, respectively, via the reactions

$$H + O_2 + M \xrightarrow{k_4} HO_2 + M \tag{15}$$

$$H + O_3 \xrightarrow{k_5} OH_{(v<9)} + O_2 \tag{16}$$

leading to the production of hydroperoxyl (HO_2) and hydroxyl (OH) radicals.

Reaction (16) is the more important one in the mesosphere. The OH and HO_2 radicals then react rapidly with O in the cyclic restoration of H and OH and catalytic destruction of O,

$$OH + O \xrightarrow{k_6} H + O_2 \tag{17}$$

$$HO_2 + O \xrightarrow{k_{7a}} OH + O_2 \tag{18}$$

This is illustrated schematically in Fig. 12. The OH and HO_2 radicals can also react with O_3, but these reactions are of lesser importance in the mesosphere:

$$OH + O_3 \xrightarrow{k_{7b}} HO_2 + O_2 \tag{19}$$

$$HO_2 + O_3 \xrightarrow{k_{7c}} 2\,O_2 + OH \tag{20}$$

The odd-hydrogen reactions have a large impact on odd-oxygen. In the mesosphere the time-dependent variation of O must be modified to

$$\begin{aligned} d[O]/dt = {} & 2(J_1 + J_2)[O_2] + (J_3 + J_4)[O_3] \\ & - \{2k_1[M][O] + k_2[M][O_2] \\ & + k_3[O_3] + k_6[OH] + k_{7a}[HO_2]\}[O], \end{aligned} \tag{21}$$

and that of O_3 to

$$d[O_3]/dt = k_2[M][O_2][O] - \{(J_3 + J_4) + k_3[O] + k_5[H]\}[O_3]. \tag{22}$$

The time-dependent equation for odd-oxygen becomes

$$\begin{aligned} d([O] + [O_3])/dt = {} & 2J_O[O_2] - \{2k_1[M][O] + 2k_3[O_3] \\ & + k_6[OH] + k_{7a}[HO_2]\}[O] - k_5[H][O_3]. \end{aligned} \tag{23}$$

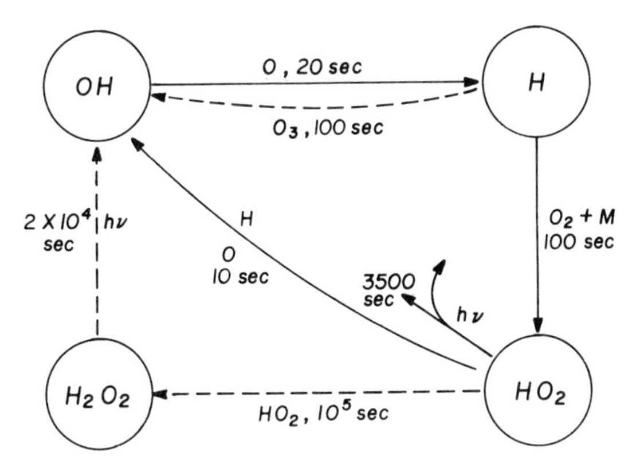

Fig. 12. Schematic illustration of the key elements of oxygen–hydrogen photochemistry and representative time constants at ~80 km.

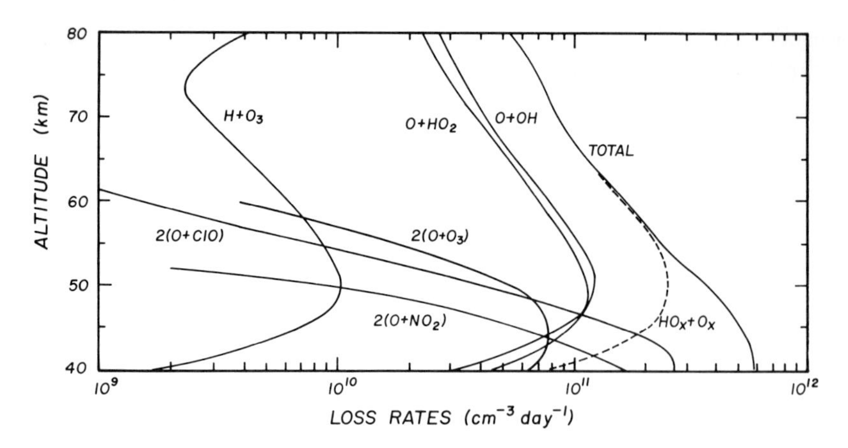

Fig. 13. Illustration of the destruction rates of atomic oxygen and ozone.

Rate coefficients for these reactions are given in Appendix II. Figure 13 shows a plot of examples of destruction rates of O and O_3. Some additional processes not discussed here are also shown. Equations (13) and (14) become, respectively,

$$[O_3] = \frac{k_2[M][O_2][O]}{J_{O_3} + k_3[O] + k_5[H]} = \frac{k_2[M][O_2][O]}{J_{O_3}(1 + \delta)}, \tag{24}$$

$$\frac{d[O]}{dt} = 2J_O[O_2] + \frac{k_2[M][O_2][O]}{(1 + \delta)} - \left\{2k_1[M][O] + k_2[M][O_2] + \frac{k_3 k_2[M][O_2][O]}{J_{O_3}(1 + \delta)} + k_6[OH] + k_{7a}[HO_2]\right\}[O]. \tag{25}$$

The time constants for O destruction (a day at ~80 km and a week at ~100 km) are too large to result in photochemical equilibrium conditions for O near the mesopause level. The vertical distribution in the lower thermosphere below the turbopause is controlled by eddy diffusion, with diffusive separation above this boundary. The diurnal variation in O in the upper mesosphere is small. In the lower mesosphere, O is converted to O_3 at night. Figure 14 illustrates altitude profiles at noon and midnight of O_x and O_3 produced by the photochemistry outlined above. The effect of OH and HO_2 on O_3 is a net decrease O_3 concentration arising from the action of atomic

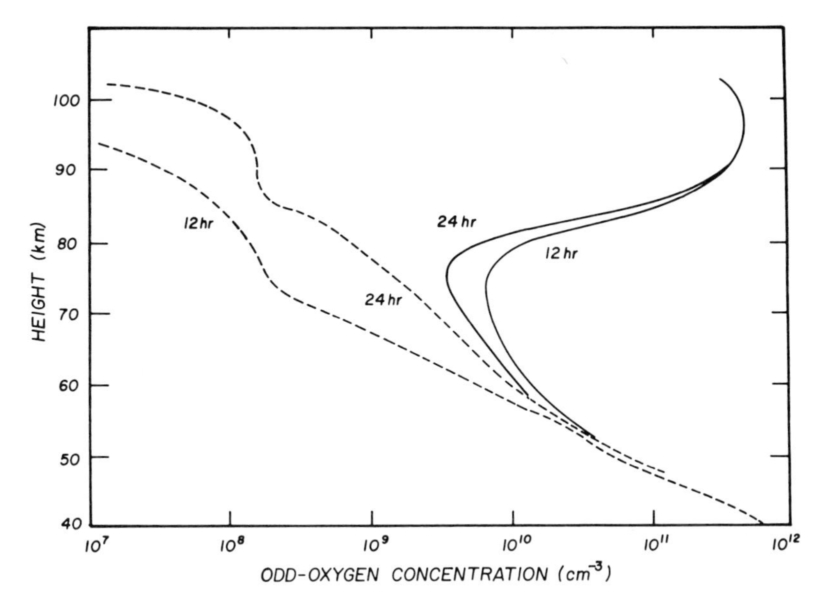

Fig. 14. Illustrative theoretical profiles of odd-oxygen (---, O_3; ——, O_x) for noontime and midnight conditions.

hydrogen. Significant discrepancies still exist between experiment and theory for mesospheric O_3. This, therefore, represents a fruitful area for further research.

3. Catalytic Destruction of Ozone by Odd-Hydrogen

The effect of odd-hydrogen radicals on O_3 can be summarized in the following well-known catalytic cycles:

$$
\begin{array}{ll}
O_3 + H \longrightarrow OH + O_2 & HO_2 + O \longrightarrow OH + O_2 \\
OH + O \longrightarrow H + O_2 & OH + O \longrightarrow H + O_2 \\
 & H + O_2 + M \longrightarrow HO_2 + M \\
\hline
O + O_3 \longrightarrow 2\,O_2 & O + O \longrightarrow O_2
\end{array}
$$

The hydrogen atoms and HO_2 are restored after each cycle, which results in the destruction of an O_3 molecule and an oxygen atom, respectively. These species are then free to participate in subsequent O_3-destruction cycles with no impact on the odd-hydrogen concentration.

a. Production and Destruction of Odd-Hydrogen Radicals. As in the case of oxygen, we can identify an odd-hydrogen family (HO_x)

comprising H, OH, and HO_2. There is one source molecule, H_2O, and there is one primary process resulting in the production of odd-hydrogen ($H + OH + HO_2$) in the mesosphere, namely, photolysis of H_2O by solar Lyα radiation,

$$H_2O + h\nu \xrightarrow{J_5} H + OH \tag{26}$$

and to a lesser extent in the lower mesosphere, by the reaction

$$H_2O + O(^1D) \xrightarrow{k_8} OH + OH^*_{(v \geq 2)} \tag{27}$$

The hydroxyl radical is transformed to H by reaction (17). Destruction of odd-hydrogen occurs through the reactions

$$OH + HO_2 \xrightarrow{k_{9a}} H_2O + O_2 \tag{28}$$

$$OH + OH \xrightarrow{k_{9b}} H_2O + O \tag{29}$$

and

$$H + HO_2 \xrightarrow{k_{10}} H_2O + O \tag{30a}$$

$$H + HO_2 \xrightarrow{k_{11}} H_2 + O_2 \tag{30b}$$

By combining the sources and sinks of H, OH, and HO_2 we can derive a time-dependent equation for HO_x:

$$\begin{aligned} d[HO_x]/dt = {} & 2(J_5 + k_8[O(^1D)])[H_2O] \\ & -2\{k_{9a}[OH][HO_2] \\ & + k_{9b}[OH]^2 + (k_{10} + k_{11})[H][HO_2]\}. \end{aligned} \tag{31}$$

The lifetime of HO_x is smaller than that of transport for heights below ~80 km. The lifetimes of H, OH, and HO_2 are relatively short in the mesosphere, so photochemical equilibrium will prevail for these constituents. Expressions for the photochemical equilibrium ratios of HO_x members can be easily derived:

$$\frac{[HO_2]}{[OH]} \simeq \left\{\frac{k_{7b}[O_3]}{k_6[O]} + \frac{k_4[M][O_2]}{k_5[O_3] + k_4[M][O_2]}\right\}\left\{\frac{k_6[O]}{k_{7c}[O_3] + k_{7a}[O]}\right\}, \tag{32}$$

which in the mesosphere reduces to ~1;

$$\frac{[H]}{[OH]} = \frac{k_6[O]}{k_4[M][O_2] + k_5[O_3]}, \tag{33}$$

which is sufficient to determine the partitioning of the HO_x family.

Below 80 km the three-body formation of HO_2 via reaction (15) proceeds rapidly and can constitute a significant source of H_2O through reactions (28) and (30a). At greater altitudes the rate of formation of HO_2 decreases rapidly, and the only significant source of H_2O is upward transport; thus its concentration drops. Hydrogen is lost by upward eddy diffusion, and the production of OH due to reaction (16) decreases. The photolysis of H_2O increases with decreasing attenuation of Lyα by O_2.

Figure 15 shows the dissociation rates of H_2O by various mechanisms for two solar zenith angles. Figure 16 summarizes the ozone–hydrogen chemistry of the mesosphere.

It should be emphasized that attempts to compare atmospheric observations of chemical species with model values must depend on the correct treatment of both dynamic and chemical coupling. Figure 17 shows the chemical and diffusive lifetimes of the O_x, HO_x, and odd-nitrogen (NO_y: NO, NO_2, NO_3, N_2O_5, and HNO_3) groups as well as of H_2O. The NO_y family has very little direct impact on the chemistry of the normal neutral mesosphere and is discussed separately in Section II,E. These results indicate that H_2O and NO_y are dependent on transport at all altitudes, and HO_x and O_x only above 85 km, but their distributions depend on the H_2O content. If the concentration of H_2O and H_2 are known, HO_x and O_x can be calculated using simple photochemical models. The H_2O abundance is sensitive to season, because of its dependence on photolytic destruction. Factors that affect the upward transport will also be important. Figure 18 shows observed and calculated H and H_2 densities.

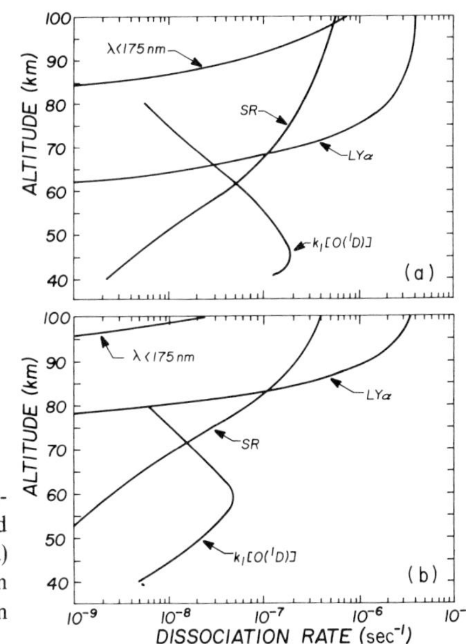

Fig. 15. Comparison of water vapor destruction rates by photodissociation and chemical paths for solar zenith angles of 0° (a) and 85° (b). From Frederick and Hudson (1980b), with permission of the American Meteorological Society.

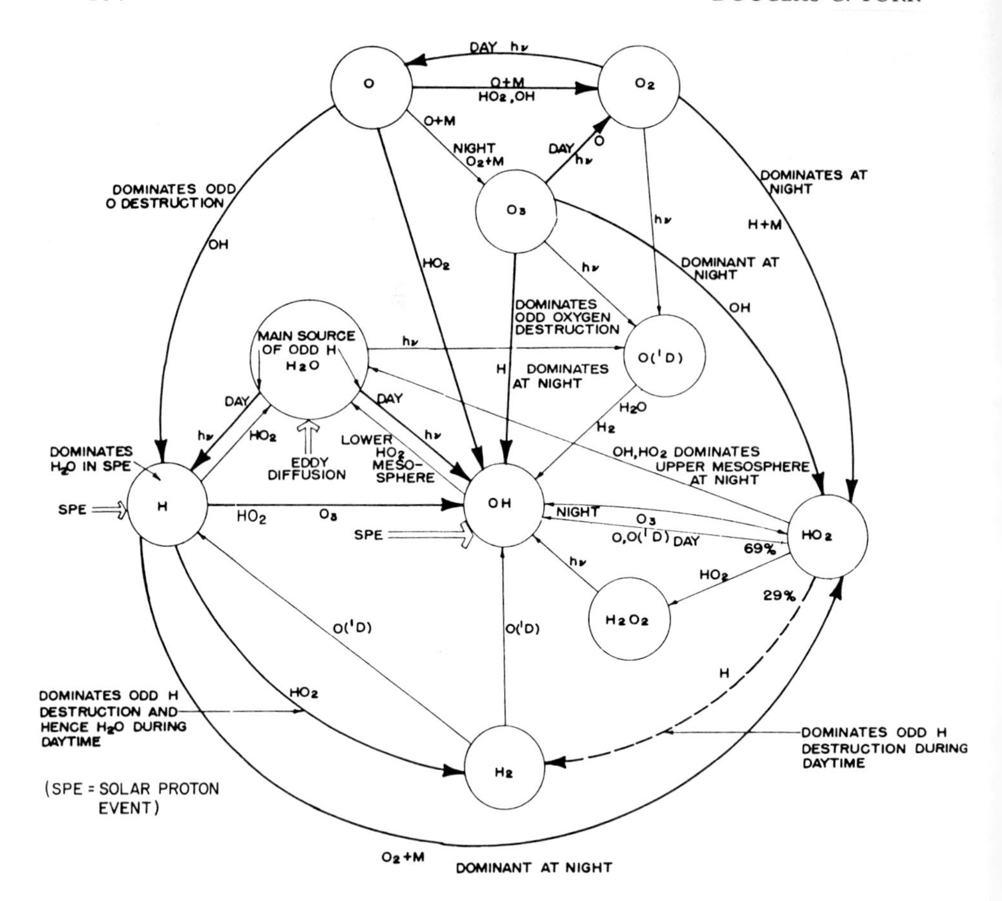

Fig. 16. Mesospheric ozone–hydrogen chemistry.

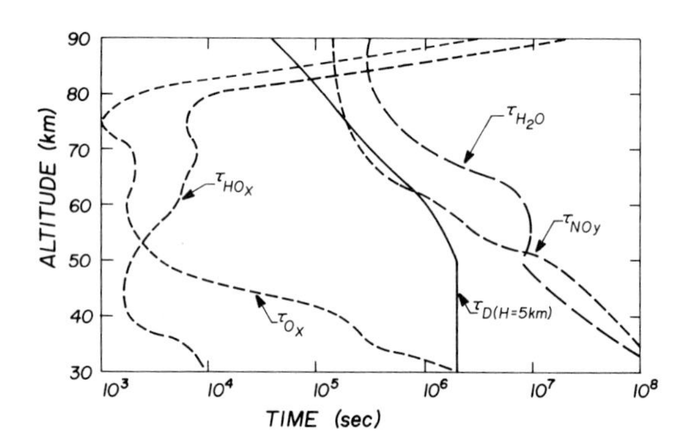

Fig. 17. Vertical distribution of odd-oxygen, odd-hydrogen, odd-nitrogen, water vapor, and eddy diffusion lifetimes. From Brasseur and Solomon (1984).

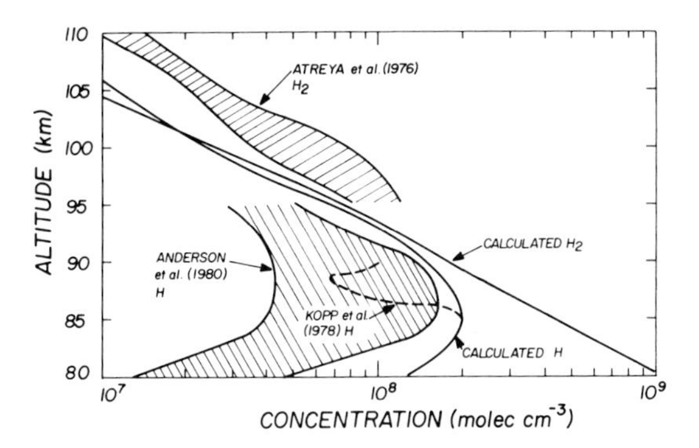

Fig. 18. Observed atomic and molecular hydrogen densities in the lower thermosphere and calculated values from the one-dimensional model for solar minimum conditions. From Solomon *et al.* (1982), with permission of Pergamon Press. References on the original figure may be found in Solomon *et al.* (1982).

b. Water Vapor. The preceding discussion on odd-hydrogen chemistry can be summarized from the perspective of H_2O chemistry as follows. The upward transport of H_2O is balanced by photochemical conversion to H and H_2 via

$$\begin{array}{c} H_2O + h\nu \longrightarrow H + OH \\ OH + O \longrightarrow H + O_2 \\ \hline H_2O + O \longrightarrow 2\,H + O_2 \end{array} \tag{34a}$$

or

$$\begin{array}{c} H_2O + h\nu \longrightarrow H + OH \\ O + OH \longrightarrow H + O_2 \\ H + O_2 + M \longrightarrow HO_2 + M \\ H + HO_2 \longrightarrow H_2 + O_2 \\ \hline H_2O + O \longrightarrow H_2 + O_2 \end{array} \tag{34b}$$

Rate coefficients are given in Appendix II. In order to illustrate current knowledge on H_2O concentrations, Fig. 19 shows experimental and theoretical H_2O mixing ratios.

c. Odd-Nitrogen. Although the photochemistry of mesospheric odd-nitrogen is important, it does not play a significant direct role in mesospheric neutral chemistry. It is an important source of NO^+ in the D region. It is also an important constituent in the chemistry of the thermosphere and stratosphere, and it appears likely that the latter two regions are coupled through the mesosphere. We discuss the chemistry of odd-nitrogen in Section II,E.

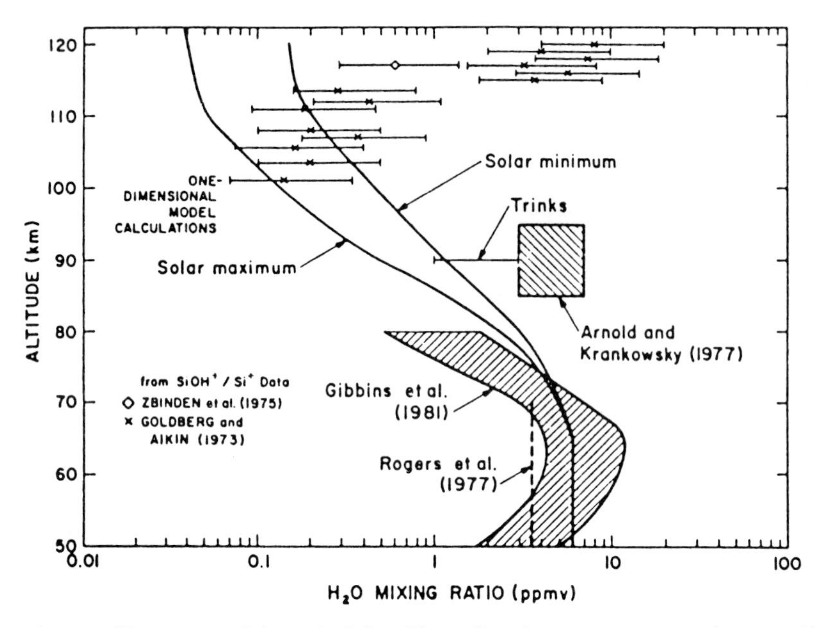

Fig. 19. Water vapor mixing ratios inferred from silicon ion measurements and measured by other techniques are shown along with the water vapor profile obtained from a one-dimensional model calculation for solar maximum and solar minimum. From Solomon *et al.* (1982), with permission of Pergamon Press. References on the original figure may be found in Solomon *et al.* (1982).

4. Effect of Energetic Particle Precipitation on Mesospheric Photochemistry

Calculations indicate that energetic particle precipitation can give rise to a series of ion reactions discussed in Section II,D,2,b that ultimately produce odd-hydrogen and odd-nitrogen. The net effect of the chain of reactions is the product reaction:

$$O_2^+ + H_2O + e^- \longrightarrow O_2 + H + OH \tag{35}$$

Odd-hydrogen is by far the most important factor in the destruction of odd-oxygen above 60 km. The odd-hydrogen produced is expected to initially decrease O_3 as discussed in the previous sections. Odd-hydrogen destruction, however, is dominated by the reaction of H with HO_2, which leads to $H_2 + O_2$ 29% of the time, $H_2O + O < 2\%$ of the time, and 2 OH 69% of the time. The latter product has no net effect on odd-hydrogen. Water vapor cannot diffuse upward rapidly enough to replace the loss estimated by the effective reaction (35). For extreme storms under sunlit conditions a large reduction in H_2O may occur, and an intermediate process in the production of odd-hydrogen in the ion chemistry chain discussed in Section II,D,2,b may be cut off. Such scenarios constitute interesting areas for study. Storm-time ion chemistry is discussed in Section II,D,2,b. The storm-time effects on the oxygen–hydrogen chemistry are indicated in Fig. 16.

C. The Ionosphere above 90 km

The ionosphere is defined loosely as the region where free electrons are present in sufficient quantity to affect the propagation of radio waves. The base of the ionosphere occurs at ~60 km, and the peak eletron density lies typically between 200 and 400 km. The ionosphere has been categorized into three major regions on the basis of maxima or inflections that occur in the electron density profile as a function of height. These are the D, E, and F regions, which are indicated in Fig. 1. The processes that give rise to these three regions will be covered in the following discussion.

1. Ionization Sources

a. Photoionization. The major source of ionization is photoionization of neutral constituents by solar extreme UV radiation at wavelengths shorter than ~1000 Å. The photoionization rate of the ith constituent q_i at a given height is proportional to the attenuated solar flux, the cross section for ionization (σ_i'), and the atmospheric concentration. Integrating over wavelength we obtain

$$q_i = n_i \int_0^\infty \sigma_i'(\lambda) F(\lambda)\, d\lambda, \tag{36}$$

where $F(\lambda)$ is given by Eq. (2). It is instructive to briefly review some of the properties of Eq. (36) using a single-constituent atmosphere. For convenience we assume the ionization and absorption cross sections are equal. Substituting for $F(\lambda)$ from Eq. (2) we get

$$q = \int_0^\infty \sigma n F(\lambda,\infty) \exp\left[-\sigma(\lambda) \int_{h_0}^\infty n \sec\chi\, dh\right] d\lambda. \tag{37}$$

To integrate Eq. (37) with respect to h, we take the log of the ideal gas law $p = nkT$ (where p is pressure, k Boltzmann's constant, and T temperature) and differentiate to obtain

$$\frac{1}{p}\frac{dp}{dh} = \frac{1}{n}\frac{dn}{dh} + \frac{1}{T}\frac{dT}{dh} \equiv -\frac{1}{H}, \tag{38}$$

which we use to define the scale height H. Hence

$$p = p_0 e^{-z}, \tag{39}$$

where

$$z = \int_{h_0}^{h} \frac{dh}{H}. \tag{40}$$

The parameter z is referred to as the reduced height. Using the ideal gas law, Eq. (39) can also be written in the form

$$n = (n_0 T_0/T)e^{-z}. \tag{41}$$

Therefore, using Eqs. (38) and (41) in (2c) and dropping the wavelength notation,

$$\tau = \sigma n_0 T_0 \int \frac{e^{-z}}{T} ds = \sigma n_0 T_0 \int e^{-z}\, dz \frac{H}{T} \sec\chi. \tag{42}$$

To express T as a function of H we differentiate the expression for the pressure at height h assuming constant g ($\int nmg\, dh$) to obtain

$$dp/dh = -nmg = -\rho g. \tag{43}$$

Comparing Eq. (43) with Eq. (38) and using the ideal gas law,

$$nmg = P/H = nkT/H, \tag{44}$$

or

$$H = kT/mg. \tag{45}$$

Hence the ratio H/T in Eq. (42) is a constant if we assume g is approximately constant. Thus

$$\tau = \frac{\sigma n_0 T_0}{T} He^{-z} \sec\chi = \sigma n H \sec\chi. \tag{46}$$

This expression is not valid for $\chi \gtrsim 80°$. It can be shown, however, that for large χ, the form of Eq. (46) can be retained if $\sec\chi$ is replaced by a function termed the Chapman function, symbolized by $\mathrm{Ch}\chi$, which can be numerically evaluated fairly readily on a computer (Smith and Smith, 1972).

To locate the height of peak production, we take the natural logarithm of Eq. (37) and differentiate to obtain the condition

$$\frac{1}{n}\frac{dn}{dh} = \frac{d\tau}{dh} = -\sigma n \sec\chi, \tag{47}$$

for $\chi < 80°$.

Substituting for $(1/n)(dn/dh)$ from Eq. (38) in Eq. (47) and using the fact that

$$\frac{1}{T}\frac{dT}{dh} = \frac{1}{H}\frac{dH}{dh}, \tag{48}$$

we find that peak production occurs at an optical depth defined by

$$\tau = \sigma n H \sec \chi = 1 + \alpha, \tag{49}$$

where α is the local vertical temperature gradient. It is common practice to choose this height as the reference level h_0 thereby setting z to zero there. Using Eq. (41) in Eq. (37) we can write

$$q = \frac{\sigma T_0 n_0 e^{-z}}{T} F(\lambda,\infty)e^{-\tau} = \sigma n F(\lambda,\infty)e^{-\tau}. \tag{50}$$

The peak value q_p occurs at $z = 0$ and $\tau = 1 + \alpha$,

$$q_p = \sigma n_0 F(\lambda,\infty)e^{-(1+\alpha)}. \tag{51}$$

Expressing q in terms of q_p,

$$q = q_p \frac{T_0}{T} \exp[1 + \alpha - z - e^{-z} \sec \chi]. \tag{52}$$

Chapman derived a classical production expression for the case $\alpha = 0$, $T = T_0$, and $\chi = 0$. If we replace q_p with its maximum value using $\sec \chi = 1$ and $\alpha = 0$, then

$$z_{peak} = \ln \sec \chi, \tag{53}$$

$$q_p = q_{max} \cos \chi. \tag{54}$$

Table A-5 lists the ionization and absorption cross sections for the major thermospheric constituents. The corresponding values for the solar flux are given in Table A-6. The data represent weighted averages over 37 selected wavelength intervals.

As mentioned above, if $\chi < 80$ then $\mathrm{Ch}\chi \simeq \sec \chi$. The Chapman function takes the curvature of the Earth into account. The total photoionization rate is obtained by summing the individual photoionization rates over all constituents. Figure 20 illustrates some typical production rate profiles. At high altitudes, where attenuation is negligible and the flux is constant with altitude, the rate of photoionization is proportional to the concentration and is characterized by the neutral constituent scale height. At low altitudes competition between increasing attenuation of the solar flux and increasing number density of the neutral ionizable constituents results in a maximum in the ionization rate. The height of the peak is determined by the magnitude of the cross section. Larger cross sections result in greater absorption and hence occurrence of the peak at higher altitudes.

b. Nocturnal Ionization Sources. Starlight (911–1026 Å) and resonance scattering of Lyβ into the night sector are the most important sources of nocturnal ionization in the E region and are capable of maintaining

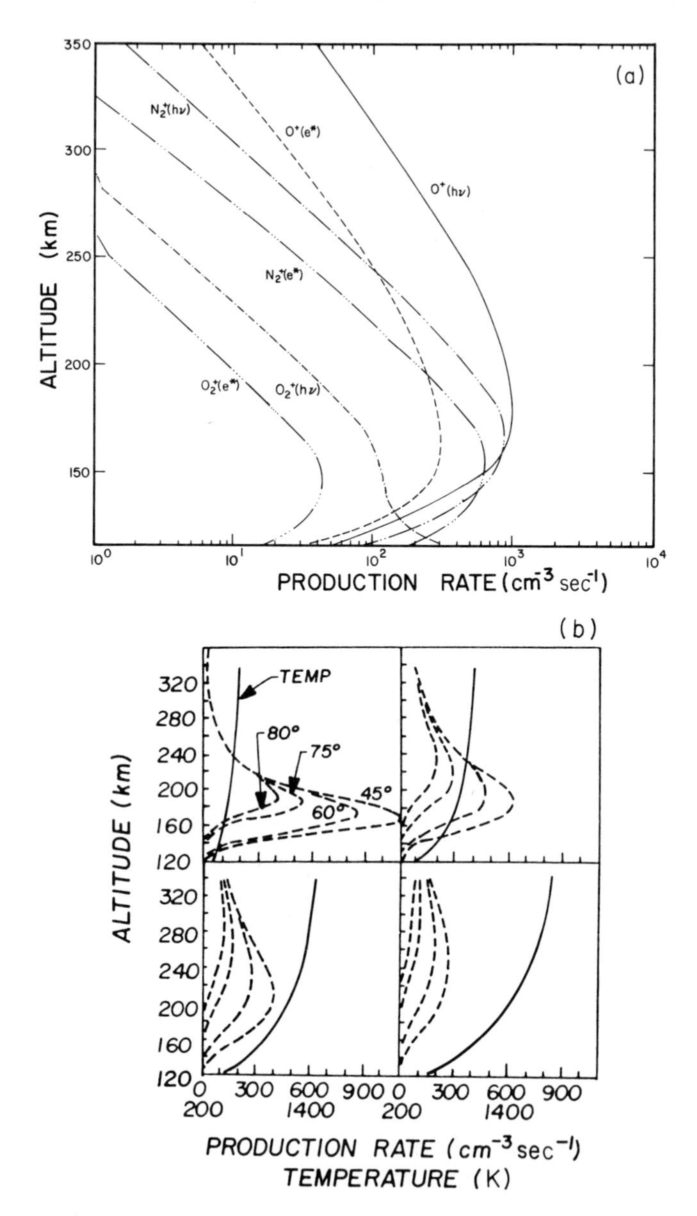

Fig. 20. (a) Total ion production rates computed for 14 February 1974 for Millstone Hill, solar zenith angle of 26°. An interesting observation that emerged from these calculations is the dominance of photoelectron impact on N_2 as an ionization source below ~150 km. The photoelectrons are produced mainly by photons of wavelengths less than 300 Å, for which the cross sections are smaller than the longer wavelengths which are largely responsible for the photoionization production rates (calculation by P. G. Richards). (b) Illustration of the effect of change in temperature T on the O^+ production rate for solar zenith angles of 45, 60, 75, and 80°.

observable electron densities of the order of 1 to 4×10^3 cm^{-3}. The source varies geographically, depending on which stars are present. In the lower F region, the major O^+ source in the equatorial ionosphere is radiation from O^+ recombination in the F_2 region, whereas away from the equatorial zones only interplanetary 584-Å radiation (resonant scattering of sunlight by intersteller helium atoms) exceeds resonance scattering of solar 584- and 304-Å radiation as the dominant source of O^+ during the month of December.

c. Ionization by Secondary Electrons. The photoionization process is illustrated by the following general process:

$$A + h\nu \longrightarrow A^{+*} + e^{-*} \tag{55}$$

where in the ionosphere A is N_2, O, or O_2. Process (55) results in the production of a flux of primary electrons, referred to as the primary photoelectron flux. The energy of the primaries depends on the difference in the energy of the absorbed photon and the ionization potential of the product ion A^{+*}. The asterisk is used to indicate that the ionization process can and generally does result in the production of excited atomic and molecular ions. Table A-7 lists photoionization branching ratios for the main states of O^+, $O_2{}^+$, and $N_2{}^+$.

The primary electrons can be energetic enough to ionize the neutral constituents, further resulting in a secondary source of ionization. The energies of the primary electrons can be of the order of the energies of soft X rays, $\sim 10^2$ eV. The primary electrons are subsequently degraded in energy in many ways. Energy is lost in inelastic collisions with neutral atoms and molecules, with ions, and with thermal electrons. In order to calculate the secondary ionization rate, it is necessary to compute the primary photoelectron and secondary (or degraded) photoelectron fluxes, which falls outside the scope of this work. Figure 21 shows the secondary photoelectron flux at three altitudes compared with a theoretical calculation. Figure 20a also shows typical ion-pair production rate height profiles due to ionization by the secondary electrons.

2. Ionospheric Photochemistry

The peak electron density of the ionosphere typicaııy lies between 10^5 and 10^6 ions cm^{-3}. The main source of ionization are the processes

$$O + h\nu \xrightarrow{J_6} O^+ + e^- \tag{56}$$

$$O_2 + h\nu \xrightarrow{J_7} O_2{}^+ + e^- \quad \text{or} \quad O + O^+ + e^- \tag{57}$$

$$N_2 + h\nu \xrightarrow{J_8} N_2{}^+ + e^- \quad \text{or} \quad N + N^+ + e^- \tag{58}$$

It was initially thought that the reverse of process (56), electron radiative

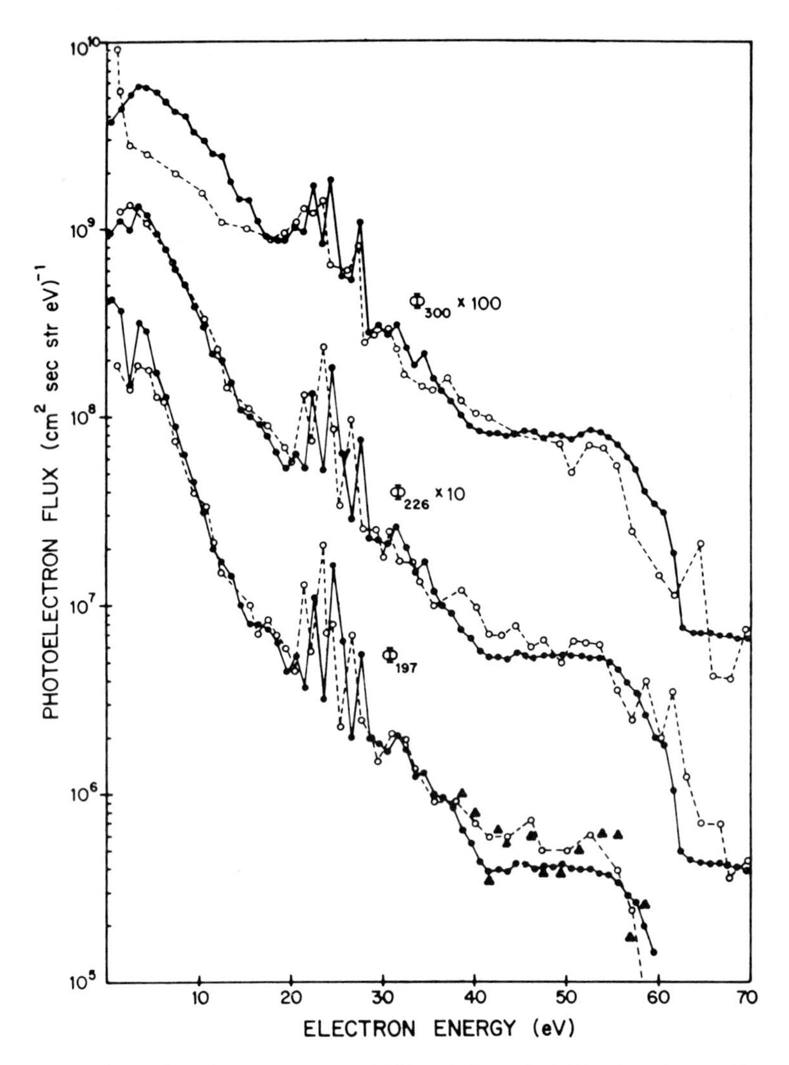

Fig. 21. Comparison between measured (○) and theoretical (●) photoelectron fluxes for three different altitudes. The photoelectron flux is the flux of secondary electrons that arise through the thermalization of primary photoelectrons produced in the process of photoionization of neutral constituents. The ϕ_{300} represents the flux at 300 km. The triangles on the lowest spectrum are also measured fluxes but from a different orbit showing the variability in this region above 35 eV. The disagreement below 15 eV at 300 km is due to the neglect of transport and the lack of simultaneous thermal electron density measurements. From the *Journal of Geophysical Research.*

recombination,

$$O^+ + e^- \longrightarrow O + h\nu \tag{59}$$

or the negative ion recombination reaction,

$$O^+ + O^- \longrightarrow O_2 + h\nu \tag{60}$$

would constitute the main loss mechanisms for electrons. However, these loss processes greatly overestimated the electron concentration. In 1947 Bates and Massey proposed dissociative recombination of O_2^+ with electrons,

$$O_2^+ + e^- \xrightarrow{\alpha_1} O + O \tag{61}$$

as the primary electron removal mechanism. This required transformation of O^+ into O_2^+, however, which they correctly suggested occurred via the ion–atom interchange reaction

$$O^+ + O_2 \xrightarrow{k_{12}} O_2^+ + O \tag{62}$$

In 1949 Biondi and Brown confirmed in laboratory studies that α_1 was large enough for the theory to work. In 1955, Bates suggested that the ion–atom interchange reaction

$$O^+ + N_2 \xrightarrow{k_{13}} NO^+ + N \tag{63}$$

followed by the dissociative recombination reaction

$$NO^+ + e^- \xrightarrow{\alpha_2} N + O \tag{64}$$

could also be important. Using ionosonde data, Bates and co-workers established that these reactions adequately described the basic photochemistry of the F_2 layer. It was subsequently found that radiative recombination does become important when the electron density exceeds 10^6 cm^{-3}. This process is believed to be the source of UV equatorial airglow emissions.

With the advent of rocket and satellite *in situ* measurements of ion composition in the 1960s and the multiparameter *Atmosphere Explorer* measurements of the 1970s, this basic picture was modified and quantified. The first major addition to the above scheme was made in 1963 by Norton *et al.*, who used measured ion density profiles and laboratory-determined values for α_1, α_2, k_{12}, and k_{13} together with rocket measurements of the solar extreme-UV flux to compare calculated and measured ion density profiles. Their results, which were confirmed later (Donahue, 1966), indicated that a major source of NO^+ was missing, which they identified as the process

$$N_2^+ + O \xrightarrow{k_{14}} NO^+ + N \tag{65}$$

Subsequent laboratory measurements confirmed the magnitude of k_{14} to be large, contrary to atomic theory considerations. Reaction (65) is the greatest source of NO^+ below $\sim$220 km. It also is mainly responsible for the destruction of N_2^+ over the same altitude range.

The next most significant modification to the photochemistry involved the addition of metastable species. The potential importance of the reaction

$$O^+(^2D) + N_2 \xrightarrow{k_{15}} N_2^+ + O \tag{66}$$

was suggested by several workers (Omholt, 1957; Dalgarno and McElroy, 1965, 1966). Once the N_2^+ loss processes had been established in the laboratory it became possible to check aeronomically the need for reaction (66). The first experimental confirmation of the effect on N_2^+ was provided by the rocket twilight measurements of N_2^+ first negative bands by Feldman (1973), who utilized the laboratory results of Mehr and Biondi (1969) on dissociative recombination of N_2^+ with electrons,

$$N_2^+ + e^- \xrightarrow{\alpha_3} 2\,N \tag{67}$$

Several minor processes subsequently identified could also affect the concentrations of the main middle F region species O^+, N_2^+, O_2^+, and NO^+. These included the reactions

$$O^+(^2P) + N_2 \xrightarrow{k_{16}} N_2^+ + O \tag{68}$$

$$N_2^+ + O \xrightarrow{k_{17}} O^+ + N_2 \tag{69}$$

$$N_2^+ + O_2 \xrightarrow{k_{18}} O_2^+ + N_2 \tag{70}$$

$$N^+ + O_2 \xrightarrow{k_{19}} O_2^+ + N \tag{71}$$

$$N^+ + O_2 \xrightarrow{k_{20}} NO^+ + O \tag{72}$$

$$O_2^+ + N(^2D) \xrightarrow{k_{21}} NO^+ + O \tag{73}$$

$$O_2^+ + N(^4S) \xrightarrow{k_{22}} NO^+ + O \tag{74}$$

$$O_2^+ + NO \xrightarrow{k_{23}} NO^+ + O_2 \tag{75}$$

The reaction of $N_2^+ + O_2$ is of major significance in the E and D regions, where it constitutes an important source of O_2^+. Reactions (73)–(75) can be significant in determining the ratio $[NO^+]/[O_2^+]$ in the E and D regions.

The introduction of metastable O^+ into the F region photochemistry resulted in several new complicating factors. It was found that since $\sim$60% of the O^+ are formed in metastable states, a considerable conversion of O^+ to N_2^+ must occur, which earlier models neglected. Computation of the rate at which this happens requires a knowledge of the $O^+(^2D)$ concentration. The chemistry of the metastable species $O^+(^2D)$ and $O^+(^2P)$ was studied during 1970s via the NASA *Atmosphere Explorer* program (see review by Torr and Torr, 1982). By the early 1980s most of the rate coefficients of the reactions listed above had been measured in the laboratory, and many aeronomical studies had also been carried out, largely with the *Atmosphere Explorer* data. Table A-8 lists current values for these coefficients; Albritton (1978) published a detailed list of rate coefficients of atmospheric interest.

It was found that several problems arose when the chemical scheme presented above was used with the laboratory-measured rate coefficients. The main problem was that the theory predicted a significant excess of N_2^+ ionization. On the basis of aeronomical studies D. G. Torr *et al.* (1980) inferred that charge exchange of vibrationally excited N_2^+ with O proceeds rapidly, thereby removing the excess N_2^+ ionization and restoring the lost O^+ ionization, namely,

$$N_2^{+*} + O \xrightarrow{k_{24}} O^+ + N_2 \tag{76}$$

This finding, however, is strictly dependent on the validity of laboratory measurements of relevant rate coefficients and particularly of the N_2^+ dissociative recombination rate coefficient.

The chemistry depicted by reaction (76) is fairly complex, since it includes all the processes that result in the production and destruction of N_2^+ vibrational excitation, which are discussed in Section II,C,5,a. Figure 22 schematically illustrates our present understandings of ionospheric chemistry. The effects of

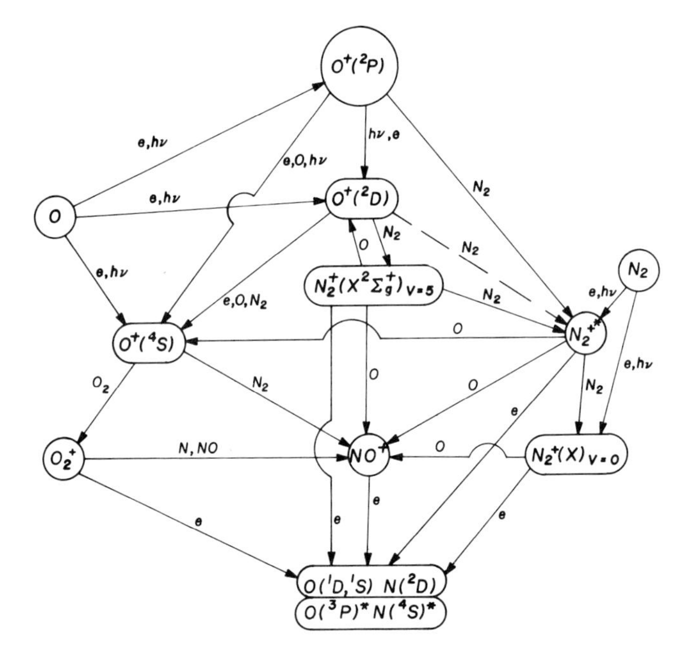

Fig. 22. Schematic illustration of our present understanding of the chemistry of the ionosphere. Considerable uncertainty still surrounds the processes involving N_2^+. The asterisks indicate vibrationally excited or translationally energetic species.

N_2^+ vibrational excitation become significant only above ~200 km because of the quenching of N_2^{+*} by N_2.

Figure 23 shows a detailed comparison of theoretical results using a chemical scheme very similar to that described above with the mean diurnal variation in electron concentration derived from *Atmosphere Explorer* data

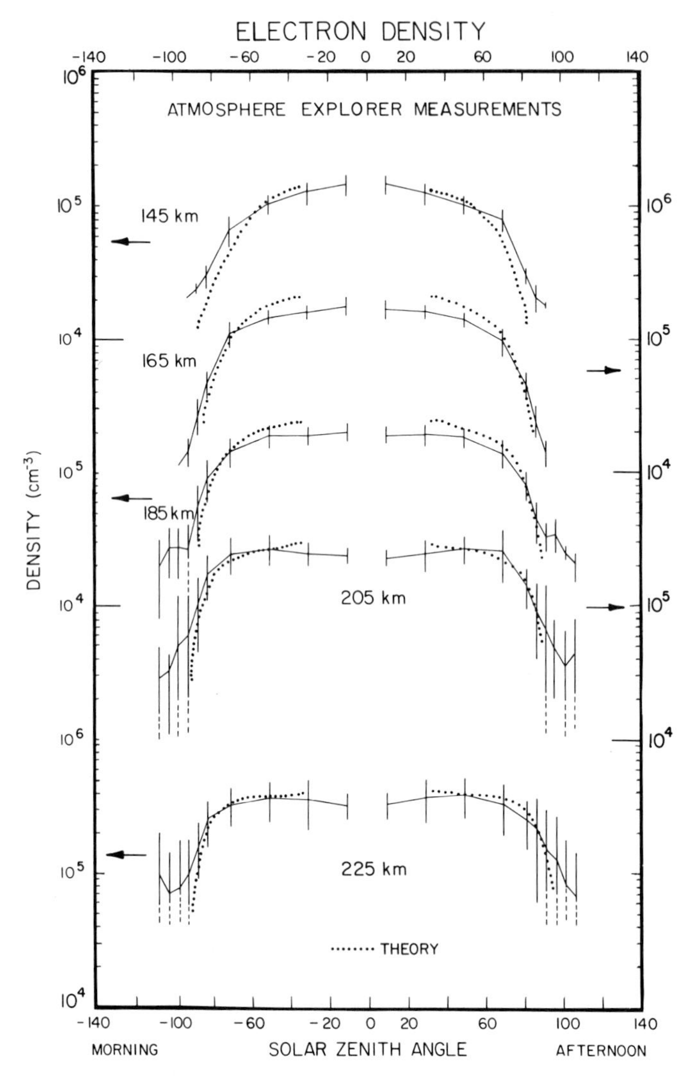

Fig. 23. Comparison of the theoretical and the mean measured diurnal variation of electron density in the F_1 layer. From Torr *et al.* (1979), with permission of Pergamon Press.

over the altitude range 150–220 km. Figure 24 shows a comparison between current theory and observations of the peak electron density made at Arecibo, Puerto Rico, on 22 March 1981 over a full diurnal cycle. This figure illustrates that the photochemistry described above, which was tested and quantified largely on the F_1 layer, also reproduces the behavior of the peak electron density of the F_2 layer, where transport phenomena become important.

3. Formation of the Ionospheric E and F Regions

In addition to separation into the main E and F regions, the F region bifurcates into the F_1 and F_2 layers. The photochemistry of the D region will be discussed in Section II,D.

The F_2 layer occurs in a regime where the rate-limiting steps are reactions (62) and (63). In this regime the O^+ and electron densities $[N_e]$ are approximately equal, so under photochemical equilibrium conditions we can write

$$[N_e] = \frac{q}{k_1[O_2] + k_2[N_2]}, \tag{77}$$

or

$$q = \beta[N_e]. \tag{78}$$

The actual peak of the layer is not photochemically produced, however, because Eq. (77) yields an electron density profile that increases indefinitely with height. In order to apply the chemistry discussed in the previous section to the F region, it is necessary to include the effects of transport of ionization, primarily by diffusion, on the electron and ion densities. The continuity

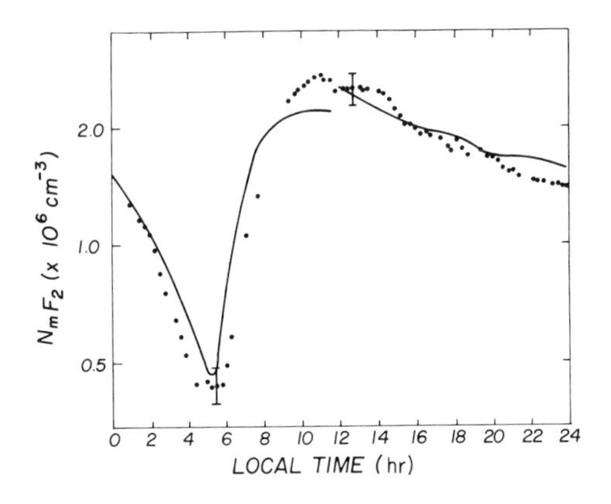

Fig. 24. Diurnal variation of maximum electron density observed at (···) and calculated for (——) Arecibo for equinox near solar maximum, 22 March 1981. From the *Journal of Geophysical Research*.

equations for ions and electrons is therefore formulated

$$\partial \mathrm{N}_i/\partial t = q_i - L_i - \nabla(\phi_i), \tag{79}$$

where q_i is the production rate, L_i the loss rate, and ϕ_i the flux of the ith constituent; $\phi_i = n_i\mathbf{v}_i$, where $\mathbf{v}_i$ is the ith constituent flow velocity and $\nabla(\phi_i)$ the divergence of the flux. A discussion of the formulation of the diffusion equation lies outside the scope of this work.

During the day, the peak of the F_2 layer occurs where transport, production, and loss processes are of comparable importance. Diffusion results in the formation of the F_2 peak at a height where the O^+ loss rate equals the diffusion rate. The formation of the F_1 layer occurs where the rate-limiting step is in transition from ion molecule reactions to dissociative recombination. When the latter becomes the rate-limiting step, a quadratic dependence of electron density on the production rate is obtained:

$$q = (\alpha_1[O_2{}^+] + \alpha_2[NO^+] + \alpha_3[N_2{}^+])[N_e]. \tag{80}$$

To illustrate the quadratic character of the dependence, we assume $\alpha_1 = \alpha_2 = \alpha_3 = \alpha$. Using the fact that in the quadratic regime O^+ is destroyed rapidly in reactions with O_2 and N_2 to form the molecular ions $O_2{}^+$, NO^+, and $N_2{}^+$, we can write

$$[N_e] \simeq [O_2{}^+] + [NO^+] + [N_2{}^+] = \frac{q(O^+) + q(O_2{}^+) + q(N_2{}^+)}{\alpha_1[O_2{}^+] + \alpha_2[NO^+] + \alpha_3[N_2{}^+]}; \tag{81}$$

hence $[N_e] \simeq \sqrt{q/\alpha}$, where $q = q(O^+) + q(O_2{}^+) + q(N_2{}^+)$.

The peak of the F_1 layer typically occurs near 150 km, and that of the F_2 between about 200 and 400 km.

The E layer peak lies between about 90 and 120 km. It is composed largely of $O_2{}^+$, formed by photoionization of O_2 by Lyβ, 977.62-Å C(III), soft X rays, and charge exchange of $N_2{}^+$ with O_2, which becomes important below ~140 km. When the NO density is large, the $O_2{}^+$ can be converted to NO^+.

4. Photochemistry of Metastable Ions

a. $O^+(^2D)$. Metastable $O^+(^2D)$ are produced by photoionization of O. The branching ratio to the ^2D state is ~34.5%. Additional minor sources include the deactivation of $O^+(^2P)$ by radiative decay and electron quenching and photoelectron impact ionization of O. The main loss processes are charge exchange with N_2 and quenching by electrons. Results from the *Atmosphere Explorer* program precluded quenching by O as a major loss process. Although the rate coefficient for quenching by O_2 is large (~10^{-9} cm^3 sec^{-1}), because the O_2 density is generally small this is not a significant loss process. Table A-9 summarizes these processes and the corresponding rate coefficients.

b. $O^+(^2P)$. Metastable $O^+(^2P)$ are produced by photoionization and photoelectron impact ionization of O. The branching ratio is ~20.2% and is not significantly different for photons and electrons. The loss processes are radiative loss and quenching by O, N_2, and electrons. Values of rate coefficients are given in Table A-9.

c. $N^+(^1D)$. The only data available on the photochemistry of metastable N^+ are based on ground-based measurements of the twilight decay of emissions. The primary source inferred is photodissociative ionization of N_2, with a small component due to photoelectrons,

$$N_2 + h\nu \xrightarrow{\beta_{^1D} J_8} N^+(^1D) + N + e^- \tag{92}$$

where the branching ratio for the production of the ^1D state, $\beta_{^1D}$, is ~10% of that for total N^+ production by photodissociative ionization. There is an additional source due to radiative decay of $N^+(^1S)$. The only significant loss mechanism is believed to be electron quenching. The photochemistry and rate coefficients are summarized in Table A-10.

d. $N^+(^1S)$. The production and loss mechanisms of $N^+(^1S)$ are believed to be photodissociative ionization of N_2 and radiative loss via decay to the ^1D state. The branching ratio for the production of ^1S ions is ~0.3% of that for the total production of N^+ by photodissociative ionization. There is also a small contribution due to ionization by photoelectrons. Table A-10 lists reactions and rate coefficients.

e. $N^+(^5S)$. $N^+(^5S)$ has been identified as the probable source of a major auroral emission at ~2150 Å, which was mistaken for several years as being due to NO γ bands. The radiative lifetime has been theoretically calculated to be ~3 msec, in agreement with laboratory measurements. The mid-latitude thermospheric source is thought to be dissociative ionization and excitation of N_2. The production of $N^+(^5S)$ was predicted to be significant in the daytime ionosphere, and this was subsequently confirmed by a rocket measurement of the emission at 2148.3 Å, although the observed value was about one-fourth the theoretical value.

The branching ratio for the production of N^+ via dissociative ionization of N_2 is 0.24. Of this ~10% is accounted for in the production of $N^+(^1D)$ and $N^+(^1S)$. The production of $N^+(^5S)$ accounts for another ~80% if the removal of each $2s\sigma_g$ orbital from the ground state of N_2 results in an $N^+(^5S_2^0)$, which means that ~10% of the N^+ are produced in the ground state, $N^+(^3P)$.

f. $NO^+(a\,^3\Sigma_g^+)$. Several reactions that result in the destruction of $NO^+(a\,^3\Sigma_g^+)$ have been measured in the laboratory. The radiative lifetime has been computed to be about 0.1–1 sec, although the uncertainties are large. Estimates have been made of likely sources, and the primary sources of

$NO^+(a\,^3\Sigma_g^+)$ are believed to be the reactions

$$N_2^+ + NO \xrightarrow{k_{25}} NO^+(a\,^3\Sigma_g^+) + N_2 \tag{100}$$

$$N^+ + O_2 \xrightarrow{k_{26}} NO^+(a\,^3\Sigma_g^+) + O \tag{101}$$

The dominant loss processes are the reactions of $NO^+(a\,^3\Sigma_g^+)$ with N_2 and O. This species probably only plays a minor role in the overall chemistry of the ionosphere. Table A-11 lists the above and additional reactions and rate coefficients.

g. $O_2^+(a^4\Pi_u)$. Approximately 40 to 70% of O_2^+ is produced in the $^4\pi$ metastable state during photoionization. The main loss mechanisms are radiative decay and charge exchange with quenching by N_2. Rate coefficients for charge exchange with O_2 and NO are large, but these reactions do not affect the density of the species significantly above 100 km altitude. Table A-11 lists reactions and rate coefficients.

5. Vibrationally Excited Ions

a. N_2^{+*}. The sources and sinks of N_2^+ vibrational excitation are reasonably well understood. The photoionization process itself will result in the creation of vibrationally excited N_2^+ directly through the production of $N_2^+(X)$ in levels for which $v > 0$. In addition to this there is significant production of A and B state ions in high-lying vibrational levels, which cascade down to the X state, populating the $v > 0$ vibrational levels. The main source of N_2^+ vibrational excitation in the daytime thermosphere, however, is resonance fluorescence of ground-state N_2^+ by pumping of the A and B states by absorption of solar radiation in the visible and near IR in the first negative and Meinel bands, respectively, as suggested by Bates (1949). Again, downward transitions to the X state result in a significant source of vibrational excitation.

The reaction of $O^+(^2D)$ with N_2 has also been suggested as a source of vibrational excitation. The products of the reaction are not well known. There are several possibilities:

$$O^+(^2D) + N_2(X\,^1\Sigma_g^+) \xrightarrow{k_{27}} O(^3P) + N_2^+(A\,^2\Pi_u)_{(v=0,1)} \tag{114}$$

followed by

$$N_2^+(A\,^2\Pi_u)_{(v=0,1)} \longrightarrow N_2^+(X)_{(v\geq 0)} + N_2^+ \text{ Meinel bands} \tag{115}$$

There is also the direct excitation process

$$O^+(^2D) + N_2(X\,^1\Sigma_g^+) \xrightarrow{k_{28}} O(^3P) + N_2^+(X\,^2\Sigma_g^+)_{(v\geq 0)} \tag{116}$$

Both of these reactions have certain channels that are accidentally energetically resonant. Reaction (114) is accidentally resonant if the product is formed in the $v = 1$ level. Reaction (116) is accidentally energetically resonant if the product is formed in the $v = 5$ level. In the case of the latter, the ground-state ions will be sufficiently long-lived to participate in subsequent chemical reactions. It has been suggested by Abdou *et al.* (1984) on the basis of aeronomical studies that the reverse of reaction (116) may account for a detectable transfer of ionization from N_2^+ to O^+. On the other hand, if reaction (114) is important one might expect to see a detectable enhancement of the N_2^+ Meinel bands.

The vibrationally excited N_2^+ are rapidly quenched in collisions with N_2. The details of the quenching process, however, are not known. Laboratory measurements have been made for $v \lesssim 3$ and have confirmed aeronomical estimates that the quenching rate is large, $\sim 5 \times 10^{-10}$ cm^3 sec^{-1}. Whether quenching occurs via a simple charge exchange process of the type

$$N_2^{+*} + N_2 \xrightarrow{k_{29}} N_2^* + N_2^+ \tag{117}$$

or whether the quenching occurs through a series of vibration–vibration exchange processes, with cascade down the vibrational ladder is not yet known. The latter process results in a Boltzmann distribution that can be characterized by an effective vibrational temperature, the magnitude of which depends on competition between processes that destroy N_2^+ and resonance fluorescence. When the latter dominate, vibrational temperatures of ~ 4500 K result, which represent radiative equilibrium with sunlight. Typical experimental temperatures in the F_2 layer (Broadfoot, 1971) lie between 2000 and 3000 K. As discussed in Section II,C,2, it is thought that the presence of vibrationally excited N_2^+ significantly affects the ion chemistry through an enhanced rate coefficient for charge exchange with O. Calculations of k_{24} (for $N_2^{+*} + O$) are discussed in Section II,C,9.

b. O_2^{+*}. The sources and sinks of O_2^+ vibrational excitation have not been studied in as much detail as those of N_2^+. However, it is expected that the basic photochemical schemes will be similar. As in the case of N_2^+, photoionization of O_2 should produce vibrationally excited O_2^+ via direct population of vibrational levels in the ground state, as well as via cascade from higher lying electronic states, such as

$$O_2 + h\nu \longrightarrow O_2^+(\mathrm{a,A,b}) + e^- \tag{118}$$

$$O_2 + e^- \longrightarrow O_2^+(\mathrm{a,A,b}) + 2\,e^- \tag{119}$$

followed, for example, by

$$O_2^+(\mathrm{A}) \longrightarrow O_2^+(\mathrm{X})_{(v>0)} + h\nu \tag{120}$$

Several reactions may also produce vibrationally excited O_2^+. For example, any one of the reactions

$$O^+(^4S) + O_2 \longrightarrow O_2^+ + O + 1.5\ \text{eV} \tag{121}$$

$$O^+(^2D) + O_2 \longrightarrow O_2^+ + O + 4.8\ \text{eV} \tag{122}$$

$$N^+ + O_2 \longrightarrow O_2^+ + N + 2.5\ \text{eV} \tag{123}$$

$$N_2^+ + O_2 \longrightarrow O_2^+ + N_2 + 3.5\ \text{eV} \tag{124}$$

could result in significant O_2^+ vibrational excitation. Since O_2^+, like N_2^+, is a homopolar molecule with long vibrational radiative lifetimes, it could react photochemically. It has been argued theoretically that collisional deactivation of O_2^{+*} would preclude the possibility of enhanced O_2^+ vibrational excitation in the ionosphere. Specifically it is thought that the atom interchange vibration–translation energy-transfer reaction

$$O_2^{+*} + O \overset{k_{30}}{} O_2^+ + O \tag{125}$$

will be rapid. In general, laboratory studies have indicated strong collisional quenching of vibrationally excited species. It appears, based on both aeronomic and laboratory data, that O_2^{+*} may be an exception. Produced vibrationally hot in many laboratory plasmas, O_2^+ does not appear to be collisionally deactivated rapidly. Studies of the terrestrial 5577-Å green line, which arises from the transition $O(^3P\text{–}^1S)$, and aeronomical studies of O_2^+ dissociative recombination have provided evidence of vibrationally excited O_2^+ in the ionosphere. The vibrational excitation appears to affect the products that arise from dissociative recombination. This is discussed further in Section II,E,2. The effect of O_2^+ vibrational excitation on the total recombination rate α_1 has not yet been established. Aeronomical studies suggest that the rate coefficient might decrease with increasing vibrational excitation. Earlier laboratory measurements failed, however, to reveal any significant dependence of the total dissociative recombination rate coefficient on vibrational temperature. This could be due to the presence of large unidentified sources of vibrational excitation in laboratory plasmas producing a hot steady state population, thereby masking the effect of minor controlled perturbations to the source functions by the experimenter.

This subject constitutes an area where much useful work remains to be done.

6. Minor Ions

a. $N^+(^3P)$. The photochemistry of ground-state N^+ proved to be problematical for several years owing to the need for a significant unidentified source mechanism in the F region. At present the most likely candidate for the missing source is the reaction

$$O^+ + N(^2D) \xrightarrow{k_{31}} N^+ + O \qquad (126)$$

At altitudes below ~200 km, photodissociative ionization of N_2 is important:

$$N_2 + h\nu \xrightarrow{\beta_{^3P} J_8} N + N^+(^3P) + e^- \qquad (127)$$

where $\beta_{^3P} \simeq 0.024$ (see Section II,C,4,c), and

$$N_2 + e^- \longrightarrow N + N^+(^3P) + 2e^- \qquad (128)$$

There are several potential additional minor sources, whose validities have not yet been firmly established. These include

$$O^+(^2P) + N_2 \xrightarrow{k_{32}} N^+ + NO \qquad (129)$$

$$O^+(^2D) + N \xrightarrow{k_{33}} N^+ + O \qquad (130)$$

$$N + h\nu \xrightarrow{J_9} N^+ + e^- \qquad (131)$$

$$He^+ + N_2 \xrightarrow{k_{34}} He + N + N^+ \qquad (132)$$

The main loss process is the reaction of N^+ with O_2:

$$N^+ + O_2 \xrightarrow{k_{35}} NO^+ + O \qquad (133)$$

$$N^+ + O_2 \xrightarrow{k_{36a}} O_2^+ + N \qquad (134)$$

$$N^+ + O_2 \xrightarrow{k_{36b}} O^+ + NO \qquad (135)$$

At altitudes above 300 km the reaction

$$N^+ + O \xrightarrow{k_{37}} O^+ + N \qquad (136)$$

becomes important. Table A-12 lists rate coefficients for these reactions. The branching ratio for the production of $N^+(^3P)$ via reaction (127) is ~2%.

b. He⁺. The primary source of He^+ is photoionization of He:

$$He + h\nu \xrightarrow{J_{10}} He^+ + e^- \qquad (137)$$

The photoionization cross section has been measured in the laboratory and accurately calculated by several different methods. The value of J_{10} is ~6×10^{-8} sec^{-1}. The photoelectron source is always close to 10% of the extreme UV source.

The primary sink for He^+ is reaction with N_2:

$$He^+ + N_2 \xrightarrow{k_{38}} N_2^+ + He \qquad (138)$$

$$He^+ + N_2 \xrightarrow{k_{34}} N + N^+ + He \qquad (139)$$

The charge exchange reaction with O,

$$He^+ + O \xrightarrow{k_{35}} He + O^+ \qquad (140)$$

is potentially a minor high-altitude sink for He^+, but the rate coefficient is not known.

c. O^{2+}. Divalent oxygen ions (O^{2+}) are important ionospheric constituents in the topside ionosphere. The primary production rates are believed to be photoionization of O^+ [reaction (141)], double photoionization of O [reaction (142)], and Auger following K-shell ionization of O by X rays [reaction (143)]:

$$O^+ + h\nu \xrightarrow{J_{11}} O^{2+} + e^- \qquad \lambda < 351 \text{ Å} \tag{141}$$

$$O + h\nu \xrightarrow{J_{12}} O^{2+} + e^- \qquad \lambda < 254 \text{ Å} \tag{142}$$

$$O + h\nu \xrightarrow{J_{13}} O^{2+} + e^- \qquad \lambda < 23 \text{ Å} \tag{143}$$

where $J_{11} = 8.7 \times 10^{-8}$ sec^{-1}, $J_{12} = 2 \times 10^{-9}$ sec^{-1}, and $J_{13} = \sim 10^{-11}$ sec^{-1}. The primary loss process was established by laboratory measurements to be a fast reaction with N_2,

$$O^{2+} + N_2 \xrightarrow{k_{41}} \text{products} \tag{144}$$

It appears on the basis of aeronomical results that O^{2+} is also destroyed in reactions with O,

$$O^{2+} + O \xrightarrow{k_{42}} 2\,O^+ \tag{147}$$

where from terrestial studies $k_{42} \simeq 6.5 \times 10^{-11}$ cm^3 sec^{-1} and from planetary work $k_{42} \simeq 2 \times 10^{-10}$ cm^3 sec^{-1}.

d. H^+. Atomic hydrogen ions (H^+) usually form the major constituent of the upper topside ionosphere and the plasmasphere. At altitudes below ~500 km, H^+ is in photochemical equilibrium. The concentration is determined almost entirely by the accidentally energetically resonant reaction

$$O^+(^4S) + H(^2S) \underset{k_{43r}}{\overset{k_{43f}}{\rightleftarrows}} H^+(1p) + O(^3P_J) + \Delta\epsilon \tag{148}$$

The energy defect $\Delta\epsilon$ is -0.00833, 0.00000, and 0.01965, respectively, for $J = 0$, 1, and 2. The ratio of the forward to reverse rate coefficients would be equal to the thermodynamic equilibrium constant if the energy defect $\Delta\epsilon$ were zero. The former is equal to the product of the product statistical weights divided by reactant statistical weights. The ground-state degeneracies for O^+, H, and H^+ are 4, 2, and 1, respectively. For $O(^3P)$ the degeneracy is given by $(2S + 1)(2L + 1) = 9$ at high temperatures. The spin orbit splitting of the ground-state levels of $O(^3P)$ is not necessarily large compared with kT, however, and the effect of the energy defect must be included, yielding

$$\frac{k_{43r}}{k_{43f}} = 8 \Bigg/ \left[\sum_{J=1}^{3} (2J + 1)\right] \exp\left(\frac{\epsilon_J - \Delta\epsilon}{kT}\right) = \frac{8}{9} \qquad \text{for large } T, \tag{149}$$

where the ϵ_J are the energies of the J levels. The forward rate coefficient has been measured at ~300 K as 3.7×10^{-10} cm^3 sec^{-1}. When the ratio of the statistical weights is taken into account, we can write the following expression relating k_{43f} and k_{43r}: $k_{43r} = \frac{8}{9}k_{43f}$.

Above ~500 km the H^+ concentration is determined by plasma transport processes and the photochemical boundary conditions at ~500 km.

7. Magnetospheric Ions

Photochemistry plays a relatively minor role in determining the ionic composition of the magnetosphere and will therefore receive only scant attention here. Defining the magnetospheric composition constitutes a vast area of research. The populations of ions of different mass and charge are described mainly by their energy characteristics in the magnetosphere, which vary greatly spatially and temporally. A key factor determining the population characteristics is the state of the magnetic field lines, whether they are open or closed. The closed field line region is the plasmasphere, where the particles are trapped by the displaced dipole-like geomagnetic field. The population energies range from thermal to several hundred mega–electron volts.

Within the plasmasphere are located the well-known inner and outer van Allen radiation belts (see Fig. 5). The lifetimes of trapped particles is determined largely by the frequency of perturbations to the magnetic field and by charge exchange collisions with neutral particles. Until recently it was thought that the energetic population of the plasmasphere was largely protons. It is now known that He^+ and O^+ can at times become the dominant constituents. In addition, N^+, N^{2+}, and O^{2+} have been observed. Reports have also been made of the detection of He^{2+} and D^+.

The only chemical reactions of significance in the magnetosphere occur in the topside ionosphere or protonosphere, at heights where the neutral densities are large enough to result in effective loss of plasmaspheric ionization. These reactions include, for example,

$$H^+ + H \longrightarrow H^* + H^+ \tag{150}$$

$$O^+ + H \longrightarrow H^+ + O \tag{151}$$

$$H^+ + O \longrightarrow O^+ + H \tag{152}$$

$$O^+ + O \longrightarrow O^* + O^+ \tag{153}$$

$$He^+ + H \longrightarrow He^* + H^+ \tag{154}$$

$$O^+ + H \longrightarrow H + O^{2+} + e^- \tag{155}$$

$$O^{2+} + H \longrightarrow O^+ + H^+ \tag{156}$$

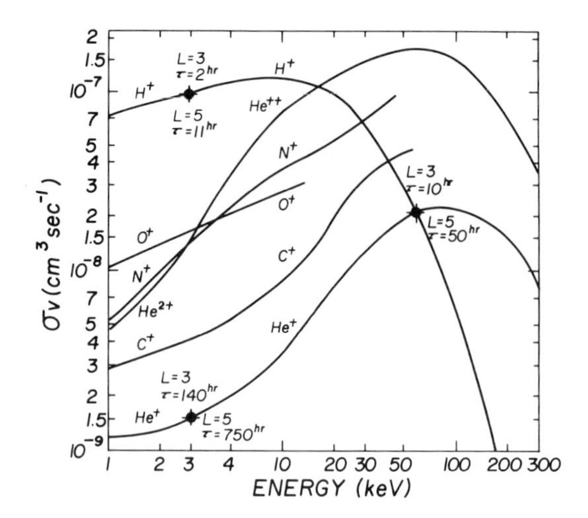

Fig. 25. Reaction rates (product of charge exchange cross section with thermal neutral hydrogen times ion velocity) for several ion species as a function of their energy. Lifetimes are shown for populations with isotropic pitch angles on L shells listed. From Tinsley (1981), with permission of Pergamon Press.

Figure 25 shows the rates for charge exchange of several species with neutral hydrogen. The asterisks indicate products that may be energetically hot. These hot atoms may have sufficient energy to escape the gravitational field of the Earth, or they may precipitate into the Earth's atmosphere. Precipitation of energetic O^+ has been observed directly through the loss cone during magnetic storms. Theoretical calculations have shown that precipitating neutral atoms can give rise to significant heating (primarily through momentum transfer) of the F region with significant dynamic effects above 300 km.

Also associated with the precipitation of kilo–electron volt O^+, which rapidly charge exchange to form energetic oxygen atoms, is a large backscattered flux of atoms in the 1- to 10-eV energy range. A significant fraction of these atoms reach escape velocity, and it has been estimated that the escape from the terrestial atmosphere could affect the total atmospheric oxygen budget over the lifetime of the Earth. Mechanisms that energize O^+ in the magnetosphere are not yet well understood. In the plasmasphere it appears that the primary source of thermal ions is outward flow from the ionosphere. Doubly charged ions in particular are transported rapidly by thermal diffusion. Recent calculations indicate that the expansion of thermal ionospheric plasma into flux tubes that have just been emptied results in the acceleration of ions to energies of the order of tens of electron volts.

8. The Escape of Gases from the Terrestrial Atmosphere

a. Jeans Escape. The Jeans theory of escape is based on the fact that the energetic component of the Maxwellian tail has sufficient energies to attain escape velocity. The average escape flux computed from the Jeans mechanism is only significant for atomic hydrogen, and it varies, for example, from about 3×10^7 to 1×10^8 as the exospheric temperature increases from 800 to 1300 K.

b. Charge Exchange of H^+ and H. Observations of the hydrogen escape flux via Lyα emissions indicated significantly larger values for the flux than is expected for Jeans escape. The additional source of flux was identified as the charge exchange mechanism

$$H^+ + H \longrightarrow H^* + H^+ \tag{157}$$

Analysis of satellite and incoherent scatter radar data yielded escape fluxes ranging between 0.5 and 6×10^8 atoms $cm^{-2}\ sec^{-1}$. Information has been acquired on the diurnal, seasonal solar cyclic, and magnetic activity variations in the charge exchange escape flux. The flux was found to have a maximum near noon and a minimum in the early morning hours.

c. Polar Wind. *Explorer 31* observations revealed upward fluxes of light ions of $\sim 5 \times 10^7\ cm^{-2}\ sec^{-1}$ in the polar regions. The observations confirmed earlier theoretical predictions of an outward flow of ionization into the tail. The source of H^+ is the charge exchange reaction

$$O^+ + H \longrightarrow H^+ + O \tag{158}$$

which determines the maximum H^+ flux. Solutions of the coupled continuity, momentum, and energy equations have yielded fluxes of $\sim 10^6\ cm^{-2}\ sec^{-1}$. Inclusion of electric fields, however, which result in Joule heating effects, increased the flow limits to about $2 \times 10^7\ cm^{-2}\ sec^{-1}$ in reasonable agreement with observations.

Similar calculations done for He^+ yielded escape fluxes of $\sim 5 \times 10^6\ cm^{-2}\ sec^{-1}$.

In the case of hydrogen, atoms that escape from the thermosphere must be replaced by upward flux from the lower atmosphere, which is determined by upward transport by eddy diffusion. Thus the step that limits the hydrogen escape flux could be the exospheric temperature, if it is low, or the diffusion rate through the lower atmosphere. Detailed calculations, which take into account the hydrogen chemistry, have yielded a limiting flux of $\sim 2 \times 10^7\ cm^{-2}\ sec^{-1}$ per part per million of stratospheric total hydrogen, which yields a net flux of between 1.0 and $3.1 \times 10^8\ cm^{-1}\ sec^{-1}$, in agreement with observed escape rates. This result is not very sensitive to the eddy diffusion coefficient.

d. Energization by Energetic Particle Precipitation. As was pointed out in Section II,C,7. theoretical calculations of the effects of the precipitation of kilo–electron volt heavy particles have indicated that thermalization of these particles results in a "backsplash" of large fluxes ($\geqslant 10$ escaping atoms per incident kilo–electron volt) of 1- to 10-eV neutral atoms. Such precipitation occurs predominantly poleward of the plasmapause during magnetically disturbed periods. The mean O escape rate due to this mechanism is estimated to be comparable with the H escape rate.

9. Ionospheric Rate Coefficients

Several of the rate coefficients discussed in the preceding sections exhibit temperature dependences that are sufficiently pronounced to affect ion composition.

a. Dissociative Recombination. The dissociative recombination rate coefficients of O_2^+ and N_2^+, which are given in Appendix II, vary by factors of about 2 and 2.5, respectively, for a change in electron temperature from 300 to 2000 K. There is some uncertainty about possible dependences on vibrational excitation. In the case of N_2^+, aeronomical studies have indicated that if α_3 has a vibrational temperature dependence of the form $\alpha_3 = 2.2 \times 10^{-7} (T_v/T_e)^{-0.4}$, then the need for reaction (76) effectively disappears. In the case of O_2^+, it appears at this stage that vibrational temperature effects are insignificant for the total recombination rate. The dissociative recombination rate coefficient of NO^+ (α_2), on the other hand, appears to have a sensitive dependence on electron temperature: $T_e^{-0.8}$ according to aeronomic studies. This results in a decrease by almost a factor of 5 in α_2 for a variation in T_e from 300 to 2000 K. Ionospheric effects of these electron temperature dependences have been observed at high latitudes in regions of plasma depletions, which are characterized by high electron temperatures. In the case of O_2^+, dissociative recombination decreases to a level where competitive loss processes such as $O_2^+ + N$ or $O_2^+ + NO$ dominate, with the result that the O_2^+ concentration shows an insensitivity to further changes in T_e. On the other hand, NO^+ is lost only by dissociative recombination, and its concentration shows a sensitive dependence on T_e under the same conditions. Figure 26 shows the temperature dependence derived for α_2 from aeronomic data. These results are of historical interest because they represent the first quantitative aeronomical determination of the temperature dependence of a rate coefficient. The result also resolved a conflict as to which laboratory measurements of α_2 should be used in ionospheric studies.

There is no laboratory evidence at present to suggest that the total dissociative recombination rates discussed above show any significant dependence on vibrational excitation. However, there is some concern that

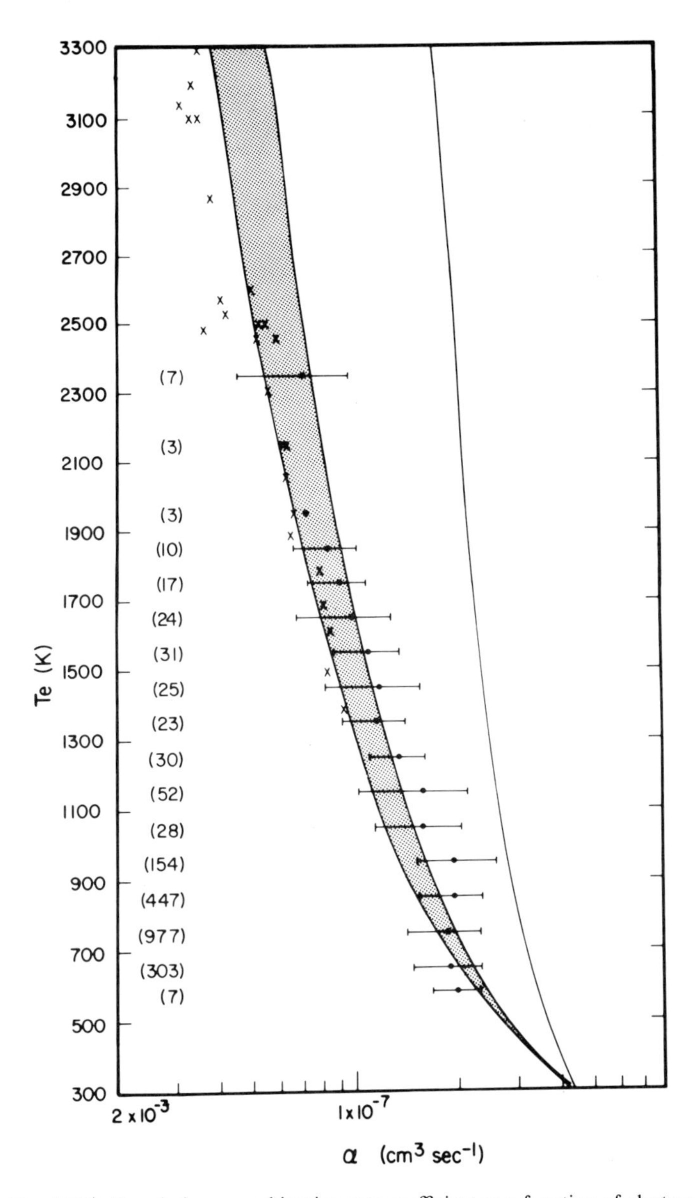

Fig. 26. NO^+ dissociative recombination rate coefficient as a function of electron temperature deduced from the *Atmosphere Explorer* data. The symbols represent results from different measurement periods. The stippled area shows the JILA results, and the line those of the Pittsburgh laboratory. The satellite results represent the first quantitative aeronomical determination of the temperature dependence of a rate coefficient. These aeronomical results decided which of the two laboratory experimental results were applicable to the ionosphere. From the *Journal of Geophysical Research.*

since laboratory plasmas exhibit unexpectedly high vibrational temperatures, processes might exist that effectively mask the sensitivity of these experiments to vibrational temperature dependences.

b. Ion–Molecule and Ion–Atom Reactions. The $O^+ + N_2$ reaction, which is of importance in the F_2 layer, exhibits strong dependences on both kinetic and vibrational temperature. Laboratory studies of the vibrational temperature dependence by Schmeltekopf *et al.* (1968) yielded the results shown in Fig. 27.

Measurements of the kinetic temperature dependence of k_{13} proved more difficult to make for direct ionospheric application, because typical ionospheric temperatures cannot be easily realized in the laboratory. High ion temperatures are therefore simulated in a flow–drift tube apparatus in which ions are accelerated to large velocities to simulate large temperatures. Although such measurements have proved to be valuable, caution is required in applying the temperature dependences directly to ionospheric problems because the ion velocity distributions under which the laboratory measurements are made will generally differ from the distributions that occur in the ionosphere. The rate coefficient for a chemical reaction is defined by

$$k = \int \sigma(E) f(v)\, dv, \tag{159}$$

where $\sigma(E)$ represents the cross section at relative energy E of the reactants and $f(v)$ their relative velocity distribution. Laboratory data have therefore been used to derive cross sections as a function of energy. These cross sections

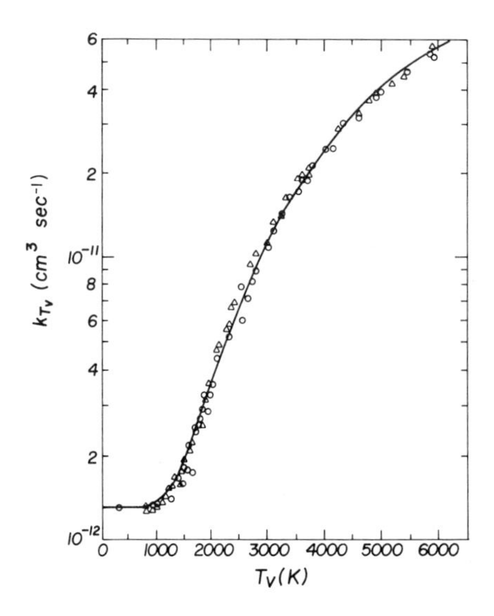

Fig. 27. Reaction rate constant k_T as a function of N_2 vibrational temperature for the reaction $O^+ + N_2 \rightarrow NO^+ + N$. From Schmeltekopf *et al.* (1968), with permission of the American Institute of Physics.

have then been folded with velocity distributions characteristic of the ionosphere to determine an appropriate expression for the nonthermal temperature dependence of k_{13} in the ionosphere.

At high latitudes the ionosphere is characterized by convection electric fields, which give rise to ion drift velocities equivalent to several thousand kelvins. It was theoretically predicted and experimentally observed that under these circumstances $[NO^+]$ becomes enhanced and at times is the dominant ionospheric constituent due to the enhanced conversion of O^+ to NO^+ via the $O^+ + N_2$ reaction.

Aeronomical studies have suggested a strong dependence of k_{24} of the $N_2^{+*} + O$ reaction [reaction (76)] on vibrational excitation, where k_{24} is given by the weighted sum of the rate coefficients for reaction of each of the $N_2^+(X)_{(v>0)}$ levels with O,

$$k_{24} = \frac{\sum k_i[N_2^+(X)_{(v=i)}]}{\sum[N_2^+(X)_{(v=i)}]}, \tag{160}$$

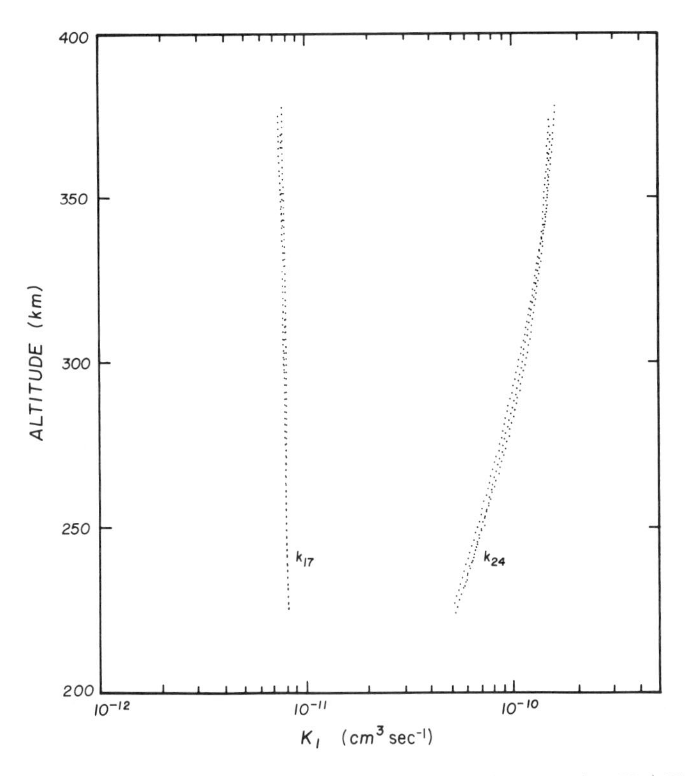

Fig. 28. Model calculations of k_{24}, the rate coefficient of the reaction $N_2^+(X)_v + O \rightarrow O^+(^4S) + N_2$, as a function of altitude. The model calculation of k_{17} for $N_2^+(X)_{(v=0)}$ is shown for comparison.

where subscript i represents the ith vibrational level. This dependence on vibrational temperature can give rise to interesting seasonal variations in ion composition. When $[N_2]$ is high, as is the case in summer at solar maximum, N_2^{+*} is quenched strongly reducing the summer production of O^+ via reaction (76). The role that this particular mechanism plays in determining the seasonal behavior of the F_2 layer has yet to be quantitatively established. To illustrate the large effect that vibrational excitation appears to have on this rate coefficient, in Fig. 28 a comparison of k_{17} $(N_2^+ + O)$ with k_{24} $(N_2^{+*} + O)$ as a function of height is shown. The large increase in k_{24} is not expected on theoretical grounds, which is a cause for concern, and suggests that problems may still exist with the F region ion chemistry.

D. The D Region of the Ionosphere

1. Observations

Our knowledge of D region ion chemistry is not as satisfactory as that of the E and F regions, although considerable progress has been made in recent years. The basic photochemistry is more complex, and the D region is not readily accessible to *in situ* satellite meaurements. Hence the available data base is smaller than for the F region.

During the daytime plasma densities are typically around 10^5 cm^3 sec^{-1} at ~ 100 km and decrease steeply at the bottom of the E layer (85–95 km). Densities usually decrease from about 10^3 cm^{-3} at 85 km to ~ 100 cm^{-3} at ~ 65 km. The concentration of free electrons decreases very rapidly relative to that of positive ions between 70 and 60 km, and negative ions become an important constituent due to the neutral density increase and rapid attachment of electrons to O_2 via three-body reactions. The positive-ion composition changes from a spectrum dominated by NO^+ and O_2^+ to one dominated by proton hydrates of the formula $H^+(H_2O)_n$. The transition height occurs between 70 and 90 km. Nocturnal densities are much less, and the region above 80 km tends to be structured.

Figure 29 shows height profiles of electrons and positive ions. Figure 30 shows ion production rates for day and night.

The primary sources of daytime D region ionization are photoionization of NO by Lyα, of metastable $O_2(a\,^1\Delta_g)$ by wavelengths in the range 1118–1027 Å, and photoionization by hard solar X rays (<10 Å). Cosmic rays become increasingly important at lower altitudes. The quiet nighttime sources are much smaller. These include scattered Lyα geocorona radiation, precipitating energetic electrons, and galactic X-ray sources. There is much uncertainty associated with the most important source, ionization of NO, because of lack of information on NO in the mesosphere. Seasonal and

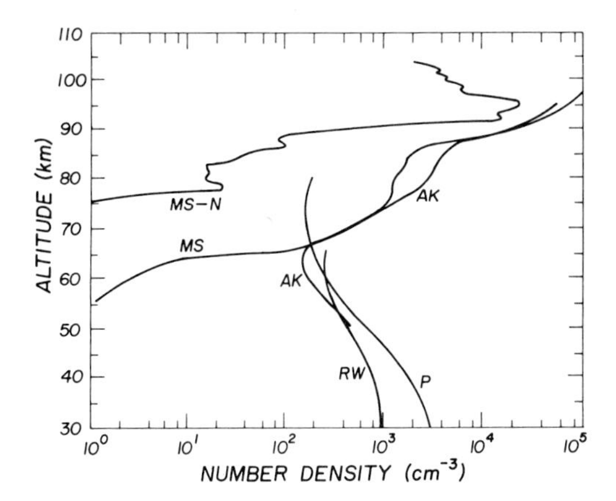

Fig. 29. Number densities of electrons during day (MS) and night (MS-N). Number densities of positive ions during day (AK, P, RW) are also shown. From Arnold and Krankowsky (1977), with permission of D. Reidel Publishing Company.

latitudinal variations in NO range from low densities of $\sim 10^6$ cm^{-3} in summer to up to 10^8 cm^{-3} in winter. Densities as high as 10^9 cm^{-3} have been inferred at high latitudes in winter from positive ion composition measurements. The main products formed by these sources are the ions N_2^+, O_2^+, N^+, O^+, and NO^+.

The primary destruction mechanism is mutual recombination. This occurs via three main processes: dissociative recombination with electrons, electron–ion recombination of complex ions, and two-body ion–ion recombination. The recombination rate coefficient varies considerably for these different processes, and the net effective rate coefficient for recombination of the plasma

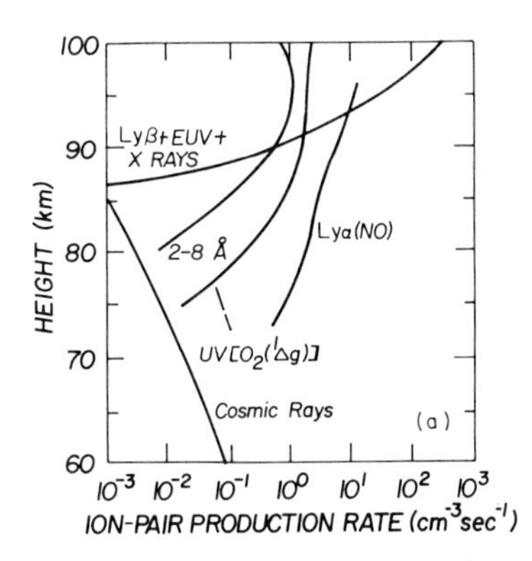

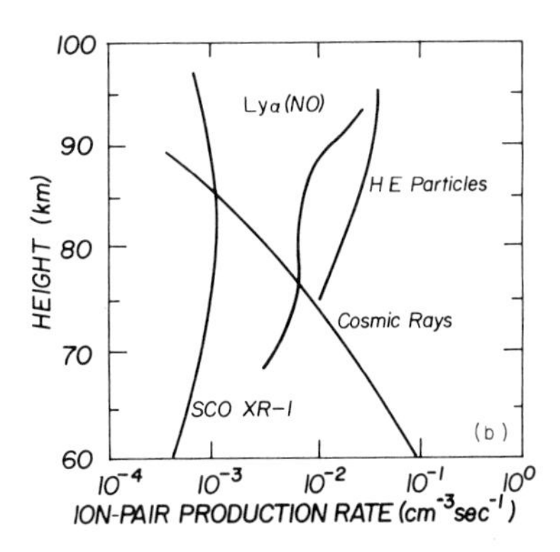

Fig. 30. Daytime (a) and nighttime (b) ionization rates. From Arnold and Krankowsky (1977), with permission of D. Reidel Publishing Company.

can vary by several orders of magnitude depending on which processes dominate. The time constant for recombination is much smaller than that for transport, and, in general, photochemical steady-state conditions prevail. Exceptions to the above scheme are positive atomic ions, which recombine via radiative recombination, a much slower process. These are discussed in Section II,D,3. Also transport processes can become significant at the base of the D region (60 km) where densities are low.

The densities of positively and negatively charged ions therefore are related by the steady-state expression

$$q = \alpha_E[N^+][N^-], \tag{161}$$

where q is the ionization rate and α_E an effective recombination rate coefficient.

The first measurements of positive and negative ions in the D region were made by Narcisi and Bailey in 1965, and in 1971 by Arnold and by Narcisi, respectively. A major finding that emerged from these measurements was the discovery of cluster ions. The cluster ions are mainly $H^+(H_2O)n$, which become dominant between 70 and 90 km. The transition from molecular to cluster ions occurs within a few kilometers. Typically the values for n, which represents the degree of hydration, lie between 2 and 4, although values as large as 20 have been observed under low temperature conditions near the mesopause.

Metallic ions also form an important component of the D region population. These are discussed separately in Section II,D,3. The transition from molecular to cluster ions occurs below ~75 km.

In addition to proton hydrates, other types of cluster species have also been observed. These include NO^+H_2O, NO^+CO_2, $O_4{}^+$, $O_2{}^+H_2O$, NO^+N_2, and others.

It is clear that, excluding the atomic ions, there are two regimes of ions: cluster and molecular. Associated with these species are their respective recombination rates, α_c and α_m. As mentioned above, these are quite different:

$$\begin{aligned} \alpha_m &\simeq 5 \times 10^{-7}\,(T/300)^{1/2} \quad \text{cm}^3\,\text{sec}^{-1}, \\ \alpha_c &\simeq (0.5\text{–}1) \times 10^{-5}\,(T/350) \quad \text{cm}^3\,\text{sec}^{-1}. \end{aligned} \tag{162}$$

This differences results in a sharp drop-off in plasma densities below the boundary.

2. Photochemistry of Cluster Ions

a. Positive Ions. The processes governing the major positive ions $O_2{}^+$, $N_2{}^+$, and NO^+ are the same as for the E region. The conversion of $O_2{}^+$ to NO^+ via charge exchange with NO and the photoionization of NO by Lyα radiation ensures that NO^+ is the major ion in the D region. It is therefore a primary precursor species for the formation of proton hydrates. The processes involving O_2^+ and NO^+ as precursors are relatively well understood, but a few inconsistencies between observations and models remain. Figure 31 summarizes the cluster formation chemistry. The corresponding rate coefficients are given in Table A-13. The two chemical paths shown are referred to as O_2^+ and NO^+ hydration. The essence of the scheme was completed in 1970 (see Ferguson, 1974).

In the O_2 path the first step is a three-body reaction forming O_4^+, which is the rate-limiting step. The time constant for this reaction increases rapidly with height because of the decrease in the densities of O_2. The time constants for the binary reactions are of the order of a second. Table A-13 shows that the

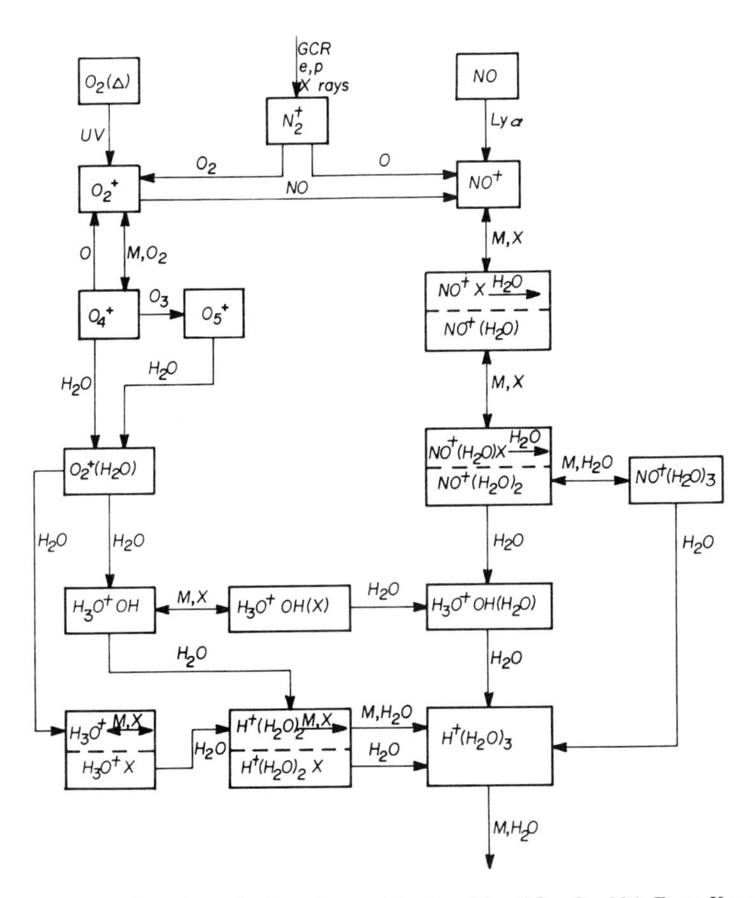

Fig. 31. Reaction scheme for D region positive ions (X = CO_2, O_2, N_2). From Kopp and Hermann (1982), as presented at the EGS Conference on Ions in the Middle Atmosphere, Leeds.

three-body reaction is strongly temperature dependent. Temperature variations in the D region therefore have a significant effect on the formation of cluster ions. Time constants for cluster ion recombination evaluated from *in situ* observations are shown in Fig. 32. The seasonal changes in temperature are largely responsible for the variations shown. Figure 31 shows a channel in which O converts $O_4{}^+$ back to $O_2{}^+$. This reaction lowers the cluster ion boundary and is the main difference between the two reaction paths.

It is possible that processes that may lead to the formation of more stable clusters also exist. For example, an ion interpreted as $O_2{}^+CO_2$ has been observed in the D region.

For the NO^+ path, the first step is clustering of CO_2, O_2, or N_2 to NO^+. Clustering of N_2 is most effective because of its high abundance. There is some uncertainty regarding competition between CO_2 exchange with the intermediate NO^+N_2 and thermal decomposition. (Generally for CO_2 the mole fraction of 3×10^{-4} at ground level is adopted for the D region.) It appears, however, that N_2 clustering will be the dominant process under all likely conditions. Problems with the NO^+ scheme also arise once the $NO^+ \cdot H_2O$ hydrate is formed because subsequent conversion to proton hydrates via a binary reaction with H_2O is slow. There is a need for rapid conversion of $NO^+ \cdot H_2O$ to higher order cluster ions because photolysis of $NO^+ \cdot H_2O$ may be rapid as well. The reactions of $NO^+(H_2O)_2$ to $H_3O^+OH(H_2O)$ and

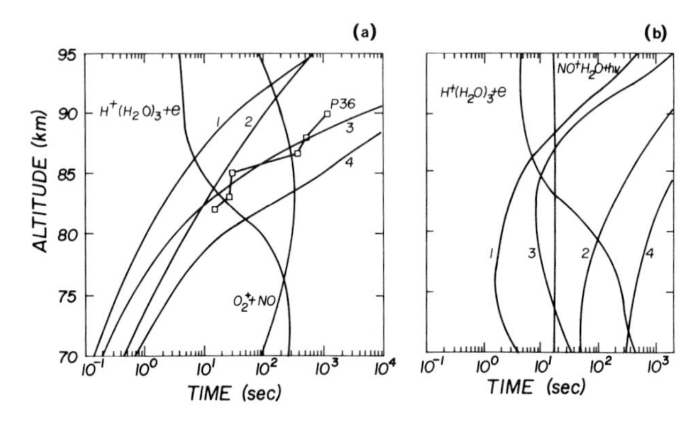

Fig. 32. (a) Time constants for $O_2{}^+$ chemistry. The numbers denote time constant for $O_2{}^+ + O_2 + M = O_4{}^+ + M$ summer (1) and winter (2); Effective time constant for $O_2{}^+ \rightarrow O_2{}^+ \cdot H_2O$ also considering the process $O_4{}^+ + O \rightarrow O_2{}^+ + O_2$, summer (3) and winter (4). Data for [M] and temperature were taken from CIRA (1972). The data points (□) denote effective time constants for $O_2{}^+$ clustering as derived from a D region positive ion composition measurement at high latitudes during summer. (b) Time constants for the NO^+ cluster ion chemistry in the D region. Numbers denote effective time constants for the conversion of NO^+ to $NO^+ \cdot H_2O$, summer (1) and winter (2); temperature enhancement by 20 K at all heights summer (3) and winter (4). From Arnold and Krankowsky (1977), with permission of D. Reidel Publishing Company.

Table II

OBSERVED MASS NUMBER AND ION IDENTITIES[a]

Mass	Tentative ion identity
14	CH_2^+, N^+
19	H^+H_2O
24	Mg^+, Na^+
27	Al^+, H^+HCN, Si^+
30	NO^+
32	O_2^+
37	$H^+(H_2O)_2$
45	Al^+H_2O; $H^+HCN \cdot H_2O$, $SiOH^+$
48	NO^+H_2O
55	$H^+(H_2O)_3$
56	Fe^+
64	O_4^+; $Al^+(H_2O)_2$; $H^+HCN(H_2O)_3$; $SiOH^+(H_2O)$
66	$NO^+(H_2O)_2$
73	$H^+(H_2O)_4$
74	NO^+CO_2
80	$Al^+(H_2O)_2$; $H^+HCN(H_2O)_3$; $SiOH^+(H_2O)_2$
95	$H^+(CH_3CN)(H_2O)_3$
103	$H^+(CH_3CN)_2H_2O$

[a] From Arnold and Viggiano (1982), with permission of Pergamon Press.

$H^+(H_2O)_3$ were proposed to take into account the discrepancy between observed and modeled $NO^+(H_2O)_3$ densities. Despite these problems the above chemical scheme appears to account reasonably well for the basic diurnal, seasonal, and solar cyclic variations observed in the D region.

The D region exhibits two anomalous characteristics in its behavior, which have become well known, namely, the D region winter anomaly and the ledge in plasma density at ~85 km in summer, which is an enhancement in D region plasma density and appears to arise from the temperature dependence of the bulk recombination coefficient. The 85-km ledge is explained by the location of the cluster ion boundary there. The rapid associated decrease in α_{eff} with increasing altitude coincides with a steep increase in ion production in summer. The solar zenith angle in winter generally reduces the solar extreme-UV and X-ray ionization rates, reducing the gradient once the boundary shifts to lower altitudes and removing the ledge. The winter anomaly appears to be associated with enhanced mesospheric NO densities in winter.

Recent improvements in rocket-borne mass spectrometer technology have resulted in the discovery of several new species. Table II lists the latest positive ions observed in the lower ionosphere between about 70 and 80 km. Theories have not been developed yet to account for the new species shown.

b. Response of the D Region to Particle Precipitation. Bombardment of the Earth's atmosphere by energetic charged particles such as solar protons or electrons can result in a significant enhancement of ion production rates in the D region. Figure 33 shows the total ion-pair production computed for the solar particle event of August 1972. During such events, the production of odd-nitrogen and odd-hydrogen increases dramatically, perturbing the ion chemistry discussed in the preceding section. Molecular nitrogen may undergo direct collisional dissociation. There is a net enhancement in odd-hydrogen which can be seen by summing the chain of cluster reactions, allowing for initial conversion of $N_2{}^+$ to $O_2{}^+$, as follows:

$$\begin{aligned}
O_2{}^+ + O_2 + M &\longrightarrow O_2{}^+O_2 + M \\
O_2{}^+O_2 + H_2O &\longrightarrow O_2{}^+H_2O + O_2 \\
O_2{}^+H_2O + H_2O &\longrightarrow H_3O^+OH + O_2 \\
H_3O^+OH + H_2O &\longrightarrow H_3O^+H_2O + OH \\
H_3O^+H_2O + e^- &\longrightarrow 2\,H_2O + H \\
O_2{}^+ + H_2O + e^- &\longrightarrow O_2 + H + OH
\end{aligned} \tag{183}$$

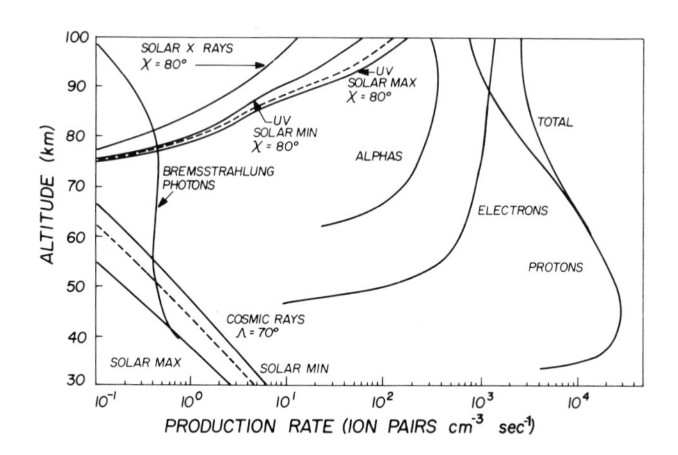

Fig. 33. Contribution of the various solar particle, solar radiations, bremsstrahlung, and cosmic rays to the total ion-pair production in the August 1972 solar particle event. The ionization rates shown were computed by a Lockheed group who used input parameters appropriate for the August 1972 event. Detailed measurements of the solar proton particle characteristics were made by the 1971-089A satellite. Electron production rate profiles were deduced from Chatanika incoherent scatter radar measurements of electron densities. From the *Journal of Geophysical Research.*

Table III

OBSERVED MASS NUMBERS AND ION IDENTITIES[a]

Mass	Tentative ion identity
16	O^-, OH^-
32	O_2^-
35	Cl^-
46	NO_2^-, SiO^- (44)
52	ClO^-; Cl^- (H_2O)(51)
60	NO_3^-, CO_3, HCO_3, SiO_2^- (60)
76	SiO_3^-, CO_4^-
78	60 (H_2O)
82	66 (16)
94	76 (H_2O)
96	78 (H_2O)
105	78 (28) = $Si_2O_3^-$
107	76 (32) = SiO_5^-

[a] From Arnold *et al.* (1982), with permission of Pergamon Press.

This scheme results in an initial increase in odd-hydrogen that causes O_3 to decrease, while the H_2O concentration is reduced. A new H distribution results after a few hours due to modified H_2O concentrations, which cannot be replenished rapidly by upward diffusion. This could affect the hydration process and break the chain, resulting in a decrease in odd-hydrogen and subsequent increase in O_3.

c. **Negative Ions.** Because of the scarcity of mass spectrometer measurements the identities of the negative ions are not firmly established. Different measurements have yielded different species in different altitude regimes. Species found to date are listed in Table III. Recent observations by Arnold *et al.* (1982) have revealed a layer of heavy ions (> 100 amu) between 80 and 90 km. The properties of the layer suggest a meteoric source. The heavy ions were also observed at stratospheric heights. Figure 34 shows a reaction scheme from Wisemberg and Kockarts (1980). (rate coefficients are given in Table A-14). The precursor negative ions in the scheme are O_2^- and O^-, which are formed by three-body attachment and the reverse of reaction (184), respectively. This is followed by rapid conversion to CO_4^- and CO_3^- and then by a slower further conversion to NO_3^- (and Cl^-). There are reversal processes that compete in the destruction of negative ions at various parts of the chain. Primary processes include photolysis, reactions with O or H, and

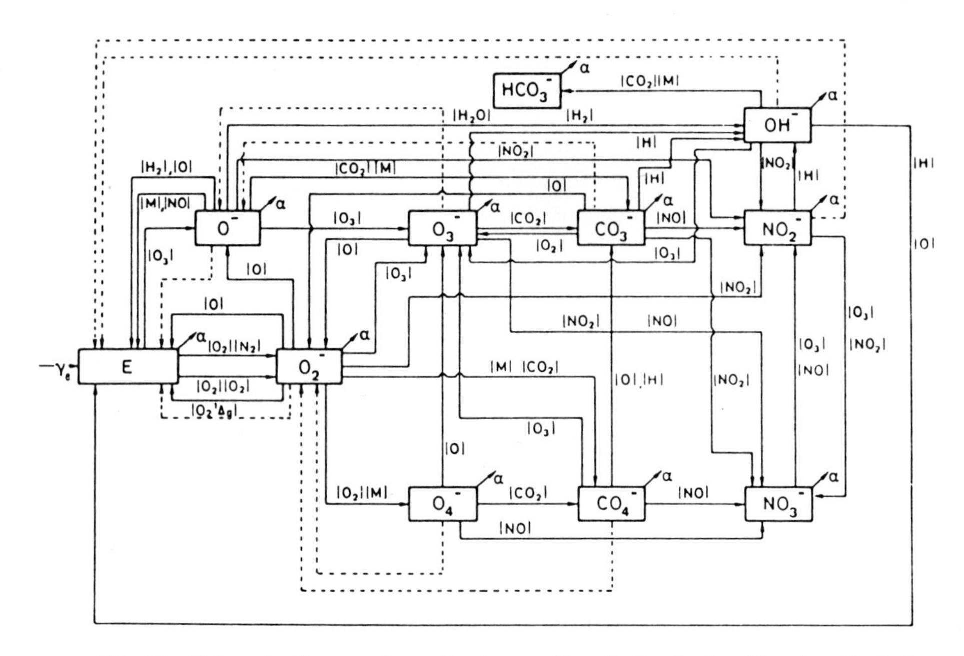

Fig. 34. Schematic diagram of negative ion reactions in the daytime D region. Neutral constituents involved in the reactions are indicated in brackets. Dashed lines correspond to photodestruction processes. Arrows labeled α correspond to dissociative recombination for electrons, and mutual neutralization for negative ions. The external production leading to electrons (E) and to positive ions is indicated by γ_e (cm^{-3} sec^{-1}). From Wisemberg and Kockarts (1980).

recombination with positive ions. Values currently in use for photodetachment rates are the following:

$$
\begin{aligned}
J_{O^-} &= 1.4 \quad \text{sec}^{-1} \\
J_{O_2^-} &= 0.38 \quad \text{sec}^{-1} \\
J_{OH^-} &= 1.1 \quad \text{sec}^{-1} \\
J_{NO_2^-} &= 8 \times 10^{-4} \quad \text{sec}^{-1} \\
J_{CO_3^-}\ (\text{for } O^- \text{ and } CO_2 \text{ products}) &= 0.47 \quad \text{sec}^{-1} \\
J_{CO_4^-}\ (\rightarrow O_2^-) &= 6.2 \times 10^{-3} \quad \text{sec}^{-1} \\
J_{O_4^-} &= 0.24 \quad \text{sec}^{-1}
\end{aligned}
\qquad (228)
$$

Various branches in the chemical scheme predominate at different heights. The increase in O with altitude limits the negative ion boundary to about 70 to 80 km. At this boundary the reaction of O_2^- with O_3 dominates, resulting in formation of CO_3^-, which is destroyed primarily by O. Hence these processes become the primary factors that determine the ratio of negative ions to electrons. NO_3^- is relatively stable, however, so that high $[NO_3^-]/[N_e]$ ratios can be realized even if the source is low.

There are many uncertainties in the negative ion chemical scheme including an unknown temperature dependence of the three-body precursor reaction. The complexity of the lower ionospheric negative ion chemistry is caused mainly by the much longer lifetimes ($\sim 10^4$ sec). Under equilibrium conditions and assuming that NO_3^- is lost primarily by recombination with positive ions, one can estimate values for the $[NO_3^-]/[CO_3^-]$ ratio as a function of the NO and positive ion concentration. Typically this amounts to ~ 270 at 70 km, which implies a negative to positive ion ratio of 1:2.

It is probable that the main sources of Cl^- are CO_3^- and CO_4^-. During the day Cl^- is destroyed rapidly by H. At night when the H concentration is low, Cl^- could become an important constituent below about 75 to 80 km.

There is not much information available on the clustering of negative ions, and if this turns out to be significant, it could modify the negative ion chemistry discussed in this chapter.

3. Metallic Ions and Neutrals

a. General Discussion. Figure 3 shows a layer of metallic ions between about 90 and 100 km. In addition to layers that are generally observed near 90 km, sporadic layers are also observed in the E region. These are usually associated with meteor showers, because of their correlation with the occurrence of showers and because the abundances of different elements relative to iron agree with meteoric abundances. Figure 35 shows observations made at Wallops Island ~ 12 hr after the maximum of the Perseid meteor shower. The ions detected include Na^+, Mg^+, Al^+, Si^+, K^+, Ca^+, Ti^+, Cr^+, Mn^+, Fe^+, CO^+, and Ni^+; $^{51}V^+$ and $^{65}Cr^+$ were observed for the first time. The metals are shown as a sum in Fig. 35, with the exception of Si^+, whose profile is distinctly different. The ions Si^+ (and Ti^+?) are depleted. The abundances of the other metal ions in the lower layer (91–99 km) and the upper layer (99–107.5 km) agree with abundances observed in carbonaceous chondrites. Figure 36 shows a chemical scheme presented by Ramseyer *et al.* (1983) at the sixth ESA Symposium on European rocket and balloon programs and represents an interesting development in metal ion chemistry. The silicon material is mainly from meteoric ablation. The neutral atoms are redistributed by ion and neutral chemical reactions of which Si, SiO, and SiO_2, the respective ions, and $HSiO^+$ are the most important. The loss processes that remove Si^+ relative to Fe^+ and Mg^+ are a reaction with H_2O, and a three-body reaction with O_2 at lower altitudes. Table A-15 gives the reactions and rate coefficients for silicon ion chemistry. Several other metallic

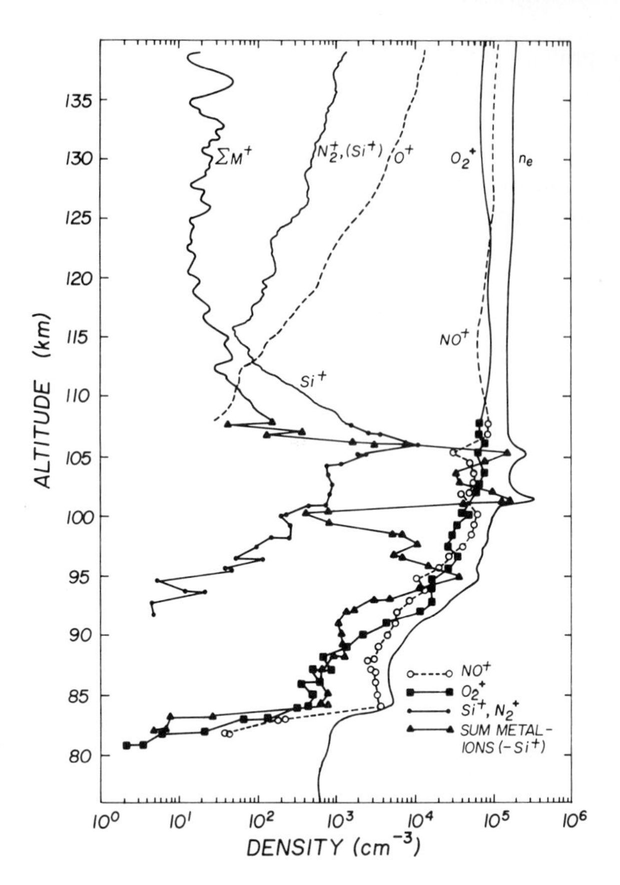

Fig. 35. Electron and ion density profiles of the lower ionosphere above Wallops Island, United States ($\chi = 28.1°$), during the Perseid meteor shower, 12 August 1975. From Kopp and Hermann (1982), as presented at the EGS Conference on Ions in the Middle Atmosphere, Leeds.

ions and neutrals also interact chemically with atmospheric constituents. We discuss the chemistry of sodium as another example, preceded by a discussion of the general chemistry of metallic ions.

The production of metals via evaporation of atoms from meteors occurs when the frictional heating reaches a sufficiently high level. The ionization energies of the metallic atoms produced are generally much smaller than those of the main atmospheric constituents. Theoretical studies indicate that because of the lower ionization energy the meteor atoms will be ionized before slowing down to energies below the ionization threshold.

The processes that result in the destruction of metallic ions in general do not appear to be well understood. Unanswered questions remain as to how metal

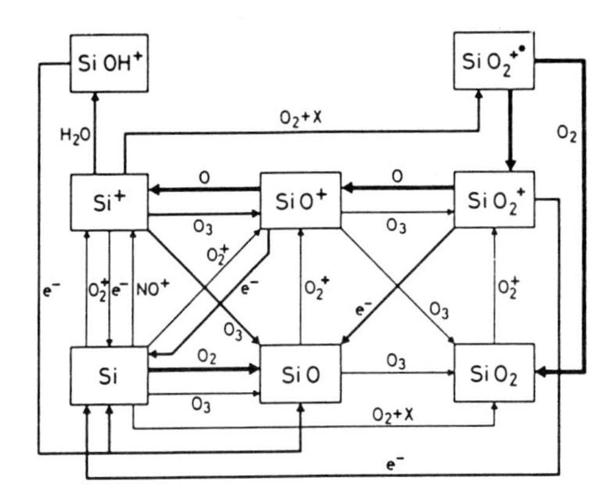

Fig. 36. Chemical reaction scheme for neutral and ionized silicon species. Broad arrows correspond to lifetimes less than 1 sec, fine arrows to lifetimes more than 10^4 sec at an altitude of 105 km. From Ramseyer *et al.* (1983), as presented at the Sixth ESA Symposium on European Rocket and Balloon Programmes.

atoms produced by meteor ablation are first ionized, arranged in layers, and then finally removed from layers into the surrounding atmosphere. In addition to charge transfer, photoionization of metallic neutrals also constitute an important ionization source, with the former becoming more important in the lower E region because of attenuation of the more energetic photons. The following kinds of reactions are the primary sources of metallic ions:

$$M + O_2^+ \xrightarrow{k_{44}} M^+ + O_2 \tag{258}$$

$$M + NO^+ \xrightarrow{k_{45}} M^+ + NO \tag{259}$$

$$M + h\nu \longrightarrow M^+ + e^- \tag{260}$$

Photoionization of and charge transfer with metal oxides MO and MO_2 can also result in the formation of metal ions:

$$MO + h\nu \longrightarrow MO^+ + e^- \tag{261}$$

$$MO_2 + h\nu \longrightarrow MO_2^+ + e^- \tag{262}$$

$$MO + O_2^+ \xrightarrow{k_{46}} MO^+ + O_2 \tag{263}$$

$$MO + NO^+ \xrightarrow{k_{47}} MO^+ + NO \tag{264}$$

Metallic ions may be lost by the reactions

$$M^+ + O_3 \xrightarrow{k_{48}} MO^+ + O_2 \tag{265}$$

$$M^+ + O_2 + X \xrightarrow{k_{49}} MO_2^+ + X \tag{266}$$

$$MO_2^+ + O \xrightarrow{k_{50}} MO^+ + O_2 \tag{267}$$

$$MO^+ + e^- \xrightarrow{k_{51}} M + O \tag{268}$$

$$MO^+ + O \xrightarrow{k_{52}} M^+ + O_2 \tag{269}$$

According to laboratory measurements k_{48} is $\sim 2 \times 10^{-10}$ cm^3 sec^{-1}, k_{51} $\sim 4 \times 10^{-7}$ cm^3 sec^{-1}, and $k_{52} \sim 10^{-10}$ cm^3 sec^{-1}. Tables A-16 and A-17 list metal–ion reactions and rate coefficients.

It is thought that the primary mechanism resulting in the formation of layers of metallic ions is transport via winds. The E and D regions are characterized by tidal winds with rapidly changing phase as a function of altitude. Winds of opposite phase result in upward and downward motion of long-lived ionization, which in turn results in a concentration buildup. It is believed that thin layers in the E region termed "sporadic E layers" are formed by this mechanism, or by similar processes, which results in wind shears. The primary prerequisite for this mechanism is that the mean lifetime of the ions be long compared with the characteristic transport lifetimes. Reactions (265) through (269) satisfy this requirement.

A parameter of interest is the ratio $[M^+]/[M]$ for which detailed information is available only for Na and K.

b. Sodium Chemistry. The presence of sodium in the atmosphere was established via a bright emission from the doublet at 5890 and 5896 Å. The peak of the emission layer is located between 88 and 95 km. The sodium peak densities lie typically between 2×10^3 and 1×10^4 cm^{-3}. Chapman proposed the mechanisms

$$Na + O + M \xrightarrow{k_{53}} NaO + M \tag{349}$$

$$Na + O_3 \xrightarrow{k_{54}} NaO + O_2 \tag{350}$$

and

$$NaO + O \xrightarrow{k_{55}} Na(^2P) + O_2 \tag{351}$$

$$NaO + O \xrightarrow{k_{56}} Na(^2S) + O_2 \tag{352}$$

An important refinement was the addition of reactions leading to NaOH, which is more stable than NaO. It is possible that ions also may play an important role, although energy considerations are more consistent with a reaction such as (351). Also, rocket measurements have shown ions to be consistently less abundant than atoms.

The rate coefficients k_{53} and k_{55} have been computed, and at 200 K: 3.4×10^{-10} and 1.0×10^{-10} cm^3 sec^{-1}, respectively. Simultaneous measurements are needed of the 5893-Å emission and $[O_3]$ to evaluate Chapman's theory quantitatively. After sunset, steady-state conditions occur, and

$$[\mathrm{Na}]/[\mathrm{NaO}] = k_{56}[\mathrm{O}]/k_{54}[\mathrm{O_3}]. \tag{353}$$

The topside scale height of the Na layer is remarkably small, and this has been interpreted as indicative of a "high-altitude" (>90 km) sink and a "low-altitude" (<90 km) source. Dust particles or aerosols have been postulated as a source, and photoionization as the high altitude sink. Redeposition on the dust particles has been recently suggested as an effective loss mechanism for atom oxides and ions. Table A-18 lists sodium reactions and rate coefficients.

E. Thermospheric Photochemistry

In this section we consider the photochemistry of the minor neutral species that exist above 120 km in the upper atmosphere. The major neutral constituents N_2 and O_2 are not discussed, since these are controlled largely by thermal and dynamic considerations, although the O_2 density can be affected by photodissociation. In the thermosphere O is a major constituent, and above 120 km it is in diffusive equilibrium. Photochemical factors that control the boundary densities at 120 km are discussed in Section II,B,1.

The major thermospheric neutral constituents have been measured and modeled. Some excellent semiempirical models provide densities and temperatures as a function of time of day, season, epoch of the solar cycle, and magnetic activity for any location on the globe at any desired altitude in the thermosphere.

The minor neutral species discussed in this section include $O(^1D)$, $O(^1S)$, $N(^4S)$, $N(^2D)$, $N(^2P)$, NO, $O_2(a\,^1\Delta_g)$, $N_2(A\,^3\Sigma_u^+)$, and vibrationally excited N_2 and O_2.

1. $O(^1D)$

The existence of $O(^1D)$ in the atmosphere was identified by the bright 6300-Å emission in the nightglow. The lifetime of the ^1D state is 147 sec, and large sources exist. The species plays an important role in channelling the flow of energy in the thermosphere. Although some controversy still exists regarding the sources of $O(^1D)$, the basic photochemistry is well understood. At night the only processes considered to be important are

$$O_2^+ + e^- \xrightarrow{\alpha_{^1D}} O(^1D) + O \tag{371}$$

and

$$O(^1D) + N_2 \xrightarrow{k_{57}} O(^3P) + N_2 \tag{372}$$

$$O(^1D) + O_2 \longrightarrow O(^3P) + O_2 \tag{373}$$

$$O(^1D) \longrightarrow O(^3P) + h\nu \tag{374}$$

where $\alpha_{1_D} = \beta_{1_D}\alpha_1$. There are two main transitions that result in radiative deactivation of $O(^1D)$. These correspond to emissions at 6300 and 6364 Å. The following additional processes become important in the daytime:

$$O + e^{-*} \longrightarrow O(^1D) + e^- \quad \text{photoelectron excitation} \tag{375}$$

$$O + e^- \longrightarrow O(^1D) + e^- \quad \text{thermal electron excitation} \tag{376}$$

$$O_2 + h\nu \xrightarrow{J_{O_2}} O(^1D) + O \tag{377}$$

Excitation cross sections for reaction (375) have been computed theoretically, and a simple expression for the excitation produced by reaction (376) as a function of electron temperature has been derived. References are given in the review by Torr and Torr (1979). In addition to these, it appears that the reaction

$$N(^2D) + O_2 \xrightarrow{k_{58}} O(^1D) + NO \tag{378}$$

may be an important process in summer.

The reaction of $O(^1D)$ with O_2 occurs at a slightly greater rate than that with N_2. Because the O_2 density is significantly less than the N_2 density, however, it is not important in the thermosphere. Figure 37 illustrates the daytime sources of $O(^1D)$. Table A-19 gives values for the rate coefficients and lifetimes.

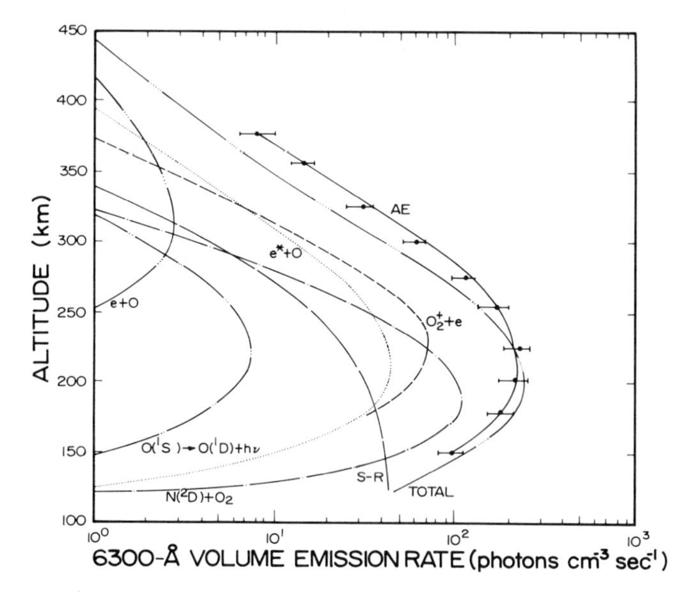

Fig. 37. Daytime OI 6300-Å volume emission rates as a function of altitude for July 1974 (four orbits). Error bars represent the spread in data for the four orbits. From Torr and Torr (1982).

2. $O(^1S)$

The transition $O(^3P-^1S)$ results in the well-known thermospheric green-line emission, the brightest feature in the visible nightglow. This emission has two components: an F region and a mesospheric component. We shall discuss the photochemistry of both components in this section. The sources of the higher altitude component are believed to be electron impact of O, dissociative recombination of O_2^+, photodissociation of O_2, and some minor sources:

$$e^{-*} + O \longrightarrow O(^1S) + e^- \tag{379}$$

$$O_2^+ + e^- \xrightarrow{\beta_{1S}\alpha_1} O(^1S) + O \tag{380}$$

$$O_2 + h\nu \longrightarrow O(^1S) + O \tag{381}$$

$$O_2^+ + N \xrightarrow{k_{59}} O(^1S) + NO^+ \tag{382}$$

Electron impact is dominant above 200 km, and photodissociation below 200 km. The only sink for $O(^1S)$ above ~ 100 km is radiative loss. At lower altitudes quenching by $O_2(a\,^1\Delta_g)$ and $O(^3P)$ become significant. Figure 38 shows the daytime sources as a function of height. A detailed discussion of the sources and sinks of $O(^1S)$ is given in the reviews by D. G. Torr and Torr (1979) and M. R. Torr and Torr (1982).

One of the main problems with regard to the production of $O(^1S)$ is the question of yield. Early laboratory and aeronomic studies had indicated the ^1S yield to be typically 0.1, although reports ranged from 2 to 10%. Subsequent laboratory studies indicated low yields for cases where O_2^+ was believed to be in low vibrational levels. Laboratory studies showed yields of 0.1 mainly from v' levels of 4 to 12, with a much lower yield from $v' < 4$. Theoretical work illustrated in Fig. 39 shows only one repulsive surface available to O_2^+ for $v < 4$ that would result in $O(^1S)$. Theoretical considerations also indicate that $O_2^+{}_{(v<4)}$ would be strongly quenched in collisions with O at a rate close to the Langevin limit. *Atmosphere Explorer* observations of the 5577-Å airglow have only indicated detectable quenching of $O_2^+{}_{(v>0)}$ by O, and interpretation of an artificial aurora below 95 km indicated values of $\sim 2\%$ for the ^1S yield, where O_2 is probably the quenching species. The O quenching rate coefficient must be significantly less than the Langevin value.

The dissociative recombination process may be described by

$$O_2^+ + e^- \longrightarrow (O_2)^* \longrightarrow 2\,O \tag{383}$$

where $(O_2)^*$ represents a repulsive state of the normal molecule. The one repulsive surface resulting in $O(^1S)$ is the $^1\Sigma_u^+$, which yields one ^1S and one ^1D atom. The calculated potential curve for $^1\Sigma_u^+$ crosses the ion potential just above the $v = 1$ vibrational level. The next available state is $5\,^3\pi_g$, which yields one ^1S and one ^3P atom and has a curve crossing near $v = 10$. In view of the

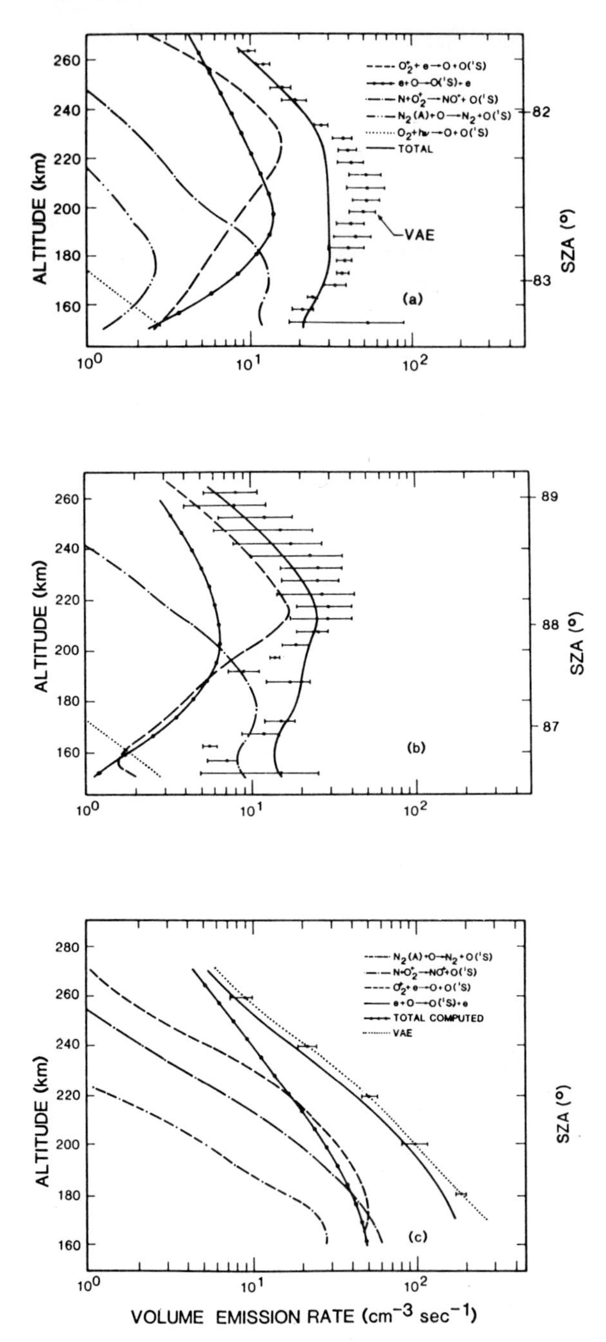

Fig. 38. Measured $O(^1S)$ volume emission rate as compared with the various sources for two *Atmosphere Explorer C* orbits: (a) 2744 upleg, (b) 2744 downleg, and (c) 606 downleg. From Torr and Torr (1982).

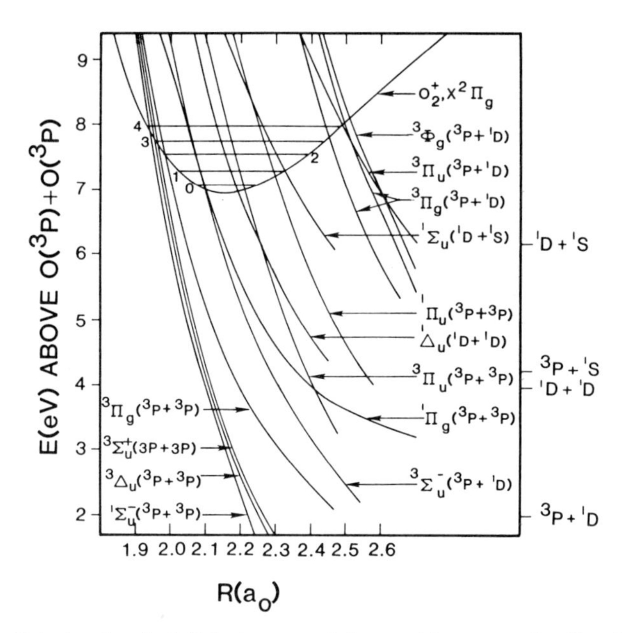

Fig. 39. Calculated excited diabatic states of O_2 providing routes for dissociative recombination of $O_2{}^+$. From Torr and Torr (1982).

experimental evidence for quenching of $O_2{}^+{}_{(v \gtrsim 1)}$ the conclusion that the $O_2{}^+$ levels $v > 1$ must cascade down the vibrational ladder to the $v = 1$ level, where recombination occurs, seems unavoidable. Data taken by the Fabry–Perot on the *Dynamics Explorer* satellite have shown that the $O(^1S) + O(^1D)$ channel is favored over the $O(^1S) + O(^3P)$ channel by a factor of 4. It is interesting to note that $O(^1D)$, for which there is an intermediate state of O_2 that crosses the ion potential near $v = 0$, does not show this kind of behavior. The F region $O(^1S)$ question should not be regarded as resolved yet.

The auroral sources of $O(^1S)$ have not yet been resolved. Direct auroral electron impact mechanisms fail to account for the observed 5577-Å emission. The reaction

$$N_2(A\,^3\Sigma_u^+) + O \longrightarrow N_2 + O(^1S) \tag{384}$$

could be the missing source, but laboratory measurements of the rate coefficient are in conflict with this suggestion. The reaction

$$N^+ + O_2 \longrightarrow NO^+ + O(^1S) \tag{385}$$

has also been suggested as the missing source. Laboratory studies, however, have shown that N^+ must be nearly thermalized for the reaction to proceed. Laboratory studies have also shown that energetic (> 1 eV) N^+ results in the products $O(^3P)$ and $NO^+(a\,^3\Sigma^+)$. Since electron impact dissociation of N_2, which results in energetic N^+ and N products, is the main source of auroral N^+, there is some uncertainty whether the reactants will cool sufficiently to guarantee that reaction (385) will proceed.

We shall now consider mesospheric $O(^1S)$. Photometric rocket and satellite measurements have shown that the lower $O(^1S)$ layer is located at 97 ± 2 km. Typically the intensity of the 5577-Å emission from this layer is ~200 R. Classically, the so-called Chapman mechanism was believed for decades to be the main source of this emission, namely,

$$3\,O \xrightarrow{k_{60}} O_2(X\,^3\Sigma_g^-)(v \leqslant 4) + O(^1S) \tag{386}$$

An alternative known as the Barth mechanism was suggested in 1962:

$$2\,O + M \xrightarrow{k_{61}} O_2{}^* + M \tag{387}$$

followed by

$$O_2{}^* + O \xrightarrow{k_{62}} O(^1S) + O_2 \tag{388}$$

$O_2{}^*$ is also quenched in competition by

$$O_2{}^* + M \xrightarrow{k_{63}} O_2 + M \tag{389}$$

where M is not an oxygen atom.

The excited O_2 may be produced in either the $A\,^3\Sigma_u^+$, the $A'\,^3\Delta_u$, or the $c\,^1\Sigma_u^-$ state. The question of which of the two mechanisms is correct appears to have been settled now, and the details of the arguments are too lengthy to present here. Briefly, Thomas (1981) used a data base of simultaneous nighttime measurements of the O density, the 5577-Å emission, and the Herzberg first bands of O_2, which arise from the $A\,^3\Sigma_u^+$ state. The rate coefficient k_{61} for the production of $O_2{}^*$ has been measured to be 4.7×10^{-33} $(300/T)^2$ cm^3 sec^{-1}. The emission intensity is given by

$$I_{Hz} = [O]^2[M]k_{61}\epsilon_1/Q_A, \tag{390}$$

where ϵ_1 is the fraction of O_2 produced in the $A\,^3\Sigma_u^+$ state and

$$Q_A = 1 + \tau_A \sum k_{Ai}[X_i], \tag{391}$$

where τ_A is the radiative lifetime of the A state and k_{Ai} the quenching rate of the ith constituent X_i. Together with the measured [O] and I_{Hz}, these parameters permitted a determination of the quenching factor as a function of height using

$$Q_A = a[O]^2[M]T^{-2}/I_{Hz}, \tag{392}$$

where T^{-2} is an assumed temperature dependence. These profiles indicated a

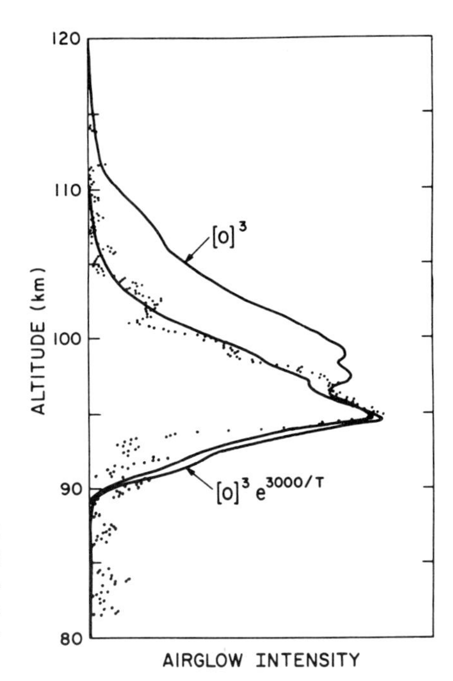

Fig. 40. Attempts to fit the Chapman mechanism to 5577-Å intensity data, showing that excess emission is predicted at high altitudes unless an unrealistic temperature dependence is added. From Torr and Torr (1982).

molecular quenching agent, which excluded O as a candidate. Attempts to fit the data with the Chapman mechanism also failed as illustrated in Fig. 40.

The data were, however, consistent with a Barth-type mechanism, but a different intermediate O_2 state with different quenching rates was required. Other facts that emerged were that, contrary to previous laboratory findings, quenching of $O(^1S)$ by O could not be the major loss process for $O(^1S)$ and that in the laboratory $O_2(a\,^1\Delta_g)$ quenching was a far more likely candidate:

$$O(^1S) + O_2(a\,^1\Delta_g) \xrightarrow{k_{64}} O + O_2 \tag{393}$$

Possibilities remaining for the identity of the intermediate state were $A'\,^3\Delta_u$ or $c\,^1\Sigma_u^-$. Emissions from both of these states from $v = 0$ have been seen on Venus, but no $O(^1S)$ emission was observed. The $v = 0$ level is endothermic for the production of $O(^1S)$ from the $c\,^1\Sigma_u^-$ state, and $A'\,^3\Delta_{u(v=0)}$ is thought to be an inefficient source of $O(^1S)$. Thus the evidence is strong that the intermediate state of O_2 is vibrationally excited, but which of the two possible states has not been established. *Spacelab 1* data suggest the intermediate state to be $O_2(A'\,^3\Delta_u)$.

An additional minor source of $O(^1S)$ between 120 and 150 km was established as photodissociation of O_2 by radiation shortward of 1334 Å. Table A-19 lists rate reactions and rate coefficients for $O(^1D)$ and $O(^1S)$.

3. Thermospheric Odd-Nitrogen

Figure 41 summarizes the processes that result in the production and destruction of odd-nitrogen in the thermosphere. Atomic nitrogen is produced primarily by ion–atom interchange of ${N_2}^+$ with O and by dissociative recombination of NO^+ at altitudes above 200 km. Above 300 km, dissociative recombination of ${N_2}^+$ and the reaction of O^+ with N_2 become significant sources. Below 200 km, dissociation of N_2 by photoelectron impact dominates.

The branching ratios for the production of $N(^2D)$ and $N(^4S)$ are critical in determining the NO and $N(^4S)$ densities in the lower thermosphere because $N(^2D)$ reacts with O_2 to produce NO and the odd-nitrogen is cannibalistically destroyed. Although $N(^4S)$ also reacts with O_2 to produce NO, the rate coefficient is very temperature dependent. The reaction proceeds slowly in the lower thermosphere where temperatures are low. In the F region, however, it

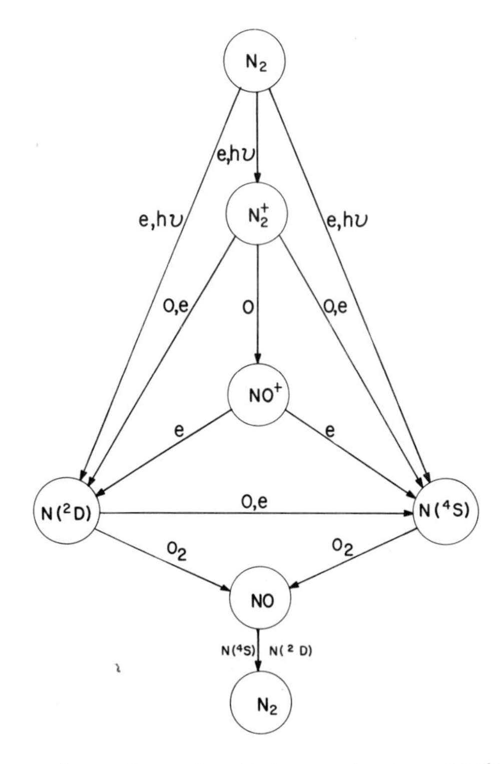

Fig. 41. Schematic illustration of the odd-nitrogen chemistry of $N(^2D)$, $N(^4S)$, and NO in the thermosphere. From Torr and Torr (1979), with permission of Pergamon Press.

becomes very important and is the major source of NO and sink for $N(^4S)$. Predominant production of $N(^2D)$ at low altitudes results in an NO-rich atmosphere. The latter in turn destroys $N(^4S)$.

The ratio $[O]/[O_2]$ increases with altitude, and O quenching of $N(^2D)$ becomes the main source of $N(^4S)$ at higher altitudes, where the latter in turn becomes a major source of NO by reacting with O_2. The buildup of $N(^4S)$ or NO tends to be self-sustaining.

a. $N(^2D)$. The sources of $N(^2D)$ in the thermosphere are

$$NO^+ + e^- \xrightarrow{\beta_2\alpha_2} N(^2D) + O \tag{397}$$

$$N_2^+ + e^- \xrightarrow{\beta_3\alpha_3} N(^2D) + N \tag{398}$$

$$N_2^+ + O \xrightarrow{\beta_4 k_{14}} N(^2D) + NO^+ \tag{399}$$

$$e^- + N_2 \longrightarrow N(^2D) + N \tag{400}$$

$$N_2 + h\nu \longrightarrow N(^2D) + N \tag{401}$$

$$N^+ + O_2 \xrightarrow{\beta_5 k_{19}} N(^2D) + O_2^+ \tag{402}$$

Originally it was found that reaction (401) is a small source of odd-nitrogen in the thermosphere. It has been suggested, however, that under optically thick conditions solar extreme-UV photons in the 800- to 1000-Å range could become entrapped, thereby increasing the probability of dissociation via predissociation of N_2.

The loss processes are

$$N(^2D) + O \xrightarrow{k_{65}} N(^4S) + O \tag{403}$$

$$N(^2D) + O_2 \xrightarrow{k_{66}} NO + O \tag{404}$$

$$N(^2D) + e^- \xrightarrow{k_{67}} N(^4S) + e^- \tag{405}$$

Table A-20 lists branching ratios and rate coefficients. These data were obtained from aeronomical studies conducted with *Atmosphere Explorer* data and from laboratory measurements (see review by Torr and Torr, 1982, for relevant references). Figures 42 and 43, respectively, show typical daytime mid-latitude sources and sinks as a function of height. At the base of the thermosphere, $N(^2D)$ is converted largely to NO.

b. $N(^4S)$. The only important sources of $N(^4S)$ above 150 km are O quenching of $N(^2D)$ and the suggested enhanced predissociation of N_2 by radiation entrapment. Below 150 km, photoelectron impact sources are important. Typical profiles are shown in Fig. 44. The importance of the $N(^2D) + O$ source is evident.

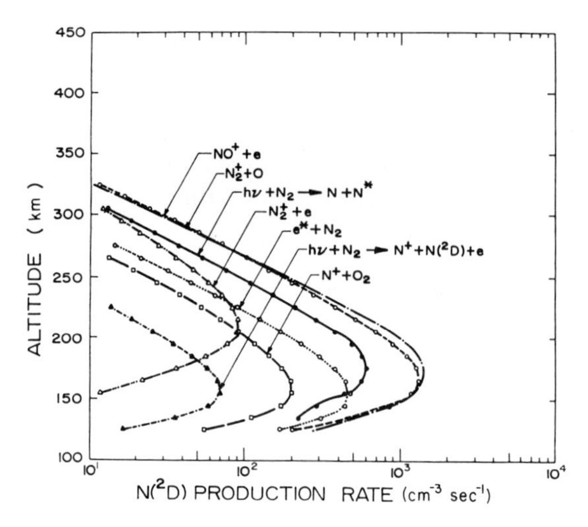

Fig. 42. Model production rates for $N(^2D)$ at 1400 hr LT on 14 February 1974 as a function of altitude. From the *Journal of Geophysical Research.*

The main loss reactions are

$$N(^4S) + O_2 \xrightarrow{k_{68}} NO + O \tag{406}$$

$$N(^4S) + NO \xrightarrow{k_{69}} N_2 + O \tag{407}$$

The ratio of the O to O_2 densities is also a major factor determining the ratio

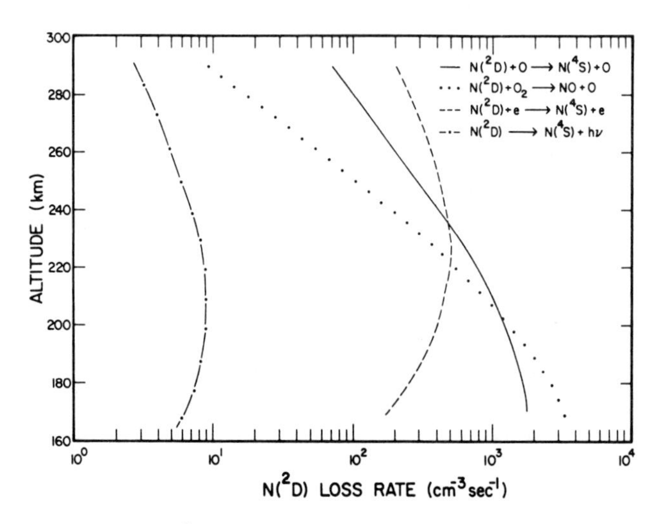

Fig. 43. Loss rates for $N(^2D)$ for *Atmosphere Explorer C* orbit 1663. From Torr and Torr (1982).

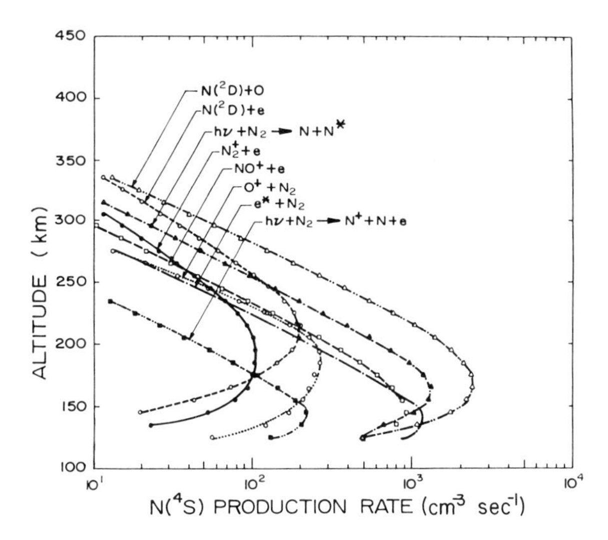

Fig. 44. Model production rates for N(^{4}S) for 1400 hr LT in midsummer 1974. From the *Journal of Geophysical Research.*

of N(^{4}S) to NO densities. At altitudes above ~200 km the N(^{4}S) sinks become small due to the rapid decrease in [NO] and [O_2], and [N(^{4}S)] is controlled by diffusion. At the base of the thermosphere the rapid conversion of N(^{2}D) to NO results in an increase in the destruction of N(^{4}S), causing it to peak at ~140 km.

c. Nitric Oxide. The reactions of N(^{2}D) and N(^{4}S) with O_2 provide the main source of NO in the thermosphere. Reaction (407) is the main sink although under certain conditions charge exchange reactions can affect the NO concentration. The diffusive lifetime becomes shorter than the chemical lifetime at the base of the thermosphere, and transport of NO into the mesosphere becomes an important removal process from the thermosphere.

A very simple expression for the NO density can be derived by including N(^{4}S) as the only source and sink, which is reasonably representative of conditions at ~200 km,

$$[\mathrm{NO}] = (k_{68}/k_{69})[\mathrm{O_2}] = 1.14[\mathrm{O_2}]\exp(-3975/T). \qquad (408)$$

d. Comparison with Measurement. Figure 45 shows a comparison of altitude profiles of model and experimental odd-nitrogen densities. Long-term variations in N(^{4}S) have been measured on the *Atmosphere Explorer* and *Aeros B* satellites. The diurnal variation has been successfully modeled. The seasonal and solar cycle variations have been interpreted in terms of photochemical effects.

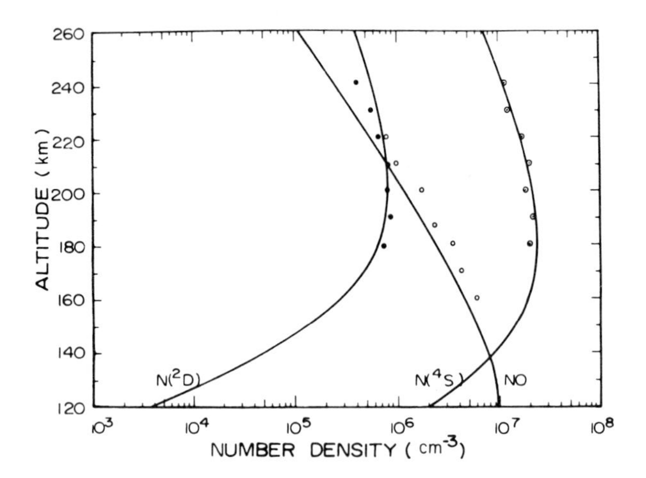

Fig. 45. Model and experimental odd-nitrogen number densities at 1400 hours LT on 14 February 1974 as a function of altitude. *AE-C* orbit 594. ⊙, $N(^4S)$ NACE; ○, NO UVNO; ●, $N(^2D)$ VAE. From the *Journal of Geophysical Research.*

Global measurements of NO concentrations were obtained by the *Atmosphere Explorer* satellites. It was found that at 200 km, knowledge of the local composition and temperature is sufficient to explain the variations of NO. The theory outlined in the previous section failed to explain the latitudinal gradient in NO. It appears that horizontal transport of auroral NO to lower latitudes is responsible for the discrepancy. This has not been quantitatively studied to date, although exploratory two-dimensional model calculations have been made.

e. Photochemical Odd-Nitrogen Coupling between the Thermosphere and Middle Atmosphere. The odd-nitrogen chemistry described in Sections II,E,3,a–d predict a global average column rate of NO production of $\sim 5 \times 10^{10}$ cm^{-2} sec^{-1} during quiet times. The global average column production rate of NO in the stratosphere is $\sim 10^{8}$ cm^{-2} sec^{-1}. Hence it is evident that a downward flux of the order of 1% of the average thermospheric column production rate through the mesosphere into the stratosphere would constitute a significant source there. This is important because of the catalytic destruction of O_3 in the stratosphere as discussed in Chapter 4.

The primary factors that affect the passage of NO through the mesosphere are the cannibalistic destruction process N + NO, photolysis of NO:

$$\mathrm{NO} + h\nu \longrightarrow \mathrm{N} + \mathrm{O} \tag{409}$$

and the rates of transport in the mesosphere. Two-dimensional models that take these factors into account suggest that, at least at high latitudes in winter, the coupling may be significant.

4. Metastable Molecules

a. $N_2(A\,^3\Sigma_u^+)$. The $A\,^3\Sigma_u^+$ state is the first electronic state of N_2, located at 6.5 eV. The radiative lifetime is $\sim$2 sec. The state, therefore, is sufficiently long-lived to undergo collisional deactivation with possible photochemical consequences. The transitions to the ground state give rise to the Vegard–Kaplan band system. In addition to direct excitation by electron (or particle) impact, sources include predominantly cascade from the higher lying $B\,^2\Pi_g$ and $C\,^3\Pi_u$ states. The Franck–Condon factors for direct excitation of the A state peak at the $v = 8$ level. Hence one would expect to see emissions from these high-lying levels. It appears, however, that inverse first positive transitions (which involve transitions from high-lying vibrational levels in the A state to nearby low-lying vibrational levels in the B state) may compete effectively with direct radiative loss from the high vibrational levels. In addition, the mechanisms that populate the A and B states are complex and require further study.

Auroral studies have related the population rate of the A state, q_A, to the N_2 O—O second positive band 3371-Å emission, η_{3371}. The value of q_A is given by

$$q_A = 5.9\eta_{3371}. \tag{410}$$

Quenching rates of N_2(A) have been determined both aeronomically and in the laboratory.

The main quencher is O:

$$N_2(A\,^3\Sigma_u^+) + O(^3P) \xrightarrow{k_{70}} N_2 + O(^3P, {}^1S) \tag{411}$$

The aeronomically determined value for the rate coefficient is $\sim 2 \times 10^{-10}$ cm^3 sec^{-1}, which is nearly an order of magnitude larger than the laboratory value. This conflict remains unresolved.

b. $O_2(a\,^1\Delta_g)$. This species was identified as a significant constituent of the lower thermosphere via its emission at 1.27 μm, which arises from the transition

$$O_2(a\,^1\Delta_g) \longrightarrow O_2(X\,^3\Sigma_g^-) + h\nu \tag{412}$$

The reaction

$$O_3 + h\nu \xrightarrow{J_3} O_2(a\,^1\Delta_g) + O(^1D) \tag{413}$$

[which is reaction (8)] is the main source. $O_2(^1\Delta_g)$ is a possible quencher of

$O(^1S)$,

$$O_2(a\,^1\Delta_g) + O(^1S) \xrightarrow{k_{71}} O + O_2(X\,^3\Sigma_g^-) \qquad (414)$$

where $k_{71} = 1.7 \times 10^{-10}$ cm^3 sec^{-1}. At night, O_2 may also be a significant quencher in the mesosphere:

$$O_2(a\,^1\Delta_g) + O_2 \xrightarrow{k_{72}} 2\,O_2 \qquad (415)$$

where $k_{72} = 2.4 \times 10^{-18}$ cm^3 sec^{-1}.

It has been pointed out that the quenching of vibrationally excited OH improves agreement between experiment and theory for nightglow emission, that is,

$$OH^* + O \xrightarrow{k_{73}} O_2(^1\Delta_g) + H \qquad (416)$$

where k_{73} has been estimated to be $\sim 4. \times 10^{-11}$ cm^3 sec^{-1}. Quenching of the intermediate O_2 state produced by the Barth mechanism [reaction (387)] by a process like reaction (388) has also been suggested as a significant source, with $O_2(c\,^1\Sigma_u^-)$ as the probable precursor.

c. Vibrationally Excited N_2. Sources of vibrationally excited N_2 in the thermosphere include photoelectron impact and chemical reactions:

$$e^- + N_2 \longrightarrow e^- + N_2^* \qquad (417)$$

$$N + NO \xrightarrow{k_{73}} N_{2(v=4)} + O \qquad (418)$$

$$O(^1D) + N_2 \xrightarrow{k_{74}} O(^3P) + N_{2(v=2)} \qquad (419)$$

The primary process that removes vibrational quanta is quenching by O:

$$N_2^* + O \xrightarrow{k_{75}} N_2 + O \qquad (420)$$

Rate coefficients are given in Table A-21.

Vibration–vibration exchange processes can play an important role in redistributing the population to yield a Boltzmann distribution that can be represented by an effective N_2 vibrational temperature. Comprehensive model calculations (see Torr and Torr, 1982) that included transport processes indicated that thermospheric N_2 vibrational temperatures are not significantly enhanced at solar minimum, because N_2^* diffuses to lower altitudes, where O quenching is rapid and the vibrational quanta are removed via vibration–translation energy transfer. At solar maximum the effects are more pronounced, because downward diffusion is inhibited by the larger neutral concentrations. Model calculations indicate that the rate coefficient for the reaction of O^+ with N_2 could increase by as much as a factor of two above 300 km.

d. Vibrationally Excited O_2. To date no quantitative study of O_2^* in the thermosphere has been carried out. Several processes could produce vibrationally excited O_2. These include

$$e^- + O_2 \longrightarrow O_2^* + e^- \quad \text{photoelectron impact} \tag{421}$$

$$O_2 + h\nu \longrightarrow O_2' \longrightarrow O_2^* + h\nu \quad \text{fluorescent scattering of solar radiation} \tag{422}$$

$$O + O + M \longrightarrow O_2^* + M \tag{423}$$

$$OH + O \longrightarrow O_2^* + H \tag{424}$$

Processes that quench vibrationally excited O_2 are uncertain. Atom–atom interchange has been suggested as an efficient process:

$$O + O_2(v) \longrightarrow O_{2(v' < v)} + O \tag{425}$$

Direct quenching may also be significant:

$$O + O_2(v) \longrightarrow O_2(v' < v) + O \tag{426}$$

Study of O_2^* in the thermosphere is an area that requires further work.

Space limitations have precluded a discussion of the chemistry of OH* and other metastable states of O_2.

III. Measurements

We shall not discuss laboratory methods for studying atmospheric processes, only the techniques used aeronomically.

A. Mesosphere

The primary constituents of interest in the mesosphere, as identified in Sections I and II, are the following:

Major neutrals: O_2, N_2

Minor constituents: O, O_3, $O(^1D)$, $O(^1S)$, $O_2(a\,^1\Delta_g)$, OH, HO_2, H_2O, H, H_2, H_2O_2, NO, NO_2, CO_2, CO

The number of measurements of these parameters made in the mesosphere is relatively small, because the region is not readily accessible to *in situ* measurements by balloons or satellites, and rocket measurements generally only give a transient snapshot picture per flight. The situation improved with the launch of the *Solar Mesosphere Explorer* (*SME*) in October 1981. Prior to this the main data sources on temperature–altitude profiles between 50 and 90 km were derived from grenade, Pitot static tubes, and falling sphere experiments conducted at several stations located around the globe. Rocket thermistor measurements of temperature have been made, but these require compensation for aerodynamic heating of the sensor and for radiational and

Table IV
INFRARED EMISSIONS

Emitting constituent	Wavelengths (μm)	Species or process observed
OH	1.5	$O_3 + HO_2$
	1.7	$O_3 + H$ and $O + HO_2$
	2.0, 3.0	$O_3 + H$
	2.7	
NO	5.1	O Interchange
	5.5	1–0 P branch
NO_2	6.2	
H_2O	6.3	
O_3	9.6	
CO_2	2.8	(Valuable monitor of mesospheric energy budget)

other nonambient heat sources. Data of this type have been analyzed harmonically for semiannual and annual cycles. Annual mean temperatures have been computed from harmonically smoothed means for use in model atmospheres. Densities of major neutral constituents are generally computed theoretically to be consistent with the measured temperature profile, since the mesosphere is mixed and the temperature uniquely defines the density profile.

Optical remote sensing from space has proved to be the most useful way of observing the mesosphere. Ozone lends itself to measurement in the near and far UV by differential absorption techniques. Much of the global data have been provided by backscatter UV spectrometers flown on *Nimbus* satellites. On *SME* the O_3 density between 50 and 70 km was determined from measurements of scattered solar radiation at two wavelengths (2650 and 2970 Å). Differential absorption by O_3 at these two wavelengths produces an absolute measurement of O_3 density. A mathematical inversion is needed to derive the O_3 density as a function of altitude. At higher altitudes, O_3 can be derived from the $O_2(a\,^1\Delta_g)$ 1.27-μm emission that arises from reaction (8). Evaluation of the O_3 concentration requires that quenching and radiative transfer of the 1.27-μm emission be taken into account. The scale height of visible wavelength scattered sunlight provides the temperature.

Potentially, information on many species can be derived from thermal emissions using cryogenically cooled radiometers. Several instruments have been flown and tested (or will be flown) on the *Upper Atmosphere Research Satellite* (*UARS*). Concentrations of NO, NO_2, CO_2, and H_2O have been successfully derived using these techniques. Table IV lists a set of constituents that either have been or could be observed using IR techniques.

Table V

NEAR-ULTRAVIOLET AND VISIBLE EMISSIONS

Emitting constituent	Wavelength	Emission processes	Species or Process observed
OH	$A\,^2\Sigma_u^+ - X\,^2\Pi$ 3080–3090 Å	Resonance Fluorescence	[OH]
NO	$A\,^2\Sigma^+ - X\,^2\Pi$ ~2150-Å γ Bands	Resonance Fluorescence	[NO]
O	1304 Å	Lamp induced Resonance Fluorescence	[O]
H	~1216	Lamp induced Resonance Fluorescence	[H]
O_3	About 2400 to 3200 Å	Absorption	[O_3]
NO_2	~4390	Absorption	[NO_2]
O_2 (atmospheric)	$b'\,\Sigma_g^+ - X\,^3\Sigma_g^-$ Visible γ and β bands	Absorption	[O_2]
$O_2(A\,^3\Sigma_u^+)$	Herzberg 1	$O + O + M \rightarrow O_2 + M$	[O]
$O(^1S)$	5577 Å	$O_2(c^1\Sigma_u^-$ or $A'\,^3\Delta_u) \rightarrow$ $O_2 + O(^1S)$	[O]
OH ($v \leqslant 9$)	Visible	Chemiluminescence	$O_3 + HO_2$
OH ($v \leqslant 5$)	Visible	Chemiluminescence	$O_3 + H$
NO_2	Continuum	O + NO	O or NO
NO	δ Bands	N + O	N or O

Spectral filtering is provided by dispersion, interferometry, or gas or interference filters. In addition, OH has recently been observed in the mesosphere via high-resolution UV spectroscopy at about 3080 to 3090 Å. Table V shows species that have or can be observed in the UV and visible.

Thermal emissions in the millimeter-wavelength spectral region can also provide valuable data on mesospheric constituents. Table VI lists potential measurement candidates proposed for the *UARS* mission.

Table VI

MICROWAVE EMISSIONS

Measurement	Spectral region (GHz)	Upper altitude limit (km)
O_3	205, 231	90
Temperature	119	100
O_2	119	120
CO	231	100
H_2O	183	90

In addition to the direct measurement techniques, in cases where measurements are particularly difficult to obtain directly, measurements of other species that are critically dependent on their concentration can be used to infer values. In particular, the formation of hydrated cluster ions and certain metallic ions involves H_2O in the chemistry chain. Measurements of key parameters in the scheme have lead to useful estimates of the H_2O concentration in the mesosphere.

B. The D and E Regions and Metallic Ions

Most of our knowledge of the bulk plasma parameters in the D and E regions have been obtained by rocket measurements of electron densities (using blunt probes), total ion densities, conductivities, and mobilities, using electrostatic probe and wave propagation methods. Partial reflection of radio waves generated by ground-based transmitters has provided useful electron density measurements. Cosmic noise absorption in the range 5–30 MHz (riometer absorption) and field strength of continuous-wave signals in the frequency range 0.01–5 MHz have also been used to infer D region electron densities.

Presently available knowledge on the positive ion composition is based on *in situ* mass spectrometer measurements. The high ambient pressures below 100 km make the measurement much more difficult than at higher altitudes. The mass spectrometer has to be mounted inside a vacuum tank which is pumped by a fast cryo- or sublimation pump. Atmospheric gases are then let in via a small orifice at low pressures. A detailed description of experimental techniques has been given by Arnold and Krankowsky (1977). Measurements have been made for the most part by three laboratories: the U.S. Air Force Geophysics Laboratory, the Max-Planck-Institut für Kernphysik (Germany), and NASA's Goddard Space Flight Center.

The discovery of large cluster ions was a surprise. It also introduced an experimental problem since ion fragmentation could occur in the detached shock wave that forms in front of the probes. Many of these problems have been solved.

Measurements of negative ions appear to be even more difficult than those of positive ions. *In situ* mass spectrometric techniques have provided the bulk of the available data, which is scarce. Whereas the sum of partial positive-ion concentrations can be normalized to a measurement of the total ion density, this is not the case for negative ions, for which more species exist, reducing their partial densities.

In the E region, in addition to *in situ* rocket mass spectrometer techniques, direct probing by coherent and incoherent radio sounding has provided useful information on electron densities. These techniques are discussed in more detail in the Section III,C.

C. Thermosphere–Ionosphere

1. High-Frequency Radio Methods

After 40 years, the pulse-sounding radio technique of Breit and Tuve continues to provide the bulk of the data acquired on a regular basis on the ionosphere. The sounder is a type of radar capable of obtaining echoes from the ionosphere over a wide range of high-frequency (HF) operating frequencies (about 1–25 MHz). The data are interpreted in terms of magnetoionic theory that relates the electron concentration to wave and plasma frequency via the refractive index of the conducting layer and the Earth's magnetic field strength. Two relevant frequencies are defined by

$$\text{Plasma frequency:} \quad (2\pi f_p)^2 = \omega_p^2 = N_e^2/m\epsilon_0,$$

$$\text{Gyro frequency:} \quad 2\pi f_H = \omega_H = Be/m,$$

where B is magnetic flux density of the Earth's field, f the radio frequency, and ω the angular frequency. These equations yield

$$N_e = 1.24 \times 10^4 f_p^2,$$

where f_p is in hertz.

With the advent of satellites, it became possible to monitor beacon signals, and the measured polarization rotation (the Faraday effect) could be used to determine the total electron content.

2. Incoherent Scatter

The sounding discussed in the previous section depends on the collective behavior of electrons, which can be described approximately in terms of refractive indices using what is known as the Appleton–Hartree equation. Two other kinds of signal are also generated when radio waves penetrate the ionospheric plasma; both involve the scattering of radiation. In one case the radiation is backscattered by irregularities or sharp gradients in the medium. In the second case the radio energy is scattered directly by the individual electrons. The latter has been termed incoherent scatter. Both techniques have been developed successfully for various ionospheric measurements. The incoherent scatter technique has proved to be very useful, since it provides data on both the neutral and ionized concentrations as well as temperatures, which are important in studying the photochemistry of the ionosphere.

A powerful radar operating in the frequency range of tens to hundreds of megahertz can be used to obtain incoherent scatter data. The amplitude of the reflected signal yields the electron density. Electron density profiles have been obtained to altitudes of several thousand kilometers with the giant Arecibo radar in Puerto Rico. The frequency broadening of the returned signal depends on the ion temperature, and the shape of the spectrum depends on the electron temperature and the ionic composition. The frequency shift yields the ion drift velocity. By rather ingenious methods the incoherent scatter data have been used to extract the neutral atmospheric density and temperature, as well as neutral wind velocities.

3. Ground-Based Optical Techniques

Measurement of optical emissions from the thermosphere is probably the earliest source of data on atmospheric photochemistry, and for many decades these data provided the only information on many thermospheric processes. Potentially, optical remote sensing of atmospheric emissions provide a powerful means of gaining information on many atmospheric processes, since many important processes generate emissions with characteristic spectral signatures. The problem with ground-based measurements, however, is lack of altitude information and limitation of the measurements to darkness when the overwhelming Rayleigh-scattered sunlight is absent. Regrettably, the major sources of ionization and excitation also cease near sunset, and nocturnal measurements can only provide very limited data on the photochemistry of the thermosphere.

4. Space-Borne Measurements

With the advent of high-altitude rockets and satellites, *in situ* measurements of many crucial thermospheric parameters became available. *In situ* mass spectrometry has provided a wealth of data on the major neutral and ionic constituents as well as the neutral temperature. *In situ* pressure-measuring devices include accelerometers, ion gauges, and capacitance manometers. *In situ* ion temperatures were recorded by retarding potential analyzers, and the electron density by cylindrical electrostatic probes. Many atmospheric emissions have been recorded by spectrometers and airglow photometers, which have scanned the limb of the Earth, providing height profiles of surface brightness that can be inverted to obtain the volume emission rates. Instruments have been developed to provide high-resolution measurements of the photoelectron flux, the solar UV flux, and the fluxes of precipitating charged and neutral particles. From the 1960s to the 1980s, large, comprehensive data bases have been acquired on nearly all key constituents of the thermosphere and ionosphere. Of particular interest is the *Atmosphere Explorer* mission of the 1970s, which was dedicated to the aeronomy of the thermosphere. The satellites were the first to provide simultaneous measurements of all important quantities. Detailed descriptions of the kind of instrumentation referred to above can be found in the April 1973 issue of *Radio Science*.

Appendix

Tabulation of Atmospheric Transmission Functions in the Spectral Region of the O_2 Schumann–Runge Bands

Table A-1a

SPECTRAL INTERVALS FOR WHICH TRANSMISSION FUNCTIONS ARE COMPUTED[a]

Label (i)	Wave-number range (cm^{-1})
1	48,600–49,000
2	49,000–49,500
3	49,500–50,000
4	50,000–50,500
5	50,500–51,000
6	51,000–51,500
7	51,500–52,000
8	52,000–52,500
9	52,500–53,000
10	53,000–53,500
11	53,500–54,000
12	54,000–54,500
13	54,500–55,000
14	55,000–55,500
15	55,500–56,000
16	56,000–56,500
17	56,500–57,000

[a] From Frederick and Hudson (1979).

Table A-1b

TRANSMISSION FUNCTIONS T FOR EACH SPECTRAL INTERVAL i[a]

z (km)[b]	N (cm^{-2})[b]	$\bar{T}$ ($i = 1$)	$\bar{T}$ ($i = 2$)
120	3.922 + 16[c]	9.998 − 1	9.998 − 1
115	7.262 + 16	9.998 − 1	9.998 − 1
110	1.555 + 17	9.998 − 1	9.998 − 1
105	3.901 + 17	9.998 − 1	9.998 − 1
100	1.060 + 18	9.998 − 1	9.998 − 1
95	2.905 + 18	9.998 − 1	9.998 − 1
90	7.718 + 18	9.997 − 1	9.997 − 1
85	1.960 + 19	9.996 − 1	9.996 − 1
80	4.716 + 19	9.991 − 1	9.991 − 1
75	1.078 + 20	9.986 − 1	9.984 − 1

(*cont.*)

Table A-1b *(cont.)*

z (km)[b]	N (cm^{-2})[b]	$\bar{T}$ ($i = 1$)	$\bar{T}$ ($i = 2$)
70	2.358 + 20	9.969 − 1	9.968 − 1
65	4.939 + 20	9.938 − 1	9.936 − 1
60	9.919 + 20	9.879 − 1	9.872 − 1
55	1.919 + 21	9.767 − 1	9.755 − 1
50	3.607 + 21	9.568 − 1	9.543 − 1
45	6.756 + 21	9.207 − 1	9.160 − 1
40	1.299 + 22	8.534 − 1	8.452 − 1
35	2.595 + 22	7.288 − 1	7.156 − 1
30	5.382 + 22	5.193 − 1	5.007 − 1
25	1.145 + 23	2.483 − 1	2.303 − 1
20	2.480 + 23	4.902 − 1	4.182 − 2

z (km)	$\bar{T}$ ($i = 3$)	$\bar{T}$ ($i = 4$)	$\bar{T}$ ($i = 5$)
120	9.998 − 1	9.999 − 1	9.999 − 1
115	9.998 − 1	9.999 − 1	9.999 − 1
110	9.998 − 1	9.999 − 1	9.999 − 1
105	9.998 − 1	9.999 − 1	9.999 − 1
100	9.998 − 1	9.999 − 1	9.999 − 1
95	9.998 − 1	9.999 − 1	9.998 − 1
90	9.997 − 1	9.998 − 1	9.996 − 1
85	9.996 − 1	9.995 − 1	9.990 − 1
80	9.991 − 1	9.992 − 1	9.997 − 1
75	9.983 − 1	9.983 − 1	9.948 − 1
70	9.967 − 1	9.962 − 1	9.890 − 1
65	9.933 − 1	9.919 − 1	9.783 − 1
60	9.865 − 1	9.835 − 1	9.602 − 1
55	9.740 − 1	9.674 − 1	9.309 − 1
50	9.512 − 1	9.377 − 1	8.846 − 1
45	9.100 − 1	8.861 − 1	8.129 − 1
40	8.350 − 1	7.996 − 1	7.056 − 1
35	7.001 − 1	6.550 − 1	5.472 − 1
30	4.805 − 1	4.333 − 1	3.342 − 1
25	2.125 − 1	1.799 − 1	1.230 − 1
20	3.548 − 2	2.683 − 2	1.513 − 2

z (km)	$\bar{T}$ ($i = 6$)	$\bar{T}$ ($i = 7$)	$\bar{T}$ ($i = 8$)
120	1.000 + 0	1.000 + 0	9.999 − 1
115	1.000 + 0	9.999 − 1	9.999 − 1
110	1.000 + 0	9.999 − 1	9.999 − 1
105	9.999 − 1	9.998 − 1	9.997 − 1
100	9.998 − 1	9.994 − 1	9.994 − 1
95	9.996 − 1	9.984 − 1	9.984 − 1
90	9.988 − 1	9.958 − 1	9.961 − 1
85	9.971 − 1	9.895 − 1	9.906 − 1
80	9.932 − 1	9.757 − 1	9.789 − 1

Table A-1b (*cont.*)

z (km)[b]	$\bar{T}(i=6)$	$\bar{T}(i=7)$	$\bar{T}(i=8)$
75	9.848 − 1	9.483 − 1	9.568 − 1
70	9.683 − 1	9.007 − 1	9.217 − 1
65	9.391 − 1	8.308 − 1	8.704 − 1
60	8.931 − 1	7.420 − 1	7.981 − 1
55	8.283 − 1	6.370 − 1	6.994 − 1
50	7.423 − 1	5.179 − 1	5.732 − 1
45	6.333 − 1	3.942 − 1	4.330 − 1
40	5.007 − 1	2.766 − 1	2.947 − 1
35	3.443 − 1	1.686 − 1	1.649 − 1
30	1.785 − 1	7.627 − 2	6.089 − 2
25	5.173 − 2	1.815 − 2	9.626 − 3
20	4.392 − 5	1.091 − 3	2.648 − 4

z (km)	$\bar{T}(i=9)$	$\bar{T}(i=10)$	$\bar{T}(i=11)$
120	9.999 − 1	9.998 − 1	9.997 − 1
115	9.998 − 1	9.997 − 1	9.994 − 1
110	9.997 − 1	9.994 − 1	9.989 − 1
105	9.994 − 1	9.987 − 1	9.974 − 1
100	9.986 − 1	9.968 − 1	9.933 − 1
95	9.964 − 1	9.918 − 1	9.829 − 1
90	9.909 − 1	9.803 − 1	9.597 − 1
85	9.786 − 1	9.586 − 1	9.176 − 1
80	9.554 − 1	9.240 − 1	8.562 − 1
75	9.195 − 1	8.733 − 1	7.775 − 1
70	8.715 − 1	8.035 − 1	6.809 − 1
65	8.102 − 1	7.115 − 1	5.635 − 1
60	7.313 − 1	5.947 − 1	4.252 − 1
55	6.296 − 1	4.496 − 1	2.739 − 1
50	5.039 − 1	3.234 − 1	1.394 − 1
45	3.686 − 1	2.050 − 1	5.573 − 2
40	2.403 − 1	1.121 − 1	1.737 − 2
35	1.273 − 1	4.463 − 2	3.220 − 3
30	4.310 − 2	8.589 − 3	1.557 − 4
25	5.509 − 3	3.536 − 4	3.497 − 7
20	8.743 − 5	6.082 − 7	9.070 − 13

z (km)	$\bar{T}(i=12)$	$\bar{T}(i=13)$	$\bar{T}(i=14)$
120	9.994 − 1	9.982 − 1	9.979 − 1
115	9.989 − 1	9.967 − 1	9.962 − 1
110	9.977 − 1	9.934 − 1	9.920 − 1
105	9.947 − 1	9.849 − 1	9.808 − 1
100	9.868 − 1	9.654 − 1	9.533 − 1
95	9.675 − 1	9.291 − 1	8.993 − 1
90	9.293 − 1	8.753 − 1	8.165 − 1
85	8.173 − 1	8.043 − 1	7.117 − 1

(*cont.*)

Table A-1b (*cont.*)

z (km)[b]	$\bar{T}(i = 12)$	$\bar{T}(i = 13)$	$\bar{T}(i = 14)$
80	7.985 − 1	7.145 − 1	5.907 − 1
75	7.091 − 1	6.053 − 1	4.561 − 1
70	5.998 − 1	4.781 − 1	3.152 − 1
65	4.717 − 1	3.404 − 1	1.818 − 1
60	3.327 − 1	2.095 − 1	7.879 − 2
55	1.998 − 1	1.028 − 1	2.182 − 2
50	9.627 − 2	3.446 − 2	2.880 − 3
45	3.567 − 2	6.752 − 3	1.354 − 4
40	9.359 − 3	6.498 − 4	1.525 − 6
35	1.158 − 3	1.615 − 5	9.508 − 10
30	2.323 − 5	1.684 − 8	8.491 − 16
25	8.472 − 9	1.387 − 14	5.186 − 28
20	4.906 − 16	1.801 − 27	0

z (km)	$\bar{T}(i = 15)$	$\bar{T}(i = 16)$	$\bar{T}(i = 17)$
120	9.974 − 1	9.947 − 1	9.919 − 1
115	9.952 − 1	9.903 − 1	9.851 − 1
110	9.904 − 1	9.807 − 1	9.690 − 1
105	9.784 − 1	9.589 − 1	9.274 − 1
100	9.509 − 1	9.183 − 1	8.324 − 1
95	9.003 − 1	8.571 − 1	6.666 − 1
90	8.287 − 1	7.697 − 1	4.512 − 1
85	7.397 − 1	6.503 − 1	2.416 − 1
80	6.331 − 1	5.011 − 1	9.677 − 2
75	5.081 − 1	3.360 − 1	2.581 − 2
70	3.648 − 1	1.819 − 1	3.283 − 3
65	2.129 − 1	7.003 − 2	8.629 − 5
60	8.447 − 2	1.465 − 2	5.951 − 8
55	1.604 − 2	9.004 − 4	1.642 − 14
50	7.737 − 4	4.498 − 6	1.301 − 27
45	5.047 − 6	3.694 − 10	0
40	2.344 − 9	7.629 − 17	0
35	8.210 − 15	1.109 − 27	0
30	5.095 − 25	1.952 − 46	0
25	1.355 − 45	0	0
20	0	0	0

[a] From Frederick and Hudson (1979), with the permission of the American Meteorological Society.

[b] Use O_2 column density N measured along the solar beam as the independent variable for interpolation purposes. The altitudes z correspond to the given N values for an overhead sun only.

[c] $3.922 + 16 \equiv 3.922 \times 10^{16}$.

Tabulation of Dissociation Rates of O_2 in the Schumann–Runge Bands

Table A-2a

SPECTRAL INTERVALS FOR WHICH DISSOCIATION RATES ARE REPORTED WHEN OZONE OPACITY IS SIGNIFICANT[a]

Label (i)	Wave-number range (cm^{-1})
1	48,600–49,000
2	49,000–49,500
3	49,500–50,000
4	50,000–50,500
5	50,500–51,000
6	51,000–51,500
7	51,500–52,000
8	52,000–52,500
9	52,500–53,000
10	53,000–53,500
11	53,500–54,000
12	54,000–54,500
13	54,500–55,000
14	55,000–55,500
15	55,500–56,000
16	56,000–56,500
17	56,500–57,000

[a] From Frederick and Hudson (1980).

Table A-2b

DISSOCIATION RATE OF O_2 IN THE SCHUMANN–RUNGE BANDS AT COLUMN DENSITIES WHERE OZONE OPACITY IS NEGLIGIBLE[a]

O_2 Column density (cm^{-2})	z_0[b] (km)	Dissociation rate (sec^{-1})
1.555 + 17[c]	110	7.76 − 8
3.901 + 17	105	6.66 − 8
1.060 + 18	100	4.96 − 8
2.905 + 18	95	3.12 − 8
7.710 + 18	90	1.78 − 8
1.960 + 19	85	9.54 − 9
4.716 + 19	80	5.31 − 9
1.078 + 20	75	3.04 − 9
2.358 + 20	70	1.80 − 9
4.939 + 20	65	1.09 − 9

[a] From Frederick and Hudson (1980).
[b] Altitude corresponding to the given O_2 column density for an overhead sun.
[c] $1.555 + 17 \equiv 1.555 \times 10^{17}$.

Table A-2c

DISSOCIATION RATE OF O_2 IN THE SCHUMANN–RUNGE BANDS AT COLUMN DENSITIES WHERE OZONE OPACITY SHOULD BE INCLUDED[a]

O_2 Column density (cm^{-2})	z_0 (km)	$J(i = 1)$[b] (sec^{-1})	$J(i = 2)$ (sec^{-1})	$J(i = 3)$ (sec^{-1})
9.919 + 20	60	1.66 − 11	1.80 − 11	1.51 − 11
1.919 + 21	55	1.65 − 11	1.79 − 11	1.52 − 11
3.607 + 21	50	1.62 − 11	1.76 − 11	1.51 − 11
6.756 + 21	45	1.55 − 11	1.68 − 11	1.43 − 11
1.299 + 22	40	1.44 − 11	1.54 − 11	1.28 − 11
2.595 + 22	35	1.23 − 11	1.30 − 11	1.06 − 11
5.382 + 22	30	8.73 − 12	9.07 − 12	7.26 − 12
1.145 + 23	25	4.17 − 12	4.17 − 12	3.20 − 12
2.480 + 23	20	8.24 − 13	7.57 − 13	5.34 − 13

O_2 Column density (cm^{-2})	$J(i = 4)$ (sec^{-1})	$J(i = 5)$ (sec^{-1})	$J(i = 6)$ (sec^{-1})
9.919 + 20	1.54 − 11	2.87 − 11	6.15 − 11
1.919 + 21	1.58 − 11	2.48 − 11	4.56 − 11
3.607 + 21	1.57 − 11	2.17 − 11	3.28 − 11
6.756 + 21	1.39 − 11	1.71 − 11	2.11 − 11
1.299 + 22	1.14 − 11	1.25 − 11	1.24 − 11
2.595 + 22	8.95 − 12	8.62 − 12	6.74 − 12
5.382 + 22	5.76 − 12	4.84 − 12	2.94 − 12
1.145 + 23	2.35 − 12	1.69 − 12	7.60 − 13
2.480 + 23	3.47 − 13	2.00 − 13	5.94 − 14

O_2 Column density (cm^{-2})	$J(i = 7)$ (sec^{-1})	$J(i = 8)$ (sec^{-1})	$J(i = 9)$ (sec^{-1})
9.919 + 20	7.65 − 11	6.59 − 11	5.92 − 11
1.919 + 21	4.76 − 11	4.85 − 11	4.13 − 11
3.607 + 21	2.87 − 11	3.29 − 11	2.69 − 11
6.756 + 21	1.44 − 11	1.76 − 11	1.39 − 11
1.299 + 22	6.50 − 12	8.32 − 12	6.16 − 12
2.595 + 22	2.82 − 12	3.59 − 12	2.45 − 12
5.382 + 22	1.05 − 12	1.10 − 12	6.84 − 13
1.145 + 23	2.28 − 13	1.52 − 13	7.81 − 14
2.480 + 23	1.32 − 14	3.75 − 15	1.14 − 15

O_2 Column density (cm^{-2})	$J(i = 10)$ (sec^{-1})	$J(i = 11)$ (sec^{-1})	$J(i = 12)$ (sec^{-1})
9.919 + 20	7.24 − 11	6.62 − 11	5.40 − 11
1.919 + 21	4.28 − 11	3.62 − 11	2.58 − 11
3.607 + 21	2.22 − 11	1.50 − 11	9.72 − 12
6.756 + 21	9.29 − 12	3.95 − 12	2.45 − 12
1.299 + 22	3.42 − 12	7.81 − 13	4.30 − 13

Table A-2c (*cont.*)

O_2 Column density (cm^{-2})	z_0 (km)	J (i = 10)[b] (sec^{-1})	J (i = 11) (sec^{-1})	J (i = 12) (sec^{-1})
2.595 + 22		1.07 − 12	1.12 − 13	4.35 − 14
5.382 + 22		1.78 − 13	4.83 − 15	7.88 − 16
1.145 + 23		6.62 − 15	1.03 − 17	2.73 − 19
2.480 + 23		1.02 − 17	2.57 − 23	1.52 − 26

O_2 Column density (cm^{-2})	J (i = 13) (sec^{-1})	J (i = 14) (sec^{-1})	J (i = 15) (sec^{-1})
9.919 + 20	4.70 − 11	2.54 − 11	2.63 − 11
1.919 + 21	1.89 − 11	5.88 − 12	5.04 − 12
3.607 + 21	5.44 − 12	6.68 − 13	2.46 − 13
6.756 + 21	7.93 − 13	2.33 − 14	1.34 − 15
1.299 + 22	5.36 − 14	1.91 − 16	4.78 − 19
2.595 + 22	1.09 − 15	1.01 − 19	1.42 − 24
5.382 + 22	1.02 − 18	8.10 − 26	7.93 − 35
1.145 + 23	8.09 − 25	4.70 − 38	1.96 − 55
2.480 + 23	1.01 − 37	0	0

O_2 Column density (cm^{-2})	J (i = 16) (sec^{-1})	J (i = 17) (sec^{-1})
9.919 + 20	6.84 − 12	1.07 − 16
1.919 + 21	4.31 − 13	3.31 − 23
3.607 + 21	2.30 − 15	2.83 − 36
6.756 + 21	1.66 − 19	0
1.299 + 22	2.73 − 26	0
2.595 + 22	3.05 − 37	0
5.382 + 22	4.42 − 56	0
1.145 + 23	0	0
2.480 + 23	0	0

[a] From Frederick and Hudson (1980), with the permission of the American Meteorological Society.

[b] $J(i) = O_2$ dissociation rate, unattenuated by O_3 for the spectral range i defined in Table A-2a.

Table A-3

SOLAR MINIMUM REFERENCE SPECTRUM FOR THE RANGE 120–190 nm[a]

Interval (nm)	Ratio of alternate irradiance measurements to the reference spectrum						Solar minimum reference spectrum: Irradiance		Flight				
	GJR 1	GJR 2	D&P	H&S	S&S	B *et al.*	10^9 photons cm^{-2} sec^{-1} nm^{-1}	mW m^{-2} nm^{-1}	A	B	C	D	E
120–121	0.81	0.75	0.51				7.68(9)	0.127	0.82	0.84	0.86	1.06	1.42
121–122	1.38	0.99	1.35				305.60	4.976	1.00	0.66	0.72	1.21	1.40
122–123	0.74	0.58	0.06				2.83	0.046	1.20	0.93	0.68	1.01	1.18
123–124	0.88	0.72	0.38	0.93			1.75	0.028	0.90	0.76	0.88	1.17	1.29
124–125	0.79	0.60	0.32	0.66			1.43	0.023	1.05	0.82	0.87	1.11	1.15
125–126	0.84	0.75	0.11	0.85			1.39	0.022	0.82	0.75	0.90	1.25	1.28
126–127	0.72	0.62	0.41	0.58			2.04	0.032	1.03	0.83	0.78	1.21	1.14
127–128	0.82	0.60	0.22	0.67			1.18	0.018	0.99	0.67	0.82	1.18	1.34
128–129	0.70	0.70	0.18	0.54			1.08	0.017	0.88	0.89	0.95	1.24	1.05
129–130	0.71	0.78	0.50	0.73			1.03	0.016	0.94	0.81	0.87	1.19	1.18
130–131	0.77	0.71	0.46	0.48			9.52	0.145	0.88	0.80	0.90	1.21	1.22
131–132	0.71	0.64	0.86	0.47			1.86	0.028	0.93	0.82	0.86	1.20	1.19
132–133	0.75	0.63	0.99	0.69			1.31	0.020	0.89	0.81	0.90	1.22	1.18
133–134	0.77	0.68	0.65	0.30			10.98	0.164	0.90	0.81	0.89	1.16	1.25
134–135	0.72	0.73	0.83	0.80			1.20	0.018	0.89	0.81	0.92	1.19	1.18
135–136	0.81	0.75	0.98	0.62			2.64	0.039	0.90	0.82	0.93	1.17	1.18
136–137	0.73	0.70	0.89	0.66			1.92	0.028	0.91	0.90	0.89	1.15	1.15
137–138	0.68	0.68	0.81	0.65			1.97	0.028	0.85	0.85	0.94	1.17	1.19
138–139	0.70	0.71	0.87	0.65			1.96	0.028	0.87	0.87	0.92	1.13	1.21
139–140	0.67	0.72	0.83	0.52			5.79	0.082	0.90	0.90	0.88	1.11	1.21
140–141	0.70	0.73	0.91	0.66			4.73	0.067	0.87	0.89	0.94	1.13	1.17
141–142	0.71	0.73	0.84	0.70			2.86	0.040	0.84	0.86	0.95	1.19	1.16
142–143	0.68	0.77	0.83	0.64			3.15	0.044	0.85	0.90	0.90	1.16	1.20
143–144	0.73	0.76	0.86	0.70			3.59	0.050	0.86	0.85	0.94	1.14	1.20
144–145	0.69	0.79	0.81	0.69			3.60	0.050	0.85	0.85	0.95	1.13	1.22
145–146	0.74	0.83	0.83	0.69			3.84	0.053	0.84	0.90	0.95	1.13	1.17
146–147	0.79	0.81	0.93	0.69			4.82	0.065	0.87	0.86	0.93	1.17	1.14
147–148	0.74	0.80	0.79	0.66			6.31	0.085	0.83	0.82	0.95	1.14	1.23
148–149	0.78	0.82	0.78	0.63			6.41	0.086	0.89	0.85	0.93	1.13	1.17
149–150	0.79	0.78	0.90	0.66			6.00	0.080	0.89	0.84	0.93	1.14	1.18
150–151	0.76	0.79	0.89	0.73			6.88	0.091	0.90	0.85	0.90	1.10	1.22
151–152	0.75	0.80	0.87	0.68	0.72		7.39	0.097	0.87	0.84	0.92	1.13	1.21
152–153	0.78	0.80	0.90	0.63	0.63		9.53	0.124	0.92	0.84	0.92	1.12	1.18

153–154	0.80	0.79	0.93	0.65	0.60		10.54	0.137	0.90	0.80	0.94	1.12	1.21
154–155	0.79	0.79	0.90	0.49	0.55		17.23	0.222	1.01	0.88	0.83	1.10	1.16
155–156	0.80	0.77	0.82	0.68	0.53		15.90	0.203	0.89	0.74	0.90	1.17	1.27
156–157	0.83	0.85	0.91	0.64	0.59		14.30	0.182	0.96	0.89	0.80	1.07	1.25
157–158	0.73	0.73	0.86	0.59	0.60		13.93	0.176	0.84	0.74	0.84	1.20	1.36
158–159	0.71	0.72	0.83	0.62	0.63		13.33	0.167	0.87	0.75	0.84	1.23	1.29
159–160	0.73	0.70	0.89	0.54	0.64		13.53	0.169	0.91	0.80	0.87	1.18	1.21
160–161	0.77	0.68	0.95	0.60	0.65		14.69	0.182	0.92	0.79	0.88	1.17	1.21
161–162	0.70	0.70	0.89	0.60	0.66		16.82	0.207	0.89	0.79	0.90	1.18	1.22
162–163	0.73	0.71	0.92	0.65	0.63		19.54	0.239	0.91	0.81	0.88	1.15	1.22
163–164	0.75	0.65	1.00	0.67	0.62		22.08	0.269	0.91	0.77	0.88	1.15	1.26
164–165	0.74	0.71	0.99	0.62	0.66		24.27	0.293	0.99	0.82	0.88	1.11	1.18
165–166	0.66	0.67	0.86	0.67	0.57		40.91	0.492	0.86	0.78	0.93	1.17	1.24
166–167	0.68	0.64	0.94	0.68	0.55		30.90	0.369	0.96	0.83	0.91	1.15	1.14
167–168	0.72	0.69	0.99	0.76	0.56		34.22	0.406	0.96	0.82	0.97	1.13	1.12
168–169	0.72	0.68	0.94	0.70	0.52		42.55	0.502	0.99	0.87	0.92	1.13	1.10
169–170	0.69	0.66	0.94	0.68	0.53		56.42	0.662	1.02	0.87	0.92	1.12	1.06
170–171	0.69	0.66	0.93	0.71	0.53		66.33	0.774	1.03	0.88	0.94	1.09	1.06
171–172	0.69	0.66	0.92	0.70	0.54		68.12	0.790	1.03	0.87	0.95	1.11	1.05
172–173	0.68	0.64	0.93	0.69	0.59		73.16	0.844	1.05	0.90	0.93	1.09	1.03
173–174	0.64	0.62	0.88	0.65	0.58		75.80	0.869	1.03	0.89	0.93	1.11	1.04
174–175	0.64	0.61	0.87	0.60	0.60	0.67	90.83	1.035	1.02	0.89	0.92	1.10	1.06
175–176	0.61	0.62	0.84	0.58	0.71	0.71	106.57	1.208	1.01	0.91	0.90	1.09	1.08
176–177	0.62	0.59	0.86	0.57	0.75	0.73	113.47	1.279	1.02	0.89	0.89	1.08	1.11
177–178	0.59	0.60	0.83	0.58	0.87	0.78	132.05	1.480	0.98	0.89	0.87	1.11	1.16
178–179	0.62	0.63	0.84	0.59	0.93	0.81	142.78	1.591	0.99	0.89	0.84	1.15	1.13
179–180	0.63	0.64	0.87	0.62	0.93	0.83	138.45	1.534	0.97	0.89	0.84	1.16	1.14
180–181	0.64	0.65	0.80	0.70	0.95	0.88	161.69	1.782	0.95	0.85	0.83	1.18	1.18
181–182	0.65	0.66	0.81	0.77	1.08	0.88	184.72	2.024	0.98	0.89	0.83	1.15	1.15
182–183	0.65	0.66	0.80	0.80	1.07	0.98	174.31	1.900	0.96	0.89	0.85	1.16	1.15
183–184	0.59	0.68	0.82	0.82	1.16	1.08	170.09	1.844	0.94	0.91	0.85	1.15	1.14
184–185	0.59	0.64	0.94	0.88	1.20	1.13	138.86	1.497	0.93	0.91	0.85	1.18	1.12
185–186			1.10	0.71	1.11	1.02	172.52	1.850	0.92	0.89	0.91	1.15	1.13
186–187			1.03	0.74	1.16	1.01	204.53	2.181	0.98	0.93	0.92	1.09	1.07
187–188			1.04	0.80	1.20	1.05	221.32	2.348	0.94	0.91	0.98	1.12	1.04
188–189			0.85	0.59	0.95	0.84	294.03	3.103	1.03	0.96	1.03	1.01	0.97
189–190			0.76	0.55	0.80	0.74	366.01	3.842	1.12	1.08	1.10	0.89	0.81

[a] From Rottman (1981). Five rocket experiments conducted between December 1972 and March 1977 were combined to give a solar minimum reference spectrum for the wavelength range 120–190 nm. Rottman also compared these data with other measurements, indicated by different identifying mnemonics. Since considerable uncertainty surrounds the absolute values of the solar UV flux, these values were included to give the reader an indication of the variability. Reference should be made to the original paper for further information.

Table A-4

ILLUSTRATIVE MEASUREMENTS OF THE SOLAR ULTRAVIOLET FLUX FOR THE WAVELENGTH RANGE 1150–3180 Å[a,b]

	Solar ultraviolet flux[c]				Solar ultraviolet flux[c]		
Wavelength	17 May 1982	15 July 1980	12 January 1983	Wavelength	17 May 1982	15 July 1980	12 January 1983
1150–1160	0.0121			1720–1730	0.643	0.999	
1160–1170	0.0193			1730–1740	0.655	1.02	
1170–1180	0.0374			1740–1750	0.867	1.05	
1180–1190	0.0186			1750–1760	1.05	1.22	
1190–1200	0.0320			1760–1770	1.13	1.39	
1200–1210	0.0854	0.236		1770–1780	1.29	1.42	
1210–1220	3.32	5.64		1780–1790	1.45	1.54	
1220–1230	0.0283	0.052		1790–1800	1.45	1.60	
1230–1240	0.0191	0.035		1800–1810	1.73	1.63	1.39
1240–1250	0.0126	0.028		1810–1820	2.09	2.01	1.71
1250–1260	0.0136	0.031		1820–1830	1.63	1.94	1.69
1260–1270	0.0157	0.041		1830–1840	1.68	2.05	1.81
1270–1280	0.0112	0.021		1840–1850	1.39	1.80	1.63
1280–1290	0.0087	0.022		1850–1860	1.73	2.05	1.82
1290–1300	0.0113	0.021		1860–1870	2.16	2.37	2.19
1300–1310	0.0733	0.154		1870–1880	2.55	2.77	2.68
1310–1320	0.0140	0.036		1880–1890	2.77	2.91	2.86
1320–1330	0.0120	0.026		1890–1900	3.20	3.31	3.36
1330–1340	0.111	0.251		1900–1910	3.50	3.41	3.56
1340–1350	0.0113	0.025		1910–1920	3.90	3.86	4.05
1350–1360	0.0245	0.053		1920–1930	4.21	4.15	4.37
1360–1370	0.0169	0.041		1930–1940	3.31	3.31	3.42
1370–1380	0.0188	0.041		1940–1950	5.14	4.85	5.16
1380–1390	0.0178	0.041		1950–1960	5.51	5.25	5.72
1390–1400	0.0547	0.128		1960–1970	5.99	5.62	6.14

1400–1410	0.0470	0.105		1970–1980	6.37	6.13	6.76
1410–1420	0.0275	0.060		1980–1990	6.17	6.07	6.48
1420–1430	0.0300	0.066		1990–2000	6.59	6.54	6.84
1430–1440	0.0366	0.076		2000–2010	7.21	7.23	7.55
1440–1450	0.0356	0.074		2010–2020	7.81	7.93	8.07
1450–1460	0.0363	0.079		2020–2030	7.82	8.26	8.29
1460–1470	0.0474	0.104		2030–2040	9.14	9.44	9.57
1470–1480	0.0590	0.125		2040–2050	10.2	10.5	10.8
1480–1490	0.0581	0.122		2050–2060	10.7	11.2	11.4
1490–1500	0.0556	0.119		2060–2070	11.0	11.5	11.7
1500–1510	0.0610	0.136		2070–2080	13.0	13.4	13.7
1510–1520	0.0654	0.143		2080–2090	14.2	15.2	15.1
1520–1530	0.0829	0.187		2090–2100	20.5	21.2	21.4
1530–1540	0.0927	0.212		2100–2110	28.2	29.1	28.1
1540–1550	0.170	0.327		2110–2120	31.6	33.3	33.3
1550–1560	0.154	0.315		2120–2130	34.3	35.6	35.7
1560–1570	0.139	0.258		2130–2140	32.8	34.6	33.7
1570–1580	0.127	0.244		2140–2150	41.2	43.4	41.6
1580–1590	0.123	0.223		2150–2160	39.8	42.2	39.8
1590–1600	0.130	0.223		2160–2170	33.3	36.1	33.6
1600–1610	0.145	0.253		2170–2180	32.9	36.6	31.9
1610–1620	0.173	0.280		2180–2190	48.2	51.7	46.8
1620–1630	0.199	0.326		2190–2200	48.9	54.4	47.3
1630–1640	0.232	0.404		2200–2210	55.3	58.1	54.2
1640–1650	0.235	0.411		2210–2220	37.5	42.9	36.9
1650–1660	0.378	0.634		2220–2230	52.3	59.9	50.5
1660–1670	0.269	0.491		2230–2240	73.7	77.9	72.1
1670–1680	0.312	0.542		2240–2250	68.3	70.7	66.7
1680–1690	0.345	0.554		2250–2260	61.2	62.9	60.0
1690–1700	0.452	0.745		2260–2270	46.5	45.1	45.6
1700–1710	0.575	0.900		2270–2280	44.0	49.1	42.5
1710–1720	0.568	0.953		2280–2290	64.3	65.3	63.0

(cont.)

Table A-4 *(cont.)*

Wavelength	Solar ultraviolet flux[c]			Wavelength	Solar ultraviolet flux[c]		
	17 May 1982	15 July 1980	12 January 1983		17 May 1982	15 July 1980	12 January 1983
2290–2300	56.7	57.5	56.4	2740–2750	184	182	182
2300–2310	63.5	68.8	61.8	2750–2760	222	214	221
2310–2320	58.0	59.4	57.1	2760–2770	352	324	352
2320–2330	59.9	65.4	59.0	2770–2780	366	350	370
2330–2340	50.9	54.2	49.6	2780–2790	254	250	271
2340–2350	38.7	48.7	37.9	2790–2800	135	133	158
2350–2360	59.9	67.8	58.2	2800–2810	141	139	136
2360–2370	53.8	61.6	52.4	2810–2820	312	303	319
2370–2380	64.7	61.5	62.8	2820–2830	441	416	444
2380–2390	48.4	50.4	47.1	2830–2840	486	458	481
2390–2400	56.7	55.2	55.5	2840–2850	344	347	326
2400–2410	51.6	48.2	49.2	2850–2860	253	241	268
2410–2420	62.7	67.0	60.6	2860–2870	526	498	528
2420–2430	92.8	88.2	90.7	2870–2880	486	476	475
2430–2440	82.4	78.7	80.2	2880–2890	520	479	540
2440–2450	78.1	71.4	76.2	2890–2900	753	684	776
2450–2460	62.3	60.5	61.1	2900–2910	921	870	912
2460–2470	63.3	63.7	62.1	2910–2920	878	828	878
2470–2480	70.6	68.2	69.8	2920–2930	761	733	767
2480–2490	55.2	51.2	55.1	2930–2940	852	800	845
2490–2500	73.3	77.6	71.7	2940–2950	815	760	776
2500–2510	75.9	72.9	75.5	2950–2960	796	744	860
2510–2520	58.2	54.4	58.6	2960–2970	892	800	776
2520–2530	53.4	54.1	52.7	2970–2980	666	689	826
2530–2540	69.8	69.1	69.0	2980–2990	791	697	677
2540–2550	74.8	80.0	74.0	2990–3000	758	720	694
2550–2560	113	108	111	3000–3010	610	567	656

2560–2570	136	142	134	3010–3020	682	666	638
2570–2580	167	165	167	3020–3030	776	725	857
2580–2590	175	157	170	3030–3040	929	876	923
2590–2600	124	117	120	3040–3050	892	847	934
2600–2610	125	117	124	3050–3060	872	826	868
2610–2620	124	127	125	3060–3070	861	828	905
2620–2630	147	133	142	3070–3080	977	900	961
2630–2640	221	252	232	3080–3090	940	872	902
2640–2650	354	323	345	3090–3100	726	668	678
2650–2660	348	343	353	3100–3110	951	928	1063
2660–2670	342	317	336	3110–3120	1062	1010	1015
2670–2680	347	330	342	3120–3130	967	922	989
2680–2690	337	319	333	3130–3140	1043	921	1077
2690–2700	336	326	337	3140–3150	933	1000	935
2700–2710	381	356	376	3150–3160	888	866	843
2710–2720	307	289	297	3160–3170	899	876	948
2720–2730	279	266	282	3170–3180			126
2730–2740	288	283	274				

[a] From Mount and Rottman (1983a,b).

[b] For three flights in 1980, 1982, and 1983. The data acquired below 1800 Å are included for completeness. The three data sets serve to indicate the variability in the measurements.

[c] In units of 10^{10} photons cm^{-2} sec^{-1} $Å^{-1}$, averaged in 10-Å intervals at 1 astronomical unit (AU) for rocket flight 33.028, 17 May 1982, rocket flight 27.004, 15 July 1980 (revised above 2400 Å), and rocket flight 33.030, 12 January 1983.

Table A-5

WEIGHTED PHOTOIONIZATION AND PHOTOABSORPTION CROSS SECTIONS σ_{eff} FOR MAJOR THERMOSPHERIC SPECIES[a]

		Species										
Interval	$\Delta\lambda$ (Å)	$O^+(^4S)$	$O^+(^2D)$	$O^+(^2P)$	$O^+(^4P)$	$O^+(^2P^*)$	Total O^+	He^+	N_2(abs)	N_2(ion)	O_2(abs)	O_2(ion)
1	50–100	0.32	0.34	0.22	0.10	0.03	1.06	0.21	0.60	0.60	1.18	1.18
2	100–150	1.03	1.14	0.75	0.34	0.27	3.53	0.53	2.32	2.32	3.61	3.61
3	150–200	1.62	2.00	1.30	0.58	0.46	5.96	1.02	5.40	5.40	7.27	7.27
4	200–250	1.95	2.62	1.70	0.73	0.54	7.55	1.71	8.15	8.15	10.50	10.50
5	256.3	2.15	3.02	1.95	0.82	0.56	8.43	2.16	9.65	9.65	12.80	12.80
6	284.15	2.33	3.39	2.17	0.89	0.49	9.26	2.67	10.60	10.60	14.80	14.80
7	250–300	2.23	3.18	2.04	0.85	0.52	8.78	2.38	10.08	10.08	13.65	13.65
8	303.31	2.45	3.62	2.32	0.91	0.41	9.70	3.05	11.58	11.58	15.98	15.98
9	303.78	2.45	3.63	2.32	0.91	0.41	9.72	3.05	11.60	11.60	16.00	16.00
10	300–350	2.61	3.98	2.52	0.93	0.00	10.03	3.65	14.60	14.60	17.19	17.19
11	368.07	2.81	4.37	2.74	0.92	0.00	10.84	4.35	18.00	18.00	18.40	18.40
12	350–400	2.77	4.31	2.70	0.92	0.00	10.70	4.25	17.51	17.51	18.17	18.17
13	400–450	2.99	4.75	2.93	0.55	0.00	11.21	5.51	21.07	21.07	19.39	19.39
14	465.22	3.15	5.04	3.06	0.00	0.00	11.25	6.53	21.80	21.80	20.40	20.40
15	450–500	3.28	5.23	3.13	0.00	0.00	11.64	7.09	21.85	21.85	21.59	21.59
16	500–550	3.39	5.36	3.15	0.00	0.00	11.91	0.72	24.53	24.53	24.06	24.06
17	554.37	3.50	5.47	3.16	0.00	0.00	12.13	0.00	24.69	24.69	25.59	25.59
18	584.33	3.58	5.49	3.10	0.00	0.00	12.17	0.00	23.20	23.20	22.00	22.00
19	550–600	3.46	5.30	3.02	0.00	0.00	11.90	0.00	22.38	22.38	25.04	25.04
20	609.76	3.67	5.51	3.05	0.00	0.00	12.23	0.00	23.10	23.10	26.10	26.10
21	629.73	3.74	5.50	2.98	0.00	0.00	12.22	0.00	23.20	23.20	25.80	25.80

22	600–650	3.73	5.50	2.97	0.00	0.00	12.21	0.00	23.22	23.22	26.02	25.94
23	650–700	4.04	5.52	0.47	0.00	0.00	10.04	0.00	29.75	25.06	26.27	22.05
24	730.36	4.91	6.44	0.00	0.00	0.00	11.35	0.00	26.30	23.00	25.00	23.00
25	700–750	4.20	3.80	0.00	0.00	0.00	8.00	0.00	30.94	23.20	29.05	23.81
26	765.15	4.18	0.00	0.00	0.00	0.00	4.18	0.00	35.46	23.77	21.98	8.59
27	770.41	4.18	0.00	0.00	0.00	0.00	4.18	0.00	26.88	18.39	25.18	9.69
28	789.36	4.28	0.00	0.00	0.00	0.00	4.28	0.00	19.26	10.18	26.66	11.05
29	750–800	4.23	0.00	0.00	0.00	0.00	4.23	0.00	30.71	16.75	27.09	9.39
30	800–850	4.38	0.00	0.00	0.00	0.00	4.38	0.00	15.05	0.00	20.87	6.12
31	850–900	4.18	0.00	0.00	0.00	0.00	4.18	0.00	46.63	0.00	9.85	4.69
32	900–950	2.12	0.00	0.00	0.00	0.00	2.12	0.00	16.99	0.00	15.54	9.34
33	977.62	0.00	0.00	0.00	0.00	0.00	0.00	0.00	0.70	0.00	4.00	2.50
34	950–1000	0.00	0.00	0.00	0.00	0.00	0.00	0.00	36.16	0.00	16.53	12.22
35	1025.72	0.00	0.00	0.00	0.00	0.00	0.00	0.00	0.00	0.00	1.60	1.00
36	1031.91	0.00	0.00	0.00	0.00	0.00	0.00	0.00	0.00	0.00	1.00	0.00
37	1000–1050	0.00	0.00	0.00	0.00	0.00	0.00	0.00	0.00	0.00	1.10	0.27

[a] From M. R. Torr *et al.* (1979). Units of cross sections are 10^{-18} cm^2.

Table A-6
SOLAR EXTREME-ULTRAVIOLET FLUX AT 1 AU

Wavelength (Å)	Ion	Solar flux period								
		(Calib.) Day 113[a,b] 1974	Day 200[a] 1976	Revised[c]	Day 348[a,d] 1978	Day 22[a,d] 1979	Day 50[a,d] 1979	Revised[c]	(Peak) Day 314[e] 1979	(Calib.) Day 226[b] 1979
50–100		0.3984	0.4382	0.323	1.0337	1.2904	1.3710	1.149	1.602	0.945
100–150		0.1497	0.1687	0.135	0.3623	0.4419	0.4675	0.343	0.455	0.284
150–200		2.3683	1.8692	1.842	4.1772	5.3708	5.7024	4.850	7.177	4.241
200–250		1.5632	1.3951	0.923	4.7953	6.6473	7.1448	3.701	5.752	3.108
256.3	He(II), Si(X)	0.4600	0.5064		0.8805	1.0331	1.0832			
284.15	Fe(XV)	0.2100	0.0773	1.212	3.2613	5.2352	5.7229	7.898	13.094	6.296
250–300		1.6794	1.3556		7.5081	11.2278	12.1600			
303.31	Si(XI)	0.8000	0.6000		2.9100	4.3380	4.6908			
303.78	He(II)	6.9000	7.7625	7.101	12.3424	13.8172	14.3956	19.412	25.790	13.997
300–350		0.9650	0.8671		4.3119	6.3164	6.8315			
368.07	Mg(IX)	0.6500	0.7394	0.951	1.2891	1.4661	1.5355	3.591	5.698	2.773
350–400		0.3140	0.2121		1.5298	2.3413	2.5423			
400–450		0.3832	0.4073	0.393	1.0922	1.4330	1.5310	0.993	1.399	0.820
465.22	Ne(VII)	0.2900	0.3299	0.474	0.6102	0.7004	0.7358	2.002	3.076	1.662
450–500		0.2851	0.3081		1.2120	1.6912	1.8228			
500–550		0.4520	0.5085	0.497	1.2303	1.5496	1.6486	1.528	2.086	1.297
554.37	O(IV)	0.7200	0.7992		1.2943	1.4537	1.5163			
584.33	He(I)	1.2700	1.5875	2.887	3.4608	4.0646	4.3005	7.528	9.014	6.290
550–600		0.3568	0.4843		0.8732	0.9985	1.0477			
609.76	Mg(X)	0.5300	0.6333		1.6782	2.3242	2.4838			
629.73	O(V)	1.5900	1.8484	2.125	3.2443	3.6938	3.8701	4.962	6.487	4.208
600–650		0.3421	0.4002		0.9606	1.2842	1.3672			
650–700		0.2302	0.2623	0.221	0.4521	0.5149	0.5388	0.452	0.548	0.387

703.31	O(III)	0.3600	0.3915	0.556	0.6363	0.7152	0.7461	1.135	1.319	0.976
700–750		0.1409	0.1667		0.3439	0.4046	0.4287			
765.15	N(IV)	0.1700	0.1997		0.3647	0.4178	0.4386			
770.41	Ne(VIII)	0.2600	0.2425	2.089	0.7760	1.1058	1.1873	4.844	6.053	4.129
789.36	O(IV)	0.7024	0.7831		1.2870	1.4501	1.5140			
750–800		0.7581	0.8728		1.8909	2.3132	2.4541			
800–850		1.6250	1.9311	1.900	3.9278	4.5911	4.8538	4.919	5.886	4.089
850–900		3.5370	4.4325	4.323	9.7798	11.5292	12.2187	12.968	15.735	10.591
900–950		3.0003	3.6994	4.366	7.9445	9.3134	9.8513	12.486	15.083	10.252
977.02	C(III)	4.4000	4.8400	7.735	8.5523	9.7478	10.2165	17.583	20.733	15.055
950–1000		1.4746	1.7155		3.3468	3.8723	4.0778			
1025.72	H(I)	3.5000	4.3750		9.5375	11.2000	11.8519			
1031.91	O(VI)	2.1000	1.9425	11.170	4.2929	5.7459	6.1049	30.812	37.099	26.474
1000–1050		2.4665	2.4775		4.7145	5.7798	6.0928			
		71.0	68.0		206.0	234.0	243.0			

[a] M. R. Torr *et al.* (1979). Units of intensity are 10^9 photons cm^{-2} sec^{-1}

[b] Rocket flight calibration.

[c] Revised by H. E. Hinteregger 16 June 1981 (Day 200, 1976 ID = SC # 21REFW; available from NSSDC at 1-Å intervals); note that line values are included in 50-Å sum.

[d] Revised values not available.

[e] Large values could represent a 27-day excursion over mean trend.

Table A-7a

PHOTOIONIZATION BRANCHING RATIOS FOR O^+

Wavelength (Å)	^4S	^2D	^2P	^4P	^2P
1025.00	0	0	0	0	0
1031.91	0	0	0	0	0
1025.72	0	0	0	0	0
975.00	0	0	0	0	0
977.02	0	0	0	0	0
925.00	1.000 + 0[a]	0	0	0	0
875.00	1.000 + 0	0	0	0	0
825.00	1.000 + 0	0	0	0	0
775.00	1.000 + 0	0	0	0	0
789.36	1.000 + 0	0	0	0	0
770.41	1.000 + 0	0	0	0	0
765.15	1.000 + 0	0	0	0	0
725.00	5.250 − 1	4.750 − 1	0	0	0
730.36	4.326 − 1	5.674 − 1	0	0	0
675.00	4.028 − 1	5.503 − 1	4.686 − 2	0	0
625.00	3.057 − 1	4.508 − 1	2.434 − 1	0	0
629.73	3.061 − 1	4.501 − 1	2.439 − 1	0	0
609.76	3.001 − 1	4.505 − 1	2.494 − 1	0	0
575.00	2.937 − 1	4.499 − 1	2.564 − 1	0	0
584.33	2.942 − 1	4.511 − 1	2.547 − 1	0	0
554.31	2.885 − 1	4.509 − 1	2.605 − 1	0	0
525.00	2.849 − 1	4.504 − 1	2.647 − 1	0	0
475.00	2.818 − 1	4.493 − 1	2.689 − 1	0	0
465.22	2.800 − 1	4.480 − 1	2.720 − 1	0	0
425.00	2.665 − 1	4.234 − 1	2.611 − 1	4.902 − 2	0
375.00	2.589 − 1	4.028 − 1	2.523 − 1	8.598 − 2	0
368.07	2.592 − 1	4.031 − 1	2.528 − 1	8.487 − 2	0
325.00	2.600 − 1	3.964 − 1	2.510 − 1	9.263 − 2	0
303.78	2.521 − 1	3.735 − 1	2.387 − 1	9.362 − 2	4.218 − 2
303.31	2.523 − 1	3.728 − 1	2.389 − 1	9.372 − 2	4.222 − 2
275.00	2.528 − 1	3.605 − 1	2.313 − 1	9.637 − 2	5.896 − 2
284.15	2.513 − 1	3.657 − 1	2.341 − 1	9.601 − 2	5.286 − 2
256.30	2.529 − 1	3.553 − 1	2.294 − 1	9.647 − 2	6.588 − 2
225.00	2.586 − 1	3.475 − 1	2.255 − 1	9.682 − 2	7.162 − 2
175.00	2.718 − 1	3.356 − 1	2.181 − 1	9.732 − 2	7.718 − 2
125.00	2.918 − 1	3.229 − 1	2.125 − 1	9.632 − 2	7.649 − 2
75.00	3.019 − 1	3.208 − 1	2.075 − 1	9.434 − 2	7.547 − 2

[a] $1.000 + 0 \equiv 1.000 \times 10^0$.

Table A-7b

PHOTOIONIZATION BRANCHING RATIOS FOR O_2^+

Wavelength (Å)	$X\,^2\Pi_g$	$a\,^4\Pi_u + A\,^2\Pi_u$	$b\,^4\Sigma_g^-$	$B\,^2\Sigma_g^-$	Remaining species
1025.00	1.000 + 0	0	0	0	0
1031.91	1.000 + 0	0	0	0	0
1025.72	1.000 + 0	0	0	0	0
975.00	1.000 + 0	0	0	0	0
977.02	1.000 + 0	0	0	0	0
925.00	1.000 + 0	0	0	0	0
875.00	1.000 + 0	0	0	0	0
825.00	1.000 + 0	0	0	0	0
775.00	1.000 + 0	0	0	0	0
789.36	1.000 + 0	0	0	0	0
770.41	9.578 − 1	4.221 − 2	0	0	0
765.15	8.960 − 1	1.040 − 1	0	0	0
725.00	5.650 − 1	4.350 − 1	0	0	0
730.36	5.650 − 1	4.350 − 1	0	0	0
675.00	3.953 − 1	4.636 − 1	1.411 − 1	0	0
625.00	2.404 − 1	3.654 − 1	3.409 − 1	5.330 − 2	0
629.73	2.421 − 1	3.592 − 1	3.522 − 1	4.647 − 2	0
609.76	2.349 − 1	3.832 − 1	3.064 − 1	7.556 − 2	0
575.00	3.542 − 1	2.809 − 1	2.149 − 1	1.501 − 1	0
584.33	3.039 − 1	3.288 − 1	2.137 − 1	1.536 − 1	0
554.31	3.551 − 1	2.354 − 1	2.166 − 1	1.930 − 1	0
525.00	3.594 − 1	2.829 − 1	1.286 − 1	2.196 − 1	9.461 − 3
475.00	2.654 − 1	1.606 − 1	7.735 − 2	8.498 − 2	4.117 − 1
465.22	2.428 − 1	1.397 − 1	6.809 − 2	6.422 − 2	4.852 − 1
425.00	4.192 − 1	2.357 − 1	1.209 − 1	1.226 − 1	1.016 − 1
375.00	3.970 − 1	2.231 − 1	1.224 − 1	1.444 − 1	1.131 − 1
368.07	3.940 − 1	2.214 − 1	1.226 − 1	1.474 − 1	1.147 − 1
325.00	3.749 − 1	2.105 − 1	1.239 − 1	1.661 − 1	1.245 − 1
303.78	3.650 − 1	2.050 − 1	1.250 − 1	5.501 − 2	2.500 − 1
303.31	3.650 − 1	2.050 − 1	1.250 − 1	5.501 − 2	2.500 − 1
275.00	3.650 − 1	2.050 − 1	1.250 − 1	5.501 − 2	2.500 − 1
284.15	3.650 − 1	2.050 − 1	1.250 − 1	5.501 − 2	2.500 − 1
256.30	3.650 − 1	2.050 − 1	1.250 − 1	5.501 − 2	2.500 − 1
225.00	3.650 − 1	2.050 − 1	1.250 − 1	5.501 − 2	2.500 − 1
175.00	3.650 − 1	2.050 − 1	1.250 − 1	5.501 − 2	2.500 − 1
125.00	3.650 − 1	2.050 − 1	1.250 − 1	5.501 − 2	2.500 − 1
75.00	3.650 − 1	2.050 − 1	1.250 − 1	5.501 − 2	2.500 − 1

Table A-7c
PHOTOIONIZATION BRANCHING RATIOS FOR N_2^+

Wavelength (Å)	$X\ ^2\Sigma_g^+$	$A\ ^2\Pi_u$	$B\ ^2\Sigma_u^+$	$D\ ^2\Pi_g$	$F\ ^2\Sigma_u$	$2S\sigma_g$
1025.00	1.000 + 0	0	0	0	0	0
1031.91	1.000 + 0	0	0	0	0	0
1025.72	1.000 + 0	0	0	0	0	0
975.00	1.000 + 0	0	0	0	0	0
977.02	1.000 + 0	0	0	0	0	0
925.00	1.000 + 0	0	0	0	0	0
875.00	1.000 + 0	0	0	0	0	0
825.00	1.000 + 0	0	0	0	0	0
775.00	1.000 + 0	0	0	0	0	0
789.36	1.000 + 0	0	0	0	0	0
770.41	1.000 + 0	0	0	0	0	0
765.15	1.000 + 0	0	0	0	0	0
725.00	5.274 − 1	4.726 − 1	0	0	0	0
730.36	6.425 − 1	3.575 − 1	0	0	0	0
675.00	3.360 − 1	6.640 − 1	0	0	0	0
625.00	3.080 − 1	5.890 − 1	1.030 − 1	0	0	0
629.73	3.080 − 1	5.890 − 1	1.030 − 1	0	0	0
609.76	3.080 − 1	5.890 − 1	1.030 − 1	0	0	0

575.00	3.320 − 1	5.683 − 1	9.975 − 2	0	0	0
584.33	3.230 − 1	5.760 − 1	1.010 − 1	0	0	0
554.31	3.519 − 1	5.511 − 1	9.706 − 2	0	0	0
525.00	3.800 − 1	5.267 − 1	9.325 − 2	0	0	0
475.00	4.058 − 1	4.696 − 1	8.291 − 2	3.929 − 2	2.465 − 3	0
465.22	4.150 − 1	4.657 − 1	8.199 − 2	3.422 − 2	3.062 − 3	0
425.00	4.398 − 1	4.464 − 1	7.853 − 2	2.856 − 2	6.687 − 3	6.621 − 5
375.00	3.618 − 1	4.801 − 1	9.938 − 2	2.668 − 2	3.007 − 2	1.944 − 3
368.07	3.467 − 1	4.790 − 1	1.011 − 1	2.968 − 2	4.077 − 2	2.741 − 3
325.00	2.674 − 1	4.626 − 1	1.073 − 1	2.263 − 2	1.147 − 1	2.542 − 2
303.78	2.561 − 1	4.405 − 1	1.038 − 1	2.114 − 2	3.903 − 2	8.942 − 2
303.31	2.558 − 1	4.400 − 1	1.038 − 1	2.108 − 2	8.824 − 2	9.108 − 2
275.00	2.353 − 1	4.044 − 1	9.559 − 2	1.158 − 2	6.949 − 2	1.836 − 1
284.15	2.421 − 1	4.161 − 1	9.835 − 2	1.470 − 2	7.809 − 2	1.507 − 1
256.30	2.214 − 1	3.805 − 1	8.994 − 2	6.278 − 3	3.767 − 2	2.642 − 1
225.00	2.070 − 1	3.559 − 1	8.411 − 2	0.000 + 0	0.000 + 0	3.530 − 1
175.00	2.048 − 1	3.520 − 1	8.320 − 2	0.000 + 0	0.000 + 0	3.600 − 1
125.00	2.048 − 1	3.520 − 1	8.320 − 2	0.000 + 0	0.000 + 0	3.600 − 1
75.00	2.048 − 1	3.520 − 1	8.320 − 2	0.000 + 0	0.000 + 0	3.600 − 1

Table A-8

RATE COEFFICIENTS FOR REGION F AND E PROCESSES

Reaction number	Reaction	Rate coefficient: Coefficient	Rate coefficient: Value (two-body: cm^3 sec^{-1}; three-body: cm^6 sec^{-1})
(59)	$O^+ + e^- \rightarrow O + h\nu$		$\sim 4 \times 10^{-12}(T/300)^{-0.7}$
(60)	$O^+ + O^- \rightarrow O_2 + e^-$		1.4×10^{-10}
(61)	$O_2^+ + e^- \rightarrow O + O$	α_1	$1.6 \times 10^{-7}(300/T_e)^{0.55}$ for $T_e \geqslant 1200$ K $2 \times 10^{-7}(300/T_e)^{0.7}$ for $T_e < 1200$ K
(62)	$O^+ + O_2 \rightarrow O_2^+ + O$	k_{12}	$1.25 \times 10^{-17} T_{eff}^2 - 3.7 \times 10^{-14} \times T_{eff} + 3.10 \times 10^{-11}$
(63)	$O^+ + N_2 \rightarrow NO^+ + N$	k_{13}	$1.533 \times 10^{-12} - 5.92 \times 10^{-13} \times (T_{eff}/300) + 8.60 \times 10^{-14} \times (T_{eff}/300)^2$ $300 \leq T_{eff} \leq 1700$ K $2.73 \times 10^{-12} - 1.155 \times 10^{-12} \times (T_{eff}/300) + 1.483 \times 10^{-13} \times (T_{eff}/300)^2$ $1700 < T_{eff} < 6000$ K
(64)	$NO^+ + e^- \rightarrow N + O$	α_2	$(4.3^{+0.3}_{-1.3}) \times 10^{-7}(T_e/300)^{(-1^{+0.1}_{-0.2})}$ (aeronomy) $2.3 \times 10^{-7}(T_e/300)^{-0.5}(\pm 15\%)$ (laboratory)
(65)	$N_2^+ + O \rightarrow NO^+ + N$	k_{14}	$1.4 \times 10^{-10}(T_i/300)^{-0.44}$ for $T_i < 1500$ K
(66)	$O^+(^2D) + N_2 \rightarrow N_2^+ + O$	k_{15}	$(8 \pm 2) \times 10^{-10}$
(67)	$N_2^+ + e^- \rightarrow N + N$	α_3	$2.2 \times 10^{-7}(300/T_e)^{0.39} \times (T_v/300)^{0.04}$ (vibrational temperature dependence could be larger)
(68)	$O^+(^2P) + N_2 \rightarrow N_2^+ + O$	k_{16}	$(4.8 \pm 1.4) \times 10^{-10}$
(69)	$N_2^+ + O \rightarrow O^+ + N_2$	k_{17}	$0.07 k_{14}(T_i/300)^{0.21}$
(70)	$N_2^+ + O_2 \rightarrow O_2^+ + N_2$	k_{18}	6×10^{-11}
(71)	$N^+ + O_2 \rightarrow O_2^+ + O$	k_{19}	4×10^{-10}
(72)	$N^+ + O_2 \rightarrow NO^+ + O$	k_{20}	2×10^{-10}
(73)	$O_2^+ + N(^2D) \rightarrow NO^+ + O$	k_{21}	$\gtrsim 1 \times 10^{-10}$
(74)	$O_2^+ + N(^4S) \rightarrow NO^+ + O$	k_{22}	$\sim 1 \times 10^{-10}$
(75)	$O_2^+ + NO \rightarrow NO^+ + O_2$	k_{23}	4.4×10^{-10}
(76)	$N_2^{+*} + O(^3P) \rightarrow O^+(^4S) + N_2$	k_{24}	$\sim 2 \times 10^{-10}$ (see Fig. 29)

Table A-9

RATE COEFFICIENTS FOR REACTIONS OF METASTABLE O^+

Reaction number	Reaction	Rate coefficient
(82)	$O^+(^2D) + N_2 \rightarrow N_2^+ + O$	$(8 \pm 2) \times 10^{-10}$ cm^3 sec^{-1}
(83)	$O^+(^2D) + O \rightarrow O^+(^4S) + O$	$\lesssim 5 \times 10^{-12}$ cm^3 sec^{-1}
(84)	$O^+(^2D) + O_2 \rightarrow O_2^+ + O$	$(7 \pm 2) \times 10^{-10}$ cm^3 sec^{-1}
(85)	$O^+(^2D) + e^- \rightarrow O^+(^4S) + e^-$	$6.6 \times 10^{-8}(300/T_e)^{0.5}$ cm^3 sec^{-1}
(86)	$O^+(^2P) \rightarrow O^+(^2D) + h\nu$	0.173 sec^{-1}
(87)	$O^+(^2P) \rightarrow O^+(^4S) + h\nu$	0.047 sec^{-1}
(88)	$O^+(^2P) + e^- \rightarrow O^+(^2D) + e^-$	$1.5 \times 10^{-7}(300/T_e)^{1/2}$ cm^3 sec^{-1}
(89)	$O^+(^2P) + e^- \rightarrow O^+(^4S) + e^-$	$4.0 \times 10^{-8}(300/T_e)^{1/2}$ cm^3 sec^{-1}
(90)	$O^+(^2P) + N_2 \rightarrow$ Products	$(4.8 \pm 1.4) \times 10^{-10}$ cm^3 sec^{-1}
(91)	$O^+(^2P) + O \rightarrow$ Products	$(5.2 \pm 2.5) \times 10^{-11}$ cm^3 sec^{-1}

Table A-10

RATE COEFFICIENTS FOR REACTIONS OF METASTABLE N^+

Reaction number	Reaction	Rate coefficient
(93)	$N^+(^1D) \rightarrow N^+(^3P) + h\nu$	4×10^{-3} sec^{-1}
(94)	$N^+(^1D) + e^- \rightarrow N^+(^3D) + e^-$	$\sim 10^{-7}$ cm^3 sec^{-1}
(95)	$N^+(^1D) + O \rightarrow$ Products	$\lesssim 5 \times 10^{-12}$ cm^3 sec^{-1}
(96)	$N^+(^1D) + N_2 \rightarrow$ Products	$\lesssim 10^{-11}$ cm^3 sec^{-1}
(97)	$N^+(^1S) \rightarrow N^+(^1D) + h\nu$	1.01 sec^{-1}
(98)	$N^+(^5S) \rightarrow N^+(^3P) + h\nu$	$\sim 2 \times 10^2$ sec^{-1}
(99)	$N^+(^5S) + M \rightarrow N^+(^3P) + M$	$\lesssim 10^{-10}$ cm^3 sec^{-1}

Table A-11

REACTIONS AND RATE COEFFICIENTS FOR REACTIONS OF NO^+ $(a^3\Sigma_g^+)$ and O_2^+ $(a^4\Pi_u)$

Reaction number	Reaction	Rate coefficient	
		Coefficient	Value (two-body: cm^3 sec^{-1})
(100)	$N_2^+ + NO \rightarrow NO^+(a\,^3\Sigma_g^+)$	k_{25}	$(0.5\text{–}2.7) \times 10^{-10}$
(101)	$N^+ + O_2 \rightarrow NO^+(a\,^3\Sigma_g^+)$	k_{26}	$(0.2\text{–}2) \times 10^{-10}$
(102)	$NO^+(a\,^3\Sigma_g^+) + N_2 \rightarrow N_2^+ + NO$		
(103)	$NO^+(a\,^3\Sigma_g^+) + N_2 \rightarrow NO^+ + N_2(A)$		$(2.5\text{–}5) \times 10^{-10}$ (102–104)
(104)	$NO^+(a\,^3\Sigma_g^+) + N_2 \rightarrow$ Other products		
(105)	$NO^+(a\,^3\Sigma_g^+) + O_2 \rightarrow O_2^+ + NO$		$(4.5 \pm 2.3) \times 10^{-10}$ (105–106)
(106)	$NO^+(a\,^3\Sigma_g^+) + O_2 \rightarrow$ Other		
(107)	$NO^+(a\,^3\Sigma_g^+) + O \rightarrow O^+ + NO$		$(1\text{–}5) \times 10^{-10}$
(108)	$NO^+(a\,^3\Sigma_g^+) \rightarrow NO^+ + h\nu$		$(1\text{–}10)$ sec^{-1}
(109)	$O_2^+(a\,^4\Pi_u) + N_2 \rightarrow N_2^+ + O_2$		4.1×10^{-10} (109–110)
(110)	$O_2^+(a\,^4\Pi_u) + N_2 \rightarrow O_2^+(X\,^2\Pi_g) + N_2$		
(111)	$O_2^+(a\,^4\Pi_u) + O_2 \rightarrow O_2^+(X\,^2\Pi_g) + O_2$		3.1×10^{-10}
(112)	$O_2^+(a\,^4\Pi_u) + NO \rightarrow NO^+ + O_2$		1.1×10^{-9}
(113)	$O_2^+(a\,^4\Pi_u) \rightarrow O_2^+(X\,^2\Pi_g) + h\nu$		(0.1–1) sec (uncertain)

Table A-12

RATE COEFFICIENTS FOR REACTIONS OF MINOR IONS $N^+(^3P)$, He^+, O^{2+}, and H^+

Reaction number	Reaction	Rate coefficient: Coefficient	Rate coefficient: Value (two-body: $cm^3\ sec^{-1}$)
(126)	$O^+ + N(^2D) \rightarrow N^+ + O$	k_{31}	$(5^{+8}_{-2}) \times 10^{-13}$
(129)	$O^+(^2P) + N_2 \rightarrow N^+ + NO$	k_{32}	$(1^{+1.0}_{-0.5}) \times 10^{-10}$
(130)	$O^+(^2D) + N \rightarrow N^+ + O$	k_{33}	$\lesssim 7.5 \times 10^{-11}$
(132)	$He^+ + N_2 \rightarrow N^+ + N + He$	k_{34}	1×10^{-9}
(133)	$N^+ + O_2 \rightarrow NO^+ + O$	k_{35}	2×10^{-10}
(134)	$N^+ + O_2 \rightarrow O_2{}^+ + N$	k_{36a}	4×10^{-10}
(135)	$N^+ + O_2 \rightarrow O^+ + NO$	k_{36b}	2×10^{-11}
(136)	$N^+ + O \rightarrow O^+ + N$	k_{37}	$\sim 10^{-12}$
(138)	$He^+ + N_2 \rightarrow N_2{}^+ + He$	k_{38}	6.5×10^{-10}
(144)	$O^{2+} + N_2 \rightarrow$ Products	k_{41}	$(1.3 \pm 0.3) \times 10^{-9}$
(145)	$O^{2+} + O_2 \rightarrow$ Products		$(1.5 \pm 0.3) \times 10^{-9}$
(146)	$O^{2+} + CO_2 \rightarrow$ Products		$(2 \pm 0.5) \times 10^{-9}$
(147)	$O^{2+} + O \rightarrow 2\,O^+$	k_{42}	$\sim 10^{-10}$
(148)	$O^+ + H \rightarrow H^+ + O$	k_{43f}	3.7×10^{-10} at 300 K
(148)	$H^+ + O \rightarrow O^+ + O$	k_{43r}	3.3×10^{-10} at 300 K

Table A-13

REACTIONS INVOLVED IN THE CONVERSION OF $O_2{}^+$ AND NO^+ TO PROTON HYDRATES IN THE D REGION

Reaction number	Reaction	Rate coefficient value (two-body: $cm^3\ sec^{-1}$; three-body: $cm^6\ sec^{-1}$)
(163)	$O_2{}^+ + 2\,O_2 \rightarrow O_4{}^+ + O_2$	$2.6 \times 10^{-30}(T/300)^{3.2}$
(164)	$O_4{}^+ + O \rightarrow O_2{}^+ + O_3$	$(3 \pm 2) \times 10^{-10}$
(165)	$O_4{}^+ + H_2O \rightarrow O_2{}^+(H_2O)$	1.5×10^{-9}
(166)	$O_2{}^+(H_2O) + H_2O \rightarrow H_3O^+(OH) + O_2$	1.0×10^{-9}
(167)	$O_2{}^+H_2O + H_2O \rightarrow H_3O + OH + O_2$	2×10^{-10}
(168)	$H_3O^+(OH) + H_2O \rightarrow H^+(H_2O)_2 + OH$	1.4×10^{-9}
(169)	$H_3O^+ + H_2O + M \rightarrow H^+(H_2O)_2 + M$	$(1.2 \pm 0.4) \times 10^{-27}$
(170)	$O_2{}^+(H_2O) + h\nu \rightarrow$ Products	$0.6\ sec^{-1}$
(171)	$NO^+ + 2\,N_2 \rightarrow NO^+(N_2) + N_2$	$2 \times 10^{-31}(300/T)^{4.4}$
(172)	$NO^+ + N_2 + M \rightarrow NO^+(N_2) + M$	$2 \times 10^{-31}(300/T)^5$ (125–180 K)
(173)	$NO^+ + N_2 + NO \rightarrow NO^+(N_2) + NO$	$(2.0 \pm 0.6) \times 10^{-31}$
(174)	$NO^+(N_2) + CO_2 \rightarrow NO^+(CO_2) + N_2$	$\sim 10^{-9}$
(175)	$NO^+(N_2) + M \rightarrow NO^+ + N_2 + M$	3×10^{-13} (210–220 K)
(176)	$NO^+ + CO_2 + N_2 \rightarrow NO^+(CO_2) + N_2$	$(2.5 \pm 1.5) \times 10^{-29}$ (200 K)
(177)	$NO^+(CO_2) + H_2O \rightarrow NO^+(H_2O) + CO_2$	$\sim 10^{-9}$
(178)	$NO^+ + H_2O + N_2 \rightarrow NO^+(H_2O) + N_2$	1.6×10^{-28}
(179)	$NO^+(H_2O) + H_2O + N_2 \rightarrow NO^+(H_2O)_2 + N_2$	1.1×10^{-27}
(180)	$NO^+(H_2O) + h\nu \rightarrow$ Products	$0.06\ sec^{-1}$
(181)	$NO^+(H_2O)_2 + H_2O + N_2 \rightarrow NO^+(H_2O)_3 + N_2$	1.9×10^{-27}
(182)	$NO^+(H_2O)_3 + H_2O \rightarrow H^+(H_2O)_3 + HNO_2$	7×10^{-11}

Table A-14

NEGATIVE ION REACTION RATES

Reaction number	Reaction	Rate coefficient value (two-body: $cm^3\ sec^{-1}$; three-body: $cm^6\ sec^{-1}$)
(184)	$O_2^- + O \rightarrow e^- + O_3$	1.5×10^{-10}
(185)	$O_2^- + O_2(^1\Delta_g) \rightarrow e^- + O_2 + O_2$	2.0×10^{-10}
(186)	$O_2^- + CO_2 + M \rightarrow CO_4^- + M$	4.7×10^{-29}
(187)	$O_2^- + O_3 \rightarrow O_3^- + O_2$	6.0×10^{-10}
(188)	$O_2^- + NO_2 \rightarrow NO_2^- + O_2$	7.0×10^{-10}
(189)	$O_2^- + O \rightarrow O^- + O_2$	1.5×10^{-10}
(190)	$O_2^- + O_2 + M \rightarrow O_4^- + M$	3.4×10^{-31}[a]
(191)	$O^- + O_2(^1\Delta_g) \rightarrow e^- + O_3$	3.0×10^{-10}
(192)	$O^- + M \rightarrow e^- +$ Neutrals	1.0×10^{-12}
(193)	$O^- + O_3 \rightarrow O_3^- + O$	8.0×10^{-10}
(194)	$O^- + H_2 \rightarrow e^- + H_2O$	5.8×10^{-10}
(195)	$O^- + H_2 \rightarrow OH^- + H$	6.0×10^{-11}
(196)	$O^- + O \rightarrow e^- + O_2$	1.9×10^{-10}
(197)	$O^- + NO \rightarrow e^- + NO_2$	2.1×10^{-10}
(198)	$O^- + CO_2 + M \rightarrow CO_3^- + M$	2.0×10^{-28}
(199)	$O^- + NO_2 \rightarrow NO_2^- + O$	1.0×10^{-9}
(200)	$O^- + H_2O \rightarrow OH^- + OH$	6.0×10^{-13}
(201)	$O_4^- + O \rightarrow O_3^- + O_2$	4.0×10^{-10}
(202)	$O_4^- + NO \rightarrow NO_3^- + O_2$	2.5×10^{-10}
(203)	$O_4^- + CO_2 \rightarrow CO_4^- + O_2$	4.3×10^{-10}
(204)	$O_3^- + O \rightarrow O_2^- + O_2$	2.5×10^{-10}
(205)	$O_3^- + NO \rightarrow NO_3^- + O$	2.6×10^{-12}
(206)	$O_3^- + CO_2 \rightarrow CO_3^- + O_2$	5.5×10^{-10}
(207)	$O_3^- + NO_2 \rightarrow NO_3^- + O_2$	2.8×10^{-10}
(208)	$O_3^- + H \rightarrow OH^- + O_2$	8.4×10^{-10}
(209)	$CO_4^- + NO \rightarrow NO_3^- + CO_2$	4.8×10^{-11}
(210)	$CO_4^- + O \rightarrow CO_3^- + O_2$	1.4×10^{-10}
(211)	$CO_4^- + O_3 \rightarrow O_3^- + CO_2 + O_2$	1.3×10^{-10}
(212)	$CO_4^- + H \rightarrow CO_3^- + OH$	2.2×10^{-10}
(213)	$NO_3^- + O_3 \rightarrow NO_2^- + O_2 + O_2$	1.0×10^{-13}
(214)	$NO_3^- + NO \rightarrow NO_2^- + NO_2$	1.0×10^{-12}
(215)	$NO_2^- + H \rightarrow OH^- + NO$	3.0×10^{-10}
(216)	$NO_2^- + O_3 \rightarrow NO_3^- + O_2$	1.2×10^{-10}
(217)	$NO_2^- + NO_2 \rightarrow NO_3^- + NO$	2.0×10^{-13}
(218)	$OH^- + H \rightarrow e^- + H_2O$	1.4×10^{-9}
(219)	$OH^- + O_3 \rightarrow O_3^- + OH$	9.0×10^{-10}
(220)	$OH^- + O \rightarrow e^- + HO_2$	2.0×10^{-10}
(221)	$OH^- + NO_2 \rightarrow NO_2^- + OH$	1.1×10^{-9}
(222)	$OH^- + CO_2 + O_2 \rightarrow HCO_3^- + O_2$	7.6×10^{-28}
(223)	$CO_3^- + O \rightarrow O_2^- + CO_2$	1.1×10^{-10}
(224)	$CO_3^- + NO_2 \rightarrow NO_3^- + CO_2$	2.0×10^{-10}
(225)	$CO_3^- + NO \rightarrow NO_2^- + CO_2$	1.1×10^{-11}
(226)	$CO_3^- + O_2 \rightarrow O_3^- + CO_2$	6.0×10^{-15}
(227)	$CO_3^- + H \rightarrow OH^- + CO_2$	1.7×10^{-10}

[a] Measured with M = He.

Table A-15

REACTIONS AND RATE CONSTANTS OF SILICON IONS[a]

Reaction number	Reaction	Rate coefficient value (two-body: $cm^3\ sec^{-1}$; three-body: $cm^6\ sec^{-1}$)
(229)	$Si + O_2^+ \rightarrow Si^+ + O_2$	$g_1 \times 10^{-9}$
(230)	$Si + O_2^+ \rightarrow SiO^+ + O$	$(1 - g_1) \times 10^{-9}$
(231)	$Si + NO^+ \rightarrow Si^+ + NO$	10^{-9}
(232)	$Si + h\nu \rightarrow Si^+ + e^-$	0
(233)	$Si + O_2 \rightarrow SiO + O(^1D)$	10^{-12}
(234)	$Si + O_3 \rightarrow SiO + O_2$	10^{-12}
(235)	$Si + O_2 + X \rightarrow SiO_2 + X$	10^{-31}
(236)	$Si^+ + H_2O \rightarrow SiOH^+ + H$	$2.3 \pm 0.9 \times 10^{-10}$
(237)	$Si^+ + O_3 \rightarrow SiO^+ + O_2$	$(g_2)8 \times 10^{-10}$
(238)	$Si^+ + O_3 \rightarrow SiO + O_2^+$	$(1 - g_2)8 \times 10^{-10}$
(239)	$Si^+ + O_2 + X \rightarrow SiO_2^{+*} + X$	3×10^{-29}
(240)	$Si^+ + e^- \rightarrow Si + h\nu$	10^{-12}
(241)	$SiO + O_2^+ \rightarrow SiO^+ + O_2$	10^{-9}
(242)	$SiO + h\nu \rightarrow SiO^+ + e^-$	0
(243)	$SiO + O_2 \rightarrow SiO_2 + O$	0
(244)	$SiO + O_3 \rightarrow SiO_2 + O_2$	10^{-12}
(245)	$SiO^+ + O \rightarrow Si^+ + O_2$	2×10^{-10}
(246)	$SiO^+ + O_3 \rightarrow SiO_2^+ + O_2$	$(g_3)4 \times 10^{-10}$
(247)	$SiO^+ + O_3 \rightarrow SiO_2 + O_2^+$	$(1 - g_3)4 \times 10^{-10}$
(248)	$SiO^+ + e^- \rightarrow Si + O$	4×10^{-7}
(249)	$SiO_2 + O_2^+ \rightarrow SiO_2^+ + O_2$	10^{-9}
(250)	$SiO_2 + h\nu \rightarrow SiO_2^+ + e^-$	0
(251)	$SiO_2^+ + O \rightarrow SiO^+ + O_2$	5×10^{-10}
(252)	$SiO_2^+ + e^- \rightarrow SiO + O$	$(g_4)3 \times 10^{-7}$
(253)	$SiO_2^+ + e^- \rightarrow Si + O_2$	$(1 - g_4)3 \times 10^{-7}$
(254)	$SiO_2^{+*} + O_2 \rightarrow SiO_2 + O_2^+$	$g_5 \times 10^{-11}$
(255)	$SiO_2^{+*} \rightarrow SiO_2^+ + h\nu$	$(1 - g_5) \times 10^{-11}$
(256)	$SiOH^+ + e^- \rightarrow SiO + H$	$(g_6)4 \times 10^{-7}$
(257)	$SiOH^+ + e^- \rightarrow Si + OH$	$(1 - g_6)4 \times 10^{-7}$

[a] From Ramseyer *et al.* (1983) as presented at the Sixth ESA Symposium on European Rocket and Balloon Programmes (ESA SP-183). X, Any third body; g_i, branching ratio; λ_0, wavelength corresponding to ionization potential.

Table A-16

RATE COEFFICIENTS FOR CHARGE TRANSFER TO NEUTRAL METAL ATOMS[a]

Reaction number	Reaction	Extrapolated thermal energy rate coefficient[b] ($cm^3\ molec^{-1}\ sec^{-1}$)
(270)	$N^+ + Na \rightarrow N + Na^+$	Small
(271)	$O^+ + Na \rightarrow O + Na^+$	Small
(272)	$N_2^+ + Na \rightarrow N_2 + Na^+$	6.2×10^{-10}[c]
(273)	$O_2^+(X\,^2\Pi_g) + Na \rightarrow O_2 + Na^+$	7.1×10^{-10}[c]
(274)	$O_2^+(a\,^4\Pi_u) + Na \rightarrow O_2 + Na^+$	2.0×10^{-9}
(275)	$H_2O^+ + Na \rightarrow H_2O + Na^+$	2.7×10^{-9}
(276)	$N_2O^+ + Na \rightarrow N_2O + Na^+$	2.0×10^{-9}
(277)	$N^+ + Mg \rightarrow N + Mg^+$	1.2×10^{-9}
(278)	$O^+ + Mg \rightarrow O + Mg^+$	Small
(279)	$N_2^+ + Mg \rightarrow N_2 + Mg^+$	$\sim 7 \times 10^{-10}$
(280)	$O_2^+ + Mg \rightarrow O_2 + Mg^+$	1.2×10^{-9}
(281)	$O_2^+(a\,^4\Pi_u) + Mg \rightarrow O_2 + Mg^+$	$>3 \times 10^{-9}$
(282)	$NO^+ + Mg \rightarrow NO + Mg^+$	8.1×10^{-10}
(283)	$H_2O^+ + Mg \rightarrow H_2O + Mg^+$	2.2×10^{-9}
(284)	$N_2O^+ + Mg \rightarrow N_2O + Mg^+$	2.2×10^{-9}
(285)	$N^+ + Ca \rightarrow N + Ca^+$	1.1×10^{-9}
(286)	$O^+ + Ca \rightarrow O + Ca^+$	7.6×10^{-10}
(287)	$N_2^+ + Ca \rightarrow N_2 + Ca^+$	1.8×10^{-9}
(288)	$O_2^+(X\,^2\Pi_g) + Ca \rightarrow O_2 + Ca^+$	1.8×10^{-9}
(289)	$O_2^+(a\,^4\Pi_u) + Ca \rightarrow O_2 + Ca^+$	3.5×10^{-9}
(290)	$NO^+ + Ca \rightarrow NO + Ca^+$	4.0×10^{-9}
(291)	$H_2O^+ + Ca \rightarrow H_2O + Ca^+$	4.0×10^{-9}
(292)	$H_3O^+ + Ca \rightarrow H_2O + H + Ca^+$	4.4×10^{-9}
(293)	$N_2O^+ + Ca \rightarrow N_2O + Ca^+$	3.7×10^{-9}
(294)	$H^+ + Fe \rightarrow H + Fe^+$	7.4×10^{-9}
(295)	$N^+ + Fe \rightarrow N + Fe^+$	1.5×10^{-9}
(296)	$O^+ + Fe \rightarrow O + Fe^+$	2.9×10^{-9}
(297)	$N_2^+ + Fe \rightarrow N_2 + Fe^+$	4.3×10^{-10}
(298)	$NO^+ + Fe \rightarrow NO + Fe^+$	9.2×10^{-10}
(299)	$O_2^+ + Fe \rightarrow O_2 + Fe^+$	1.1×10^{-9}
(300)	$H_2O^+ + Fe \rightarrow H_2O + Fe^+$	1.5×10^{-9}

[a] From McEwan and Phillips (1975), with permission of Edward Arnold Publishers, Ltd.

[b] These rate coefficients were extrapolated to 300 K from higher energy results in the range 1–500 eV.

[c] From Albritton (1978).

Table A-17

THERMAL ENERGY ATMOSPHERIC METAL–ION REACTIONS[a]

Reaction number	Reaction	Rate coefficient value (two-body: $cm^3\ sec^{-1}$; (three-body: $cm^6\ sec^{-1}$)	Comments
Singly charged species			
(301)	$N_2^+ + Na \rightarrow Na^+ + N_2$	6.2×10^{-10}	
(302)	$NO^+ + Na \rightarrow Na^+ + NO$	7.0×10^{-11}	
(303)	$O_2^+ + Na \rightarrow Na^+ + O_2$	7.1×10^{-10}	
(304)	$O_2^+ + Na \rightarrow NaO^+ + O$	7.7×10^{-11}	
(305)	$Mg^+ + O_3 \rightarrow MgO^+ + O_2$	2.3×10^{-10}	
(306)	$Ca^+ + O_3 \rightarrow CaO^+ + O_2$	1.6×10^{-10}	
(307)	$Fe^+ + O_3 \rightarrow FeO^+ + O_2$	1.5×10^{-10}	
(308)	$Na^+ + O_3 \rightarrow NaO^+ + O_2$	$<1 \times 10^{-11}$	
(309)	$K^+ + O_3 \rightarrow KO^+ + O_2$	$<1 \times 10^{-11}$	
(310)	$MgO^+ + O \rightarrow Mg^+ + O_2$	$\sim 1 \times 10^{-10}$	
(311)	$SiO^+ + O \rightarrow Si^+ + O_2$	$\sim 2 \times 10^{-10}$	
(312)	$SiO^+ + N \rightarrow Si^+ + NO$	$\sim 2 \times 10^{-10}$	
	$\rightarrow NO^+ + Si$	$\sim 1 \times 10^{-10}$	
(313)	$Na^+, K^+, Ba^+ + O_2(NO) \rightarrow$ Products	$<1 \times 10^{-13}$	For ion kinetic energies up to ~ 5 eV
(314)	$Mg^+ + O_2 + Ar \rightarrow MgO_2^+ + Ar$	$\sim 2.5 \times 10^{-30}$	
(315)	$Ca^+ + O_2 + Ar \rightarrow CaO_2^+ + Ar$	$\sim 6.6 \times 10^{-30}$	
(316)	$Ca^+ + O_2 + He \rightarrow CaO_2^+ + He$	$\sim 2 \times 10^{-30}$	
(317)	$Fe^+ + O_2 + Ar \rightarrow FeO_2^+ + Ar$	$\sim 1.0 \times 10^{-30}$	
(318)	$Na^+ + O_2 + Ar \rightarrow NaO_2^+ + Ar$	$<2 \times 10^{-31}$	
(319)	$K^+ + O_2 + Ar \rightarrow KO_2^+ + Ar$	$<2 \times 10^{-31}$	
(320)	$Na^+ + 2CO_2 \rightleftharpoons Na^+ \cdot CO_2 + CO_2$	2×10^{-29}	$k_{rev} = 1 \times 10^{-14}$
(321)	$Na^+ \cdot CO_2 + 2CO_2 \rightleftharpoons Na^+(CO_2)_2 + CO_2$	2×10^{-29}	$k_{rev} = 5 \times 10^{-13}$
(322)	$Na^+ + 2O_2 \rightleftharpoons Na^+ \cdot O_2 + O_2$	5×10^{-32}	$k_{rev} = 8 \times 10^{-13}$

(323)	$K^+ + 2CO_2 \rightleftharpoons K^+\cdot CO_2 + CO_2$	4×10^{-30}	$k_{rev} = 2.5 \times 10^{-13}$
(324)	$Na^+ + H_2O + He \rightarrow Na^+\cdot H_2O + He$	4.7×10^{-30}	
(325)	$Na^+ + 2H_2O \rightarrow Na^+\cdot H_2O + H_2O$	1.0×10^{-28}	
(326)	$K^+ + H_2O + He \rightarrow K^+\cdot H_2O + He$	2.6×10^{-30}	
(327)	$K^+ + 2H_2O \rightarrow K^+\cdot H_2O + H_2O$	4.5×10^{-29}	
(328)	$Ba^+ + CO_2 + He \rightarrow Ba^+\cdot CO_2 + He$	2.8×10^{-30}	
(329)	$Ca^+ + CO + He \rightarrow Ca^+\cdot CO + He$	2.7×10^{-30}	
(330)	$Ca^+ + O_2 + He \rightarrow Ca^+\cdot O_2 + He$	$\sim 2 \times 10^{-30}$	
(331)	$Ba^+ + CO_2 + He \rightarrow Ba^+\cdot CO_2 + He$	2.8×10^{-30}	
Doubly charged species			
(332)	$Mg^{2+} + Ar + He \rightarrow Mg^{2+}\cdot Ar + He$	3.1×10^{-30}	
(333)	$Mg^{2+} + N_2 + He \rightarrow Mg^{2+}\cdot N_2 + He$	1.9×10^{-29}	
(334)	$Mg^{2+} + CO + He \rightarrow Mg^{2+}\cdot CO + He$	4.7×10^{-29}	
(335)	$Mg^{2+} + CO_2 + He \rightarrow Mg^{2+}\cdot CO_2 + He$	3.1×10^{-27}	
(336)	$Ca^{2+} + Ar + He \rightarrow Ca^{2+}\cdot Ar + He$	$\sim 1 \times 10^{-30}$	
(337)	$Ca^{2+} + N_2 + He \rightarrow Ca^{2+}\cdot N_2 + He$	6.2×10^{-30}	
(338)	$Ca^{2+} + O_2 + He \rightarrow Ca^{2+}\cdot O_2 + He$	8.9×10^{-30}	
(339)	$Ca^{2+} + CO + He \rightarrow Ca^{2+}\cdot CO + He$	2.0×10^{-29}	
(340)	$Ca^{2+} + CO_2 + He \rightarrow Ca^{2+}\cdot CO_2 + He$	1.1×10^{-27}	
(341)	$Ca^{2+} + N_2O + He \rightarrow Ca^{2+}\cdot N_2O + He$	2.5×10^{-27}	
(342)	$Ca^{2+} + H_2O + He \rightarrow Ca^{2+}\cdot H_2O + He$	$\sim 5 \times 10^{-28}$	
(343)	$Ba^{2+} + N_2 + He \rightarrow Ba^{2+}\cdot N_2 + He$	$\sim 1.6 \times 10^{-30}$	
(344)	$Ba^{2+} + O_2 + He \rightarrow Ba^{2+}\cdot O_2 + He$	3×10^{-30}	
(345)	$Ba^{2+} + CO + He \rightarrow Ba^{2+}\cdot CO + He$	$\sim 5 \times 10^{-30}$	
(346)	$Ba^{2+} + CO_2 + He \rightarrow Ba^{2+}\cdot CO_2 + He$	1.1×10^{-28}	
(347)	$Ba^{2+} + N_2O + He \rightarrow Ba^{2+}\cdot N_2O + He$	$\sim 1.9 \times 10^{-28}$	
(348)	$Ba^{2+} + H_2O + He \rightarrow Ba^{2+}\cdot H_2O + He$	1.1×10^{-28}	

[a] From McEwan and Phillips (1975), with permission of Edward Arnold Publishers, Ltd.

Table A-18

SODIUM CHEMISTRY[a]

Reaction number	Reaction	Rate coefficient value (two-body: $cm^3\ sec^{-1}$; three-body: $cm^6\ sec^{-1}$)
(350)	$Na + O_3 \rightarrow NaO + O_2$	3.4×10^{-10}
(351)	$NaO + O \rightarrow Na^* + O_2$	1.2×10^{-11}
(352)	$NaO + O \rightarrow Na + O_2$	2.8×10^{-11}
(354)	$Na + O_2 + M \rightarrow NaO_2 + M$	$6.7 \times 10^{-31} \exp(290/T)$
(355)	$Na + HO_2 \rightarrow NaOH + O$	1.4×10^{-11}
(356)	$NaO + O_3 \rightarrow NaO_2 + O_2$	3.5×10^{-11}
(357)	$NaO + O_3 \rightarrow Na + 2\,O_2$	1.0×10^{-10}
(358)	$NaO + HO_2 \rightarrow NaOH + O_2$	1.0×10^{-11}
(359)	$NaO + H_2O \rightarrow NaOH + OH$	1.0×10^{-10}
(360)	$NaO + H \rightarrow Na + OH$	1.0×10^{-14}
(361)	$NaO_2 + h\nu \rightarrow Na + O_2$	1.0×10^{-4}
(362)	$NaO_2 + O \rightarrow NaO + O_2$	1.0×10^{-13}
(363)	$NaO_2 + OH \rightarrow NaH + O_2$	1.0×10^{-11}
(364)	$NaO_2 + H \rightarrow NaOH + O$	1.0×10^{-13}
(365)	$NaOH + h\nu \rightarrow Na + OH$	2.0×10^{-3}
(366)	$NaOH + H \rightarrow Na + H_2O$	1.4×10^{-12}
(367)	$NaOH + O(^1D) \rightarrow NaO + OH$	1.0×10^{-10}
(368)	$NaOH + HCl \rightarrow NaCl + H_2O$	1.0×10^{-11}
(369)	$NaCl + h\nu \rightarrow Na + Cl$	1.0×10^{-2}

[a] From Kirchoff (1983).

Table A-20

REACTIONS AND RATE COEFFICIENTS FOR REACTIONS OF ODD-NITROGEN

Reaction number	Reaction	Rate coefficients: Coefficient	Rate coefficients: Value (two-body: $cm^3\ sec^{-1}$; three-body: $cm^6\ sec^{-1}$)
(397)	$NO^+ + e^- \rightarrow N(^2D) + O$	$\beta_2\alpha_2$	$\beta_2 \simeq 0.76$
(398)	$N_2^+ + e^- \rightarrow N(^2D) + N$	$\beta_3\alpha_3$	$\beta_3 \simeq 1$
(399)	$N_2^+ + O \rightarrow N(^2D) + NO^+$	$\beta_4 k_{14}$	$\beta_4 \simeq 1$
(402)	$N^+ + O_2 \rightarrow N(^2D) + O_2^+$	$\beta_5 k_{19}$	$\beta_5 \simeq 1$
(403)	$N(^2D) + O \rightarrow N(^4S) + O$	k_{65}	$\sim 7 \times 10^{-13}$
(404)	$N(^2D) + O_2 \rightarrow NO + O$	k_{66}	6×10^{-12}
(405)	$N(^2D) + e^- \rightarrow N(^4S) + e^-$	k_{67}	$(3.6\text{–}6.5) \times 10^{-10}(T_e/300)^{1/2}$
(406)	$N(^4S) + O_2 \rightarrow NO + O$	k_{68}	$4.4 \times 10^{-12} \exp(-3220/T)$
(407)	$N + NO \rightarrow N_2 + O$	k_{69}	3.4×10^{-11}

Table A-19

REACTIONS AND RATE COEFFICIENTS RELEVANT TO $O(^1D)$ AND $O(^1S)$ CHEMISTRY

Reaction number	Reaction	Rate coefficient: Coefficient	Rate coefficient: Value (two-body: $cm^3 sec^{-1}$; three-body: $cm^6 sec^{-1}$)
(371)	$O_2^+ + e^- \rightarrow O(^1D) + O$	α_{1D}	$\simeq (0.9\text{–}1.3)\alpha_1$
(372)	$O(^1D) + N_2 \rightarrow O(^3P) + N_2$	k_{57}	2.3×10^{-11}
(373)	$O(^1D) + O_2 \rightarrow O(^3P) + O_2$		$(3.6\text{–}7.0) \times 10^{-11}$
(374)	$O(^1D) \rightarrow O(^3P) + h\nu$		$A_{1D} = 0.068\ sec^{-1}$ $[A_{6300\,A} = (5.15 \pm 1.25) \times 10^{-3}\ sec^{-1}$ $A_{6364\,A} = (1.66 \pm 0.42) \times 10^{-3}\ sec^{-1}]$
(378)	$N(^2D) + O_2 \rightarrow O(^1D) + NO$	k_{58}	$\simeq 5 \times 10^{-12}$
(380)	$O_2^+ + e^- \rightarrow O(^1S) + O'$	$\beta_{15}\alpha_1$	$\beta_{15} \simeq 2\text{–}10\%$
(382)	$O_2^+ + N \rightarrow O(^1S) + NO^+$	k_{59}	$(1.2\text{—}5) \times 10^{-11}$
(386)	$3\,O \rightarrow O_2(X\,^3\Sigma_g^-)(v \leqslant 4) + O(^1S)$	k_{60}	6.3×10^{-33}
(387)	$2\,O + M \rightarrow O_2^* + M$	k_{61}	$4.7 \times 10^{-33}(300/T)^2$
(388)	$O_2^* + O \rightarrow O(^1S) + O_2$	k_{62}	3×10^{-13} (uncertain)[a]
(389)	$O_2^* + N_2 \rightarrow O_2 + N_2$	k_{63}	2.8×10^{-13}[b]
(389)	$O_2^* + O_2 \rightarrow 2\,O_2$	k_{63}	5×10^{-13}[c]
(393)	$O(^1S) + O_2(a\,^1\Delta_g) \rightarrow O + O_2$	k_{64}	1.7×10^{-10}
(394)	$O(^1S) + O_2 \rightarrow O(^3P) + O_2$		$4.9 \exp(-1730/RT)$
(395)	$O(^1S) + N_2 \rightarrow O(^3P) + N_2$		$< 10^{-17}$
(396)	$O(^1S) + O(^3P) \rightarrow 2\,O$		2×10^{-14} (theoretical)

[a] The uncertainty involves the efficiency for the production of $O_2(A'\,^3\Delta_u)$ by reaction (387), and $O(^1S)$ by $O_2^* + O$; k_{61} is the total rate coefficient for production of all intermediate states, and k_{62} includes the combined efficiencies.

[b] Here O_2^* is $O_2(A'\,^3\Delta_u)$ (*Spacelab 1* data suggest this to be O_2^*).

[c] Here O_2^* is $O_2(A\,^3\Sigma)$ [there are no data on O_2 quenching of $O_2(A'\,^3\Delta_u)$].

Table A-21

REACTIONS OF VIBRATIONALLY EXCITED N_2

Reaction number	Reaction	Rate coefficients: Coefficient	Rate coefficients: Value (two-body: $cm^3\ sec^{-1}$; three-body: $cm^6\ sec^{-1}$)
(418)	$N + NO \rightarrow N_{2(v=4)} + O$	k_{73}	$\lesssim 3.4 \times 10^{-11}$
(419)	$O(^1D) + N_2 \rightarrow N_{2(v=2)} + O(^3P)$	k_{74}	$\sim 7 \times 10^{-12}$
(420)	$N_2^* + O \rightarrow N_2 + O$	k_{75}	4.4×10^{-14}

References

Abdou, W. A., Torr, D. G., Richards, P. G., Torr, M. R., and Breig, E. (1984). Results of a comprehensive study of the photochemistry of N_2^+ in the ionosphere. *JGR, J. Geophys. Res.* **89,** 9069–9079.

Albritton, D. L. (1978). Ion-neutral reaction-rate constants measured in flow reactors through 1977. *At. Data Nucl. Data Tables* **22,** 1–101.

Arnold, F., and Krankowsky, D. (1977). Ion composition and electron- and ion-loss processes in the Earth's atmosphere. *In* "Dynamical and Chemical Coupling Between the Neutral and Ionized Atmosphere" (B. Grandal and J. A. Holtet, eds.), pp. 93–127. Reidel Publ., Dordrecht, Netherlands.

Arnold, F., and Viggiano, A. A. (1982). Combined mass spectrometric composition measurements of positive and negative ions in the lower ionosphere—I. Positive ions. *Planet. Space Sci.* **30,** 1295.

Arnold, F., Viggiano, A. A., and Ferguson, E. E. (1982). Combined mass spectrometric composition measurements of positive and negative ions in the lower ionosphere—II. Negative ions. *Planet. Space Sci.* **30,** 1307.

Bates, D. R. (1949). The intensity distribution in the nitrogen band systems emitted from the Earth's upper atmosphere. *Proc. R. Soc. London, Ser. A* **196,** 217.

Bates, D. R. (1955). Charge transfer and ion–atom interchange collisions. *Proc. Phys. Soc. London, Sect. A* **68,** 344.

Bates, D. R., and Massey, H. S. W. (1947). *Proc. R. Soc. London, Ser. A* **192,** 1.

Biondi, M. A., and Brown, S. C. (1949). *Phys. Rev.* **76,** 1697.

Brasseur, G., and Solomon, S. (1984). "Aeronomy of the Middle Atmosphere." Reidel Publ., Dordrecht, Netherlands.

Breig, E. L., Torr, M. R., and Kayser, D. C. (1982). "Observations and Photochemistry of O^{2+} in the Daytime Thermosphere," Aerospace Rep. No. ATR-82(7871)-2.

Broadfoot, A. L. (1971). Dayglow nitrogen band system. *In* "The Radiating Atmosphere" (B. McCormac, ed.), p. 33. Reidel, Dordrecht, Netherlands.

Dalgarno, A., and McElroy, M. B. (1965). The fluorescence of solar ionizing radiation. *Planet. Space Sci.* **13,** 947.

Dalgarno, A., and McElroy, M. B. (1966). Twilight effects of solar ionizing radiation. *Planet. Space Sci.* **14,** 1321.

Donahue, T. M. (1966). Ionospheric reaction rates in the light of recent measurements in the ionosphere and the laboratory. *Planet. Space Sci.* **14,** 33.

Feldman, P. D. (1973). Daytime ion chemistry of N_2^+. *J. Geophys. Res.* **78,** 2010.

Ferguson, E. E. (1974). Laboratory measurements of ionospheric ion–molecule reaction rates. *Rev. Geophys. Space Phys.* **12,** 703.

Frederick, J. E., and Hudson, R. D. (1980a). Atmospheric opacity in the Schumann–Runge bands and the aeronomic dissociation of water vapor. *J. Atmos. Sci.* **37,** 1088.

Frederick, J. E., and Hudson, R. D. (1980b). Dissociation of molecular oxygen in the Schumann–Runge bands. *J. Atmos. Sci.* **37,** 1099–1106.

Garcia, R. R., and Solomon, S. (1983). A numerical model of the zonally averaged dynamical and chemical structure of the middle atmosphere. *JGR, J. Geophys. Res.* **88,** 1379.

Kirchhoff, V. W. J. (1983). Atmospheric sodium chemistry and diurnal variations: an up-date. *Geophys. Res. Lett.* **10,** 721.

Kopp, E., and Herrmann, U. (1982). Ion composition in the lower ionosphere. *EGS Conf. Ions Middle Atmos., Leeds 23–27 Aug. 1982.*

McEwan, M. J., and Phillips, L. F. (1975). "Chemistry of the Atmosphere." Arnold, London.

Mehr, F. J., and Biondi, M. A. (1969). Electron temperature dependence of recombination of O_2^+ and N_2^+ ions with electrons. *Phys. Rev.* **181,** 264.

Mount, G. H., and Rottman, G. J. (1983a). The solar absolute spectral irradiance 1150–3173 Å: May 17, 1982. *JGR, J. Geophys. Res.* **88,** 5403.

Mount, G. H., and Rottman, G. J. (1983b). The solar absolute spectral irradiance at 1216 Å and 1800–3173 Å: January 12, 1983. *JGR, J. Geophys. Res.* **88,** 6807.

Nicolet, M. (1974). An overview of aeronomic processes in the stratosphere and mesosphere. *Can. J. Chem.* **52,** 1381.

Norton, R. B., Van Zandt, T. E., and Denison, J. S. (1963). *In* "Proceedings of International Conference on the Ionosphere" (A. C. Strickland, ed.), p. 26. Inst. Phys. and Phys. Soc., London.

Omholt, A. (1957). The red and near-infra-red auroral spectrum. *J. Atmos. Terr. Phys.* **10,** 320.

Ramseyer, H., Eberhardt, P., and Kopp, E. (1983). Silicon chemistry and inferred water vapor in the lower ionosphere. *ESA Symp. Eur. Rocket Balloon Programmes, 6th* ESA SP-183, pp. 465–472.

Rottman, G. J. (1981). Rocket measurements of the solar spectral irradiance during solar minimum, 1972–1977. *JGR, J. Geophys. Res.* **86,** 6697.

Rottman, G. J. (1983). 27-Day variations observed in solar UV (120–300 nm) irradiance. *Planet. Space Sci.* **31,** 1001.

Schmeltekopf, A. L., Ferguson, E. E., and Fehsenfeld, F. C. (1968). Afterglow studies of the reactions He^+, $He(2^3S)$, and O^+ with vibrationally excited N_2. *J. Chem. Phys.* **48,** 2966.

Smith, F. L., III, and Smith, C. (1972). Numerical evaluation of Chapman's grazing incidence integral ch(X, χ). *J. Geophys. Res.* **77,** 3592.

Solomon, S., Ferguson, E. E., Fahey, D. W., and Crutzen, P. J. (1982). On the chemistry of H_2O, H_2, and meteoritic ions in the mesosphere and lower thermosphere. *Planet. Space Sci.* **30,** 1117.

Thomas, L. (1980). The composition of the mesosphere and lower thermosphere. *Philos. Trans. R. Soc. London, Ser. A* **296,** 243.

Thomas, R. J. (1981). Analyses of atomic oxygen, the green line and Herzberg bands in the lower thermosphere. *JGR, J. Geophys. Res.* **86,** 206.

Tinsley, B. A. (1981). Neutral atom precipitation—a review. *J. Atmos. Terr. Phys.* **43,** 617.

Torr, D. G., and Torr, M. R. (1979). Chemistry of the thermosphere and ionosphere. *J. Atmos. Terr. Phys.* **41,** 797.

Torr, D. G., Richards, P. G., and Torr, M. R. (1980). Ionospheric composition: the seasonal anomaly explained. *AGARD Conf. Proc.* **AGARD-CP-295,** 18-1.

Torr, M. R., and Torr, D. G. (1982). The role of metastable species in the thermosphere. *Rev. Geophys. Space Phys.* **20,** 91.

Torr, M. R., Torr, D. G., Ong, R. A., and Hinteregger, H. E. (1979). Ionization frequencies for major thermospheric constituents as a function of Solar Cycle 21. *Geophys. Res. Lett.* **6,** 771.

Torr, M. R., Torr, D. G., and Hinteregger, H. E. (1980). Solar flux variability in the Schumann–Runge continuum as a function of Solar Cycle 21. *JGR, J. Geophys. Res.* **85,** 6063.

Walker, J. C. G. (1977). "Evolution of the Atmosphere." Macmillan, New York.

Wisemberg, J., and Kockarts, G. (1980). Negative ion chemistry in the terrestrial D region and signal flow graph theory. *JGR, J. Geophys. Res.* **85,** 4642.

II

The Other Planets

6

The Photochemistry of the Atmosphere of Venus

RONALD G. PRINN

Department of Earth, Atmospheric, and Planetary Sciences
Massachusetts Institute of Technology
Cambridge, Massachusetts

I. Introduction

Venus is our nearest planetary neighbor and has often been described in the popular literature as Earth's twin. In fact, while Venus is similar to the Earth in some respects, it is surprisingly dissimilar in others. It is very similar in mass (0.81 Earth masses), radius (0.95 Earth radii), mean density (95% that of

ISBN 0-12-444920-4

Table 1
SOME BASIC DATA FOR EARTH AND VENUS

	Venus	Earth
Mass (kg)	4.87×10^{24}	5.98×10^{24}
Radius (km)	6051	6378
Mean density (g cm^{-3})	5.24	5.52
Distance from Sun (km)	1.08×10^{8}	1.50×10^{8}
Gravitational acceleration (cm sec^{-2})	887	981
Orbital period (days)	225	365
Spin period (days)	243	1
Obliquity (deg)	2.6	23.4
Surface pressure (bars)	92	1
Surface temperature (K)	735	297
Albedo	0.77	0.30

Earth), and gravity (90% that of Earth). On the other hand, its average surface temperature exceeds by some 450 K the average surface temperature on Earth, it has no oceans, it has no biota, and it is completely shrouded by concentrated sulfuric acid (H_2SO_4) clouds—an exceedingly inhospitable planet to be called the Earth's twin! A summary of some basic physical parameters for both planets is provided in Table I.

When close to the Earth (specifically when at its large-crescent phase) Venus is the brightest object in the night sky after the moon. Indeed, humans have been aware of the existence of this bright nearby planet for a very long time. As reviewed by Cruikshank (1983), the presence of Venus as an entity distinct from the bright stars was evident even to the Babylonians around 3000 BC. Very much later, in AD 1610, Galileo Galilei observed the crescent and other phases of Venus with his newly invented telescope and concluded that this familiar celestial body is not self-luminous; despite its frequent designation as a morning or evening star, it was therefore definitely a planet as opposed to a star! The transit of Venus across the disk of the sun in AD 1761 enabled M. V. Lomonosov to observe the halo around Venus and to deduce correctly that this planet possessed an atmosphere. These early telescopic observations also showed Venus to be a nearly featureless planet to the human eye. With no visible signs of a solid surface, the conclusion was reached naturally and correctly that this planet was shrouded totally and continuously in clouds. What these featureless clouds were composed of and what was hidden beneath them (an arid, dusty, and lifeless desert? a humid, foggy, vital swamp?) were then matters of great interest, but the answers were purely conjecture.

At invisible wavelengths, distinctive cloud features would later become apparent. In the 1920s W. H. Wright and F. E. Ross took the first photographs of Venus in ultraviolet (UV) light and observed the now well-documented

UV dark markings. In the 1930s the first clue to atmospheric composition appeared with the discovery of CO_2 absorption bands in the spectra of Venus near 800 nm taken by W. S. Adams and T. Dunham. The subsequent three decades of observations would witness the use of successively higher and higher spectral and spatial resolution, the extension of the accessible wavelengths well into the UV, infrared (IR), and microwave regions, and the use of balloons, aircraft, and rockets as instrumental platforms. Then in 1962, *Mariner 2* (U.S.A.) flew by Venus at a distance of 35,000 km, and in 1967, *Venera 4* (U.S.S.R.) entered its atmosphere and *Mariner 5* (U.S.A.) approached to within 4100 km of its surface. The modern space-age era of investigation of our enigmatic neighbor had begun.

By the late 1960s five gases had been unambiguously identified in the atmosphere above the Venus clouds using Earth-based spectroscopy: CO_2, H_2O, CO, HCl, and HF. The latter three gases, present at levels of about 50, 0.5, and 0.01 parts per million by volume (ppmv), respectively, were discovered by Connes *et al.* (1967, 1968) using a very high resolution Fourier transform spectrometer; the detection of such miniscule amounts of these gases in the atmosphere of another planet was a remarkable demonstration of the power of this (then novel) instrument in planetary astronomy. Connes *et al.* (1967) had also argued (correctly) that CO_2 was the dominant gas in the atmosphere. Water vapor (H_2O), first discovered by Bottema *et al.* (1965), was subsequently shown by Schorn *et al.* (1969) to be highly variable. Mixing ratios for H_2O between 1 and 100 ppmv have since been seen at various times and places in the Venusian clouds. These are very low H_2O mixing ratios compared to those in clouds on Earth, where H_2O may comprise up to a few percent of the air by volume.

Microwave emission from Venus was expected and first reported by Mayer *et al.* (1958). The strength of the emission was, however, very much larger than expected and corresponded to a blackbody temperature approaching 600 K! Subsequent microwave measurements from *Mariner 2* indicated that the emission was almost certainly from a very hot surface and not from hot electrons in a dense ionosphere. From the intensity and polarization of the microwave emission, Kuzmin and Clark (1965) were later able to deduce a surface temperature of 760 K. Earlier Wildt (1940) had made the remarkably prescient suggestion that the then known presence of CO_2 in the atmosphere of Venus would lead to surface temperatures exceeding 400 K. Specifically, while CO_2 enables sunlight to readily penetrate to ground level, it also strongly absorbs and reemits the IR radiation emanating from the surface. As a result, both the surface and lower atmospheric temperatures become elevated by the so-called greenhouse effect. We now know that the greenhouse effect is the essential mechanism leading to the observed elevated temperatures; the precise values for the ground and air temperatures are, however, governed ultimately by dynamic as much as radiative processes.

Table II
SUMMARY OF SUCCESSFUL AMERICAN AND RUSSIAN MISSIONS TO VENUS

Mission	Date of encounter	Mission type
Mariner 2	14 December 1962	Flyby
Venera 4	18 October 1967	Bus, four entry probe/hard-landers
Mariner 5	19 October 1967	Flyby
Venera 5	16 May 1969	Bus, entry probe/hard-lander
Venera 6	17 May 1969	Bus, entry probe/hard-lander
Venera 7	15 December 1970	Bus, entry probe/soft-lander
Venera 8	22 July 1972	Bus, entry probe/soft-lander
Mariner 10	5 February 1974	Flyby
Venera 9	22 October 1975	Orbiter, entry probe/soft-lander
Venera 10	25 October 1975	Orbiter, entry probe/soft-lander
Pioneer Venus 1	4 December 1978	Orbiter
Pioneer Venus 2	9 December 1978	Bus, four entry probes/hard-landers
Venera 11	25 December 1978	Flyby, entry probe/soft-lander
Venera 12	21 December 1978	Flyby, entry probe/soft-lander
Venera 13	1 March 1982	Flyby, entry probe/soft-lander
Venera 14	5 March 1982	Flyby, entry probe/soft-lander
Venera 15	10 October 1983	Orbiter
Venera 16	15 October 1983	Orbiter

The complete cloud cover on Venus had naturally thwarted any global-scale studies of the surface of Venus using traditional techniques. Consequently radars became the major tool for probing the morphology and physical properties of the Venusian surface. Beginning in the 1960s Earth-based radars were aimed at Venus (e.g., see Goldstein, 1964) and provided the first reliable estimates of its radius (6051 km) and its spin period (243 days) and direction (retrograde). Images of major surface features such as Maxwell Montes were also obtained (e.g., see Pettengill, 1978).

Ultraviolet images of Venus taken from Earth-based telescopes over several years had revealed a retrograde rotation period for the observed dark markings in the clouds of about 4 or 5 days (Boyer and Carmichel, 1961). When combined with the radar-derived retrograde rotation period for the surface of 243 days, it was immediately apparent that the atmosphere at the cloud level (~ 65 km altitude) was superrotating with a velocity of ~ 100 m sec^{-1}. The precise dynamic mechanism producing this superrotation is still being hotly debated today (e.g., see Schubert, 1983).

Partly as a result of the relative inaccessibility of the lower atmosphere and surface and partly due to a focusing of Soviet planetary missions on Venus, there have been a remarkably large number of satellite missions to Venus, namely, 22 to date (see Table II). Results obtained from these satellites have collectively played a major role in furthering our understanding of Venus. A few of the relevant highlights of these missions are worthy of specific mention.

The inaugural Venusian entry proble, *Venera 4*, provided the first direct evidence that the atmosphere was principally CO_2. This probe and the later Soviet entry probes *Venera 5* to *8* would indicate decisively that CO_2 is some 97% by volume of the atmosphere.

The *in situ* measurements by *Venera 4* to *6* and radio occultation measurements by *Mariner 5* also provided the first accurate assessments of atmospheric temperatures between the altitudes of 20 and 80 km at several locations over the planet. Later, *Venera 7* landed successfully on the surface and obtained the first direct measurement of the surface temperature, namely 747 K. By 1970 it was thus known that the temperature lapse rate below 50 km altitude was remarkably close to the adiabatic rate (i.e., 8 K km^{-1}) and that horizontal temperature gradients were very small compared to those in the Earth's atmosphere. *Mariner 5* had also provided the first measurements of electron densities in the Venusian ionosphere.

Mariner 10, which flew by Venus in 1973, obtained a remarkable set of near-UV images that would prove invaluable in the study of atmospheric dynamics at the cloud-top level (e.g., see Fig. 1). The same satellite yielded additional measurements of atmospheric temperatures and ionospheric electron densities as well as detailed UV spectra in the 20- to 170-nm wavelength region. The latter spectra verified dramatically the more tentative conclusions reached from previous observations, namely, that CO and O had surprisingly low concentrations in the Venusian thermosphere, considering the continuous rapid production of these two species from CO_2 photodissociation. Associated with this puzzling result was the very low CO mixing ratio (50 ppmv) observed at cloud-top level and the fact that O_2 had never (and still has never) been detected in the visible atmosphere. How are CO, O, and O_2 recycled back to CO_2 with such efficiency? This dilemma, also common to the Martian atmosphere, remains a subject of great interest to this day.

Venera 9 and *10* entered the Venusian atmosphere in 1975 carrying extensive scientific payloads, which included nephelometers. The latter instruments demonstrated that the cloud base was at an altitude of ~50 km and that cloud particle number densities were 50–500 cm^{-3}. A few years earlier, Sill (1972) had suggested that the clouds of Venus were composed of concentrated H_2SO_4 droplets, and theoretical models of these putative H_2SO_4 clouds (Prinn, 1973a, 1975; Wofsy and Sze, 1975) had implied particle number densities and cloud-base positions remarkably similar to those subsequently observed. Compelling, although indirect, evidence for the H_2SO_4 hypothesis was beginning to emerge: the spherical shape and high refractive index (1.44) of the particles (Hansen and Arking, 1971), the observed low H_2O vapor pressures, and the near-IR reflectivity of the clouds (Young and Young, 1973; Pollack *et al.*, 1975).

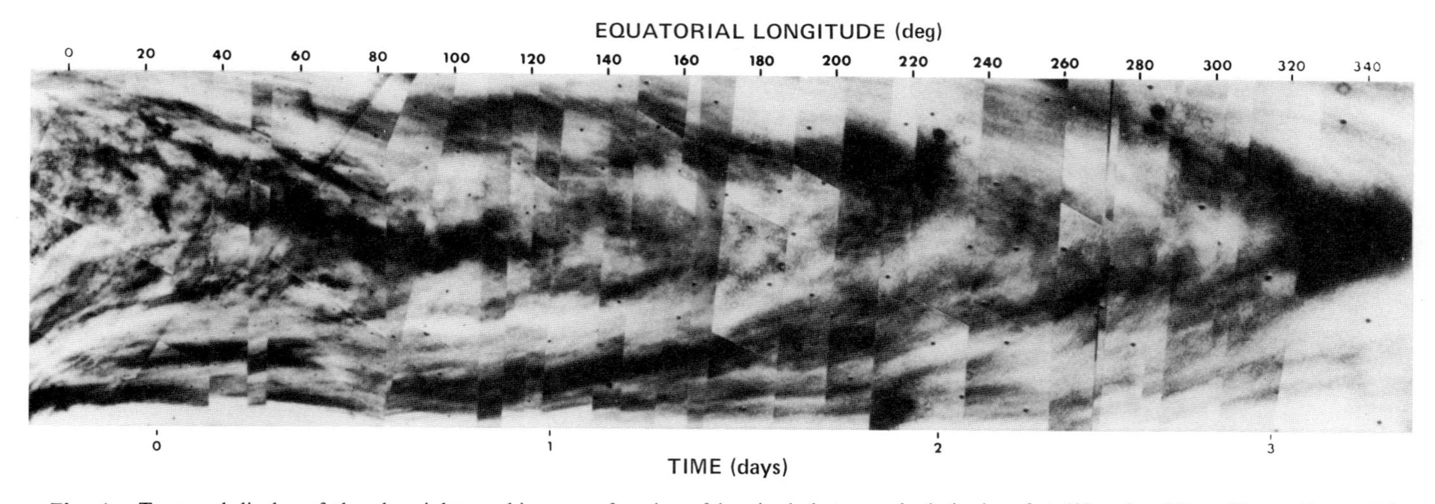

Fig. 1. Temporal display of the ultraviolet markings as a function of longitude between the latitudes of $+40^{\circ}$ and -50° on Venus. Equatorial longitude is based on 4.0^{d} rotation at the equator. The display is constructed using successive television images taken by *Mariner 10*, using a 300- to 400-nm wavelength filter. From Murray *et al.* (1974).

Table III

EXPERIMENTS ON THE 1978 *PIONEER VENUS* AND *VENERA 11* AND *12* MISSIONS

Pioneer Venus	
Pioneer Venus orbiter	*Pioneer Venus* entry probes
Neutral mass spectrometer	Large probe mass spectrometer
Ultraviolet spectrometer	Large probe gas chromatograph
Cloud photopolarimeter	Large/small probes atmospheric structure detectors
Infrared radiometer	Large/small probes nephelometers
Ion mass spectrometer	Large probe cloud particle size spectrometer
Electron temperature probe	Large probe solar flux radiometer
Retarding potential analyzer	Large probe infrared radiometer
Magnetometer	Small probes net flux radiometers
Plasma analyzer	Differential long baseline interferometric tracking experiment
Electric field detector	Doppler tracking experiment
Dual-frequency occultation experiment	Atmospheric turbulence experiment
Radar mapper	Atmospheric propagation experiment
Gamma burst detector	
Internal density distribution experiment	
Celestial mechanics experiment	
Atmospheric drag experiment	
Pioneer Venus bus	
Ion mass spectrometer	
Neutral mass spectrometer	
Venera 11 and 12	
Probes	Flybys
Atmospheric structure detectors	Ultraviolet spectrometer
Probe tracking experiments	
Mass spectrometers	
Gas chromatograph (*Venera 12*)	
Solar radiation spectrophotometers	
Nephelometers	
Thunderstorm detectors	
Cloud composition X-ray fluorescence detector (*Venera 12*)	

The year 1978 saw the most intensive exploration of Venus by spacecraft in history. In this single year, the orbiter, main probe, three miniprobes, and bus vehicle, which collectively comprised NASA's *Pioneer Venus* mission, plus the Soviet *Venera 11* and *12* entry probes all encountered Venus and sampled its atmosphere and surface with an impressive array of *in situ* and remote-sensing instruments. A summary of the scientific payloads on the vehicles of this veritable Venusian invasion force is given in Table III.

The radar on the *Pioneer Venus* orbiter provided the first comprehensive picture of the surface topography of Venus (see Fig. 2). In contrast to the

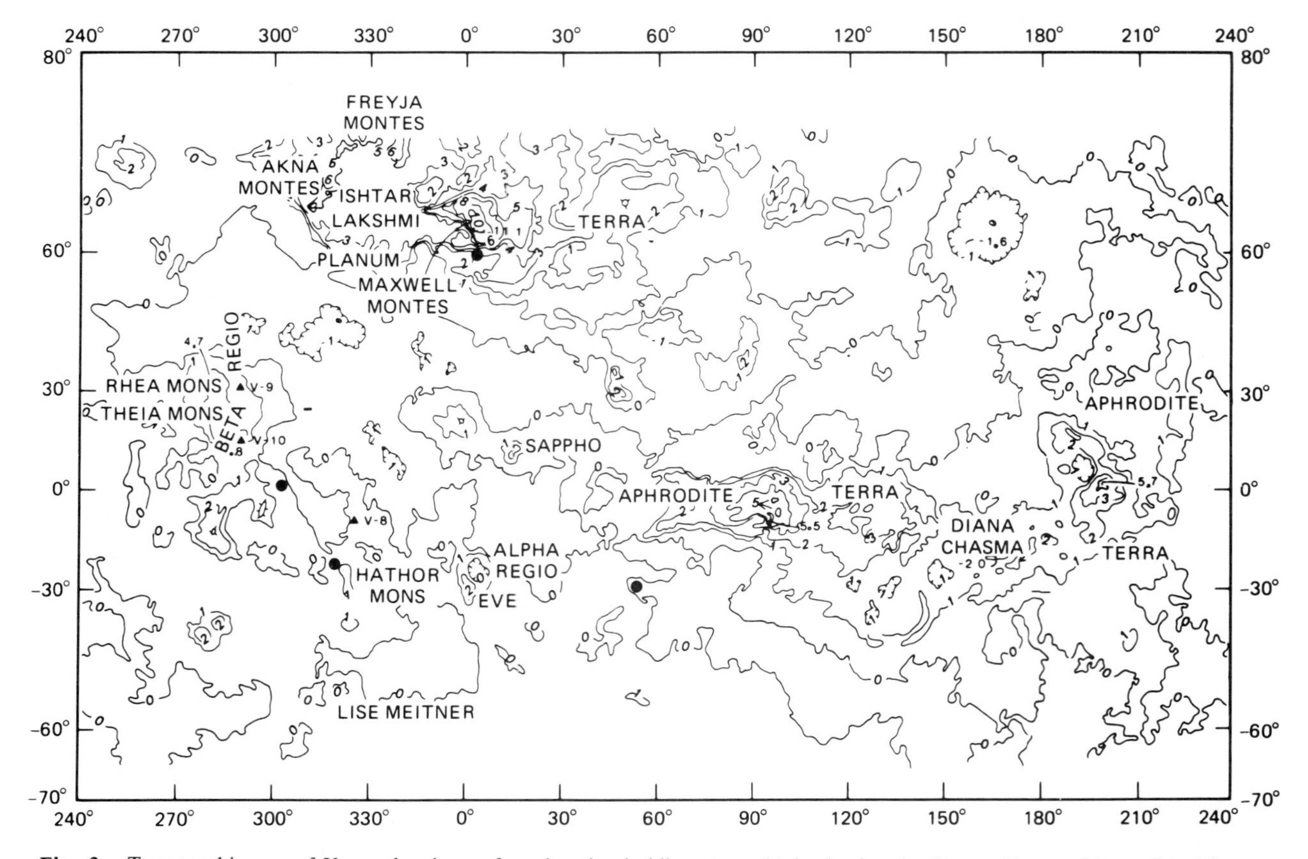

Fig. 2. Topographic map of Venus showing surface elevation in kilometers, obtained using the *Pioneer Venus* orbiter radar altimeter. From Masursky *et al.* (1980).

Earth, which has a distinctly bimodal distribution for surface elevation (i.e., continents and ocean floors), Venus has an essentially unimodal distribution. There are only two features that could be compared in any serious way to our continents, namely, Ishtar Terra in the northern hemisphere and Aphrodite Terra in the equatorial region. The range of surface elevations is, however, quite similar on the two planets. The highest mountain on Venus, Maxwell Montes, is 13.1 km higher than the lowest valley, Diana Chasma (higher spatial resolution, however, may yet reveal an even greater total relief). There is also evidence, albeit indirect, that volcanoes exist on Venus; in particular, Rhea Mons and Theia Mons appear to be similar to large shield volcanoes on Earth (Masursky *et al.*, 1980).

Both the *Pioneer Venus* and *Venera 11* and *12* missions incorporated mass spectrometers and gas chromatographs that sampled the atmosphere above and below the clouds. Gases identified were N_2, CO, SO_2, He, Ne, Ar, Kr, and H_2O. There was more tentative evidence for the presence of S_2, S_3, H_2S, OCS, and O_2. Two particular surprises were the large $^{36}Ar/^{40}Ar$ ratio (about unity on Venus compared to 0.003 on Earth) and the predominance of SO_2 over OCS as the principal sulfur-containing gas (prior theoretical considerations had favored OCS).

Using the signals received at several radio observatories on Earth, the four *Pioneer Venus* probes were tracked very accurately as they descended into the atmosphere, yielding the most precise estimates of wind velocities below 70 km ever obtained for this planet (Counselman *et al.*, 1979). The previously inferred 100 m sec^{-1} zonal winds at altitudes of $\sim$65 km were very apparent in the new data, and the first evidence was obtained for extensive equatorward motion at the 50-km level. The meteorology of Venus was further elucidated by the IR radiometer on the *Pioneer Venus* orbiter, which showed that equatorial temperatures at altitudes of 75 to 95 km were some 15–20 K *cooler* than polar temperatures (F. Taylor *et al.*, 1979); an occurrence not too different from that in the lower stratosphere on Earth. At altitudes of 140 to 190 km in the Venusian thermosphere, analyses of drag on the *Pioneer Venus* orbiter indicated daytime temperatures up to 300 K and nightside temperatures down to 100 K (Keating *et al.*, 1980); these very low nightside temperatures imply that the thermosphere of Venus is quite different in its dynamics from the Earth's thermosphere.

A novel laser cloud-particle analyzer on the *Pioneer Venus* main probe showed a remarkable trimodal particle size distribution (only one mode of which is visible from Earth) and a complex layered structure to the clouds (Knollenberg and Hunten, 1980). The ionosphere was also investigated in detail during the *Pioneer Venus* mission. The ion mass spectrometers observed O^+ as the dominant ion above 200 km and O_2^+ below 180 km. Large diurnal variations in ion densities were observed, for example, at 200 km altitude the densities of O^+ and O_2^+ at noon were ~100 times those at midnight.

An important series of Soviet landers, *Venera 8, 9, 10, 13*, and *14*, have provided the only *in situ* observations of the chemical and physical properties of the Venusian surface. The two latest of these landers, *Venera 13* and *14*, encountered Venus in March 1982. Images of the surface taken by successive landers show a landscape varying from site to site in smoothness and in relative abundances of rocks, rock fragments, and dust. Albedos of the generally very dark surface materials varied from <3 to almost 12%. γ-Ray spectrometers on *Venera 8* through *10* showed Earth-like abundances of the radioactive elements K, U, and Th. X-Ray fluorescence experiments on *Venera 13* and *14* provided a detailed analysis of the composition of the surface rocks at two sites (−7°, 305° and −13°, 310° in Fig. 2). The gas chromatographs on these latest two *Venera* craft also provided a new twist. The abundances of H_2S and OCS appeared to be comparable to that of SO_2, in contrast to the earlier *Pioneer Venus* measurements: Do the abundances of the various sulfur gases change on a 4-year time scale (1978–1982)?

Our knowledge of the planet Venus has now reached a level of detail where many important scientific questions can be asked and then satisfactorily answered using the available evidence. This chapter focuses on important problems related to the chemistry and photochemistry of the atmosphere. Our discussion of atmospheric chemistry must, however, also take into account the interactions between chemical, radiative, and dynamic processes in the atmosphere and between the atmosphere and surface. We shall therefore begin in the next section with a brief review of the structure of the atmosphere and clouds and the circulation of the atmosphere. This will be followed by a summary of the composition of the atmosphere and surface. A detailed look at the chemistry and photochemistry of the atmosphere will form the core of this chapter. We shall finish with a discussion of theories for the evolution and maintenance of the atmosphere and some comments on outstanding unanswered questions and future directions in research.

II. Atmospheric Structure, Circulation, and Composition

A. Thermal Structure

A reference model for the variation of temperature and pressure with altitude on Venus has been recently constructed using available data by Seiff (1983), and the profile for 30° latitude is shown in Fig. 3. The detailed temperature variations with both latitude and altitude obtained with the

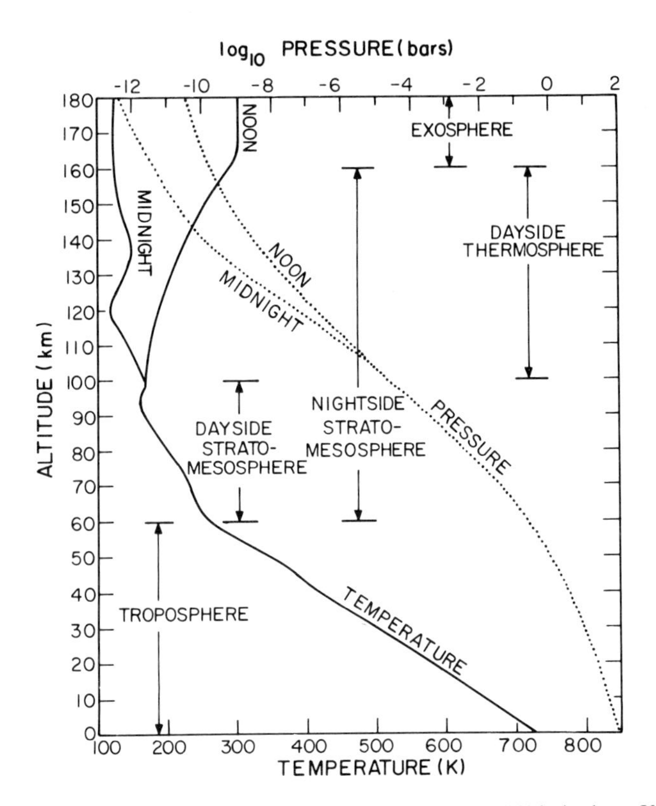

Fig. 3. Variation of temperature and pressure with altitude at 30° latitude on Venus in the standard atmosphere of Seiff (1983). Also shown are the various regions of the atmosphere defined in the text.

Pioneer Venus orbiter infrared radiometer (F. Taylor *et al.*, 1983) for the 55- to 110-km altitude region are shown in Fig. 4. To facilitate comparison with the Earth's atmosphere, we shall refer to the region below 60 km as the troposphere, the 60- to 100-km region (dayside) or 60- to 160-km region (nightside) as the strato–mesosphere, the 100- to 160-km region on the dayside as the thermosphere, and the region above 160 km as the exosphere.

Temperatures in the Venusian lower troposphere (Fig. 3) differ from those on Earth in three important ways. First, the vertical temperature gradient on Venus is quite close to the adiabatic gradient for CO_2 (which is 8–11 K km^{-1}, depending on temperature) whereas our own troposphere is distinctively subadiabatic (specifically 6.5 K km^{-1} compared to the N_2/O_2 adiabatic gradient of 10 K km^{-1}). Second, the equator-to-pole temperature change on

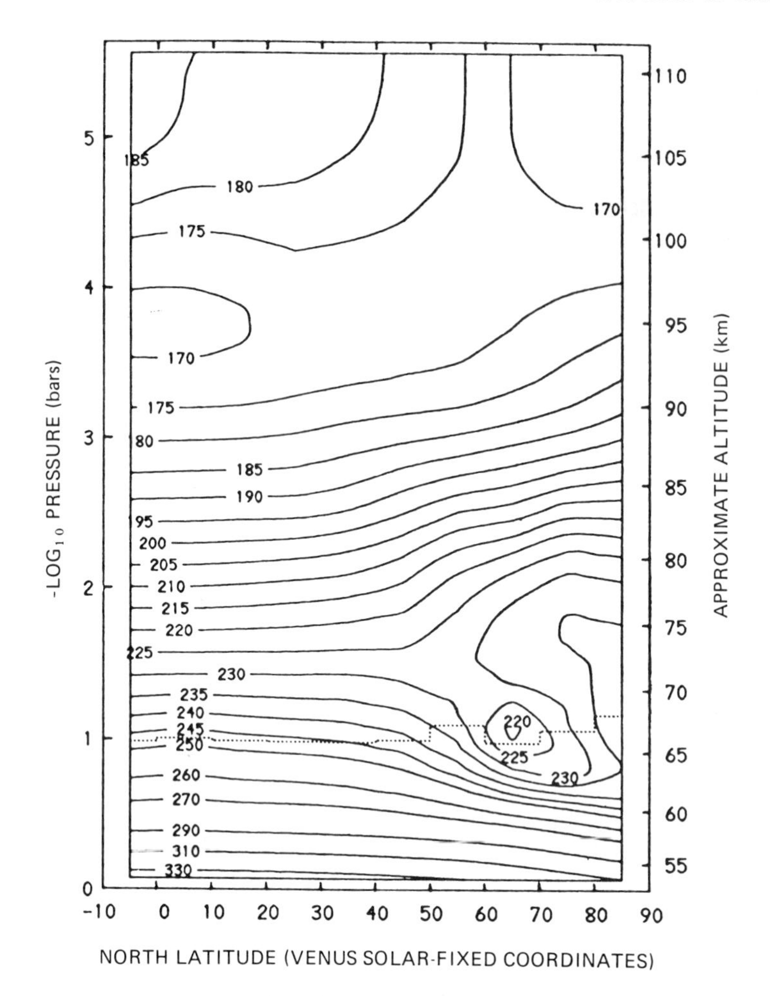

Fig. 4. Meridional cross section of the temperature (K) above the 1-bar pressure level on Venus. The dashed line denotes the level of unit optical depth for 11.5-μm radiation. From Taylor *et al.* (1983).

the Venus surface is at best a few kelvins compared to 50 K on Earth. Third, and most obvious, the surface temperature on Venus is high enough to preclude the existence of even very saline oceans and to permit significant vapor pressures of compounds of the normally lithophilic elements, Cl, F, and S to exist in equilibrium with surface rocks.

In the Venusian upper troposphere (Fig. 4), which shows more Earth-like horizontal temperature gradients, temperatures drop by some 20 K from the equator to the pole. The lower strato–mesosphere on Venus has a horseshoe-shaped region of anomalously cold air centered at midmorning at 65° latitude and 67 km altitude. There is no analogous phenomenon in the Earth's stratosphere. The upper strato–mesosphere between 70 and 95 km is characterized by a 10- to 20-K increase in temperature from equator to pole; such a dynamically induced temperature gradient is seen on Earth in the lower stratosphere and lower thermosphere but not in the upper stratosphere or mesosphere, where local heating due to absorption of solar UV radiation by ozone (O_3) plays a major role in determining the temperatures. This heating due to O_3 on Earth also produces the well-known inversion in the upper stratosphere that serves to divide the stratosphere from the mesosphere. As is evident from Fig. 3 there is no such inversion on Venus, and given the absence of O_3, no such inversion is expected.

A distinct thermosphere on Venus is present only on the dayside where temperatures increase with altitude, largely because of absorption of solar UV radiation by CO_2. The exosphere, the isothermal region above 160 km, has a temperature of ~300 K at noon and ~125 K at midnight; on Earth the exospheric temperatures are very much higher (typically 1000 K).

B. Cloud Structure

The clouds of Venus are best described as a stratified low-density haze. They can be considered to extend upwards to altitudes of 65–67 km, where unit cloud optical depth for IR radiation is reached (see the dashed line in Fig. 4) and downwards to a usually well-defined cloud base in the 46- to 50-km region. Three particle types have been identified: aerosols (~0.3 μm diameter), designated as mode 1; spheres (~2 μm diameter), designated as mode 2; and finally large particles (~7 μm diameter), which may be crystalline, designated as mode 3. The vertical mass distributions of these three types of particles as determined by the *Pioneer Venus* particle-size spectrometer is given in Fig. 5 (Knollenberg and Hunten, 1980). The total extinction optical depth of the clouds in visible light is ~29 with modes 1, 2, and 3 contributing approximately 3, 10 and 16 optical depths, respectively (Esposito *et al.*, 1983). The extinction of visible light is due almost totally to scattering.

The opaque, whitish clouds on Venus are sufficiently reflective at yellow and red wavelengths to enable Venus to possess the highest albedo of any planet in our solar system. Its albedo of 0.77 is so high that despite its greater proximity to the Sun Venus actually absorbs much less solar energy than the Earth (150 versus 240 W m^{-2}); the magnitude of the greenhouse effect on Venus is sufficient to produce the high surface temperatures despite this relative deficit in solar input.

The Venusian clouds do absorb solar violet, near-UV, and near-IR photons to a sufficient extent to cause (with some aid from CO_2) ~70% of the solar

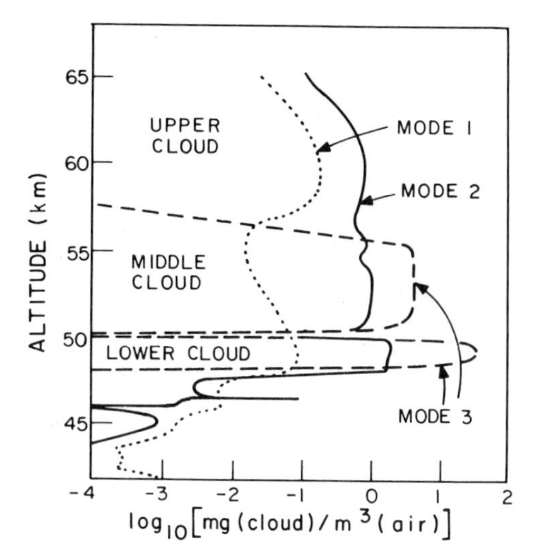

Fig. 5. Density as a function of altitude for the three classes of cloud particles detected by the *Pioneer Venus* particle size spectrometer. Profiles shown are smoothed versions of the actual profiles. From Knollenberg and Hunten (1980).

energy input to the planet to be deposited in the clouds and atmosphere above 50 km. The remaining 30% is deposited about equally in the 20- to 40-km region and at the surface (Tomasko *et al.*, 1980). This deposition of the bulk of the solar heating in the upper troposphere on Venus is quite different from the Earth, where the bulk (66%) of the solar input occurs at the surface (Liou, 1980).

It is important to note that one-half of the solar energy deposited in the Venusian clouds (i.e., 35% of the total solar input) results from absorption at wavelengths less than 500 nm by an UV absorber in the clouds themselves (Tomasko *et al.*, 1980). As we shall discuss later, current ideas indicate that both the highly reflective whitish clouds (which produce the high albedo) and the UV-absorbing material (which contributes to the high-altitude heating) are of photochemical origin; the photochemistry of the atmosphere therefore plays a major role in determining the meteorology and climate of Venus (Prinn, 1984). Titan is the only other body in the solar system in which photochemistry is involved in such an intimate way with bulk atmospheric dynamics and thermodynamics.

C. Ionospheric Structure

Venus possesses a well-developed ionosphere, with a peak in electron density of $\sim 3 \times 10^5$ cm^{-3} occurring around 150 km altitude on the dayside;

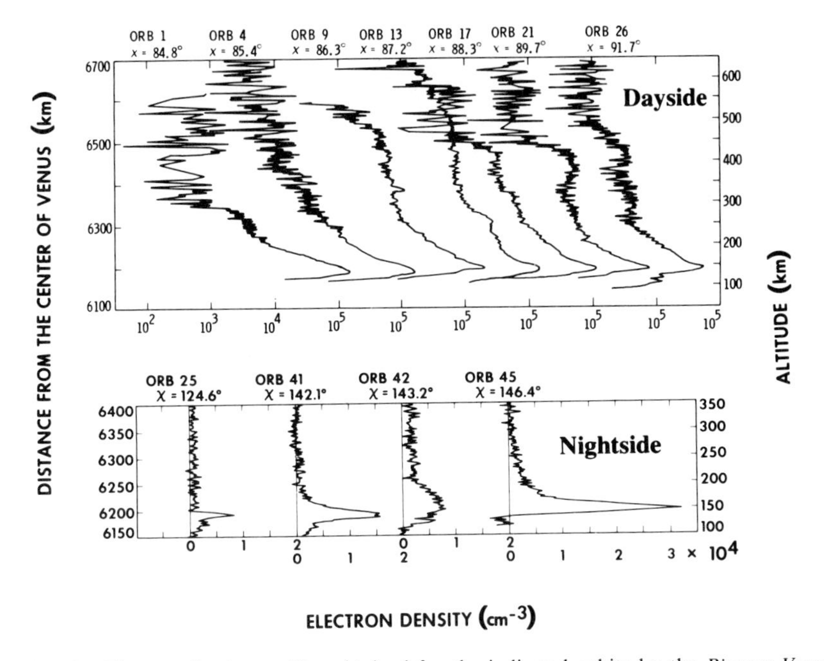

Fig. 6. Electron density profiles obtained for the indicated orbits by the *Pioneer Venus* orbiter radio occultation experiment. Dayside (polar) profiles are from Kliore *et al.* (1979b), and nightside profiles are from Kliore *et al.* (1979a). Dayside profiles are staggered by one decade, and the 10^5 cm^{-3} level is shown for each profile. Nightside profiles are in units of 10^4 cm^{-3}. The solar zenith angle χ for each orbit is indicated.

nightside electron densities at 150 km are about 30 times less than those on the dayside. A major difference between the ionospheres of Venus and Earth is caused by the lack of an intrinsic magnetic field on Venus of sufficient strength to form an extensive magnetosphere to ward off the solar wind. Hence the Venusian ionosphere is affected not only by the variations in the solar UV flux that produce most of the ionization of the neutral atmosphere, but also by variations in the solar wind which affect in particular the upper boundary of the ionosphere (the ionopause). It is therefore not surprising that successive spacecraft missions have recorded substantially different structures, especially for the Venusian upper ionosphere. Figure 6 shows representative samples of the dayside and nightside ionospheres observed by the *Pioneer Venus* radio occultation experiment (Kliore *et al.*, 1979a,b). An extensive review of the ionospheric structure as a function of both space and time is beyond the scope of this treatise; such a review has, however, been made by Brace *et al.* (1983).

D. Atmospheric Circulation

The dominant feature of the Venusian circulation is the retrograde zonal wind present at essentially all altitudes below 100 km. Because the surface rotation is also retrograde, the atmosphere is "superrotating" relative to the surface. A sketch of the altitude dependence of the zonal wind velocity u is given in Fig. 7. These velocities are based on tracking of the *Pioneer Venus* probes below 60 km (Counselman *et al.*, 1980) and on the measured temperature–pressure data, assuming cyclostrophic balance above 60 km (Seiff *et al.*, 1980).

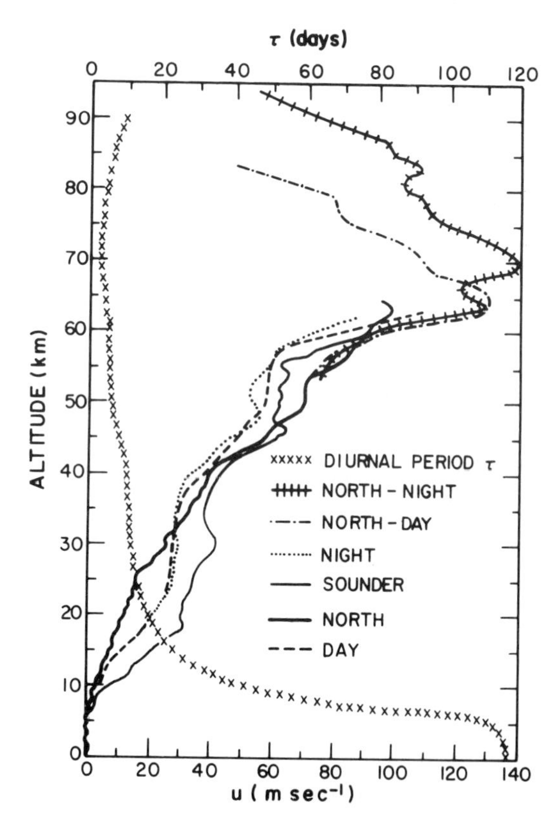

Fig. 7. Retrograde zonal wind velocity u as a function of altitude on Venus obtained by tracking the four *Pioneer Venus* entry probes (designated night, sounder, north, and day) and by combining pressure and temperature data from the north and night or north and day entry probes assuming cyclostrophy (Counselman *et al.*, 1980; Seiff *et al.*, 1980; Schubert, 1983). Also shown is the diurnal period τ at each altitude (see text).

Photochemistry is driven by solar photons, and it is clear that the period τ of the diurnal photon cycle (i.e., the length of the day) experienced by a parcel of air embedded in the above retrograde zonal flow will vary widely with altitude z. The diurnal period is given by

$$1/\tau = \bar{u}/2\pi(a + z) + 1/\tau(0), \tag{1}$$

where a is the radius of Venus and $\tau(0)$ the length of the day at the surface (= 117 Earth days). The altitude dependence of τ is also shown in Fig. 7; τ varies from 117 days at the surface to ~4 days at 70 km.

Information on the equator-to-pole (meridional) circulation below 100 km comes from tracking of entry probes and UV cloud features (Counselman *et al.*, 1980; Rossow *et al.*, 1980; Limaye and Suomi, 1981) and from the observed IR emission (Taylor *et al.*, 1983). This information is useful but insufficient to define unambiguously the meridional flow field. A straightforward interpretation of available data suggests the picture given in Fig. 8 (Schubert *et al.*, 1980). Solar energy deposited at the surface and at the cloud level drives two thermally direct semi-global-scale Hadley cells; in the cloud-level cell the meriodional velocity v is typically $\leqslant 10$ m sec^{-1}, corresponding to equator-to-pole transport times of $\geqslant 11$ days.

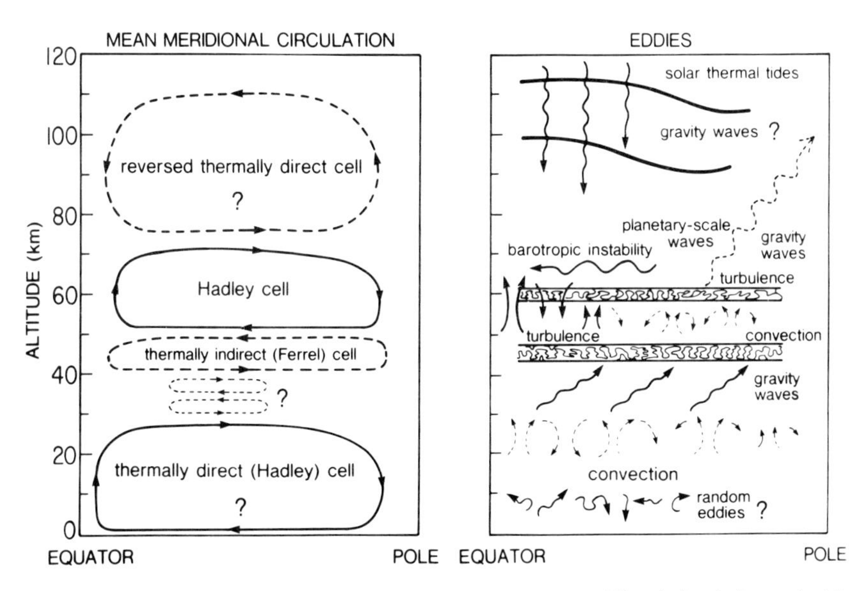

Fig. 8. Sketch summarizing current ideas about the mean meridional circulation and eddy (or wave) phenomena in the Venus atmosphere. From Schubert *et al.* (1980).

Waves on Venus (i.e., motions not having zonal symmetry) are readily observed in UV cloud images, and their dynamic properties have been analyzed (e.g., Belton *et al.*, 1976; Rossow *et al.*, 1980). The most obvious planetary-scale wave feature is the dark Y shape evident in Fig. 1. Waves on a wide variety of space and time scales are expected, which are due to thermal forcing by the Sun (producing tides, etc.), to mechanical forcing caused by overshooting air parcels in two recognized convective regions at 45 and 60 km, and to mechanical forcing caused by flow over mountains. Some ideas are summarized in Fig. 8.

For understanding the photochemistry of the atmosphere we are particularly interested in waves that cause vertical transport of species in the region above the clouds. Our current knowledge of the meteorology of Venus is insufficient to enable the true three-dimensional flows of atmospheric species to be accurately portrayed. Instead, a one-dimensional model in which the globally averaged vertical transport due to both waves and the zonally symetric circulation is parameterized as a Fickian diffusion process is utilized (albeit as a model of last resort). In this model the vertical flux of a species is simply $-K[M]\,\partial f/\partial z$, where K is the "eddy diffusion" coefficient, [M] the total atmospheric molecular number density, and f the volume or number mixing ratio of the species. Of course, such a model is not useful for simulating species with large horizontal gradients (e.g., those with large equator-to-pole or day-to-night variations) for which models with at least two dimensions must be used (e.g., Dickinson and Ridley, 1977).

The advantage of the simple eddy diffusion formulation is that K can be inferred directly, for example, from trace species distributions. In the 60- to 80-km region, K values of approximately 10^4 to 10^5 cm^2 sec^{-1} are compatible with the cloud-particle scale height (Prinn, 1974a), the SO_2 scale height (Winick and Stewart, 1980), and analyses of radio signal scintillations (Woo and Ishimaru, 1981). It is reasonable to assume that transient internal waves (as opposed to the zonally averaged circulations) are the principal mode of vertical transport above 80 km (Prinn, 1975). In this case K will increase upward as $[M]^{-0.5}$ (Lindzen, 1971), and indeed, the formula

$$K = 1.4 \times 10^{13}[M]^{-0.5} \quad \text{cm}^2\,\text{sec}^{-1} \tag{2}$$

provides a good fit to the vertical distribution of neutral constituents in the 100- to 150-km region (von Zahn *et al.*, 1980) as well as to the SO_2 and cloud particle scale height data below 80 km cited above.

The molecular diffusion coefficient for, say, H_2 in CO_2 is given approximately by (McElroy and Hunten, 1969)

$$D = 1.85 \times 10^{19}(T/500)^{0.5}[M]^{-1} \quad \text{cm}^2\,\text{sec}^{-1}. \tag{3}$$

Using the noontime atmospheric model in Fig. 3, we deduce that the *turbopause* for H_2, defined where $D = K$ (i.e., where molecular and eddy diffusion are of equal importance), lies at 130 km altitude. The turbopause levels for other species (e.g., He, N_2) will lie within a few kilometers of this level.

E. Atmospheric Composition

The abundances of the known chemically active gases in the atmosphere of Venus below 90 km are shown as a function of altitude in Fig. 9. The solid curves denote volume mixing ratios based on the extensive review and analysis by Von Zahn *et al.* (1983), who addressed all of the relevant measurements prior to and including the *Pioneer Venus* and *Venera 11* and *12* entry probes. The dashed lines summarize mixing ratio measurements by the spectrophotometers and gas chromatographs on the more recent *Venera 13* and *14* entry probes (Moroz, 1983).

There are particular lacunae and contradictions concerning the chemically active gases of which the reader should be aware. In view of the fact that 80% of the atmospheric mass lies below 22 km, our present lack of knowledge of the concentrations of several chemically important gases hampers our understanding of atmosphere–lithosphere interactions. The absolute

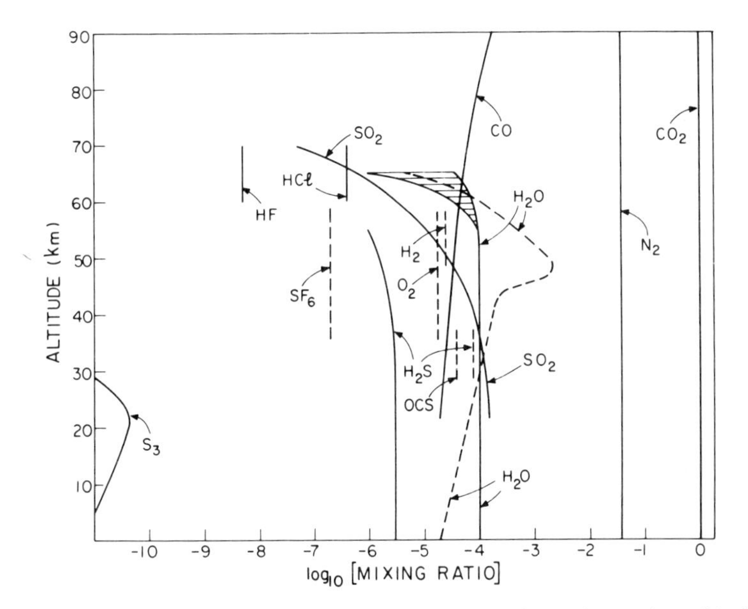

Fig. 9. Volume mixing ratios of chemically active gases observed at various altitudes on Venus. Solid curves summarize pre-1982 data reviewed by von Zahn *et al.* (1983). Dashed curves refer to data from the 1982 *Venera 13* and *14* entry probes (Moroz, 1983).

abundance of H_2O in the lower atmosphere is obviously poorly defined, and it is not yet clear whether the H_2O concentration is constant below the clouds or decreases toward the surface (see the solid and dashed H_2O curves in Fig. 9). The *Venera 13* and *14* gas chromatographic data imply that the reduced sulfur gases (H_2S, OCS) are of comparable abundance to SO_2, while the data available prior to these missions suggest only a minor role for these reduced gases. The presence of significant amounts of O_2 below the clouds is very difficult to reconcile with results from other probe instruments and with the stringent upper limit to the O_2 mixing ratio at cloud level of 3×10^{-7} (Trauger and Lunine, 1983). The evidence for significant amounts of H_2 below the clouds provided by *Venera 13* and *14* must also be regarded as tentative: the possibility that *both* O_2 and H_2 could have mixing ratios exceeding 10^{-5} in the lower atmosphere is chemically implausible.

Above 90 km, microwave data imply that CO mixing ratios continue to increase, reaching values $\sim 10^{-3}$ at 110 km (Wilson *et al.*, 1981). The *Pioneer Venus* mission provided detailed data on the atmospheric composition between 140 and 200 km, and the deduced dayside and nightside abundances of various species are summarized in Fig. 10 (Niemann *et al.*, 1980). Note the very large diurnal variations in the densities of all species at these high altitudes, which are related to the large-amplitude thermospheric diurnal temperature cycle noted earlier. The heavy species CO_2, CO, O, N_2, and N have daytime maxima while the light species H (not shown) and He have nighttime maxima.

In addition to He, which as noted above is readily observable in the thermosphere, there are three other inert gases detected thus far in the Venusian atmosphere: Ne with a mixing ratio $\sim 7 \times 10^{-6}$, Ar with a mixing ratio $\sim 3.1 \times 10^{-5}$, and Kr with two contradictory estimates of its abundance ($\sim 6 \times 10^{-6}$ or $\sim 2.5 \times 10^{-8}$; see Von Zahn *et al.*, 1983, for a review). Isotopic information is also available for certain volatile elements. In particular, as reviewed by Von Zahn *et al.* (1983), we have

$$^{2}H:{}^{1}H = 1.6 \times 10^{-2}:1,$$

$$^{13}C:{}^{12}C = 1.1 \times 10^{-2}:1,$$

$$^{15}N:{}^{14}N = 3.7 \times 10^{-3}:1,$$

$$^{18}O:{}^{16}O = 2 \times 10^{-3}:1,$$

$$^{37}Cl:{}^{35}Cl \simeq 3.2 \times 10^{-1}:1,$$

$$^{38}Ar:{}^{36}Ar:{}^{40}A = 0.18:0.94:1,$$

$$^{80}Kr:{}^{86}Kr:{}^{83}Kr:{}^{82}Kr:{}^{84}Kr = 0.15:0.17:0.29:0.48:1.$$

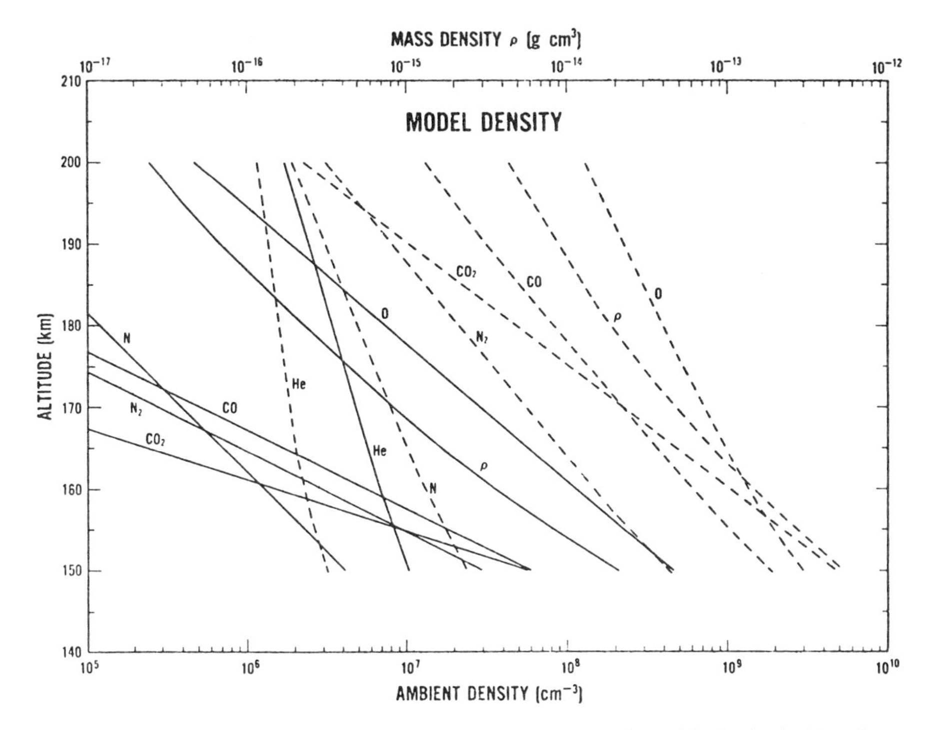

Fig. 10. Number densities of observed neutral species at various altitudes in the Venusian upper atmosphere at noon (---) and midnight (——) at the equator as deduced by Niemann *et al.* (1980). Also shown is the total atmospheric mass density ρ at each altitude at the equator for exospheric temperatures of 294 K (noon) and 118 K (midnight).

Of particular interest here are the very large enrichments of deuterium (^{2}H) and ^{36}Ar and ^{38}Ar on Venus relative to their terrestrial abundances.

F. Ionospheric Composition

As noted earlier, the Venusian ionosphere has a very variable structure, and the electron density variability is mirrored by variations in positive-ion densities and compositions. Sample daytime ion composition profiles are given in Fig. 11 (H. Taylor *et al.*, 1979a). The dominant ions are O_2^+ below 180 km and O^+ above this level. The large diurnal ionospheric variation alluded to earlier is very evident when comparing the samples of the daytime and nighttime ion densities, which are also shown in Fig. 11 (H. Taylor *et al.*, 1979b).

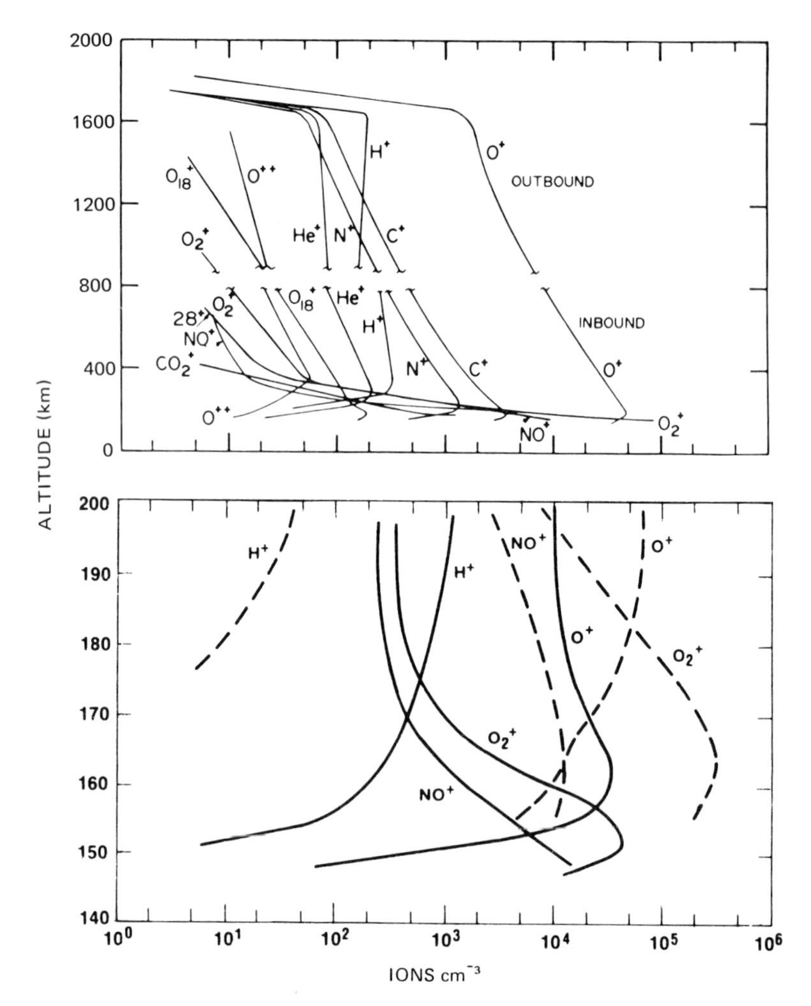

Fig. 11. Number densities of ions observed during orbit 5 (top) and orbits 12 (---, bottom) and 59 (——, bottom) of the *Pioneer Venus* orbiter are shown as functions of altitude. Data from orbits 5 and 12 pertain to the late afternoon, while orbit 59 data was taken at night. From Taylor *et al.* (1979a,b).

G. Cloud Composition

There is now considerable evidence, albeit indirect, indicating that the dominant constituent of the Venusian upper clouds (i.e., the "mode 2" particles in Fig. 5, which are the ones visible from Earth) are composed of concentrated sulfuric acid (H_2SO_4) droplets. Specifically, the data indicate that the droplets contain 75% by mass H_2SO_4 corresponding roughly to the

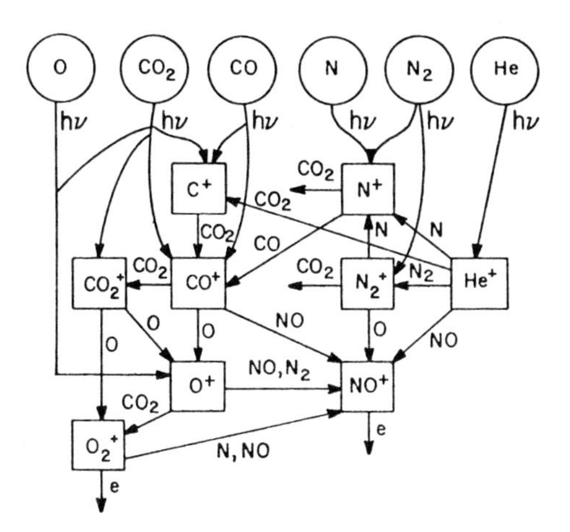

Fig. 12. Flow diagram illustrating the major species and transformations involved in the maintenance of the Venusian ionosphere. Parent neutral species are enclosed in circles and ions in squares. From Nagy *et al.* (1983).

formula $H_2SO_4 \cdot 2H_2O$. The cloud reflectivity (Pollack *et al.*, 1975), the spherical particle shape and high refractive index (~ 1.44; Hansen and Hovenier, 1974), the observed position of the cloud base, and the rapid coherent decrease noted earlier in SO_2 and H_2O within the clouds are all compatible with this identification. As reviewed by Esposito *et al.* (1983), the other two components (modes 1 and 3) may also be composed of H_2SO_4, but evidence is lacking. Alternative suggestions for the mode 1 particle composition are elemental sulfur and meteoritic dust. If the mode 3 particles are crystals, a prime candidate for their composition would be perchloric acid ($HClO_4$) crystals, $HClO_4 \cdot H_2O$ (see Von Zahn *et al.*, 1983). Indeed, $HClO_4$ itself has properties remarkably similar to H_2SO_4, including low H_2O vapor pressures and a refractive index in the range 1.38–1.42, depending on temperature and concentration. Because $HClO_4$ also forms stable mixtures with H_2SO_4, it is a feasible additional component of the H_2SO_4 droplets in the clouds.

H. Surface Composition

We complete this section with a brief summary of the composition of the surface of Venus determined by the X-ray fluorescence experiment on *Venera 13* and *14* (see Table IV; Surkov *et al.*, 1983). Note that the only volatile element detected thus far in surface rocks is sulfur and that the measured abundance of sulfur (expressed as SO_3) is much lower than that of calcium (expressed as CaO). Thus most of the calcium in the rocks must be in a form

Table IV

CHEMICAL COMPOSITION OF VENUSIAN ROCKS DETERMINED BY THE X-RAY FLUORESCENCE EXPERIMENTS ON *VENERA* *13* AND *14*[a]

Element (oxide)	*Venera 13* site	*Venera 14* site
MgO	11.4 ± 6.2	8.1 ± 3.3
Al_2O_3	15.8 ± 3.0	17.9 ± 2.6
SiO_2	45.1 ± 3.0	48.7 ± 3.6
K_2O	4.0 ± 0.63	0.2 ± 0.07
CaO	7.1 ± 0.96	10.3 ± 1.2
TiO_2	1.59 ± 0.45	1.25 ± 0.41
MnO	0.2 ± 0.1	0.16 ± 0.08
FeO	9.3 ± 2.2	8.8 ± 1.8
SO_3	1.62 ± 1.0	0.88 ± 0.77
Cl	<0.3	<0.4
Na_2O[b]	2.0 ± 0.5	2.4 ± 0.4
	~98.1	~98.7

[a] From Surkov *et al.* (1983). Each detected element is assumed to be present as its oxide.
[b] Estimated value; sodium not actually detected.

other than anhydrite ($CaSO_4$). Indeed, the implicit assumption made in Table IV is that sulfur is in the form of a sulfate; if this is not true [e.g., the sulfur is in pyrite (FeS_2)] then the above conclusion is made even firmer.

III. Chemistry and Photochemistry

A chemist taking even a cursory look at the current information on the composition of the Venusian atmosphere is led to ask a number of obvious questions:

1. The atmosphere is predominantly CO_2, and yet CO_2 is photochemically less stable than its photodissociation products CO and O_2. What prevents solar UV radiation from producing a predominantly CO, O_2 atmosphere?
2. The ionosphere is presumably produced by photoionization of CO_2. Why is the dominant ion O_2^+, and what produces the ionosphere on the nightside where there are obviously no solar photons?
3. The putative concentrated H_2SO_4 clouds are produced presumably by oxidation of the observed SO_2, H_2S, and OCS gases, yielding SO_3 and thus

H_2SO_4. The abundance of O_2 in the clouds, however, is so low that it has never been detected! How do we cause oxidation in such an O_2-poor atmosphere, and is this oxidation related to the recycling of CO and O_2 to CO_2 (question 1)?

4. What compounds cause the UV markings in the clouds, and how are they produced?

5. The atmosphere contains remarkably large concentrations of compounds of the normally lithophilic elements S, Cl, and F. How are these large concentrations maintained?

6. In seeking an explanation for the massive atmosphere on Venus, it is natural to explore the consequences of heating the surface of the Earth to a Venus-like value, that is, 750 K. We would then produce indeed a massive atmosphere. It would have an H_2O partial pressure at the surface of ~300 bars (from evaporation of the oceans), a CO_2 partial pressure of ~55 bars (from thermal decomposition of crustal carbonates, etc.), and a N_2 partial pressure of about 1 to 3 bars (from present atmosphere plus degassing of crustal nitrogen). For comparison, at the surface of Venus we presently have partial pressures of ~89 bars for CO_2, ~3.2 bars for N_2, but only ~0.01 bar for H_2O! Was Venus formed with very little H_2O relative to the Earth, or did it lose its H_2O, and if so how?

7. Why are the D/H and $^{36}Ar/^{40}Ar$ ratios on Venus about 100 and 277 times their values on Earth, and does the answer to this question help to answer question 6?

In this section we discuss in succession the chemistry and photochemistry of the thermosphere, ionosphere, strato–mesosphere, and troposphere. In doing so, current ideas on answers to questions 1 through 5 above (and to related queries) will be addressed. Questions 6 and 7 relate to the chemical evolution of the atmosphere, which will be addressed in Section IV.

A. Thermosphere

Photodissociation of CO_2 is very facile in the daytime thermosphere, where it has a photochemical lifetime ~1 week. The specific reactions are

$$\begin{aligned} CO_2 + h\nu &\longrightarrow CO + O(^3P) \qquad \lambda < 2240\ \text{Å} \\ &\longrightarrow CO + O(^1D) \qquad \lambda < 1650\ \text{Å} \\ &\longrightarrow CO + O(^1S) \qquad \lambda < 1273\ \text{Å} \end{aligned} \tag{4}$$

with the $O(^1D)$ and $O(^1S)$ atoms being rapidly quenched to $O(^3P)$ by collisions. The observations in Fig. 10 indicate that, despite this decomposition, CO_2 remains the dominant species up to ~160 km. Thus CO_2 must be replenished at and below this altitude on a time scale of ~1 week.

If we use the three-body reaction

$$O(^3P) + CO + M \longrightarrow CO_2 + M \quad (5)$$

to replenish CO_2, we find, however, that three-body reactions are efficient only at the higher pressures occurring at and below 100 km altitude. Also, at these lower altitudes the reaction

$$O(^3P) + O(^3P) + M \longrightarrow O_2 + M \quad (6)$$

is 10^4–10^5 times faster than reaction (5), which is spin forbidden. Thus essentially all of the photolyzed CO_2 produces CO and O_2. We are therefore led to three apparent dilemmas: photolyzed CO_2 must be replenished in the thermosphere in ~1 week; photolysis of CO_2, which occurs at a column rate of about 10^{13} molec cm^{-2} sec^{-1} on Venus, will build up *above the cloud tops* an amount of O_2 equal to its observed upper limit of 5×10^{17} molec cm^{-2} in only ~1 day and an amount of CO equal to its observed abundance of 8×10^{19} mole cm^{-2} in only ~3 months; finally, photolysis will convert the entire 1.4×10^{27} molec cm^{-2} of CO_2 present on Venus to CO and O_2 on the geologically very short time scale of $\sim 4 \times 10^6$ years. Dilemmas analogous to these three apply also to the CO_2 atmosphere on Mars.

Early studies of Venusian photochemistry focused on the above three dilemmas. In the thermosphere, while novel *in situ* schemes for catalyzing the CO, O reaction [reaction (4)] were proposed (e.g., Reeves *et al.*, 1966; McElroy and Hunten, 1969), these did not stand the test of time. It became apparent that the only way to maintain low O and CO (and high CO_2) concentrations in the 100- to 150-km region was by rapid downward transport of CO and O, balanced by upward transport of CO_2 (Donahue, 1968; Shimizu, 1969; Dickinson and Ridley, 1972). Transport over an average vertical distance of ~25 km in 1 week (6×10^5 sec) is obviously required. If the transport process is eddy diffusive, we have

$$K \simeq (25 \times 10^5)^2 (6 \times 10^5)^{-1} = 10^7 \quad \text{cm}^2\,\text{sec}^{-1}. \quad (7)$$

Values for K of this large magnitude are consistent with Eq. (2). One-dimensional models made with the benefit of the *Pioneer Venus* data confirm the need for large K values and indicate that Eq. (2) provides a satisfactory fit to all available information (Von Zahn *et al.*, 1980; Krasnopolsky and Parshev, 1983). If the transport process is advective, the two-dimensional chemical–dynamic model of Dickinson and Ridley (1975, 1977) predicted upward velocities of ~0.4 m sec^{-1} at the subsolar point and dayside-to-nightside winds of ~300 m sec^{-1}. The predicted dayside O and CO concentrations, however, were too low. Either their predicted circulation is too intense, or there exist reverse circulations not realized in their model, or eddy diffusive transports are important.

Mayr *et al.* (1980) have used the *Pioneer Venus* compositional data to infer properties of the actual three-dimensional thermospheric circulation. Day-to-nightside winds of ~ 200 m sec^{-1} are necessary to transport O, He, and H at sufficient rates to explain the nightside concentrations of these species. In addition, the observed diurnal variations in CO_2, O, H, He, and temperature demand two things: first, that the thermosphere like the atmosphere below it be superrotating with a period of 5 to 10 days (i.e., 50–100 m sec^{-1} zonal wind), and second, that mass exchange with the mesosphere commensurate with $K = 3 \times 10^7$ cm^2 sec^{-1} must occur. The superrotation is needed to explain the observed shifts in phases from precise midnight or noon peaks. The large K values are required to maintain the observed large amplitudes in density variations against the ameliorating effects of horizontal transport. In particular, for negligibly small K values and no superrotation [conversant with the Dickinson and Ridley (1977) model], the predicted cycle in O density peaks at midnight, instead of noon as observed, and also has a smaller amplitude than observed. The Dickinson and Ridley (1977) model is a laudable *ab initio* attempt at modeling the Venusian thermosphere, but it is clear that the real thermosphere will unfortunately not succumb so easily to theoretical analysis.

Hydrogen is a component of the thermosphere in the form of H and H_2. Information on H was obtained very early in Venusian investigations due to the readily observable resonant scattering of solar Lyα by the hydrogen atom "corona" surrounding the planet. A most puzzling result was obtained by *Mariner 5*. While a portion of the observed Lyα from Venus could be attributed to H with a scale height corresponding to an exospheric temperature of 650 K, the larger portion (which originated closer to the planet) implied a source with a scale height about one-half of that for H at 650 K. Early explanations included an isothermal 650 K mixture of H and H_2 (Barth, 1968), an isothermal 650 K H and D mixture (Donahue, 1968; McElroy and Hunten, 1969), and H at two temperatures (A. Stewart, 1968; Barth, 1968; Kumar and Hunten, 1974). As reviewed by von Zahn *et al.* (1983), the model that now appears to best fit the facts is the two-temperature model with an exospheric temperature of ~ 350 K and with hot hydrogen atoms (designated H*) produced by various ion–electron, ion–molecule, charge exchange, and energy exchange reactions. In particular, the reactions

$$
\begin{aligned}
O^+ + H_2 &\longrightarrow OH^+ + H^* \\
OH^+ + e^- &\longrightarrow O + H^* \\
H + H^{+*} &\longrightarrow H^+ + H^* \\
O + H^{+*} &\longrightarrow O^+ + H^* \\
O^* + H &\longrightarrow O + H^*
\end{aligned}
\tag{8}
$$

are the dominant exospheric sources of H* based on *Pioneer Venus* data (Cravens *et al.*, 1980; Kumar *et al.*, 1981; Hodges and Tinsley, 1981; McElroy *et al.*, 1982).

The photochemistry of H_2 in the thermosphere was investigated theoretically by Sze and McElroy (1975), Liu and Donahue (1975), and more recently by Kumar *et al.* (1981) using *Pioneer Venus* data. The major dissociation mechanism for H_2 is

$$CO_2^+ + H_2 \longrightarrow CO_2H^+ + H \tag{9}$$

and this reaction followed by

$$CO_2H^+ + e^- \longrightarrow CO_2 + H \tag{10}$$

also provides the major source for thermospheric H (photodissociation of HCl becomes important at lower levels). The equator-to-pole winds of about 50 to 100 m sec^{-1} carry H from its dayside source to the nightside (Mayr *et al.*, 1980). Transport of both H and He by these winds above the turbopause is much more efficient than the transport of the heavier thermospheric constituents (O, CO, N_2, N, CO_2) due to their much larger scale heights. This causes peaks in the H and He densities near midnight as opposed to near noon for the heavier species. For $K \gg 3 \times 10^7$ cm^2 sec^{-1} both light and heavy species would have noontime peaks (Mayr *et al.*, 1980), which is clearly not the case.

Nitrogen is also observed in the thermosphere, and its photochemistry was considered by McElroy and Strobel (1969) and Cravens *et al.* (1978) prior to *Pioneer Venus*, and by Rusch and Cravens (1979) with the benefit of the *Pioneer Venus* observations. According to Rusch and Cravens (1979) nitrogen atoms are produced mainly by

$$\begin{aligned} N_2 + e^{-*} &\longrightarrow N + N + e^- \\ NO^+ + e^- &\longrightarrow [N(^2D) \text{ or } N] + O \\ N(^2D) + M &\longrightarrow N + M \end{aligned} \tag{11}$$

and removed mainly by

$$N + O_2^+ \longrightarrow NO^+ + O \tag{12}$$

$$N + NO \longrightarrow N_2 + O \tag{13}$$

Nitric oxide (NO) is produced by the unusual occurrence of an efficient two-body recombination reaction

$$N + O \longrightarrow NO + h\nu \tag{14}$$

which leads to NO emission in its δ and γ systems. This "airglow" is readily observable on the nightside and has a peak at 111 km (Stewart *et al.*, 1980). Destruction of NO is principally due to reaction (13). The morphology of the NO nightglow depends sensitively on atmospheric transport since the source of N and O required in reaction (14) is downward mixing. Thus Gerard *et al.* (1981) were able to derive a value for K on the nightside which was $\sim 60\%$ of that deduced on the dayside for the same density levels [see Eq. (2)].

B. Ionosphere

Early attempts at explaining the Venusian ionosphere were made by R. Stewart (1968) and McElroy (1969) for the dayside and by McElroy and Strobel (1969) for the nightside. The principal proposed electron source on the dayside was CO_2 photoionization at wavelengths less than 90 nm,

$$CO_2 + h\nu \longrightarrow CO_2{}^+ + e^- \tag{15}$$

and the principal proposed ion was $CO_2{}^+$. These early models had to be changed, first because the observed large thermospheric O densities enable the rapid charge exchange reaction

$$O + CO_2{}^+ \longrightarrow O_2{}^+ + CO \tag{16}$$

to occur, so $O_2{}^+$ is the dominant ion (Stewart, 1972; Donahue, 1971), and second because the dayside exospheric temperature is actually one-half of that assumed in these early models (Kumar and Hunten, 1974). The modern theory of the Venusian ionosphere, extensively reviewed by Nagy *et al.* (1983), is based on the detailed observations made by the *Pioneer Venus* instruments. Photoionization of CO_2 below 170 km altitude and of O above 170 km provides the primary electron sources on the dayside. Below 170 km, ion transport is unimportant and $O_2{}^+$ is the dominant ion. Loss of $O_2{}^+$ is largely due to

$$O_2{}^+ + e^- \longrightarrow 2\,O \tag{17}$$

While reactions (15)–(17) serve largely to explain the major features of the lower ionosphere, nonnegligible concentrations of the ions CO^+, $CO_2{}^+$, $N_2{}^+$, and NO^+ are also produced. A summary of the dayside ion chemistry is given in Fig. 12.

Above 170 km, the dominant ion produced is O^+:

$$\begin{aligned} O + h\nu &\longrightarrow O^+ + e^- \\ O + e^{-*} &\longrightarrow O^+ + 2\,e^- \\ O + CO_2{}^+ &\longrightarrow O^+ + CO_2 \end{aligned} \tag{18}$$

which is destroyed by

$$O^+ + NO \longrightarrow NO^+ + O$$
$$O^+ + CO_2 \longrightarrow O_2^+ + CO \qquad (19)$$
$$O^+ + H \longrightarrow H^+ + O$$

Although not explicitly stated by Nagy *et al.* (1983), a net downward diffusion of O^+ to lower levels where reactions (19) are efficient must occur and be balanced by an upward neutral O flux. Minor ions in this upper ionospheric region are C^+, N^+, He^+, and H^+, and their production and removal mechanisms are summarized in Fig. 12.

There are two currently viable suggestions for production of the nightside ionosphere: either weakly energetic electrons precipitating into the nightside or thermal O^+ transported from the dayside and subsiding into the nightside thermosphere. Observational evidence for both types of fluxes exists (Intriligator *et al.*, 1979; Knudsen *et al.*, 1980), but two-dimensional modeling by Cravens *et al.* (1981) indicates that the day-to-night transport of O^+ is alone sufficient to provide the observed nightside ionization.

Space limitations do not permit a detailed review here of the energetics of the ionosphere or the physics of its interaction with the solar wind. For coverage of these topics, the reader is referred instead to Nagy *et al.* (1983), Russell and Vaisberg (1983), and Cloutier *et al.* (1983).

C. Strato–Mesosphere

This portion of the Venusian atmosphere has been the object of intensive theoretical study not only because of its importance in answering the fundamental chemical questions introduced at the beginning of this section (particularly questions 1, 3, and 4) but also because of its relevance to important chemical problems on Earth and Mars. Indeed, research on the photochemistry of Venus has provided prescience of important developments in our own atmosphere. For example, the photochemistry of HCl, Cl, ClO, and ClOO was considered in the Venusian atmosphere (Prinn, 1971) before the importance of chlorine photochemistry in our own O_3 layer was appreciated (Stolarski and Cicerone, 1974). In addition, formation of H_2SO_4 aerosols from OCS was also studied in the Venusian atmosphere (Prinn, 1973a), before the very same reactions were shown to be a major contributor to the terrestrial stratospheric H_2SO_4 (or "Junge") layer (Crutzen, 1976).

First, let us address the question of the stability of CO_2 and the low abundance of O_2 in the Venusian atmosphere. While rapid downward transport of O and CO proved capable of resolving this problem in the thermosphere, it merely transferred the ultimate problem of CO_2 recycling to the levels below 100 km. While this problem is also common to Mars, its resolution on Venus actually depends on chlorine and sulfur compounds not present on Mars.

The thermosphere on Venus effectively absorbs all solar photons with wavelengths less than ~ 1700 Å, leaving longer wavelengths to pump the photochemistry of the strato–mesosphere. Photodissociation of CO_2 proceeds down to the cloud tops at 70 km and in the process removes essentially all of the photons in the 1700- to 2000-Å region. Water vapor is also dissociated by 1700- to 2000-Å photons, but it is effectively shielded by the abundant overlying CO_2. In contrast, HCl is readily photodissociated at wavelengths up to 2200 Å and is therefore not shielded by CO_2 (Prinn, 1971); as a result it plays a central role in the chemistry of the strato–mesosphere.

The photochemistry of HCl on Venus was first studied by Prinn (1971, 1972, 1973b). The important reactions destroying and reforming HCl were predicted to be

$$\begin{aligned} HCl + h\nu &\longrightarrow H + Cl \qquad \lambda < 2200 \text{ Å} \\ Cl + H_2 &\longrightarrow HCl + H \end{aligned} \tag{20}$$

This early work emphasized both the importance of HCl photodissociation as a source and sink of H_2 and the possible roles played by reactive chlorine (Cl, ClO, and ClO_2) reactions in the catalysis of the CO, O_2 reaction. The work also suggested photochemically produced Cl_2, HOCl, Cl_3^-, and OCl^- in acidic cloud droplets as candidates for the UV and blue absorbers in the clouds.

Later McElroy *et al.* (1973) reported a chemical model for the Venusian upper atmosphere that emphasized the roles of HCl and Cl in the production of odd-hydrogen (H, OH, and HO_2). This odd-hydrogen was used to recombine CO and O_2 using schemes similar to those already employed successfully for Mars (McElroy and Donahue, 1972). Then Sze and McElroy (1975) combined what appeared to be the most plausible chlorine and odd-hydrogen reactions from these earlier papers. They concluded that reactions of CO with OH and, to a lesser degree, with ClO were the principal modes of recyling CO_2.

While chlorine compounds were clearly important, their precise roles were predicated by the discovery, discussed earlier, that the clouds of Venus contained H_2SO_4 and by the suggestion by Prinn (1973a) that photochemical oxidation of reduced sulfur gases was the probable source of these clouds. In the region of the visible clouds a new sink for O_2, namely, the reaction

$$S + O_2 \longrightarrow SO + O \tag{21}$$

followed by further reactions yielding SO_3 and thus H_2SO_4, was proposed as the major O_2 sink in the lower strato–mesosphere (Prinn, 1973a, 1975, 1978). The recycling of CO_2 was accomplished in this scheme by thermochemical reactions in the hot lower atmosphere, beginning with

$$SO_3 + CO \longrightarrow SO_2 + CO_2 \tag{22}$$

After the 1978 American and Russian missions to Venus, Winick and Stewart (1980) and Krasnopolsky and Parshev (1980) readdressed the coupled photochemistry of HCl and sulfur compounds in the light of the new data derived from these missions. Winick and Stewart (1980) were able to model successfully the vertical distributions of SO_2 and cloud densities above 60 km, but their chemical model produced O_2 and CO abundances 6 and > 300 times, respectively, the observed cloud-top values. This latter discrepancy is related to two factors. First, they deduced odd hydrogen and reactive chlorine concentrations more than an order of magnitude less than those deduced by Sze and McElroy (1975). This difference was caused by new kinetic data that indicated that the reaction

$$HO_2 + Cl \longrightarrow HCl + O_2 \tag{23}$$

which destroys both odd-hydrogen and reactive chlorine, was fast but the alternative reaction (Prinn, 1971)

$$HO_2 + Cl \longrightarrow ClO + OH \tag{24}$$

was negligibly slow. The lower odd-hydrogen predictions led in turn to much higher O_2 concentrations and downward fluxes in the Winick and Stewart model (1980) relative to the Sze and McElroy model (1975). Second, Winick and Stewart (1980) used *Pioneer Venus* data to argue that OCS was absent from the cloud region. Thus the potent photochemical source of sulfur atoms,

$$OCS + h\nu \longrightarrow CO + S \qquad \lambda < 3000 \text{ Å} \tag{25}$$

which drives the removal of O_2 by reaction (21) in the Prinn (1973a) model, was absent in their model.

Krasnopolsky and Parshev (1980) were also able to model the vertical SO_2 distribution successfully, and in addition, they predicted CO and O_2 abundances that agreed much better with observations. Their scheme for recombining CO and O_2, however, utilized the Prinn (1971) reaction

$$ClO + CO \longrightarrow Cl + CO_2 \tag{26}$$

and assumed a rate constant for this reaction several orders of magnitude greater than that adopted by Winick and Stewart (1980). In addition, they included formation of COCl (Prinn, 1971); the subsequent putative rapid reaction

$$COCl + O_2 \longrightarrow CO_2 + ClO \tag{27}$$

in fact dominates recombination of CO and O_2 in their model. This latter reaction was, however, very speculative. With reaction (27) included, these authors predicted O_2 concentrations, O_2 downward fluxes, and reactive chlorine concentrations similar to those of Sze and McElroy (1975). Their odd-hydrogen concentrations and consequently their oxidation rates of CO by OH, however, are much lower than those of Sze and McElroy (1975) and instead resemble those of Winick and Stewart (1980).

The most extensive investigation to date of the coupled chlorine–sulfur chemistry of the Venusian strato–mesosphere has been carried out by Yung and DeMore (1982; see also DeMore and Yung, 1982). Their overall cycles for CO_2 and sulfur species are easy to summarize. Carbon dioxide is photodissociated at an average rate of 8.5×10^{12} molec cm^{-1} sec^{-1} in the Venusian strato–mesosphere, yielding CO and mostly O_2. About three of four CO and O_2 molecules recombine above the 70-km level through various reactions involving odd-hydrogen, reactive chlorine, or perhaps even nitrogen oxides. This leaves downward fluxes for CO and O_2 at 70 km of 2.2×10^{12} and 1.1×10^{12} molec cm^{-2} sec^{-1}, respectively. The O_2 serves to oxidize an upwardly diffusing SO_2 flux of 2.2×10^{12} molec cm^{-1} sec^{-1} to produce SO_3 and thus H_2SO_4. The H_2SO_4 falls down, evaporates to yield SO_3, and the SO_3 reacts with CO [reaction (22)] to recycle the CO_2 and SO_2.

The problem lies not in the definition of these overall cycles, however, but in the details of the catalytic subcycles involved. Yung and DeMore (1982) emphasized the enormous importance of the H_2 mixing ratio in differentiating between various possible catalytic subcycles; as noted earlier, all that is definitely known about H_2 is that it has a mixing ratio of $\leqslant 10^{-5}$. These workers therefore considered three possible H_2 mixing ratios: 2×10^{-5} (model A), 5×10^{-7} (model B), and 1×10^{-13} (model C). Their reaction scheme included HCl photodissociation and the previously recognized roles played by various products of HCl photodissociation (Prinn, 1971; McElroy *et al.*, 1973; Krasnopolsky and Parchev, 1980; Winick and Stewart, 1980). They also added some potentially important new reactants and reactions. Their HCl photochemical scheme is summarized in Fig. 13. Note particularly the central role played by H_2 in both the reformation of photodissociated HCl and in the production of odd-hydrogen [the relevant reaction is reaction (20), discussed earlier].

The rates of the major reactions producing and consuming oxygen in model A are shown in Fig. 14. The high odd-hydrogen concentrations accompanying the high H_2 concentrations in this model cause the reaction

$$CO + OH \longrightarrow CO_2 + H \qquad (28)$$

to dominate CO_2 recycling, with a lesser role played by SO_2 oxidation. The

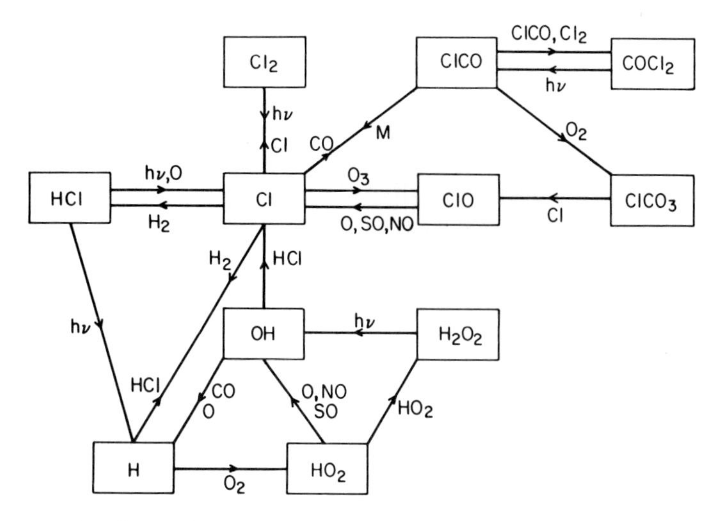

Fig. 13. Flow diagram illustrating the major sources, sinks, and recycling pathways for odd-hydrogen and reactive chlorine species in the Venusian strato–mesosphere. From Yung and DeMore (1982).

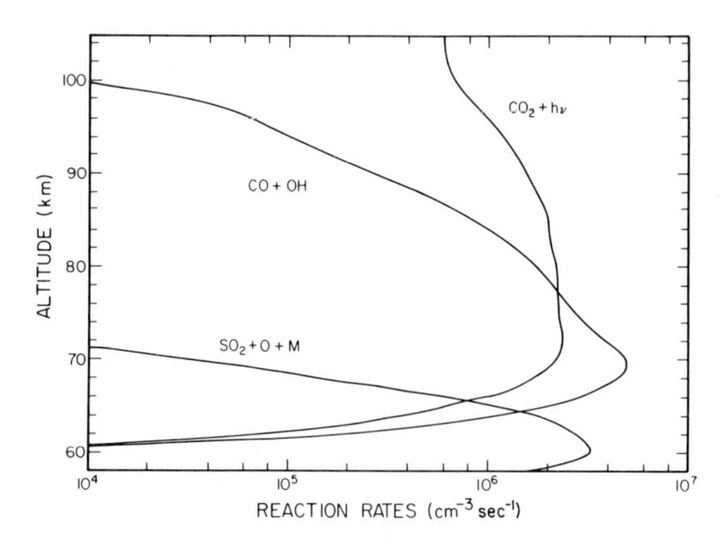

Fig. 14. The major reactions producing and consuming oxygen in model A of Yung and DeMore (1982) for the Venusian strato–mesosphere. From Yung and DeMore (1982).

high odd-hydrogen concentrations also lead to a net *production* of H_2O in this model due to

$$OH + HCl \longrightarrow H_2O + Cl \tag{29}$$

and an accompanying net *destruction* of H_2 due to

$$Cl + H_2 \longrightarrow HCl + H \tag{30}$$

These net tendencies for H_2O and H_2 must be offset by the occurrence of the net reaction

$$CO + H_2O \longrightarrow CO_2 + H_2 \tag{31}$$

elsewhere in the atmosphere at a rate sufficient to yield an H_2 flux of $\sim 10^{11}$ molec cm^{-2} sec^{-1}. Yung and DeMore (1982) compared this latter number to the production rate of CH_4 by the Earth's biota, which is $\sim 3 \times 10^{10}$ molec cm^{-2} sec^{-1}. Methane (CH_4) plays on Earth a role similar to that played on Venus by H_2; in particular, the abundances of reactive chlorine in the Earth's ozone layer depend critically on the reaction

$$Cl + CH_4 \longrightarrow HCl + CH_3 \tag{32}$$

which can be compared to its Venusian analogy, reaction (30).

A necessary part of any catalytic scheme to recombine CO and O_2 is the breaking of the O—O bond. In model A this is dominated by two reactions, with reaction (34) occurring to a lesser extent:

$$SO + HO_2 \longrightarrow SO_2 + OH \tag{33}$$

$$S + O_2 \longrightarrow SO + O \tag{34}$$

The dominant pathway for breaking the O—O bond, converting CO to CO_2, and oxidizing SO_2 to SO_3 in model A is thus

$$\begin{aligned}
SO_2 + h\nu &\longrightarrow SO + O \\
SO + HO_2 &\longrightarrow SO_2 + OH \\
OH + CO &\longrightarrow CO_2 + H \\
H + O_2 + M &\longrightarrow HO_2 + M \\
SO + h\nu &\longrightarrow S + O \\
S + O_2 &\longrightarrow SO + O \\
3\,O + 3\,SO_2 + 3\,M &\longrightarrow 3\,SO_3 + 3\,M
\end{aligned} \tag{35}$$

Except for the need for an unspecified H_2O-to-H_2 recycling mechanism, model A provides satisfactory fits to available data on CO and O_2 concentrations as well as to the observed 1.27-μm Venusian airglow, which is caused by the relaxation of chemically produced excited O_2, namely,

$$O_2(^1\Delta) \longrightarrow O_2 + h\nu \tag{36}$$

Oxygen consumption and production in model B of Yung and DeMore (1982) involve largely the same reactions as in model A (Fig. 14) and also lead to the same demand for an unspecified H_2 source of $\sim 10^{11}$ molec cm^{-2} sec^{-1}. Agreement with the relevant CO and O_2 data is not as good as in model A but nevertheless is still satisfactory. The assumed lower H_2 concentrations in model B, however, lead to lower odd-hydrogen concentrations, and reaction (33) is no longer efficient in breaking the O—O bond. Thus model B postulates the existence of odd-nitrogen species (N, NO, NO_2, NO_3, HNO, HNO_2, HNO_3) in the strato–mesosphere with a total mixing ratio of $\sim 3 \times 10^{-8}$, which enables the reaction

$$NO + HO_2 \longrightarrow NO_2 + OH \tag{37}$$

to replace reaction (33) as the dominant O—O bond-breaking method. There are at least two problems with the odd-nitrogen hypothesis, however: first, odd-nitrogen is destroyed in model B at a rate of $\sim 1.2 \times 10^9$ molec cm^{-2} sec^{-1}, and this production rate cannot be provided even by the putative lightning discharges on Venus (Borucki *et al.*, 1981); second, the *Venera 11* and *12* spectrophotometer established an *upper limit* for NO_2 in the lower atmosphere of $\sim 5 \times 10^{-10}$ (Moroz, 1981).

The reactions producing and consuming oxygen in model C are summarized in Fig. 15. Agreement between theory and observation for this model is about the same as that for model B. In this model the sequential reactions

$$\begin{aligned} COCl + O_2 + M &\longrightarrow ClCO_3 + M \\ ClCO_3 + (Cl, O) &\longrightarrow CO_2 + ClO + (Cl, O) \end{aligned} \tag{38}$$

replace not only reaction (28) as the dominant CO oxidation mechanism but also reactions (33) or (37) as the dominant O—O bond-breaking mechanism. The very low assumed H_2 abundance at 60 km in model C leads to very low odd-hydrogen concentrations, thus eliminating the large H_2O production and H_2 removal rates characteristic of models A and B. The H_2 concentrations are so low in model C, however, that reaction (30), which recycles HCl, is no longer efficient enough to maintain a constant mixing ratio for this compound in the strato–mesosphere. Large mixing ratios for Cl (>70 km) and Cl_2 (<70 km) result; in particular, the mixing ratio of Cl_2 in the clouds is predicted to be $\sim 10^{-7}$.

Yung and DeMore (1982) preferred model C for three principal reasons: first, it does not require a significant lower atmospheric source for H_2; second, Pollack *et al.* (1980) had earlier concluded that sulfur could not be the UV absorber in the clouds and that amounts of Cl_2 not much greater than those in model C are therefore the most likely cause of the absorption; third, microwave observations of CO (Wilson *et al.*, 1981) indicate a depletion of CO in the nightside at 80–90 km, and CO destruction in model C is large in the 80- to 90-km region (Fig. 15) while negligible in models A and B (Fig. 14).

The apparently large H_2 input required in model A (and B), however, is not entirely improbable. In particular, as we shall discuss shortly, photochemical reactions driven by thiozone (S_3) and S_4 photodissociation in the hot lower atmosphere may be a sufficiently potent source of H_2. Also, Toon *et al.* (1982) have reinstated sulfur as the most viable candidate for the UV absorber, thus obviating the need for large Cl_2 concentrations. The principal remaining argument for model C is thus the apparent nightside CO depletion; however, a definitive study showing that even model C is capable of quantitatively explaining this depletion is unfortunately not yet available. An important argument *against* model C is that it requires what appear to be unrealistically *small* H_2 concentrations in the lower atmosphere. In thermochemical equilibrium at the surface, the predicted H_2 mixing ratio is 6.7×10^{-7} (Lewis, 1970). Without a major upper atmospheric sink for H_2 it is difficult to see how the lower atmospheric H_2 mixing ratio could be depleted by nearly seven

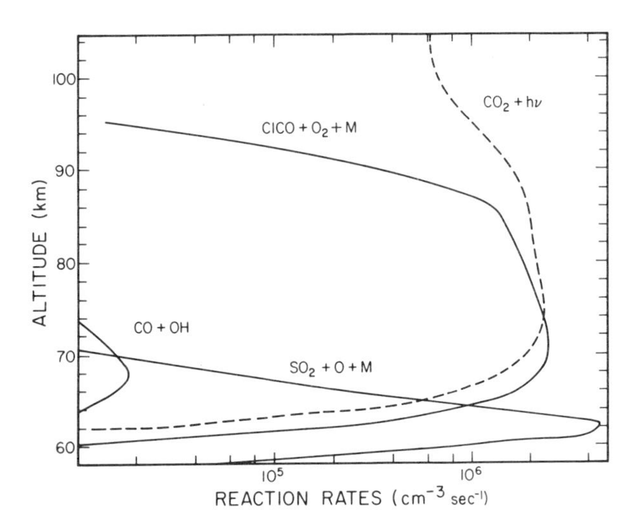

Fig. 15. The major reactions producing and consuming oxygen in model C of Yung and DeMore (1982) for the Venusian strato–mesosphere. From Yung and DeMore (1982).

orders of magnitude below its equilibrium value, which is what model C requires. We are left therefore at the present time with no clearly preferred model for the recombination of CO and O_2 in the Venusian strato–mesosphere. The importance of determining the lower atmospheric H_2 mixing ratio and of further documenting and then modeling the nightside CO depletion is, however, quite apparent from the seminal work of Yung and DeMore (1982).

Sulfur dioxide (SO_2) removal and H_2SO_4 production is facile in all three of the above models; each model serves to convert an upward SO_2 flux of $\sim 10^{12}$ molec cm^{-2} sec^{-1} into an equal downward flux of $H_2SO_4 \cdot 2H_2O$. Since the Venusian clouds contain $\sim 7 \times 10^{18}$ H_2SO_4 molec cm^{-2}, the time for chemical regeneration of the clouds is ~ 80 days. This chemical regeneration process should be carefully distinguished from the simple reversible evaporation–recondensation process that undoubtedly occurs at the cloud base:

$$H_2SO_4 \cdot 2H_2O \longrightarrow H_2SO_4 + 2\,H_2O,\ SO_3 + 3\,H_2O \tag{39}$$

and that could result in a recycling of particles in the cloud-base region on a much shorter time scale (e.g., see Rossow, 1978). We reiterate, however, a point made earlier that the clouds of Venus are a direct product of the rich photochemistry of this planet and are not a simple condensation phenomenon.

It remains for us to address the nature and maintenance of the blue- and UV-absorbing component of the clouds. The nature of this absorber has been a topic of considerable debate; concentrated H_2SO_4 particles do not absorb at the relevant UV wavelengths, and SO_2 absorbs at only wavelengths less than 0.32 μm. While many candidates for the absorbers have been proposed, only elemental sulfur (Prinn, 1971, 1974b; Hapke *et al.*, 1974) is presently considered viable, although even this candidate has received very mixed reviews over the years. Toon *et al.* (1982) showed that elemental sulfur particles containing traces of the metastable blue- and UV-absorbing sulfur allotropes S_3 and S_4 and constituting only $\sim 3\%$ by mass of the clouds in the 55- to 70-km region are capable of explaining the relevant data on the refractive index and reflectivity of the clouds.

If S_3 and S_4 contained in elemental sulfur particles are the required absorber, then how are they formed? Prinn (1975) proposed that both elemental sulfur and H_2SO_4 are formed in a sulfur "bicycle" that begins with OCS and H_2S in the lower atmosphere. Later work indicated that the allotropes S_2, S_3, and S_4 are natural products of OCS and H_2S decomposition (Prinn, 1978, 1979). The failure of the 1978 *Pioneer Venus* and *Venera 11* and *12* missions to positively detect OCS, however, placed the sulfur bicycle in abeyance. Since then, the *Venera 13* and *14* missions have indicated OCS and H_2S concentrations are much larger than was indicated by the 1978 missions, which serves to revive the importance of the sulfur bicycle.

The basic requirements for the sulfur bicycle (see Fig. 16) are the presence of reduced sulfur gases (OCS, H_2S) and a spatially and/or temporally variable O_2 concentration. The reduced sulfur gases are converted to H_2SO_4 in O_2-rich regions and to elemental sulfur in O_2-poor regions. From Section II,E we note that the reduced sulfur gases may comprise anywhere from approximately 3% (pre–*Venera 13* and *14* results) to 44% (*Venera 13* and *14*) of the total sulfur below the clouds. If all of the OCS and H_2S transported up into the clouds produces S (rather than H_2SO_4), then even the 3% value would be almost sufficient to yield the amounts of sulfur required by Toon *et al.* (1982). The bicycle begins with the production of sulfur atoms. There are a number of suggested sources of S in the clouds. From Prinn (1975, 1978) we have

$$\begin{aligned} &\mathrm{OCS} + h\nu \longrightarrow \mathrm{CO} + \mathrm{S} && \lambda < 300\ \mathrm{nm} \\ &\mathrm{H_2S} + h\nu \longrightarrow \mathrm{SH} + \mathrm{H} && \lambda < 300\ \mathrm{nm} \\ &\mathrm{SH} + \mathrm{SH} \longrightarrow \mathrm{S} + \mathrm{H_2S} && \end{aligned} \tag{40}$$

with rates depending on the assumed OCS and H_2S concentrations. From Yung and DeMore (1982) we have S from SO photodissociation, which produces $\sim 5 \times 10^{11}$ atoms $\mathrm{cm}^{-2}\ \mathrm{sec}^{-1}$ in the high H_2 case and somewhat less in the low H_2 cases. As discussed by Prinn (1975), we expect any S at cloud

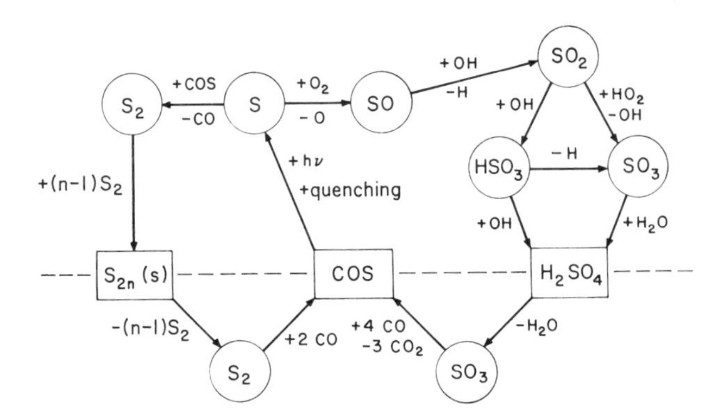

Fig. 16. Flow diagram illustrating the bicycle for carbonyl sulfide in the Venus atmosphere. Transient species are enclosed in circles, and stable species in rectangles. The dashed curve represents the cloud base. Reactions above this curve are photochemical, and below this line they are thermochemical. From Prinn (1975).

level to be converted to S_2 through the reaction

$$S + OCS \longrightarrow S_2 + CO \tag{41}$$

rather than to SO (and ultimately to H_2SO_4) through the reaction

$$S + O_2 \longrightarrow SO + O \tag{42}$$

provided that the O_2 mixing ratio in the air parcel is less than 0.07% of the OCS mixing ratio. If this condition is met, then the S produced from SO photodissociation alone would be sufficient to convert all of the OCS in an upward-moving air parcel to S_2, even if the upward OCS flux is as great as 23% of the SO_2 flux of 2.2×10^{12} molec cm^{-2} sec^{-1}. In addition, the reaction (Prinn, 1979)

$$OCS + SO \longrightarrow CO_2 + S_2 \tag{43}$$

may further augment the OCS conversion to S_2.

An attractive aspect of the bicycle scheme is that it provides a natural explanation for the occurrence of unstable sulfur allotropes in the clouds. Once S_2 is formed, the subsequent reactions

$$\begin{aligned} OCS + S_2 &\longrightarrow CO + S_3 \\ OCS + S_3 &\longrightarrow CO + S_4 \\ S_2 + S_2 + M &\longrightarrow S_4 + M \end{aligned} \tag{44}$$

provide an effective path in OCS-rich, O_2-poor air parcels for producing the near-UV, blue, and green absorbers S_3 and S_4 (Prinn, 1979). As noted by Toon *et al.* (1982), S_3 and S_4 that condensed onto particles would be metastable and thus provide an explanation for the ephemeral nature of the UV absorbers in the clouds.

Another aspect of the bicycle scheme (Prinn, 1975) is that upward-moving air parcels are expected to be poor in O_2 and rich in OCS and H_2S. Thus production of the blue- and UV-absorbing sulfur is expected to be favored in upward moving air parcels and conversely to be suppressed in downward moving parcels. About 35% of the solar energy deposited on Venus involves absorption at wavelengths less than 500 nm (Tomasko *et al.*, 1980). Since this absorption is concentrated in the putative sulfur-rich regions, then a positive correlation is expected between upward motion and solar heating, which may be important in the dynamics of the visible atmosphere.

D. Troposphere and Surface

The chemistry of sulfur, chlorine, carbon, and hydrogen compounds on Venus appears to be driven by two competing influences (Prinn, 1973a, 1978;

Florensky *et al.*, 1978). A photochemical region exists (as discussed above) within and above the clouds, where the chemistry is driven by UV light, resulting (except for carbon) in a net increase in the oxidation state of species. This photochemical regime may also extend downward to within 10 km of the surface because of photochemistry induced by near-UV and visible light. In contrast, a thermochemical region exists at and near the surface, where high temperatures and pressures enable the action of purely thermochemical reactions, which will result usually in a net decrease in the oxidation state of species (except for carbon).

A number of studies have been carried out that explicitly hypothesize chemical equilibrium in the lower atmosphere and between the atmosphere and surface minerals and then explore the consequences of this hypothesis for atmospheric and surface composition (Lewis, 1970; Barsukov *et al.*, 1980; Lewis and Kreimendahl, 1980; Khodakovsky, 1982).

For fluorine and chlorine compounds, the seminal paper by Lewis (1970) showed that the following equilibria at the Venusian surface:

$$2\,NaCl + CaAl_2Si_2O_8 + SiO_2 + H_2O \longleftrightarrow 2\,NaAlSiO_4 + CaSiO_3 + 2\,HCl$$

$$2\,NaCl + Al_2SiO_5 + 3\,SiO_2 + H_2O \longleftrightarrow 2\,NaAlSi_2O_6 + 2\,HCl \qquad (45)$$

$$2\,CaF_2 + MgSiO_3 + SiO_2 + 2\,H_2O \longleftrightarrow Ca_2MgSi_2O_7 + 4\,HF$$

lead to predicted mixing ratios for HCl and HF of $\sim 10^{-6}$ and 2×10^{-8}, respectively, which are only slightly greater than the values observed above the clouds. There is now little doubt that the much higher atmospheric concentrations of HCl and HF on Venus relative to those on Earth is simply a result of the high surface temperature on Venus. The agreements between the predicted and observed mixing ratios for HCl and HF also strongly suggest that thermochemical equilibrium is closely attained between the atmosphere and both halite (NaCl) and fluorite (CaF_2) on the surface. Apparently, the putative *photo*chemical conversions of HCl to Cl_2, chlorine oxides, and chlorine oxyacids discussed earlier are not efficient enough relative to the reverse *thermo*chemical reactions in the hot lower atmosphere to significantly disturb this thermochemical equilibrium.

For sulfur compounds the picture is much more complex. It is educational to first consider what sulfur chemistry on Venus would look like if we were able to deactivate totally the photochemical regime. In this case, the lower atmosphere and surface would be in thermochemical equilibrium. The abundance of O_2 in the atmosphere, and thus the oxidation state of all other species, should then be determined by the pyrite–anhydrite equilibrium (Lewis and Kreimendahl, 1980; Barsukov *et al.*, 1980):

$$FeS_2 + 2\,CaCO_3 + \tfrac{7}{2}\,O_2 \longleftrightarrow 2\,CaSO_4 + FeO + 2\,CO_2 \qquad (46)$$

The predicted equilibrium O_2 mixing ratio is $\sim 3 \times 10^{-25}$, and at this low O_2 level the dominant sulfur-bearing gas is OCS (mixing ratio of $\sim 6 \times 10^{-4}$), with smaller amounts of H_2S and SO_2 (mixing ratios of 1.3×10^{-4} and 1.6×10^{-5}, respectively). If we increase the O_2 mixing ratio from this equilibrium value (e.g., by reactivating the photochemical regime), the mixing ratios of *all* sulfur gases decrease while the activity of FeO and $CaSO_4$ in surface rocks increases. Maintenance of the total sulfur ($OCS + H_2S + SO_2$) mixing ratio at its presently observed value of 1.5×10^{-4}, however, requires that the activity of FeO not exceed one-tenth of its value on Earth (Lewis and Kreimendahl, 1980). In other words, even a mildly oxidized Venusian crust is incompatible with significant amounts of *any* sulfur gases in the atmosphere. Equally important, in order for the total sulfur mixing ratio to equal 1.5×10^{-4}, the dominant sulfur gases *must* be OCS and H_2S; the maximum SO_2 mixing ratio in equilibrium under *any* conditions is $\sim 1.6 \times 10^{-5}$ (Lewis and Kreimendahl, 1980).

At the opposite extreme to equilibrium lies the atmospheric state ensuing when we deactivate the thermochemical regime but maintain all of the photochemical processes. Sulfur will be driven toward its highest stable oxidation state and thus exist entirely as H_2SO_4 above the cloud base and as SO_3 gas below this base. Since neither chemical regime alone can explain the observed composition of the Venusian atmosphere as discussed in Section II,E, it is apparent that the contemporary atmosphere represents a balance between these two regimes. If we accept the *Venera 13* and *14* data on OCS and H_2S, then the two regimes compete remarkably well; if we accept the much lower OCS and H_2S abundances inferred prior to these two probes, then the thermochemical regime is more subdued but still influential (e.g., the thermochemical reduction of SO_3 to SO_2 is still very efficient).

While the existence of the above balance is easily inferred, the detailed cycling of sulfur compounds that leads to the balance is not. In discussing the overall sulfur cycle on Venus it is useful to refer to a triarchy involving three subcycles (Prinn, 1984; von Zahn *et al.*, 1983). These subcycles are shown in Fig. 17, in which the regions of stability and relative abundances of the major sulfur compounds are qualitatively indicated by the position and size, respectively, of the areas assigned to each (the question marks in Fig. 17 denote the uncertainty in the OCS and H_2S abundances). In the atmosphere above 50 km the net photochemical reactions

$$\begin{aligned} CO_2 &\longrightarrow CO + \tfrac{1}{2} O_2 \\ \tfrac{1}{2} O_2 + SO_2 &\longrightarrow SO_3 \\ SO_3 + 3\, H_2O &\longrightarrow H_2SO_4 \cdot 2H_2O \end{aligned} \tag{47}$$

initiate a so-called fast atmospheric cycle. To complete this cycle the net

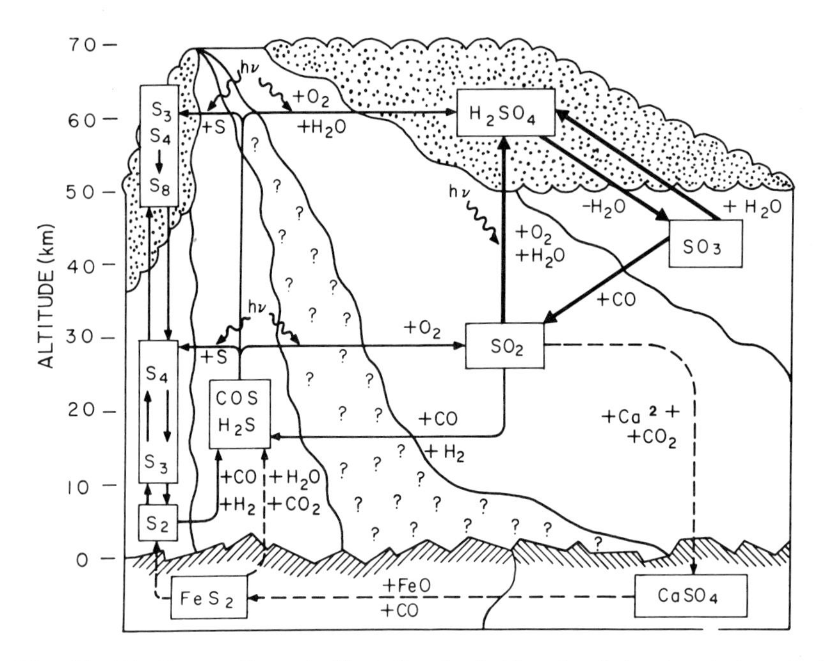

Fig. 17. The Venus sulfur cycle, illustrating the fast (——) and slow (——) atmospheric subcycles and the geological subcycle (---). The regions of stability and relative abundances of the major sulfur compounds are qualitatively indicated by the position and size of the shaded area assigned to each compound. From Prinn (1984).

thermochemical reactions

$$\begin{aligned} H_2SO_4 \cdot 2H_2O &\longrightarrow 3\,H_2O + SO_3 \\ SO_3 + CO &\longrightarrow SO_2 + CO_2 \end{aligned} \tag{48}$$

proceed in the hot lower (< 50 km) atmosphere. The formation of the H_2SO_4 clouds and the partial recycling of the photodissociation products of CO_2 (i.e., CO, O_2) are direct products of this fast cycle.

A "slow atmospheric cycle" begins with the following net photochemical reactions in the atmosphere above 50 km:

$$\begin{aligned} 6\,CO_2 &\longrightarrow 6\,CO + 3\,O_2 \\ \tfrac{3}{2}\,O_2 + H_2S &\longrightarrow SO_3 + H_2 \\ \tfrac{3}{2}\,O_2 + OCS &\longrightarrow SO_3 + CO \\ 2\,SO_3 + 6\,H_2O &\longrightarrow 2(H_2SO_4 \cdot 2H_2O) \\ H_2S &\longrightarrow H_2 + (1/x)S_x \\ COS &\longrightarrow CO + (1/x)S_x \end{aligned} \tag{49}$$

This cycle (essentially the "sulfur bicycle") is closed in the lower atmosphere by the net thermochemical reactions

$$\begin{aligned} 2(H_2SO_4 \cdot 2H_2O) &\longrightarrow 6\,H_2O + 2\,SO_3 \\ SO_3 + 4\,CO &\longrightarrow OCS + 3\,CO_2 \\ H_2 + SO_3 + 3\,CO &\longrightarrow H_2S + 3\,CO_2 \\ H_2 + (1/x)S_x &\longrightarrow H_2S \\ CO + (1/x)S_x &\longrightarrow OCS \end{aligned} \tag{50}$$

Important products of this cycle are various allotropes of elemental sulfur (S_x) that can account for the UV absorptivity of the clouds. If the high OCS and H_2S levels inferred from *Venera 13* and *14* are correct, then this cycle is also an important contributor to the H_2SO_4 clouds. As discussed later, the photochemical portion of this cycle may be aided by reactions in the lower atmosphere instigated by S_3 and S_4 photodissociation, which convert OCS and H_2S to S_x (and perhaps also to SO_2, using H_2O as the oxygen source).

The "geological cycle" completes the triarchy. While it is the slowest of the three cycles, it is in a sense the most important for it determines both the total mixing ratio of atmospheric sulfur compounds and the balance between the photochemical and thermochemical tendencies. The geological cycle operates on the time scale of crustal weathering and recycling. It begins with crustal outgassing of the reduced sulfur gases (OCS, H_2S) through net reactions involving pyrite (FeS_2):

$$\begin{aligned} FeS_2 + 2\,H_2O + CO &\longrightarrow FeO + 2\,H_2S + CO_2 \\ FeS_2 + CO_2 + CO &\longrightarrow FeO + 2\,OCS \end{aligned} \tag{51}$$

with the FeO produced being sequestered in surface silicates. The total sulfur mixing ratio in the emanating gases must equal or exceed 1.5×10^{-4}, which in turn demands that the degassing crustal rocks have low FeO activities, as already noted. The possibility that the required FeS_2-rich deposits may be exposed at high elevations on Venus (e.g., through weathering or volcanism) has some observational support. In particular, Pettengill *et al.* (1982) have deduced the presence of highly conductive minerals at high elevations (e.g., Theia Mons) based on radar-reflectivity data; FeS_2 is one of the very few feasible surface materials with the required high electrical conductivity.

Once degassed, the H_2S and OCS are pumped many times through the fast and slow atmospheric cycles. Conclusions about the relative times spent in these two cycles are confused by the conflicting observational data on OCS and H_2S; if these latter two gases are minor then sulfur will flux largely through the fast cycle. In any case, continual outgassing of OCS and H_2S will build up SO_2 in the atmosphere through the net photochemical oxidations

$$\begin{aligned} 2\,H_2S + 6\,CO_2 &\longrightarrow 2\,H_2O + 2\,SO_2 + 6\,CO \\ 2\,OCS + 4\,CO_2 &\longrightarrow 2\,SO_2 + 6\,CO \end{aligned} \tag{52}$$

These net oxidations lead in turn to an SO_2 concentration that exceeds its equilibrium value, and various weathering reactions involving the leaching of calcium from surface silicates to form $CaSO_4$,

$$4\,Ca^{2+} + 4\,O^{2-} + 4\,SO_2 + 4\,CO_2 \longrightarrow 4\,CaSO_4 + 4\,CO \tag{53}$$

will proceed in an attempt to restore equilibrium. There is ample available Ca^{2+} at the surface for this purpose; from Table IV it can be seen that CaO comprises 7 and 10%, respectively, of the rocks analyzed at the *Venera 13* and *14* landing sites, while SO_3 (in sulfates) comprises only 1.6 and 0.9%, respectively, of the same rocks (Surkov *et al.*, 1983). Thus there is no doubt that SO_2 is being removed from the atmosphere by Ca^{2+} at the present time. Of course, the current lack of data on the sulfur gases below 22 km prevents any firm conclusions on the exact degree to which Ca^{2+} acts to restore completely equilibrium concentrations of SO_2 at the surface.

The geological cycle must be ultimately completed beneath the surface by the conversion of $CaSO_4$ to FeS_2:

$$4\,CaSO_4 + 2\,FeO + 14\,CO \longrightarrow 2\,FeS_2 + 4\,Ca^{2+} + 4\,O^{2-} + 14\,CO_2 \tag{54}$$

As already noted, we currently lack the observations required to determine whether the restorative reactions in the geological cycle are efficient enough to maintain atmosphere–surface equilibrium. We also have no evidence to suggest that this geological cycle is in a steady state. Sporadic volcanic activity and cometary impacts may govern the outgassing of OCS and H_2S, while the incorporation of SO_2 into the crust and regeneration of FeS_2 may depend on equally unsteady crustal recycling processes. An exact balance between outgassing and reburial rates of sulfur species at any one time on Venus may be an exception rather than a rule. The total atmospheric sulfur mixing ratio and the mean oxidation state of atmospheric sulfur may have undergone significant changes. For example, in periods of intense volcanic activity the total sulfur mixing ratio and the abundance of OCS relative to SO_2 would be expected to be distinctly higher than at present; the clouds would therefore be more dense, extend to lower altitudes, and have a significantly higher S/H_2SO_4 ratio than the clouds evident today. Such changes would clearly impact the meteorology and climate during these volcanically active periods.

Of particular interest in this respect is the recent remarkable report by Esposito (1984) that SO_2 in the visible atmosphere has decreased by a factor of 10 from 1978 to 1983, accompanied by a similar decline in the amount of submicrometer-sized haze particles above the clouds. As Esposito (1984) noted, it is very tempting to ascribe these (and other) secular changes observed over the past several decades on Venus to sporadic volcanic eruptions.

Finally, let us address tropospheric and surface carbon and hydrogen compounds. Lewis (1970) predicted CO, H_2O, and H_2 mixing ratios for chemical equilibrium in the lower atmosphere between the atmosphere and surface of 5×10^{-5}, 5×10^{-4}, and 6.7×10^{-7}, respectively. These predicted CO and H_2O values are somewhat higher than the observed values reviewed earlier (Fig. 9); nevertheless this does not rule out the possibility that their abundances in the lower atmosphere are determined by equilibrium reactions at the surface, such as

$$\begin{aligned} &3\,\mathrm{FeMgSiO_4} + \mathrm{CO_2} \longleftrightarrow 3\,\mathrm{MgSiO_3} + \mathrm{Fe_3O_4} + \mathrm{CO} \\ &\mathrm{Ca_2Mg_5Si_8O_{22}(OH)_2} \longleftrightarrow 3\,\mathrm{MgSiO_3} + 2\,\mathrm{CaMgSi_2O_6} + \mathrm{SiO_2} + \mathrm{H_2O} \end{aligned} \tag{55}$$

Since we have also identified photochemical sources and sinks for CO, H_2, and H_2O, there is therefore ambiguity to the question of photochemical or thermochemical control of their abundances in the lower atmosphere.

We finish this section devoted to the troposphere with a speculative note on the possibility of active photochemistry below the clouds of Venus. In particular, Prinn (1979) noted that photodissociation of the observed S_3 and S_4 in the 10- to 40-km region by UV, blue, and yellow light was probable, and would thus lead to nonequilibrium concentrations for S and S_2 in the lower troposphere. This would provide an important sink for OCS through the chain reactions

$$\begin{aligned} \mathrm{S} + \mathrm{OCS} &\longrightarrow \mathrm{CO} + \mathrm{S_2} \\ \mathrm{S_2} + \mathrm{OCS} &\longrightarrow \mathrm{CO} + \mathrm{S_3} \\ \mathrm{S_3} + \mathrm{OCS} &\longrightarrow \mathrm{CO} + \mathrm{S_4} \end{aligned} \tag{56}$$

Also, the photochemical sulfur atoms produced are expected to be "hot" with excitation energies ranging up to about 32–48 kcal $\mathrm{mol^{-1}}$ for, say, 350-nm photons (Prinn, 1979). The illustrative sequence

$$\begin{aligned} \mathrm{S_3} + h\nu &\longrightarrow \mathrm{S_2} + \mathrm{S^*} \qquad \lambda < 500\ \mathrm{nm} \\ \mathrm{S^*} + \mathrm{HCl} &\longrightarrow \mathrm{SH} + \mathrm{Cl} \\ \mathrm{Cl} + \mathrm{H_2O} &\longrightarrow \mathrm{OH} + \mathrm{HCl} \\ \mathrm{OH} + \mathrm{CO} &\longrightarrow \mathrm{CO_2} + \mathrm{H} \\ \mathrm{H} + \mathrm{SH} &\longrightarrow \mathrm{H_2} + \mathrm{S} \\ \mathrm{S} + \mathrm{S_2} + \mathrm{M} &\longrightarrow \mathrm{S_3} + \mathrm{M} \\ \hline \mathrm{CO} + \mathrm{H_2O} + h\nu &\longrightarrow \mathrm{CO_2} + \mathrm{H_2} \end{aligned} \tag{57}$$

may be the potent source of H_2 and sink for H_2O in the lower atmosphere required, for example, to balance the reverse net reaction occurring in the Yung and DeMore (1982) models A and B.

IV. Atmospheric Origin and Evolution

The origin of the planets and the evolution of their atmospheres is the challenging subject of a book by Lewis and Prinn (1984) and several papers (e.g., Prinn, 1982, and nine following papers; Donahue and Pollack, 1983). The atmospheres of Venus, Earth, and Mars appear to be largely derived from outgassing of the solid planet, and the initial compositions are therefore intimately connected to the primordial volatile content and thermal history of each planet. Once outgassed, the atmosphere is then chronically or episodically affected by evolutionary processes such as escape of certain light elements, cometary and meteoritic infall, atmospheric chemistry and photochemistry, atmosphere–surface reactions, and, for the Earth (but not Venus), the influence of living organisms. Space does not permit us to address here the origin and evolution of the atmosphere of Venus in a very comprehensive way; the reader is referred instead to the book and papers cited above. In this review we shall concentrate instead on the two "obvious questions" 6 and 7, cited at the beginning of Section III.

In a crude sense, it is not unreasonable to assume that the primordial volatile contents of Venus and Earth were similar: their proximity would suggest that they accreted from closely related material in the primitive solar nebula, and today they have similar sizes and similar mean densities. As noted earlier, such a model would suggest that Venus and Earth contain similar amounts of nitrogen and carbon, which indeed appears to be the case. They would also be expected to contain similar amounts of the radioactive elements ^{40}K, ^{235}U, ^{238}U, and ^{232}Th and thus of ^{40}Ar and ^{4}He, which are products of the decay of these elements. The radioactive elemental abundances do appear to be similar in crustal rocks on both planets, and the data on ^{40}Ar and ^{4}He are compatible with Earth-like abundances (provided we are willing to accept a somewhat smaller outgassing rate on Venus compared to Earth as ^{40}Ar is somewhat underabundant on Venus relative to Earth).

The above similarities form the basis of one of the two competing hypotheses for the origin and evolution of the Venusian atmosphere (e.g., see Walker *et al.*, 1970). This hypothesis (the "wet Venus" hypothesis) states that Venus and Earth began with roughly comparable endowments of the volatile elements or of their radioactive precursors. As noted in questions 6 and 7 in Section III, Venus must therefore have subsequently lost some 300 bars of H_2O. This is accomplished by the so-called runaway greenhouse, in which outgassed H_2O on Venus enters the atmosphere and thus contributes to a steadily increasing atmospheric opacity, whereas on Earth it condenses into the oceans, and the oceans remove CO_2 into carbonates. The principal reason for this difference between the two planets is hypothesized to be simply the greater proximity of Venus to the Sun. Facile irreversible dissociation of the H_2O-rich Venusian atmosphere to H_2 and O with subsequent escape of the H_2 and escape and/or reaction of the O with the crust then leads to the present dehydrated atmosphere. Photochemistry thus plays a central role in this hypothesis.

The other major hypothesis for the evolution of the Venusian atmosphere arises from a combination of specific physical and chemical models of the primitive solar nebula (e.g., see Lewis, 1972). In these theoretical models, Venus and Earth do *not* accrete from precisely similar material, rather, the accreting Earth receives significantly more of the hydrated silicates (tremolite, serpentine, talc, etc.) than Venus because the hydrated silicates only become thermodynamically stable in the cool nebular region outside of the Earth's orbit. In these models both planets are predicted to receive comparable amounts of carbon, nitrogen, and sulfur (see, e.g., Prinn, 1982), but the paucity of the hydrated silicates on Venus means that this planet *never* possessed substantial amounts of H_2O; the need for massive H escape and O burial is thus obviated. We shall call this the "dry Venus" hypothesis.

Which of these two general hypotheses (if either) is correct? First, let us address the question of how H (and O) can be removed from the Venus atmosphere and at what rate. Venus today is unique among the terrestrial planets in that it possesses a "corona" of nonthermal hydrogen atoms as discussed in Section III,A. These nonthermal hydrogen atoms (H*) *dominate* hydrogen escape from Venus. In particular, if the *lower* atmosphere is rich in H_2 (mixing ratio $\sim 2 \times 10^{-5}$) the exospheric OH^+ recombination reactions

$$\begin{aligned} O^+ + H_2 &\longrightarrow OH^+ + H^* \\ OH^+ + e^- &\longrightarrow O + H^* \end{aligned} \tag{58}$$

provide $\sim 7 \times 10^7$ cm^{-2} sec^{-1} H* that can escape (Kumar *et al.*, 1981). Alternatively, if the lower atmosphere is poor in H_2, then the abundance of exospheric OH^+ becomes small, and the sequential reactions

$$\begin{aligned} O_2^+ + e^- &\longrightarrow O^* + O^* \\ O^* + H &\longrightarrow H^* + O \end{aligned} \tag{59}$$

should dominate exospheric H* production, providing an H* escape flux of $\sim 8 \times 10^6$ cm^{-2} sec^{-1} (McElroy *et al.*, 1982). For comparison with these numbers, the escape flux of *thermal* hydrogen atoms from the 300 K Venusian exosphere by the classical Jeans escape mechanism is only $\sim 10^5$ cm^{-2} sec^{-1}

(Earth and Mars possess far more impressive Jeans escape fluxes for H of $\sim 10^8$ and 2×10^8 cm^{-2} sec^{-1}, respectively; Hunten and Donahue, 1976).

At a rate of 10^7 cm^{-2} sec^{-1}, the hydrogen present in only a 20-cm-deep layer of H_2O on Venus could be depleted over geological time. In fact, because atmospheric H_2O and H_2 abundances would be larger in the past than at present, H escape fluxes in the past are expected to exceed 10^7 cm^{-2} sec^{-1}. Assuming that the H escape flux in the past varied *linearly* with the H_2O abundance, McElroy *et al.* (1982) deduced that an 8-m-deep layer of H_2O could be depleted from Venus over geological time; this is only about one-three-hundredth of the amount of H_2O on Earth. In addition, since escape of deuterium is negligible compared to the escape of hydrogen, the D/H ratio on Venus is predicted by McElroy *et al.* (1982) to have steadily increased over geological time; specifically it begins at an Earth-like value of 5×10^{-5} some 4.6×10^9 years ago and rises to a value of $\sim 10^{-2}$ today. Indeed, there is evidence from the *Pioneer Venus* mass spectrometer (Donahue *et al.*, 1982) that the D/H ratio on Venus today is $\sim 1.6 \times 10^{-2}$. We thus have some reasonably self-consistent arguments suggesting that Venus was once endowed with water, but only a small fraction of that on Earth; we shall call this the "damp Venus" hypothesis.

Donahue *et al.* (1982) have argued that their measured D/H ratio not only favors the damp-Venus hypothesis but it is also consistent with the wet-Venus hypothesis. In particular, they note that for H_2O mixing ratios in the early Venusian atmosphere exceeding $\sim 2 \times 10^{-2}$ the loss of hydrogen to space no longer involves essentially collisionless atomic or molecular exospheric fluxes. Instead a fluid hydrodynamic upward flow occurs in which deuterium is carried outward along with H. Thus enhancement of the D/H ratio can only begin once the H_2O mixing ratio becomes less than $\sim 2 \times 10^{-2}$; the latter mixing ratio is equivalent to about a 9-m-deep layer of water. Depletion of H but not D in a 9-m-deep layer of water on Venus would yield a D/H ratio that is rather close to the putative present-day value. In this scenario Venus could begin with the equivalent of an Earth-sized ocean of water. As this H_2O is outgassed along with CO_2, the combined greenhouse effects of these two gases prevent the condensation of H_2O on the surface. Photodissociation of H_2O is rapid while it is the dominant atmospheric constituent unshielded by CO_2, and Watson *et al.* (1982) envisage H_2 escape fluxes by hydrodynamic outflow of up to 10^{12} cm^{-2} sec^{-1}; such a flux would dehydrogenate the initial Earth-like endownment of H_2O on Venus in only 0.3×10^9 years, leaving plenty of time for nonhydrodynamic escape mechanisms to enhance the D/H ratio to its present high value.

We should also address the fate of oxygen in the above scenarios. This is, in fact, a complicated matter because it depends on the modes of accretion of carbon and H_2O onto the primordial Venus. As discussed by Lewis and Prinn

(1984), retention of these volatiles probably occurred through quite different mechanisms. If carbon arrived in a reduced form (e.g., as elemental carbon in an iron–nickel alloy or as bound carbon in complex hydrocarbons), then some of the O_2 produced from H_2O dissociation could be used to oxidize the accreted C to CO_2 to produce the present atmosphere. The amount of O_2 present in CO_2 today is equivalent to that contained in a 900-m-deep layer of water on the planet; clearly carbon oxidation could be a potent sink for oxygen.

An additional sink is provided by oxidation of crustal iron. The average oxidation state of iron in the primordial Venusian crust, however, is a complicated function of the ratio of metallic iron to Fe^{2+} to Fe^{3+} in accreting material and of the subsequent degree of differentiation of the early Venusian interior. Burial of the O_2 content of a 300-bar primordial H_2O Venusian atmosphere through oxidation of Fe to FeO or of FeO to Fe_3O_4 would require $\sim 5 \times 10^{24}$ g of Fe or $\sim 2 \times 10^{25}$ g of FeO. The mass of Venus is 5×10^{27} g, and it is not unreasonable to assume that $\sim 10\%$ of its mass is Fe or FeO. Thus 1–4% of the Venusian interior must be exposed to the atmosphere in the first 0.3×10^9 years if iron oxidation provides the oxygen sink for the wet-Venus hypothesis; this is not an easy task (Lewis and Prinn, 1984).

Two further pieces of available evidence are important in the "wet" versus "damp" versus "dry Venus" controversy. We have already alluded to the remarkable decrease in the abundance of ^{36}Ar as we proceed from Venus to Earth to Mars. Present theories explain this gradient by postulating that the ^{36}Ar is derived from the solar wind, which impregnates the grains from which the planets accreted (Wetherill, 1980; McElroy and Prather, 1981). Since Venus is closest to the early Sun, it accreted from ^{36}Ar-rich grains. This theory, however, *requires* that gravitational stirring of grains in the inner solar system be sufficiently weak to prevent the ^{36}Ar-rich grains from being mixed in any important amounts out to the orbits of Earth and Mars. This argues against a common origin for the grains which formed Venus and Earth. The second piece of evidence concerns the oxidation state of the present crust. In Section III,D we noted that even a mildly oxidized Venusian crust is incompatible with significant amounts of any sulfur gases in the atmosphere. This argues against the crust having provided a sink for the large amount of O_2 associated with the wet-Venus hypothesis. These latter two pieces of evidence therefore make the wet-Venus hypothesis less tenable but do not disprove it. If the D/H ratio is indeed $\sim 10^{-2}$, then this certainly argues against the dry-Venus hypothesis. Thus if there is any compromise or preferred model at the present time, it is the damp-Venus model.

How can we make further progress in understanding the evolution of the Venusian atmosphere? It is clear that we need to obtain more knowledge of the composition *and* oxidation state of the present Venusian surface and of its volcanic emanations. It is also clear that the suggestions of an enhanced D/H ratio need to be further corroborated. Finally, the evolution of the Venusian atmosphere has and must continue to be a subject developed not in isolation but with a full appreciation of its interrelationships with the evolution of its neighboring terrestrial planets.

Acknowledgments

This work was supported in part by the Atmospheric Chemistry Program of the National Science Foundation through a grant to the Massachusetts Institute of Technology. I thank Elizabeth Manzi for her part in transforming my scattered writings into a legible final manuscript.

References

Barsukov, V., Volkov, V., and Khodakovsky, I. (1980). The mineral composition of surface rocks: a preliminary prediction. *Geochim. Cosmochim. Acta, Suppl.* **14,** 765–773.

Barth, C. (1968). Interpretation of the Mariner 5 Lyman alpha measurements. *J. Atmos. Sci.* **25,** 564–567.

Belton, M., Smith, G., Schubert, G., and Del Genio, A. (1976). Cloud patterns, waves, and convection in the Venus atmosphere. *J. Atmos. Sci.* **33,** 1394–1417.

Borucki, W., Dyer, J., Thomas, G., Jordan, J., and Comstock, D. (1981). Optical search for lightning on Venus. *Geophys. Res. Lett.* **8,** 233–238.

Bottema, M., Plummer, W., and Strong, J. (1965). A quantitative measurement of water vapor in the atmosphere of Venus. *Ann. Astrophys.* **28,** 225–230.

Boyer, C., and Carmichel, H. (1961). Observation photographique de la planete Venus. *Ann. Astrophys.* **24,** 531–535.

Brace, L., Taylor, H., Gombosi, T., Kliore, A., Knudsen, W., and Nagy, A. (1983). The ionosphere of Venus: Observations and their interpretation. *In* "Venus" (D. Hunten, L. Colin, T. Donahue, and V. Moroz, eds.), pp. 779–840. Univ. of Arizona Press, Tucson.

Cloutier, P., Tascione, T., Daniell, R., Taylor, H., and Wolff, R. (1983). Physics of the interaction of the solar wind with the ionosphere of Venus: flow/field models. *In* "Venus" (D. Hunten, L. Colin, T. Donahue, and V. Moroz, eds.), pp. 941–979. Univ. of Arizona Press, Tucson.

Connes, P., Connes, J., Benedict, W., and Kaplan, L. (1967). Traces of HCl and HF in the Venus atmosphere. *Astrophys. J.* **147,** 1230–1237.

Connes, P., Connes, J., Kaplan, L., and Benedict, W. (1968). Carbon monoxide in the Venus atmosphere. *Astrophys. J.* **152,** 731–743.

Counselman, C., Gourevitch, S., King, R., Loriot, G., and Prinn, R. (1979). Venus winds are zonal and retrograde below the clouds. *Science* **205,** 85–87.

Counselman, C., Gourevitch, S., King, R., Loriot, G., and Ginsberg, E. (1980). Zonal and meridional circulation of the lower atmosphere of Venus determined by Radio Interferometry. *JGR, J. Geophys. Res.* **85,** 8026–8030.

Cravens, T., Nagy, A., Chen, R., and Stewart, A. (1978). The ionosphere and airglow of Venus: Prospects for Pioneer Venus. *Geophys. Res. Lett.* **5,** 613–616.

Cravens, T., Gombosi, T., and Nagy, A. (1980). Hot hydrogen in the exosphere of Venus. *Nature (London)* **283,** 178–180.

Cravens, T., Nagy, A., and Crawford, S. (1981). *Pap., 4th IAGA Sci. Assembly, 1981.*

Cruikshank, D. (1983). The development of studies of Venus. *In* "Venus" (D. Hunten, L. Colin, T. Donahue, and V. Moroz, eds.), pp. 1–9. Univ. of Arizona Press, Tucson).

Crutzen, P. (1976). The possible importance of CSO for the sulfate layer of the statosphere. *Geophys. Res. Lett.* **3,** 73–76.

DeMore, W., and Yung, Y. (1982). Catalytic processes in the atmospheres of Earth and Venus. *Science* **217,** 1209–1213.

Dickinson, R., and Ridley, C. (1972). Numerical solution for the composition of a thermosphere in the presence of a steady subsolar-to-antisolar circulation with application to Venus. *J. Atmos. Sci.* **29,** 1557–1570.

Dickinson, R., and Ridley, C. (1975). A numerical model for the dynamics and composition of the Venusian thermosphere. *J. Atmos. Sci.* **32,** 1219–1231.

Dickinson, R., and Ridley, C. (1977). Venus mesosphere and thermosphere temperature structure. II: day–night variations. *Icarus* **30,** 163–178.

Donahue, T. (1968). The upper atmosphere of Venus: a review. *J. Atmos. Sci.* **25,** 568–573.

Donahue, T. (1971). Aeronomy of CO_2 atmospheres: a review. *J. Atmos. Sci.* **28,** 895–900.

Donahue, T., and Pollack, J. (1983). Origin and evolution of the atmosphere of Venus. *In* "Venus" (D. Hunten, L. Colin, T. Donahue, and V. Moroz, eds.), pp. 1003–1036. Univ. of Arizona Press, Tucson.

Donahue, T., Hoffman, J., Hodges, R., and Watson, A. (1982). Venus was wet: a measurement of the ratio of deuterium to hydrogen. *Science* **216,** 630–633.

Esposito, L. (1984). Sulfur dioxide shows evidence for Venus volcanism. *Science* **223,** 1072–1074.

Esposito, L., Knollenberg, R., Marov, M., Toon, O., and Turco, R. (1983). The clouds and hazes of Venus. *In* "Venus" (D. Hunten, L. Colin, T. Donahue, and V. Moroz, eds.), pp. 484–564. Univ. of Arizona Press, Tucson.

Florensky, C., Volkov, V., and Nikolaeva, O. (1978). A geochemical model of the Venus troposphere. *Icarus* **33,** 537–553.

Gerard, J., Stewart, A., and Bougher, S. (1981). The altitude distribution of the Venus ultraviolet nightglow and implications on vertical transport. *Geophys. Res. Lett.* **8,** 633–636.

Goldstein, R. (1964). Venus characteristics by Earth-based radar. *Astron. J.* **69,** 12–18.

Hansen, J., and Arking, A. (1971). Clouds of Venus: evidence for their nature. *Science* **171,** 669–672.

Hansen, J., and Hovenier, J. (1974). Interpretation of the polarization of Venus. *J. Atmos. Sci.* **31,** 1137–1160.

Hapke, B., Nelson, R., Woodman, J., and Barker, T. (1974). UV spectroscopy on Venus: evidence for an elemental sulfur component of the clouds. *Bull. Am. Astron. Soc.* **6,** 368–369.

Hodges, R., and Tinsley, B. (1981). Charge exchange in the Venus ionosphere as the source of the hot exospheric hydrogen. *JGR, J. Geophys. Res.* **86,** 7649–7656.

Hunten, D., and Donahue, T. (1976). Hydrogen loss from the terrestrial planets. *Annu. Rev. Earth Planet. Sci.* **4,** 265–292.

Intriligator, D., Collard, H., Mihalov, J., Whitten, R., and Wolfe, J. (1979). Electron observations and ion flows from the *Pioneer Venus* orbiter plasma analyser experiment. *Science* **205,** 116–119.

Keating, G., Nicholson, J., and Lake, L. (1980). Venus upper atmosphere structure. *JGR, J. Geophys. Res.* **85,** 7941–7956.

Khodakovsky, I. (1982). Atmosphere–surface interactions on Venus and implications for atmospheric evolution. *Planet. Space Sci.* **30,** 803–818.

Kliore, A., Woo, R., Armstrong, J., and Patel, I. (1979a). The polar ionosphere of Venus near the terminator from early *Pioneer Venus* orbiter occultations. *Science* **203,** 765–768.

Kliore, A., Patel, I., Nagy, A., Cravens, T., and Gombosi, T. (1979b). Initial observations of the nightside ionosphere of Venus from *Pioneer Venus* orbiter radio occultations. *Science* **205,** 99–102.

Knollenberg, R., and Hunten, D. (1980). The microphysics of the clouds of Venus: results of the *Pioneer Venus* particle size spectrometer experiment. *JGR, J. Geophys. Res.* **85,** 8039–8058.

Knudsen, W., Spenner, K., Miller, K., and Novak, V. (1980). Transport of ionospheric O^+ ions across the Venus terminator and implications. *JGR, J. Geophys. Res.* **85,** 7803–7810.

Krasnopolsky, V., and Parshev, V. (1980). Photochemistry of the Venus atmosphere down to 50 km (results of calculations). *Kosm. Issled.* **19,** 261–278.

Krasnopolsky, V. and Parshev, V. (1983). Photochemistry of the Venus atmosphere. *In* "Venus" (D. Hunten, L. Colin, T. Donahue, and V. Moroz, eds.), pp. 431–458. Univ. of Arizona Press, Tucson.

Kumar, S., and Hunten, D. (1974). Venus: an ionosphere model with an exospheric temperature of 350°K. *JGR, J. Geophys. Res.* **79,** 2529–2532.

Kumar, S., Hunten, D., and Taylor, H. (1981). H_2 abundance in the atmosphere of Venus. *Geophys. Res. Lett.* **8,** 237–239.

Kuzmin, A., and Clark, B. (1965). The polarization measurement and the brightness temperature of Venus at 10.6 cm wavelength. *Astron. J.* **43,** 595–617.

Lewis, J. (1970). Venus: atmospheric and lithospheric composition. *Earth Planet. Sci. Lett.* **10,** 73–80.

Lewis, J. (1972). Metal-silicate fractionation in the solar system. *Earth Planet. Sci. Lett.* **15,** 286–290.

Lewis, J., and Kreimendahl, F. (1980). Oxidation state of the atmosphere and crust of Venus from *Pioneer Venus* observations. *Icarus* **42,** 330–337.

Lewis, J., and Prinn, R. (1984). "Planets and Their Atmospheres: Origin and Evolution." Academic Press, New York.

Limaye, S., and Suomi, V. (1981). Cloud motions on Venus: global structure and organization. *J. Atmos. Sci.* **38,** 1220–1235.

Lindzen, R. (1971). Tides and gravity waves in the upper atmosphere. *In* "Mesospheric Models and Related Experiments" (G. Fiocco, ed.), pp. 122–130. Reidel Publ., Dordrecht, Netherlands.

Liou, K.N. (1980). "An Introduction to Atmospheric Radiation." Academic Press, New York.

Liu, S., and Donahue, T. (1975). The aeronomy of the upper atmosphere of Venus. *Icarus* **24,** 148–156.

Masursky, H., Eliason, E., Ford, P., McGill, G., Pettengill, G., Schaber, G., and Schubert, G. (1980). *Pioneer Venus* radar results: geology from images and altimetry. *JGR, J. Geophys. Res.* **85,** 8232–8260.

Mayer, C., McCullough, T., and Sloanaker, R. (1958). Observations of Venus at 3.15 cm wavelength. *Astrophys. J.* **127,** 1–10.

Mayr, H., Harris, I., Niemann, H., Brinton, H., Spencer, N., Taylor, H., Hartle, R., Hoegy, W., and Hunten, D. (1980). Dynamic properties of the thermosphere inferred from *Pioneer Venus* mass spectrometer measurements. *JGR, J. Geophys. Res.* **85,** 7841–7848.

McElroy, M. (1969). Structure of the Venus and Mars atmospheres. *J. Geophys. Res.* **74,** 29–41.

McElroy, M., and Donahue, T. (1972). Stability of the Martian atmosphere. *Science* **177,** 986–988.

McElroy, M., and Hunten, D. (1969). The ratio of deuterium to hydrogen in the Venus atmosphere. *J. Geophys. Res.* **74,** 1720–1739.

McElroy, M., and Prather, M. (1981). Noble gases in the terrestrial planets: clues to evolution. *Nature (London)* **293,** 535–539.

McElroy, M., and Strobel, D. (1969). Models for the nighttime Venus ionosphere. *J. Geophys. Res.* **74,** 1118–1127.

McElroy, M., Sze, D., and Yung, Y. (1973). Photochemistry of the Venus atmosphere. *J. Atmos. Sci.* **30,** 1437–1447.

McElroy, M., Prather, M., and Rodriguez, J. (1982). Escape of hydrogen from Venus. *Science* **215,** 1614–1615.

Moroz, V. (1981). The atmosphere of Venus. *Space Sci. Rev.* **29,** 3–127.

Moroz, V. (1983). Summary of preliminary results of the *Venera 13* and *Venera 14* missions. *In* "Venus" (D. Hunten, L. Colin, T. Donahue, and V. Moroz, eds.), pp. 45–68. Univ. of Arizona Press, Tucson.

Murray, B., Belton, M., Danielson, G., Davies, M., Gault, D., Hapke, B., O'Leary, B., Strom, R., Suomi, V., and Trask, N. (1974). Venus: atmospheric motion and structure from Mariner 10 pictures. *Science* **183,** 1307–1315.

Nagy, A., Cravens, T., and Gombosi, T. (1983). Basic theory and model calculations of the Venus ionosphere. *In* "Venus" (D. Hunten, L. Colin, T. Donahue, and V. Moroz, eds.), pp. 841–872. Univ. of Arizona Press, Tucson.

Niemann, H., Kasprzak, W., Hedin, A., Hunten, D., and Spencer, N. (1980). Mass spectrometric measurements of the neutral gas composition of the thermosphere and exosphere of Venus. *JGR, J. Geophys. Res.* **85,** 7817–7827.

Pettengill, G. (1978). Physical properties of the planets and satellites from radar observations. *Annu. Rev. Astron. Astrophys.* **16,** 265–292.

Pettengill, G., Ford, P., and Nozette, S. (1982). Venus: global surface radar reflectivity. *Science* **217,** 640–642.

Pollack, J., Erickson, E., Goorvitch, D., Baldwin, B., Strecker, D., Witteborn, F., and Augason, G. (1975). A determination of the composition of the Venus clouds from aircraft observations in the near-infrared. *J. Atmos. Sci.* **32,** 1140–1150.

Pollack, J., Toon, O., Whitten, R., Boese, R., Ragent, B., Tomasko, M., Esposito, L., Travis, L., and Weideman, D. (1980). Distribution and source of the UV absorption in Venus atmosphere. *JGR, J. Geophys. Res.* **85,** 8141–8150.

Prinn, R. (1971). Photochemistry of HCl and other minor species in the atmosphere of Venus. *J. Atmos. Sci.* **28,** 1058–1068.

Prinn, R. (1972). Venus atmosphere: structure and composition of the ClOO radical. *J. Atmos. Sci.* **29,** 1004–1007.

Prinn, R. (1973a). Venus: composition and structure of the visible clouds. *Science* **182,** 1132–1135.

Prinn, R. (1973b). The upper atmosphere of Venus: a review. *In* "Physics and Chemistry of the Upper Atmosphere" (B. McCormac, ed.), pp. 335–344. Reidel Publ., Dordrecht, Netherlands.

Prinn, R. (1974a). Venus: vertical transport rates in the visible atmosphere. *J. Atmos. Sci.* **31,** 1691–1697.

Prinn, R. (1974b). Some aspects of the chemistry and dynamics of the upper atmosphere. *In* "The Atmosphere of Venus: NASA SP-382" (J. Hansen, ed.), pp. 155–156. Natl. Tech. Inf. Serv, Springfield, Illinois.

Prinn, R. (1975). Venus: chemical and dynamical processes in the stratosphere and mesosphere. *J. Atmos. Sci.* **32,** 1237–1247.

Prinn, R. (1978). Venus: Chemistry of the lower atmosphere prior to the *Pioneer Venus* mission. *Geophys. Res. Lett.* **5,** 973–976.

Prinn, R. (1979). On the possible roles of gaseous sulfur and sulfanes in the atmosphere of Venus. *Geophys. Res. Lett.* **6,** 807–810.

Prinn, R. (1982). Origin and evolution of planetary atmospheres: an introduction to the problem. *Planet. Space Sci.* **30,** 741–754.

Prinn, R. (1984). The sulfur cycle and clouds of Venus. *In* "Recent Advances in Planetary Meteorology" (G.E. Hunt, ed.), pp. 15–30. Cambridge Univ. Press, London and New York.

Reeves, R., Harteck, P., Thompson, B., and Waldron, R. (1966). Photochemical equilibrium studies of CO_2 and their significance for the Venus atmosphere. *J. Phys. Chem.* **70,** 1637–1640.

Rossow, W. (1978). Cloud microphysics: analysis of the clouds of Earth, Venus, Mars, and Jupiter. *Icarus* **36,** 1–50.

Rossow, W., Del Genio, A., Limaye, S., Travis, L., and Stone, P. (1980). Cloud morphology and motions from *Pioneer Venus* images. *JGR, J. Geophys. Res.* **85,** 8107–8128.

Rusch, D., and Cravens, T. (1979). A model of the neutral and ionic nitrogen chemistry in the daytime thermosphere of Venus. *Geophys. Res. Lett.* **6,** 791–794.

Russell, C., and Vaisberg, O. (1983). The interaction of the solar wind with Venus. *In* "Venus" (D. Hunten, L. Colin, T. Donahue, and V. Moroz, eds.), pp. 873–940. Univ. of Arizona Press, Tucson.

Schorn, R., Barker, E., Gray, L., and Moore, R. (1969). High-precision spectroscopic studies of Venus. II: the water vapor variations. *Icarus* **10,** 98–104.

Schubert, G. (1983). General circulation and the dynamical state of the Venus atmosphere. *In* "Venus" (D. Hunten, L. Colin, T. Donahue, and V. Moroz, eds.), pp. 681–765. Univ. of Arizona Press, Tucson.

Schubert, G., Covey, C., Del Genio, A., Elson, L., Keating, G., Seiff, A., Young, R., Apt, J., Counselman, C., Kliore, A., Limaye, S., Revercomb, H., Sromovsky, L., Suomi, V., Taylor, F., Woo, R., and von Zahn, U. (1980). Structure and circulation of the Venus atmosphere. *JGR, J. Geophys. Res.* **85,** 8007–8025.

Seiff, A. (1983). Models of Venus's atmospheric structure. *In* "Venus" (D. Hunten, L. Colin, T. Donahue, and V. Moroz, eds.), pp. 1045–1048. Univ. of Arizona Press, Tucson.

Seiff, A., Kirk, D., Young, R., Sommer, S., Blanchard, R., Findlay, J., and Kelly, G. (1979). Thermal contrast in the atmosphere of Venus: initial appraisal from *Pioneer Venus* probe data. *Science* **205,** 46–49.

Seiff, A., Kirk, D., Young, R., Blanchard, R., Findlay, J., Kelly, G., and Sommer, S. (1980). Measurements of thermal structure and thermal contrasts in the atmosphere of Venus and related dynamical observations: results from the four *Pioneer Venus* probes. *JGR, J. Geophys. Res.* **85,** 7903–7933.

Shimizu, M. (1969). A model calculation of the Cytherean upper atmosphere. *Icarus* **10,** 11–25.

Sill, G. (1982). Sulfuric acid in the Venus clouds. *Commun. Lunar Planet. Lab.* **9,** 191–198.

Stewart, A. (1968). *Pap., Ariz. Conf. Planet. Atmos., 2nd, 1968.*

Stewart, A. (1972). *Mariner 6* and *7* ultraviolet spectrometer experiment: An interpretation of the intense CO_2^+, CO, and O airglow emissions. *J. Geophys. Res.* **77,** 54–64.

Stewart, A., Gerard, J., Rusch, D., and Bougher, S. (1980). Morphology of the Venus ultraviolet night airglow. *JGR, J. Geophys. Res.* **85,** 7861–7870.

Stewart, R. (1968). Interpretation of *Mariner 5* and *Venera 4* data on the upper atmosphere of Venus. *J. Atmos. Sci.* **25,** 578–582.

Stolarski, R., and Cicerone, R. (1974). Stratospheric chlorine: a possible sink for ozone. *Can. J. Chem.* **52,** 1610–1618.

Surkov, Y., Moskalyeva, L., Sheglov, O., and Gromov, V. (1983). New data on the composition, structure, and properties of Venus rocks obtained by *Venera 13* and *14*. *Pap., Lunar Planet. Sci. Conf., 14th, 1983.*

Sze, D., and McElroy, M. (1975). Some problems in Venus aeronomy. *Planet. Space Sci.* **23,** 763–786.

Taylor, F., Diner, D., Elson, L., McCleese, D., Martonchik, J., Delderfield, J., Bradley, S., Schofield, J., Gille, J., and Coffey, M. (1979). Temperature, cloud structure, and dynamics of Venus middle atmosphere by infrared remote sensing from *Pioneer Orbiter*. *Science* **205,** 65–67.

Taylor, F., Hunten, D., and Ksanfomaliti, L. (1983). The thermal balance of the middle and upper atmosphere of Venus. *In* "Venus" (D. Hunten, L. Colin, T. Donahue, and V. Moroz, eds.), pp. 650–680. Univ. of Arizona Press, Tucson.

Taylor, H., Brinton, H., Bauer, S., Hartle, R., Cloutier P., Michel, F., Daniells, R., Donahue, T., and Maehl, R. (1979a). Ionosphere of Venus: first observations of the effects of dynamics on the dayside ion composition. *Science* **203,** 755–757.

Taylor, H., Brinton, H., Bauer, S., Hartle, R., Cloutier, P., Daniell, R., and Donahue, T. (1979b). Ionosphere of Venus: first observations of day-night variation of the ion composition. *Science* **205,** 96–99.

Tomasko, M., Doose, L., Smith, P., and Odell, A. (1980). Measurements of the flux of sunlight in the atmosphere of Venus. *JGR, J. Geophys. Res.* **85,** 8167–8186.

Toon, B., Turco, R., and Pollack, J. (1982). The ultra-violet absorber on Venus: amorphous sulfur. *Icarus* **51,** 358–373.

Trauger, J., and Lunine, J. (1983). Spectroscopy of molecular oxygen in the atmospheres of Venus and Mars. *Icarus* **55,** 272–281.

von Zahn, U., Fricke, K., Hunten, D., Krankowsky, D., Mauersberger, K., and Nier, A. (1980). The upper atmosphere of Venus during morning conditions. *JGR, J. Geophys. Res.* **85,** 7829–7840.

von Zahn, U., Kumar, S., Niemann, H., and Prinn, R. (1983). Composition of the Venus atmosphere. *In* "Venus" (D. Hunten, L. Colin, T. Donahue, and V. Moroz, eds.), pp. 299–430. Univ. of Arizona Press, Tucson.

Walker, J., Turekian, K., and Hunten, D. (1970). An estimate of the present-day deep mantle degassing rate from data on the atmosphere of Venus. *J. Geophys. Res.* **75,** 3558–3568.

Watson, A., Donahue, T., and Walker, J. (1982). The dynamics of a rapidly escaping atmosphere: Applications to the evolution of Earth and Venus. *Icarus* **48,** 150–166.

Wetherill, G. (1980). Ar^{36} on Venus. *Geochim. Cosmochin. Acta, Suppl.* **14,** 1239–1242.

Wildt, R. (1940). Note on the surface temperature of Venus. *Astrophys. J.* **91,** 266–268.

Wilson, W., Klein, M., Kakar, R., Gulkis, S., Olsen, E., and Ho, P. (1981). Venus. I. Carbon monoxide distribution and molecular line searches. *Icarus* **45,** 624–637.

Winick, J., and Stewart, A. (1980). Photochemistry of SO_2 in the Venus upper cloud layers. *JGR, J. Geophys. Res.* **85,** 7849–7860.

Wofsy, S., and Sze, D. (1975). Venus cloud models. *In* "Atmospheres of Earth and the Planets" (B. McCormac, ed.), pp. 369–384. Reidel Publ., Dordrecht, Netherlands.

Woo, R., and Ishimaru, A. (1981). Eddy diffusion coefficient for the atmosphere of Venus from radio scintillation measurements. *Nature* (*London*) **289,** 383–384.

Young, A., and Young, L. (1973). Comments on the composition of the Venus cloud tops in light of recent spectroscopic data. *Astrophys. J.* **179,** 39–43.

Yung Y., and DeMore, W. (1982). Photochemistry of the stratosphere of Venus: implications for atmospheric evolution. *Icarus* **51,** 199–247.

7

The Photochemistry of the Atmosphere of Mars

CHARLES A. BARTH

Department of Astrophysical, Planetary, and Atmospheric Sciences
and Laboratory for Atmospheric and Space Physics
University of Colorado
Boulder, Colorado

I. Introduction

The atmosphere of Mars is a natural laboratory of photochemistry. The atmosphere over the winter polar cap is the cleanest atmosphere in the solar system, consisting of almost pure carbon dioxide (CO_2). In the spring and summer in the polar regions and throughout the year elsewhere on the planet, water vapor (H_2O) is injected into and removed from the atmosphere. There is also a small amount of molecular nitrogen (N_2) and argon (Ar) in the atmosphere. Mars has distinct seasons, with the temperature varying from 290 K in the southern summer to 148 K in the polar winter.

The annual sublimation–condensation cycle in the polar regions produces a planet-wide rise and fall of the total atmospheric pressure. The study of the

ISBN 0-12-444920-4

photochemistry of the Mars atmosphere is a study of the photochemistry of a CO_2 atmosphere with varying amounts of H_2O added under varying conditions of pressure, temperature, and solar illumination. The photochemistry of this $CO_2/H_2O/N_2$ system produces carbon monoxide (CO), molecular oxygen (O_2), and oxides of nitrogen (NO_x). Surprisingly, the CO_2 is relatively undissociated. Climatic change occurs on Mars; there is clear-cut evidence that the climate in the past differed greatly from the climate today. The atmosphere is still evolving; atoms are escaping from the top of the atmosphere today as they have been, no doubt, since the origin of the Mars atmosphere. This chapter describes the astronomical setting of the planet, the history of the spacecraft exploration of Mars, the composition of the atmosphere, the photochemistry of both the lower and upper atmosphere, the evolution of the atmosphere, climatic change, and the directions for future research.

The reference section contains a number of papers that review the photochemistry of Mars (McElroy, 1973; Barth, 1974a,b; Hunten, 1974; McElroy *et al.*, 1977) as well as a list of references to the observations and theories described in the text.

II. The Astronomical Setting

Table I contains the astronomical data that determine the physical conditions pertinent to the Mars atmosphere. Mars is about half the size of Earth; its mass is about one-tenth that of Earth. These two properties result in

Table I

ASTRONOMICAL DATA FOR MARS AND EARTH

	Mars	Earth
Radius (km)	3394	6371
Mass (kg)	6.42×10^{23}	5.98×10^{24}
Gravitational acceleration (m sec^{-2})	3.72	9.82
Mass of atmosphere	1.71×10^2 kg m^{-2}	1.03×10^4 kg m^{-2}
Heliocentric distance	1.52 AU	1.0 AU
Obliquity (deg)	25.1	23.5
Eccentricity of orbit	0.093	0.017
Ratio of aphelion to perihelion distance squared	45%	7%
Equilibrium temperature (K)	216	253
Mean surface temperature (K)	220	288
Period of revolution (days)	686.98	365.26
Period of rotation	24 hr 37 min	23 hr 56 min

a gravitational acceleration about four-tenths the gravitational acceleration of Earth. This means that for the same temperature, the Mars atmosphere is more extended than is Earth's. The mass of the Mars atmosphere (per unit area) is about one-hundredth of the mass (per unit area) of the Earth's atmosphere. This means that conditions in the lower atmosphere of Mars are more similar to conditions in the stratosphere of Earth rather than the Earth's troposphere. The obliquity of Mars is nearly the same as the obliquity of Earth; thus Mars has seasons just as Earth does. Because of the greater eccentricity of the orbit, however, the seasons are accentuated on Mars. The solar flux falling on the Mars atmosphere at perihelion is 45% greater than that incident at aphelion.

The mean distance of Mars from the Sun is half again as great as the mean distance of Earth from the Sun. Consequently, the amount of solar energy

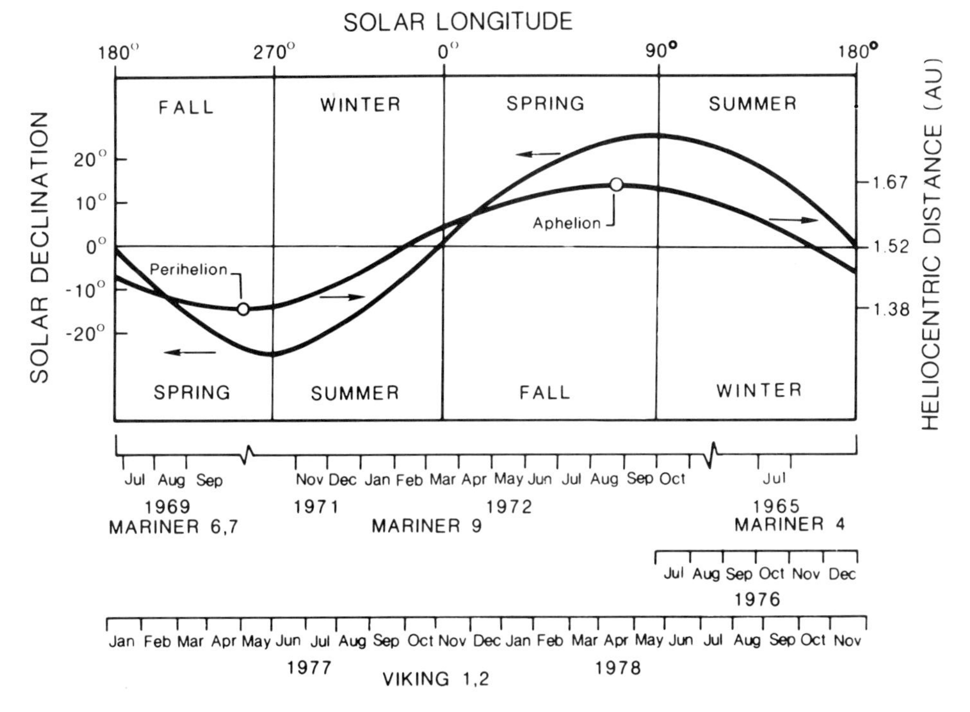

Fig. 1. Mars seasonal calendar. The solar declination and the heliocentric distance are plotted as a function of the solar longitude as viewed from Mars. The northern spring equinox occurs at a solar longitude of 0°. The seasons in the two hemispheres are marked. The Earth calendar dates of the *Mariner* and *Viking* missions are shown so that observations made at different times may be compared according to seasonal data. From Barth (1974b). Reproduced with permission from the *Annual Review of Earth and Planetary Sciences*, Vol. 2, copyright 1974 by Annual Reviews, Inc.

falling on the Mars atmosphere per unit area is less than half of the solar energy per unit area falling on the Earth's atmosphere. The equilibrium temperature on Mars (the temperature determined by balancing outgoing thermal energy to the incoming solar energy) is 216 K compared to 253 K on Earth. There is only a small greenhouse effect on Mars, so the average surface temperature is only 220 K.

The period of revolution of Mars around the Sun is almost twice as long as the period of revolution of Earth. The period of rotation of Mars about its axis is just 2% longer than the period of rotation of Earth. Thus the Mars year is much longer than the Earth year, but the length of the day on Mars is similar to the length of the day on Earth. The heliocentric distance and the solar declination are plotted in Fig. 1 as a function of the solar longitude as seen from Mars. This Mars calendar illustrates the characteristics of the seasons. The southern spring and summer are short compared to the northern spring and summer, but since perihelion occurs in the late southern spring, the solar flux falling on the southern hemisphere during the spring and summer is more intense than that incident upon the northern hemisphere during its spring–summer season. Dust storms occur on Mars usually in the late southern spring near the time of perihelion.

III. Space Exploration

Almost all of our knowledge of the Mars atmosphere has been acquired during the period of space exploration. Before the beginning of space exploration, ground-based astronomers had detected CO_2 in the Mars atmosphere, but the total pressure of the atmosphere was not known nor was the percentage abundance of CO_2. It was not known whether CO_2 was a major or minor constituent of the atmosphere.

Preparations for the first spacecraft mission to Mars began in the early 1960s. In anticipation of that mission, ground-based astronomers renewed their spectroscopic observations of Mars. In 1963, H_2O was discovered in the Mars atmosphere by spectroscopic observations in the near infrared (Spinrad *et al.*, 1963). In July of 1965, the first "flyby" mission to Mars was carried out by *Mariner 4*. The radio occultation experiment, which measured the index of refraction of the atmosphere directly above the surface, determined that the pressure of the atmosphere was 5.5 mbars, a value much lower than anticipated by planetary astronomers (Kliore *et al.*, 1965). The measurement of the low atmospheric pressure on Mars led to the important prediction that the Mars polar cap should be composed of frozen CO_2 (Leighton and Murray, 1966).

At the time of the 1967 opposition of Mars, high-resolution spectroscopic observations (using Earth-based telescopes) of infrared (IR) bands of CO_2

showed that the composition of the Mars atmosphere was nearly completely CO_2 (Belton *et al.*, 1968). In addition, during the 1967 opposition, very high resolution IR spectra of Mars that were obtained using the Mt. Palomar 5-m telescope, and the powerful technique of Fourier spectroscopy showed that the partial pressure of CO_2 in the Mars atmosphere was 5.2 mbars (Connes *et al.*, 1969). These same observations also showed that the amount of CO in the atmosphere was very small, only 0.2%. Telescope observations of the H_2O content of the Mars atmosphere over several Mars seasons showed that the amount of H_2O is variable, sometimes dropping below the detection limit of the spectroscopic technique (Barker *et al.*, 1970).

In July and August of 1969, the two flyby missions *Mariner 6* and *7* made close-up spectroscopic, occultation, and television observations of the Mars atmosphere. The *Mariner 6* and *7* IR spectrometer detected only three atmospheric constituents, CO_2, CO, and H_2O, even though these spectrometers had the sensitivity to measure a large number of different molecules, even if they had been present in only parts per million (Horn *et al.*, 1972). Measurements of the southern polar cap by the *Mariner 7* IR radiometer showed the temperature to be 148 K, the temperature of subliming CO_2 (Neugebauer *et al.*, 1971). This observation was evidence that the Mars polar cap during the spring season is indeed frozen CO_2. Ozone (O_3) was discovered over the south polar cap by the *Mariner* ultraviolet (UV) spectrometer experiment (Barth and Hord, 1971). The amount of O_3 in the atmosphere elsewhere on the planet was much less, below the detection limit of the UV spectrometer. The UV spectrometer experiment also determined that the temperature of the upper atmosphere of Mars was 350 K, a value much less than anticipated from theoretical models (Barth *et al.*, 1971; Stewart, 1972). Analysis of the UV spectra of the upper atmosphere showed that CO_2^+ is not the major ion even though the major neutral constituent is CO_2 (Barth *et al.*, 1971; Stewart, 1972). Atomic hydrogen was discovered in the upper atmosphere of Mars by the *Mariner 6* UV spectrometer experiment, and the amount escaping from the atmosphere was determined (Barth *et al.*, 1969, 1972). Atomic oxygen was also measured in the Mars upper atmosphere; the amount was found to be very small, only 1% (Strickland *et al.*, 1972). During the 1971 opposition of Mars, O_2 was discovered in the Mars atmosphere using ground-based telescopes and high-resolution spectroscopy. The amount was very small, corresponding to a composition of 0.13% (Barker, 1972; Carleton and Traub, 1972). The combination of spacecraft and ground-based telescope observations showed that Mars has an essentially undissociated atmosphere. From the surface to the top of the atmosphere, CO_2 is the major constituent. In spite of the photodissociation produced by solar UV radiation, the amount of CO, O_2, O, and O_3 is small. The development of the photochemical theory of Mars later in this chapter will show how this is possible.

Mariner 9 started orbiting Mars in November 1971 and conducted a systematic set of scientific observations until October 1972. This observation period covered four seasons, winter and spring in the north, and summer and fall in the south. Figure 1 shows the Earth dates for the period of *Mariner 9* observations and the corresponding seasonal periods on Mars. This figure also shows that the *Mariner 4* observations were made during northern summer and the *Mariner 6* and *7* observations during southern spring–northern fall. The *Mariner 9* IR interferometer spectrometer mapped the temperature structure of the atmosphere during these four seasons. These remote soundings showed that temperature inversions occurred in the winter polar regions, with very cold air trapped under a warmer layer of air above (Hanel *et al.*, 1972). The *Mariner 9* UV spectrometer mapped the abundance of O_3 as a function of latitude and time over these four seasons. The largest amount of O_3 occurred in the cold, dry atmosphere in the winter polar regions (Barth *et al.*, 1973). The escape of atomic hydrogen from the top of the Mars atmosphere was monitored by the *Mariner 9* UV spectrometer. The escape flux was constant over the period of the *Mariner 9* observations and had nearly the same value as it did 2 years earlier at the time of the *Mariner 6* and *7* observations.

On 20 July 1976 and then again on 3 September 1976, *Viking* spacecraft descended through the Mars atmosphere and landed on the surface. Mass spectrometers on the entry probes and on the lander showed that the Mars atmosphere contained N_2 (Nier and McElroy, 1977; Biemann *et al.*, 1976). Argon was also discovered to be present in the Mars atmosphere. These same instruments measured the isotopic abundances of carbon, nitrogen, and oxygen in the Mars atmosphere. The structure of the Mars atmosphere was measured from the atmospheric probes during the descent through the atmosphere (Seiff and Kirk, 1977). The composition and structure of the Mars ionosphere were also measured (Hanson *et al.*, 1977). The major ion in the ionosphere was O_2^+, with CO_2^+ present with an abundance of $\sim 10\%$. The temperature of the upper atmosphere was measured to be less than 200 K.

The *Viking* orbiters conducted global observations of H_2O and temperature from July 1976 until November 1978. The H_2O observations showed that the surface and north polar caps are sources of H_2O for the Mars atmosphere (Farmer *et al.*, 1977). The observations of surface temperature showed a maximum of 290 K in the southern hemisphere at the time of the southern summer solstice (Kieffer *et al.*, 1977). The lowest temperatures were found over the winter polar caps. In Fig. 1, the times of the *Viking* observations are shown so that they may be compared to *Mariner* observations taken at the corresponding seasonal date.

A very important discovery of the *Viking* mission came from the gas chromatograph–mass spectrometer experiment (Biemann *et al.*, 1977), namely, that there are no organic compounds on the surface of Mars. There is no methanol, formaldehyde, propane, octane, benzene, acetone, acetonitrile, or thiophene. Photochemical processes can produce some of these simple molecules from a $CO_2/CO/H_2O/N_2$ system, and photochemical processes can destroy the organic compounds. The absence of organic compounds on the surface is evidence that there is no widespread biological activity on Mars today.

Another important discovery of the *Mariner* mission resulted from obtaining images of nearly the entire surface of Mars. First the *Mariner* images and later the higher resolution *Viking* images show channels distributed widely on the Martian surface (Masursky, 1973; Carr, 1981). Figure 2 shows a

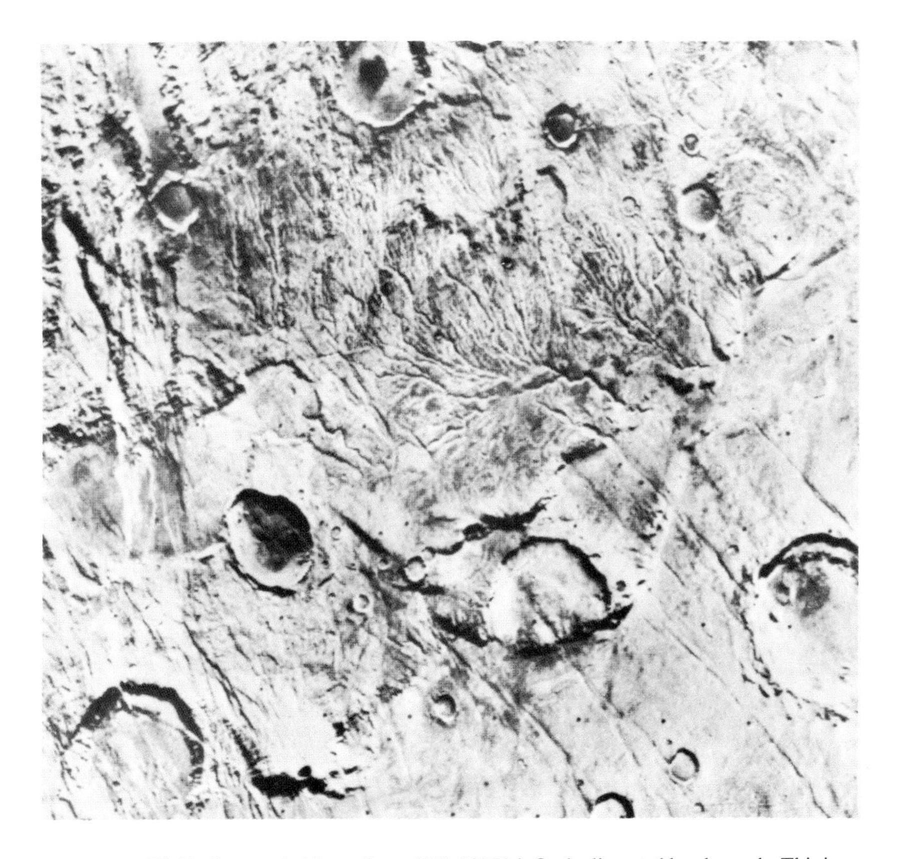

Fig. 2. Highly fractured old terrain at 48°S, 98°W, is finely dissected by channels. This is one of the densest drainage networks observed on the planet. The picture is 250 km across. From Carr (1981).

region covered with runoff channels. Many of these channels connect with other channels to form tributary networks. Figure 3 shows a large channel flowing to the north, intersecting another large channel flowing to the east. Further north in the figure, large channels flow out of large craters and flow to the east. Figure 4 shows streamlined islands that lie at the mouths of several valleys converging on this region. The flow is from the lower left around the crater, with the long tapering tail pointing downstream toward the upper right corner of the figure. These are examples of widespread fluid erosion that took place at a time when water was able to erode the surface more easily than it can today.

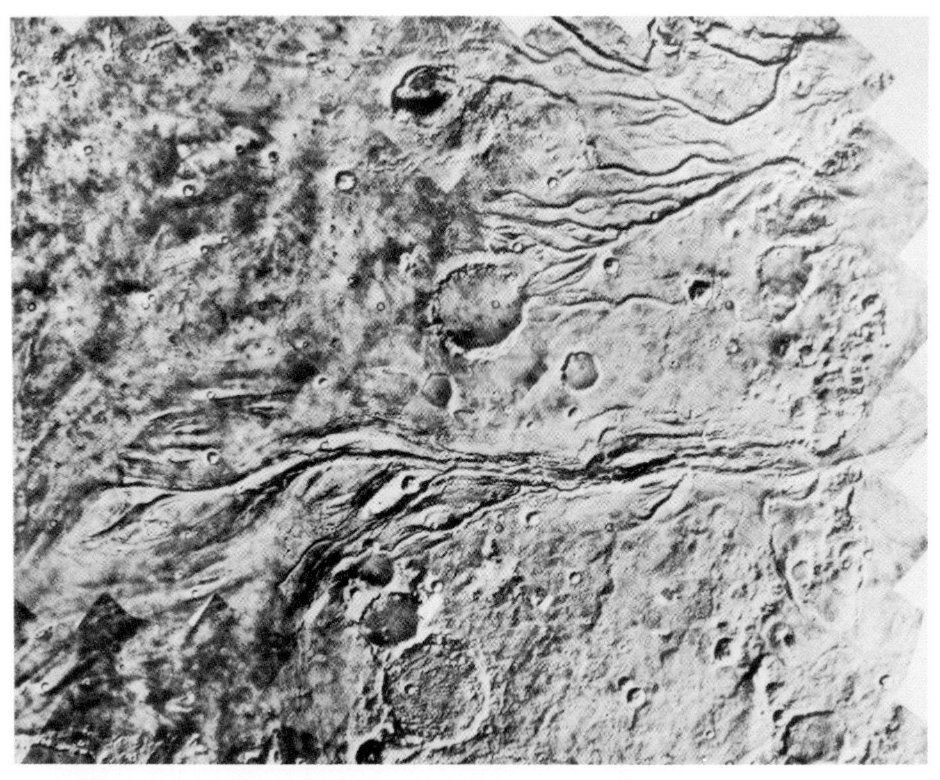

Fig. 3. Maja Vallis to the south and Vedra Vallis to the north are deeply incised into the old cratered terrain between Lunae Planum to the left and Chryse Planitia to the right. Juventae Chasma, several hundred kilometers to the south of the area shown, gives rise to a large channel that extends northward along the eastern boundary of Lunae Planum into this region. Flow apparently converged on Maja Vallis to cut its gorge. The scene is 300 km across. From Carr (1981).

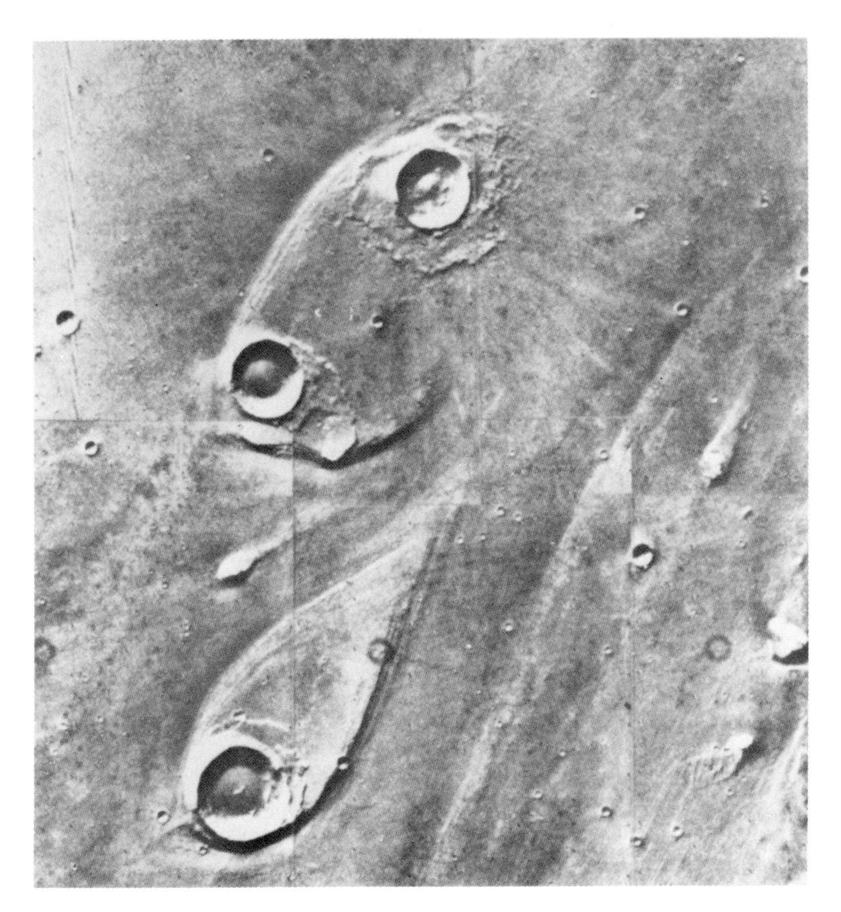

Fig. 4. Streamlined islands at the mouth of Ares Vallis, at 20°N, 31°W. The islands appear to be eroded remnants of a former plateau. The flow from the lower left diverged around two craters to form islands, with a sharp-pointed prow upstream and a long tapering tail downstream. The adjacent channel floor has a faint longitudinal scour. The upper crater postdates formation of the channel. Each island is ~40 km long. From Carr (1981).

IV. Composition and Structure of the Atmosphere

The Mars atmosphere is composed of CO_2 with a small amount of N_2 and Ar. The rare gases Ne, Kr, and Xe are present in parts per million. All of these gases are the result of outgassing from the interior of Mars; ^{40}Ar is produced by the radioactive decay of ^{40}K. The composition of the Mars atmosphere as measured by instruments on the *Viking* landers and entry probes is given in Table II. Carbon monoxide and O_2 are present in the atmosphere as a result of

Table II
COMPOSITION OF THE MARS ATMOSPHERE

Gas	Abundance (%)
CO_2	95.32
N_2	2.7
Ar	1.6
O_2	0.13
CO	0.07

photochemical reactions described in the next section. Water vapor and O_3 are also present in the atmosphere. The amount and distribution of these two constituents, which are highly variable, will be described later in this section.

The temperature structure of the Mars lower atmosphere has been measured by two *Viking* spacecraft during their descent through the atmosphere; *Viking 1* landed at a latitude of 22°N, *Viking 2* at 48°N. The mean of these two temperature profiles is shown in Fig. 5 which is taken from the Mars Reference Atmosphere. At the surface, the temperature is 214 K. In this profile, the temperature remains nearly constant for the first 5 km. Above this altitude, the temperature decreases with increasing height until it reaches a temperature of 140 K at 70 km. From 70 to 100 km the temperature profile is isothermal.

The major constituents of the Mars lower atmosphere are CO_2, N_2, and Ar (95.5, 2.7, and 1.6%, respectively), yielding a mean molecular weight of 43.4. The atmosphere is well mixed, and the relative composition of the major constituents remains constant until above 100 km. Using the three physical parameters surface pressure, composition, and the temperature profile, the pressure and density may be calculated as a function of altitude using the barometric equation,

$$dp/p = -(Mg/kT)\,dz = -dz/H,$$

where p is the atmospheric pressure, M the mean molecular weight, g the acceleration due to gravity, k the Boltzmann constant, T the temperature, z the altitude, and H the scale height. Pressure at an altitude h is calculated using the barometric equation in its integrated form,

$$p(h) = p(h_0)\exp\left[-\int_0^h \frac{dz}{H(z)}\right],$$

recognizing that H is a function of altitude. The density is calculated from the

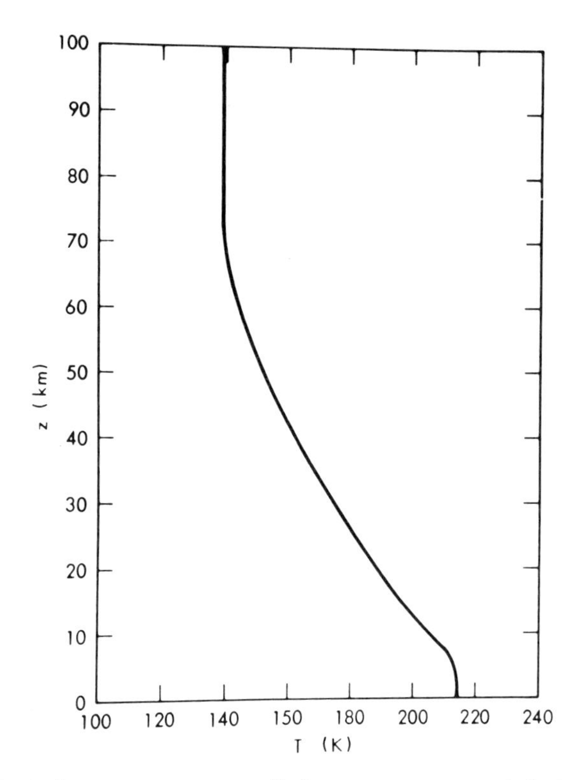

Fig. 5. Nominal mean temperature profile for summer temperate latitudes (Kliore, 1982). Copyright 1982 Pergamon Press Ltd.

pressure and temperature using the perfect gas law,

$$p = nkT,$$

where n is the number density. The mass density is equal to the number density times the mean molecular mass. Table III lists the physical properties of the Mars atmosphere.

The structure of the upper atmosphere of Mars is determined by the thermospheric temperature, the level of the turbopause, and the mixing ratio of atomic oxygen at the turbopause. The thermosphere is the part of the upper atmosphere from approximately 100 to 220 km, where the temperature increases with increasing altitude. The temperature variation as a function of altitude may be calculated with the following empirical equation:

$$T = T_\infty + (T_0 - T_\infty)\exp[-s(z - z_0)],$$

where T_∞ is the thermospheric temperature, T_0 the temperature at altitude z_0,

Table III

PHYSICAL PROPERTIES OF THE MARS ATMOSPHERE

z (km)	T (K)	p (mbars)	z (km)	T (K)	p (mbars)
0	214.0	6.36	52	150.3	2.45×10^{-2}
2	213.8	5.30	54	148.7	1.90×10^{-2}
4	213.4	4.41	56	147.2	1.47×10^{-2}
6	212.4	3.68	58	145.7	1.14×10^{-2}
8	209.2	3.06	60	144.2	8.79×10^{-3}
10	205.0	2.54	62	143.0	6.76×10^{-3}
12	201.4	2.10	64	142.0	5.20×10^{-3}
14	197.8	1.73	66	141.0	3.98×10^{-3}
16	194.6	1.42	68	140.0	3.04×10^{-3}
18	191.4	1.16	70	139.5	2.33×10^{-3}
20	188.2	9.47×10^{-1}	72	139.0	1.78×10^{-3}
22	185.2	7.70×10^{-1}	74	139.0	1.36×10^{-3}
24	182.5	6.25×10^{-1}	76	139.0	1.04×10^{-3}
26	180.0	5.06×10^{-1}	78	139.0	7.96×10^{-4}
28	177.5	4.08×10^{-1}	80	139.0	6.09×10^{-4}
30	175.0	3.28×10^{-1}	82	139.0	4.66×10^{-4}
32	172.5	2.63×10^{-1}	84	139.0	3.57×10^{-4}
34	170.0	2.11×10^{-1}	86	139.0	2.73×10^{-4}
36	167.5	1.68×10^{-1}	88	139.0	2.09×10^{-4}
38	164.8	1.33×10^{-1}	90	139.0	1.60×10^{-4}
40	162.4	1.06×10^{-1}	92	139.0	1.23×10^{-4}
42	160.0	8.33×10^{-2}	94	139.0	9.41×10^{-5}
44	158.0	6.56×10^{-2}	96	139.0	7.21×10^{-5}
46	156.0	5.15×10^{-2}	98	139.0	5.53×10^{-5}
48	154.1	4.03×10^{-2}	100	139.0	4.24×10^{-5}
50	152.2	3.15×10^{-2}			

and s a parameter related to the lapse rate. The thermospheric temperature T_∞, which is the temperature near the top of the thermosphere, is determined by measuring the neutral or plasma scale height of the atmosphere above the ionospheric peak (near 135 km). The thermospheric temperatures measured by remote sounding techniques on *Mariner 4*, *6*, *7*, and *9* and *Viking 1* and *2* and by atmospheric instruments on the *Viking* entry probes were found to vary from less than 200 K at the time of the *Viking* missions to 350 K at the time of *Mariner 9*. Figure 6 shows the result of the radio occultation measurements of the plasma scale height for these missions. The temperature scale deduced from plasma scale height is shown on the left. Also plotted is the 10.7-cm solar radio flux, which is used as an indicator of the solar UV radiation. The apparent correlation suggests that the varying temperatures are caused by the changing solar flux. Figure 7 shows a plot of thermospheric

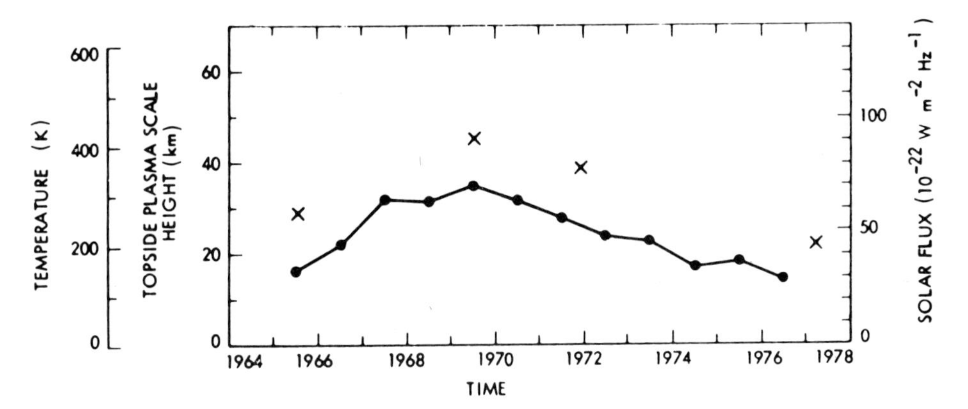

Fig. 6. Changes in the temperature and the topside plasma scale height (marked by crosses) of the Martian upper atmosphere during the solar cycle. The trace on the graph represents the 10.7-cm solar flux. The data, which apply to the 150- to 200-km altitude region of the sunlit side of Mars, were acquired by conducting radio occultation measurements with *Mariner 4, 6, 7* and *9* and with *Viking*. Also shown are the 10.7-cm solar flux data adjusted to the range of Mars. From Fjeldbo *et al.* (1977). Copyright 1977 the American Geophysical Union.

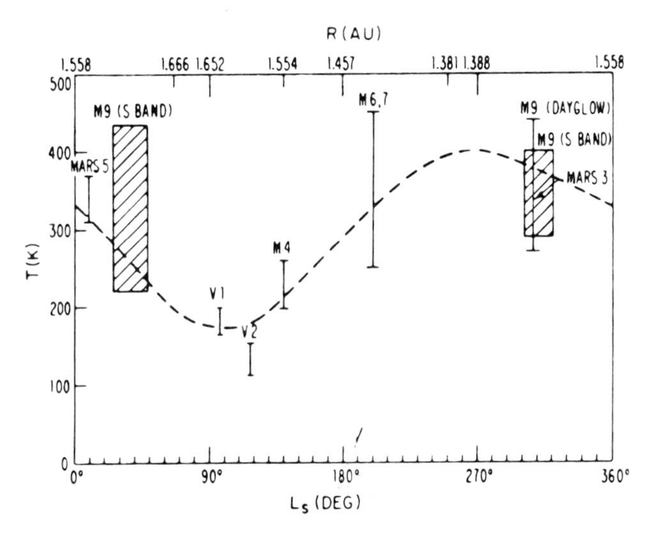

Fig. 7. Temperature of Mars upper atmosphere obtained during the missions of *Mariner 4, 6, 7,* and *9*, *Mars 3* and *5*, and *Viking 1* and *2*. Data (each uncertainty being indicated by an error bar) are plotted against the planetocentric solar longitude of Mars L_s and the radial distance from the Sun R during the period of each mission. Hatched regions represent temperatures derived from topside plasma scale heights obtained by the *Mariner 9* radio occultation experiment. Dashed curve indicates a possible seasonal variation for temperature in the Martian upper atmosphere. From McElroy *et al.* (1977). Copyright 1977 the American Geophysical Union.

temperatures as a function of the Mars heliocentric distance and the Mars seasonal date. Also included in this plot are the UV dayglow measurements from *Mariner 6*, *7*, and *9* and measurements from the Soviet *Mars 3* and *5* missions.

Figure 8 shows a model atmosphere for conditions at the times of the *Mariner 6* and *7* and *Mariner 9* missions. Two temperature profiles are used, at 300 and 350 K, to show the range of temperatures. The atomic oxygen mixing ratio at 135 km is 1%. Carbon dioxide and atomic oxygen densities are shown as a function of altitude. Figure 9 shows a model atmosphere appropriate to the time of the *Viking* missions.

Water vapor in the atmosphere of Mars is variable. The amount in the atmosphere and its location is controlled by the temperature of the surface and of the atmosphere. The northern polar cap is a source of H_2O during the northern summer; the southern polar cap is not, at least not during the time of *Viking* observations in 1977. The surface of Mars is a source of H_2O, depending on the location and the season. Figure 10 shows a time–latitude plot of the H_2O abundance in the Mars atmosphere as measured by the *Viking* orbiters. The seasonal time is indicated by L_s, the solar longitude as measured from Mars. This figure shows the five seasons of the *Viking* observations, $L_s = 90$–$180°$ northern summer, $L_s = 180$–$270°$ northern fall, $L_s = 270$–$360°$ northern winter, $L_s = 360$–$90°$ northern spring, and $L_s = 90$–$180°$ northern

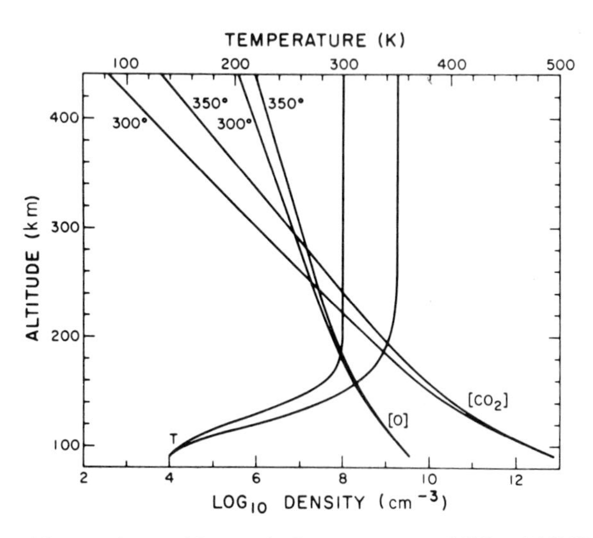

Fig. 8. Model atmospheres with exospheric temperatures of 300 and 350 K for Mars. The atomic oxygen density is ~1% in each model at the altitude (~135 km) where the CO_2 column density is 4×10^{16} cm^{-2}. From Strickland *et al.* (1973). Copyright 1973 the American Geophysical Union.

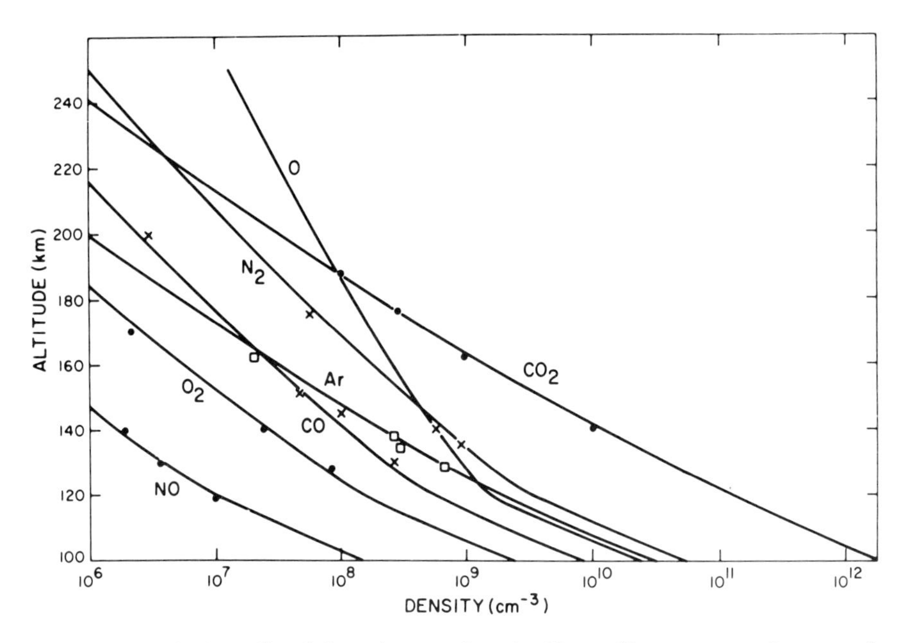

Fig. 9. Altitude profiles of the major neutral species. The profiles are compared to some of the densities measured by *Viking 1*. The carbon dioxide, molecular oxygen, and nitric oxide measurements are represented by solid circles, the carbon monoxide and molecular nitrogen densities by crosses, and the argon densities by open squares. From Fox and Dalgarno (1979). Copyright 1979 the American Geophysical Union.

summer again. The maximum (50 pr μm)* in the north in the summer (1976) shows the release of H_2O from the north pole after the CO_2 has sublimed away. This behavior is repeated the following Mars year (1978) with even a larger release (90 pr μm). A comparable release in the south during southern summer ($L_s \sim 300°$) did not occur. There is a local maximum (15 pr μm), however, at $L_s = 265°$ and a latitude of 60°S. The surface of Mars is the source of H_2O during this southern spring and summer. The surface is also a source of H_2O during the northern spring. Contours in the north starting at $L_s = 30°$ show an increase in H_2O content of the atmosphere at all northern latitudes. The H_2O content of the atmosphere drops below 1 pr μm during the approach and duration of the polar night in both hemispheres. For the entire planet, the surface is a larger source of H_2O than the polar cap. The total amount in the atmosphere varies seasonally between the equivalent of 1 and 2 km^3 of liquid water, with the maximum occurring in the northern summer and the minimum in the northern winter.

* 1 pr μm $\equiv$ 1 precipitable micrometer of H_2O, which is equivalent to 1×10^{-4} g cm^{-2} of H_2O or a column of 3.3×10^{18} molec cm^{-2}.

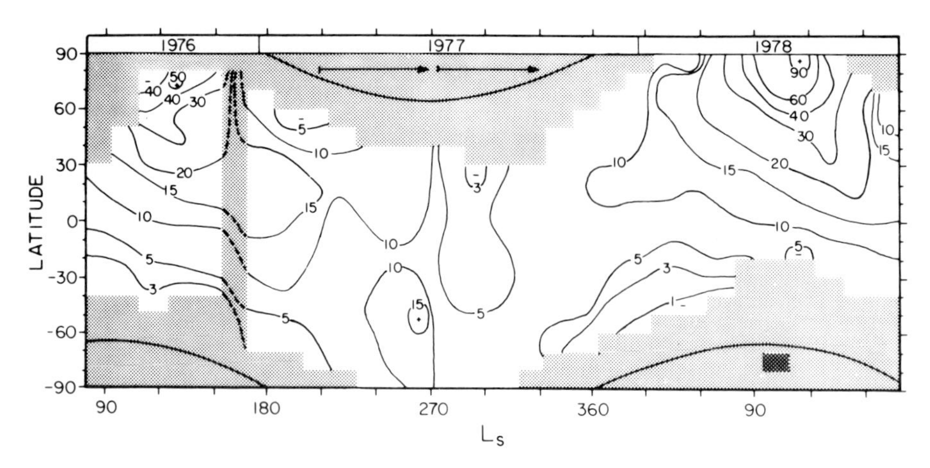

Fig. 10. Contour plot of vapor abundance (in precipitable μm) as a function of season and latitude. Shaded areas represent regions of no data. The arcs bound the regions of polar night. From Jakosky and Farmer (1982). Copyright 1982 the American Geophysical Union.

Ozone is highly variable as a constituent of the Mars atmosphere. It was first measured by the *Mariner 7* UV spectrometer over the southern polar cap during the southern spring. *Mariner 9* observations showed that the amount of O_3 over the northern polar cap is a maximum in the winter, decreases during the spring, and disappears during the summer. Ozone is present in the Mars atmosphere when the atmosphere is cold and dry. Figure 11 shows O_3, temperature, and cloud observations made during the northern winter. Clouds appear when the temperature falls below 185 K northward of 46°N latitude. This temperature is sufficiently low for ice crystals to form. The polar atmosphere above the clouds is exceedingly dry. *Mariner* observations show that O_3 appears under these conditions, increasing as the observations went poleward and the atmosphere became colder and colder. During the entire period of *Mariner* observations, O_3 was not detected equatorward of 45° latitude.

The annual sublimation and precipitation of CO_2 out of and into the polar caps produces a planet-wide pressure change of 2.4 mbars, a pressure change of 37% when compared to the Mars mean pressure of 6.36 mbars. Figure 12 shows pressure measurements made by the two *Viking* landers, one at 22°N, the other at 48°N. Two maxima and two minima occur during the Mars year. The highest pressure occurs at the time of the summer solstice in the south ($L_s = 270°$) while the southern cap is still subliming. The lowest pressure occurs toward the end of the southern winter ($L_s = 150°$) while precipitation of CO_2 is still occurring. The annual pressure change of 2.4 mbars is equivalent to a layer of CO_2 ice (density 1.5 g cm^{-3}) 1 m thick covering the winter polar cap ($\sim 5\%$ of the surface area of the planet).

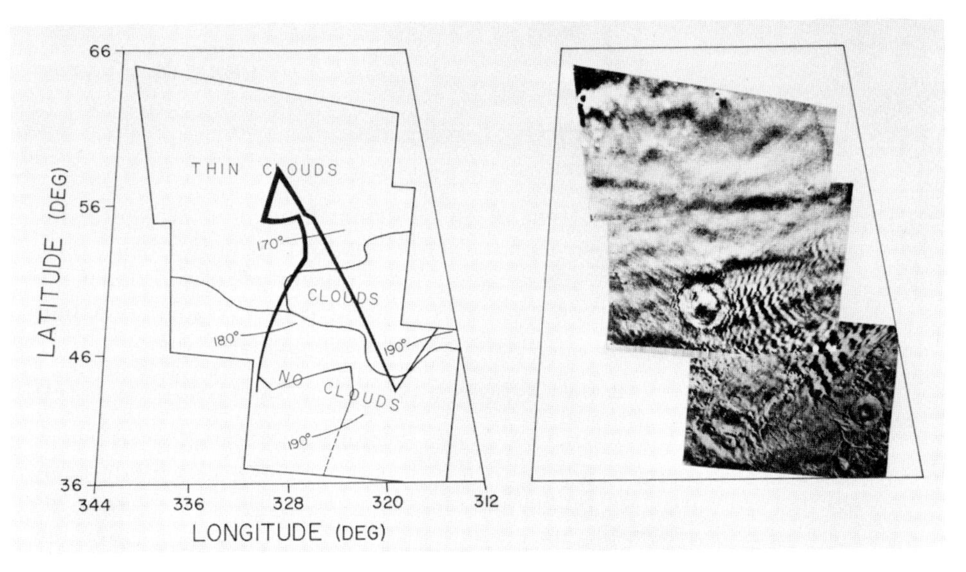

Fig. 11. Photomosaic of ground and cloud formations around the crater Lyot, together with ozone concentrations and atmospheric temperatures. The left-hand side outline is a rectilinear projection of the photographs on the right. The amount of ozone measured is proportional to the thickness of the viewing track; temperatures are given in kelvins. From Barth and Dick (1974).

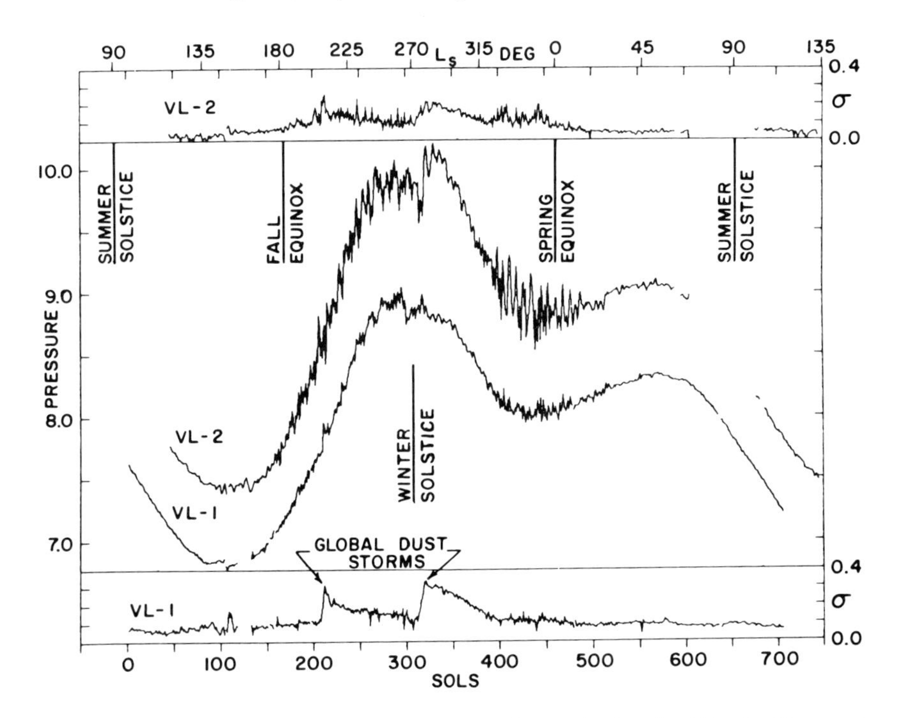

Fig. 12. (Center block) Daily mean pressures at the two landers for 700 sols. (Upper and lower blocks) Standard deviations of pressure within each sol. All pressures and standard deviations are in mbars. The abscissa is time measured in *VL-1* sols. The scale labeled L_s is the areocentric longitude of the Sun. The dates of onset of two global dust storms are marked. Gaps are due to irretreivably lost data. From Hess *et al.* (1980). Copyright 1980 the American Geophysical Union.

V. Photochemistry of the Lower and Upper Atmosphere

On Mars, photodissociation of CO_2 occurs throughout the lower atmosphere all the way to the surface. For comparison, the 6.36-mbar surface pressure of the Mars atmosphere corresponds to an altitude of ~33 km in the Earth's atmosphere. Carbon dioxide is dissociated by UV solar radiation shortward of 2270 Å into ground-state atomic oxygen and CO:

$$CO_2 + h\nu \xrightarrow{J_1} CO + O(^3P) \qquad \lambda < 2270 \text{ Å} \tag{1}$$

The absorption cross section of CO_2 as a function of wavelength is shown in Fig. 13. For wavelengths shorter than 1670 Å, oxygen atoms are produced in the excited ^1D state. The production rate of atomic oxygen in the ground ^3P state that results from the photodissociation of CO_2 is calculated from the equation

$$P(O) = \sum F_\lambda \sigma_\lambda [CO_2] \exp\left(-\sigma_\lambda \int_h^\infty [CO_2]\, dz \sec\chi\right), \quad 1670 \text{ Å} < \lambda < 2270 \text{ Å},$$

where F_λ is the solar flux, which is a function of wavelength, σ_λ the absorption cross section, $[CO_2]$ the density of CO_2, h the altitude, and χ the solar

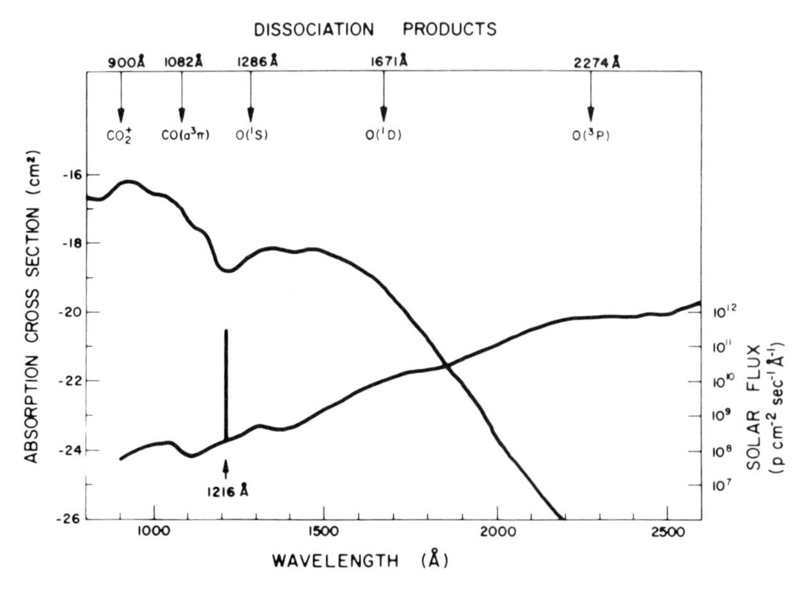

Fig. 13. Solar flux and carbon dioxide absorption cross sections, 900–2300 Å. Threshold wavelengths of several photodissociation products are shown, as well as the photoionization threshold. From Barth (1974a). Copyright 1974 Verlag Chemie GmbH.

zenith angle. An example of such a calculation is shown in Fig. 14. Above ~25 km, the solar UV radiation is unattenuated. Below 20 km, where attenuation is occurring, the production rate of atomic oxygen is $\sim 1 \times 10^6$ atoms cm^{-2} sec^{-1}.

The discussion that follows describes in a simplified way the photochemistry of the first 10 km of the Mars atmosphere. Although there are horizontal winds and possibly strong vertical mixing, the discussion will assume local photochemical equilibrium. First, the most important reactions will be described and equilibrium relationships between the various species will be derived. Then additional reactions will be added, still retaining the assumption of photochemical equilibrium. Finally, the results of model calculations involving many reactions, vertical mixing, and temporal variations will be described.

Initially, the Mars lower atmosphere will be assumed to be a pure CO_2 atmosphere with the addition of a small amount of H_2O. In the first 10 km of the Mars atmosphere, the H_2O content is variable and depends on the temperature of the atmosphere, the temperature of the surface, and the previous history of the air parcel.

Recombination of atomic oxygen and CO occurs principally through reactions with hydroperoxyl (HO_2) and hydroxyl (OH) radicals:

$$HO_2 + O \xrightarrow{k_4} OH + O_2 \tag{2}$$

$$OH + CO \xrightarrow{k_5} H + CO_2 \tag{3}$$

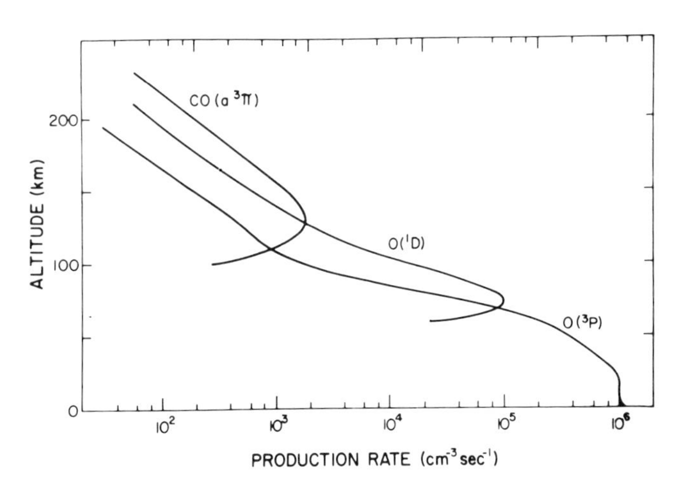

Fig. 14. Photoproduction rates of $O(^3P)$, $O(^1D)$, and $CO(a\ ^3\Pi)$ states in Mars atmosphere. A solar zenith angle of 60° was used in these calculations. Adapted from McElroy and McConnell (1971). Copyright 1971 the American Meteorological Society.

A third reaction converts H to HO_2, forming a cycle that replenishes the odd hydrogen consumed in the reaction with atomic oxygen and CO:

$$H + O_2 + M \xrightarrow{k_6} HO_2 + M \tag{4}$$

The net effect of this cycle of odd-hydrogen reactions is the catalytic recombination of odd-oxygen. The sum of the three reactions of the cycle is

$$CO + O \longrightarrow CO_2$$

The density of atomic oxygen in the lower atmosphere is directly proportional to the photodissociation rate of CO_2 and inversely proportional to the HO_2 density:

$$[O] = P(O)/k_4[HO_2]. \tag{5}$$

Since the odd-hydrogen reactions are rapid, the odd-hydrogen densities are in photochemical equilibrium with each other. The ratio of the density of one odd-hydrogen species to another is determined by the atomic oxygen and CO densities. For example, the ratio of the density of OH to the density of HO_2 is controlled by the densities of atomic oxygen and CO:

$$[OH]/[HO_2] = k_4[O]/k_5[CO]. \tag{6}$$

The source of the odd hydrogen is primarily from H_2O, either from the photodissociation by UV radiation shortward of 2420 Å,

$$H_2O + h\nu \xrightarrow{J_2} H + OH \qquad \lambda < 2420 \text{ Å} \tag{7}$$

or from the reaction of $O(^1D)$ with H_2O when sufficient numbers of these excited atoms are present,

$$H_2O + O(^1D) \xrightarrow{k_7} OH + OH \tag{8}$$

Odd-hydrogen is converted back to even hydrogen by odd–odd hydrogen reactions such as

$$HO_2 + OH \xrightarrow{k_8} H_2O + O_2 \tag{9}$$

While this is a fast reaction, the two reactants are minor constituents of the atmosphere, hence the loss rate of odd-hydrogen is relatively slow. The relationship of odd-hydrogen to even-hydrogen may be determined from these reactions:

$$[HO_2][OH] = k_7[O(^1D)][H_2O]/k_8, \tag{10}$$

and by introducing the relationship between [OH] and $[HO_2]$:

$$[HO_2]^2 = k_5k_7[O(^1D)][H_2O][CO]/k_4k_8[O]. \tag{11}$$

The source of the $O(^1D)$ in the lower atmosphere of Mars is the photodissociation of O_3 by solar UV radiation shortward of 3100 Å:

$$O_3 + h\nu \xrightarrow{J_3} O_2 + O(^1D) \qquad \lambda < 3100 \text{ Å} \tag{12}$$

The major loss process for the excited oxygen atoms is deactivation by the major constituent of the atmosphere, CO_2:

$$O(^1D) + CO_2 \xrightarrow{k_9} O + CO_2 \tag{13}$$

This is a rapid reaction, and consequently the equilibrium density of $O(^1D)$ is small:

$$[O(^1D)] = J_3[O_3]/k_9[CO_2]. \tag{14}$$

Ozone is formed from the three-body reaction of atomic oxygen with O_2:

$$O + O_2 + M \xrightarrow{k_2} O_3 + M \tag{15}$$

The principal reaction destroying O_3 is photodissociation from solar radiation, especially in the UV shortward of 3100 Å [Eq. (12)]. The photochemical equilibrium relationship between these two odd-oxygen species is

$$[O_3] = \frac{k_2[O_2][M]}{J_3}[O]. \tag{16}$$

The time constants associated with these reactions are short, measured in hundreds of seconds.

Substituting into the expression relating odd-oxygen and odd-hydrogen, the equilibrium relationships for odd–even hydrogen, odd–odd oxygen, and $O(^1D)$ gives the expressions for the two odd-oxygen species in terms of even-hydrogen:

$$[O] = P(O)\left(\frac{k_8k_9}{k_2k_4k_5k_7}\right)^{1/2}\frac{1}{[O_2]^{1/2}[CO]^{1/2}[H_2O]^{1/2}}, \tag{17}$$

$$[O_3] = \frac{P(O)}{J_3}\left(\frac{k_2k_8k_9}{k_4k_5k_7}\right)^{1/2}\frac{[O_2]^{1/2}[CO_2]}{[CO]^{1/2}[H_2O]^{1/2}}. \tag{18}$$

These two equations show that both forms of odd-oxygen (O and O_3) are inversely proportional to the square root of the density of H_2O: the more H_2O, the less O_3. The maximum amount of O_3 should occur when the least amount of H_2O is in the atmosphere.

Striking confirmation of this model comes from the *Mariner 9* UV spectrometer measurements of O_3. Ozone was measured over the polar cap during the winter and spring, when the atmosphere above the cap was very dry. With arrival of summer, H_2O was released from the polar cap, and the O_3 disappeared. Measurements from the water detector instrument on the *Viking*

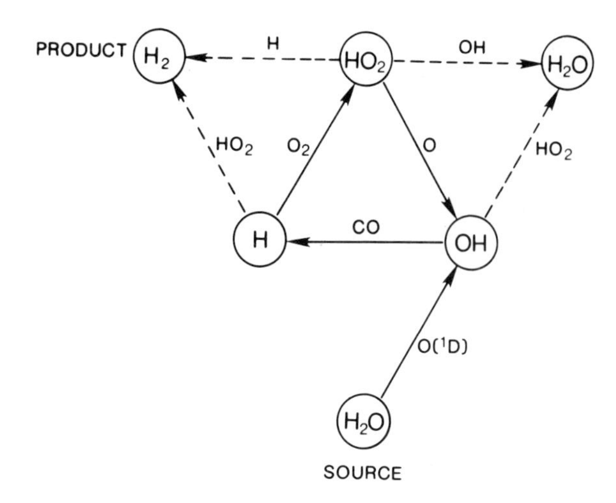

Fig. 15. Flow diagram of odd–even hydrogen reactions in a wet atmosphere. The chemical symbols in the circles represent the equilibrium density of that hydrogen species. The chemical symbols next to the arrows represent the chemical reactions that convert one species to another. The inner triangle is the catalytic cycle that converts odd-oxygen to even-oxygen.

orbiters showed that H_2O is released from the north polar cap at the beginning of summer. A comparison of the *Mariner* and *Viking* observations confirms the theory that when H_2O is present in the Mars atmosphere, O_3 is not observable and when the atmosphere is dry, O_3 appears.

Although the principal source of odd-hydrogen in the Mars lower atmosphere is the dissociation of H_2O, a steady-state abundance of H_2 should be present because of the reaction

$$H + HO_2 \xrightarrow{k_{10}} H_2 + O_2 \tag{19}$$

The conversion of H_2O to H_2 is illustrated in Fig. 15. In a wet atmosphere, the dissociation of H_2O is the source of odd-hydrogen (indicated at the bottom of the diagram), with the products moving upward to the right supplying the odd-hydrogen for the catalytic cycle. The inner triangle illustrates the catalytic conversion of odd-oxygen to even-oxygen. The odd-hydrogen densities are steady state. In the upper right and upper left corners, odd-hydrogen reactions produce even-hydrogen, namely, H_2O and H_2. In a wet atmosphere devoid of H_2, H_2O is the source and H_2 the product.

Molecular hydrogen is dissociated by its reaction with $O(^1D)$, which in turn is produced by the photodissociation of O_3 [Eq. (12)],

$$H_2 + O(^1D) \xrightarrow{k_{11}} H + OH \tag{20}$$

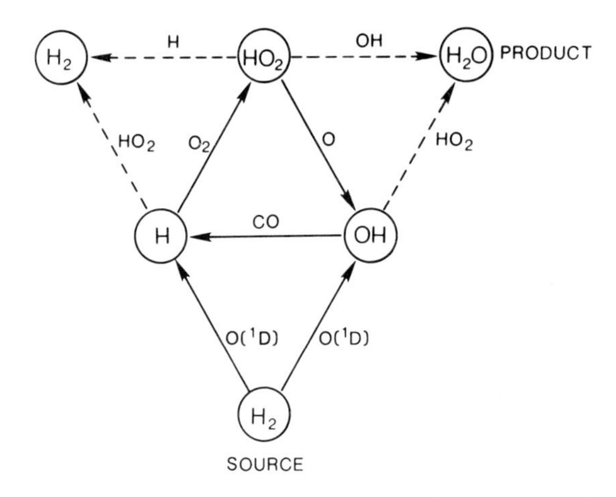

Fig. 16. Flow diagram of odd–even hydrogen reactions in a dry atmosphere.

In photochemical equilibrium, the steady-state density of H_2 is

$$[H_2] = k_{10}[H][HO_2]/k_{11}[O(^1D)]. \tag{21}$$

When the atmosphere is completely dry, as over the winter polar caps, the source of odd-hydrogen is H_2. Under these conditions, the equilibrium relationship between O_3 and even-hydrogen is

$$[O_3] = \frac{P(O)}{J_3}\left(\frac{k_2 k_8 k_9}{k_4 k_5 k_{11}}\right)^{1/2} \frac{[O_2]^{1/2}[CO_2]}{[CO]^{1/2}[H_2]^{1/2}}. \tag{22}$$

The conversion of H_2 to odd-hydrogen in a dry atmosphere is illustrated in Fig. 16. At the bottom of the diagram, H_2 is the source of odd-hydrogen. The dissociation by $O(^1D)$ produces OH and atomic hydrogen. The odd-hydrogen steady-state densities catalyze the odd-oxygen recombination in the inner triangle. Odd-hydrogen reactions produce H_2 (upper left corner) and H_2O (upper right corner). In a dry atmosphere, H_2 is the source and H_2O the product.

Additional odd-hydrogen reactions catalyze the recombination of odd-oxygen. Particularly, when there are large amounts of odd-hydrogen in the atmosphere (such as after the release of H_2O in the summer hemisphere), a catalytic cycle involving hydrogen peroxide (H_2O_2) becomes important:

$$\begin{aligned} HO_2 + HO_2 &\longrightarrow H_2O_2 + O_2 \\ H_2O_2 + h\nu &\xrightarrow{J_4} OH + OH \end{aligned} \tag{23}$$

Two HO_2 react to produce H_2O_2, which in turn is photodissociated into two OH. These reactions are followed by two each of the reaction of OH with CO

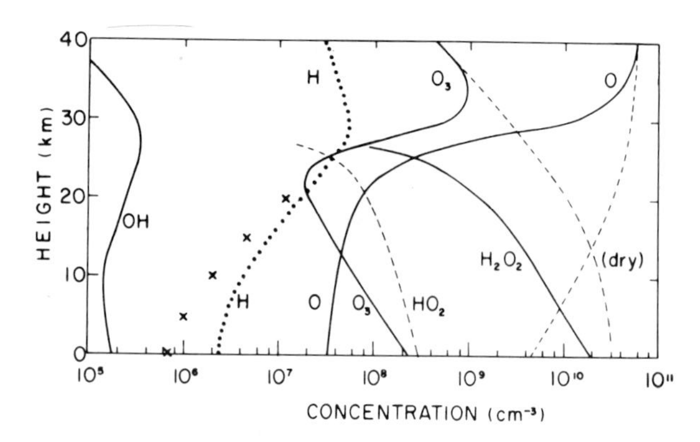

Fig. 17. The wet atmospheric model: the components of the odd-hydrogen are shown separately. From Parkinson and Hunten (1972). Copyright 1972 the American Meteorological Society.

and conversion of H to HO_2:

$$OH + CO \longrightarrow H + CO_2 \qquad (2\times)$$

$$H + O_2 + M \longrightarrow HO_2 + M \qquad (2\times)$$

The sum of these four (actually six) reactions forms a cycle that catalytically causes CO and O_2 to combine to form CO_2:

$$2\,CO + O_2 \quad \to 2\,CO_2$$

In this reaction cycle, H_2O_2 actually serves as a reservoir of odd hydrogen, which is released when photodissociation occurs. The H_2O_2 may freeze out of the atmosphere during the night and sublime during the day, directly injecting odd-hydrogen into the atmosphere. Another unique characteristic of this cycle is that it converts O_2 to CO_2 without the involvement of atomic oxygen.

The results of a model atmosphere calculation using both of the odd-hydrogen catalytic cycles is shown in Fig. 17. For a dry atmosphere, the O_3 density near the surface is 3×10^{10} molec cm^{-3}. For a wet atmosphere (15 pr μm H_2O), the calculated O_3 density decreases by over two orders of magnitude to 2×10^8 molec cm^{-3}. In this model, the large amount of H_2O_2 near the surface (3×10^8 molec cm^{-3}) plays a significant role.

Vertical mixing occurs throughout the Mars atmosphere. In the middle atmosphere (10–70 km) eddy diffusion is assumed to be the dominant mechanism for vertical mixing. Calculations of the vertical density distribution of chemically reacting species may be made by introducing the concept of an eddy diffusion coefficient K. The diffusive flux of an individual species is

$$\Phi_i = -K(dn_i/dz + dn_i/dz_{eq}).$$

The term dn_i/dz is the gradient of the actual density distribution of species n_i in the atmosphere, while the term dn_i/dz_{eq} represents the gradient of the density in hydrostatic equilibrium. For an isothermal atmosphere.

$$dn_i/dz_{eq} = -n_i/H_{ave},$$

where H_{ave} is the scale height for the mean molecular weight, 43.4. The expression for diffusive flux for each species,

$$\Phi_i = -K(dn_i/dz + n_i/H_{ave}),$$

and the continuity equation for each species,

$$d\Phi_i/dz = P_i - L_i n_i,$$

are solved simultaneously; P_i is the production rate and L_i the loss frequency for species n_i. The results of the model atmosphere calculation using an H_2O abundance of 12.5 pr μm and strong vertical mixing ($K = 3 \times 10^8$ cm^2 sec^{-1}) are shown in Fig. 18. This calculation shows a column abundance of 2.7 μm atm of O_3, which is in agreement with measurements made by the *Mariner 9* UV spectrometer. Ozone is more abundant than atomic oxygen below 30 km. Hydroperoxyl is the dominant form of odd-hydrogen. Atomic

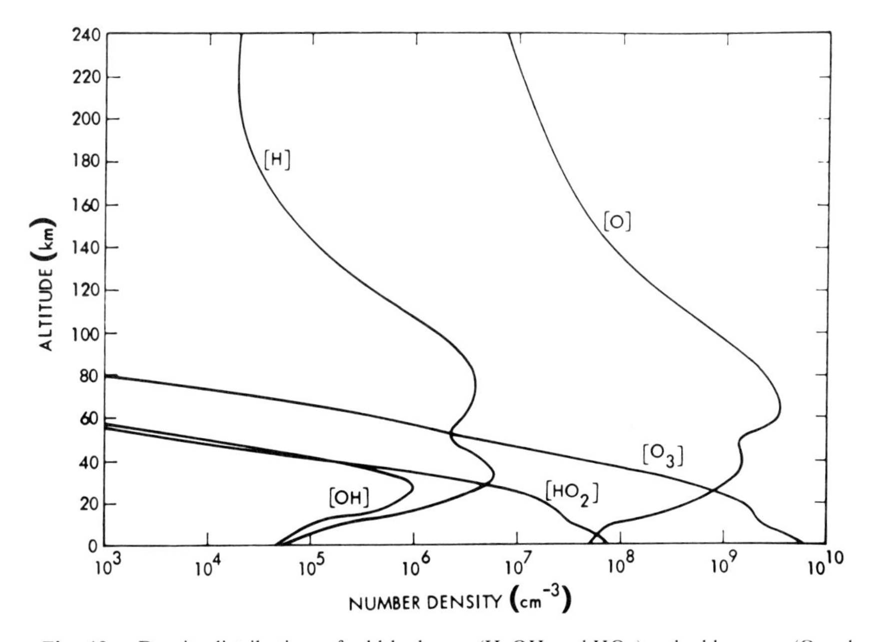

Fig. 18. Density distributions of odd-hydrogen (H, OH, and HO_2) and odd-oxygen (O and O_3). $T_{exobase} = 365$ K [K_z (III)]. From Kong and McElroy (1977).

hydrogen and OH densities are nearly equal and decrease with decreasing altitude. Figure 19 shows the production and loss rates for OH, illustrating the relative role of the odd-hydrogen reactions. The reactions of the catalytic cycle,

$$O + HO_2 \longrightarrow O_2 + OH$$

$$CO + OH \longrightarrow CO_2 + H$$

are clearly the dominant reactions below 30 km. Figure 20 shows the production and loss rates for total odd-hydrogen. The dissociation of H_2O by $O(^1D)$ is the major source below 10 km. Under the assumptions of this model, the contribution from $O(^1D)$ dissociation of molecular hydrogen is over two orders of magnitude less. Below 30 km, the dominant loss of odd-hydrogen is the reaction

$$OH + HO_2 \longrightarrow H_2O + O_2$$

The photoequilibrium relationships derived in the previous section may be used to check the results of this detailed model calculation.

Surface reactions may also occur on Mars. The very reactive odd-hydrogen and odd-oxygen species are in direct contact with the surface. Ultraviolet radiation of wavelengths as short as 2000 Å penetrates to the surface. A

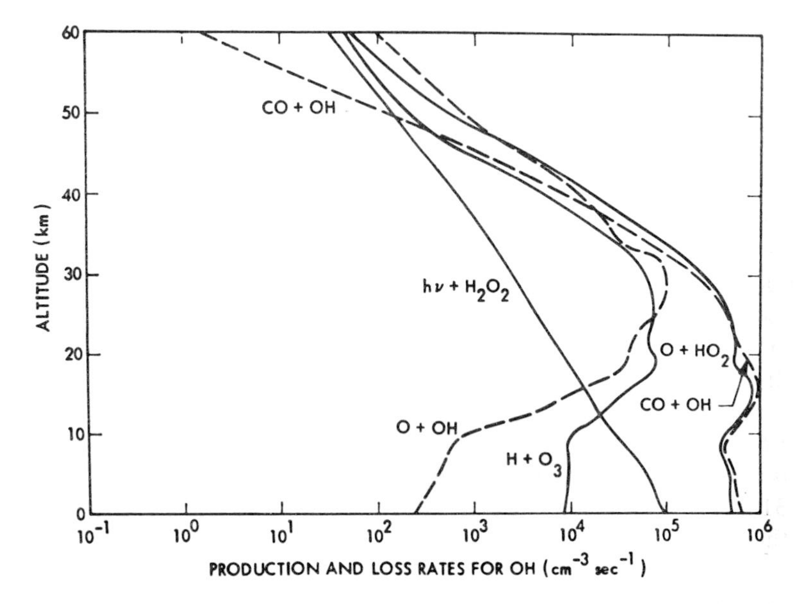

Fig. 19. Rates of various production (——) and loss (– – –) processes of hydroxyl in the lower Martian atmosphere. From Kong and McElroy (1977).

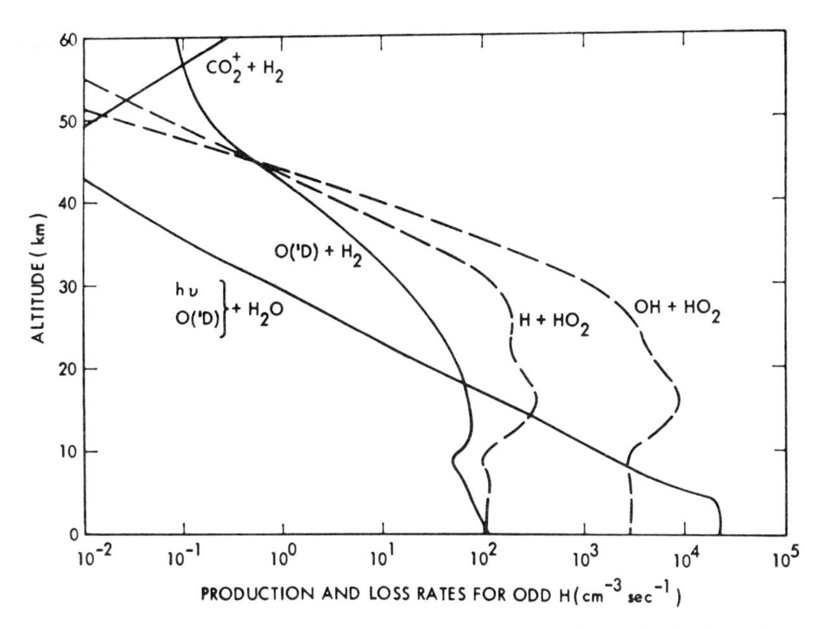

Fig. 20. Production (——) and loss (– – –) rates for odd-hydrogen in the lower Martian atmosphere. Conditions are same as in previous figures. From Kong and McElroy (1977).

particularly important reaction that may occur on Mars is the surface catalysis of the recombination of CO and O_2 to form CO_2:

$$2\,CO + O_2 \xrightarrow{\text{surface}} 2\,CO_2$$

The results of a model calculation that includes surface recombination are shown in Fig. 21. The vertical mixing in this model has been adjusted so that the calculated odd-oxygen and odd-hydrogen remain in agreement with observations. To achieve this result, the eddy diffusion coefficient K has been reduced by a factor of 30 ($K = 1 \times 10^7$ cm^2 sec^{-1}) compared to the model shown in Fig. 18. A comparison of these two models shows the O_3 density near the surface to be nearly the same, but the atomic oxygen and atomic hydrogen densities above 30 km are dramatically different. The result shows that atmospheric photochemistry, surface chemistry, and vertical mixing are intimately linked on Mars.

Both products of the photodissociation of O_3 by solar radiation shortward of 3100 Å are in excited states, the O_2 in the a $^1\Delta_g$ state and the atomic oxygen in the ^1D state:

$$O_3 + h\nu \longrightarrow O_2(a\,^1\Delta_g) + O(^1D) \qquad \lambda < 3100\ \text{Å} \tag{24}$$

Both products may radiate, the $O_2(a\,^1\Delta_g)$ state at 1.27 μm and the $O(^1D)$ at

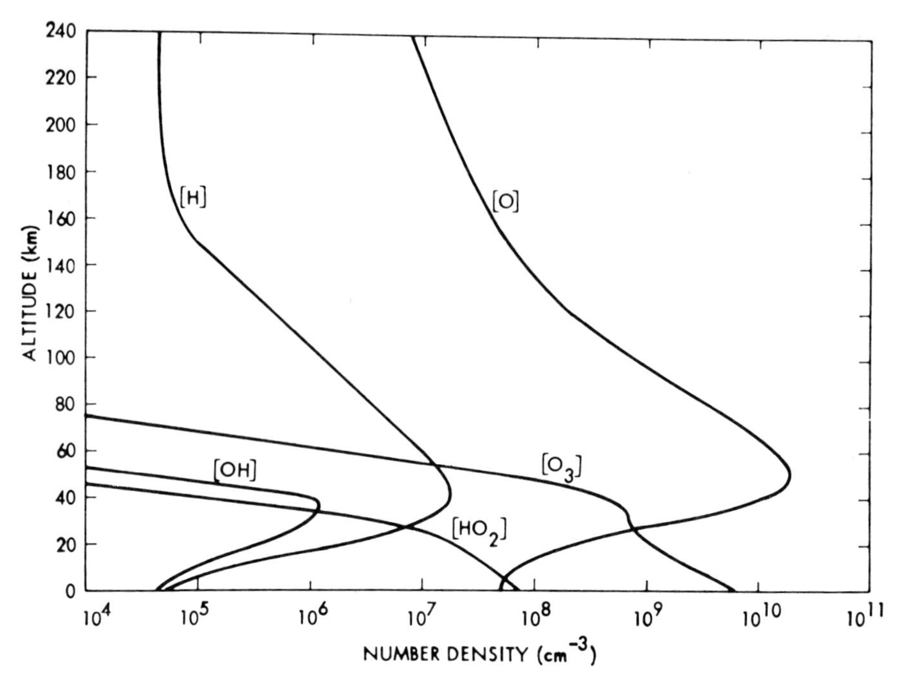

Fig. 21. Density distributions of H, OH, HO_2, O, and O_3 of a model calculation that includes surface reactions. Eddy coefficient K_z is reduced by a factor of 30 near an altitude of 30 km. $T_{\text{exobase}} = 315$ K [K_z (VII)]. From Kong and McElroy (1977).

6300 Å. The measurement of this airglow radiation, particularly the 1.27-μm emission, is an excellent method of determining the density and distribution of O_3:

$$O_2(a\,{}^1\Delta_g) \longrightarrow O_2(X\,{}^3\Sigma_g^-) + h\nu \qquad \lambda = 1.27\ \mu\text{m}$$

The transition probability A of this state is 2.6×10^{-4} sec^{-1}. The $O_2(a\,{}^1\Delta_g)$ excited state may also be deactivated by collisions with CO_2, the principal constituent of the atmosphere:

$$O_2(a\,{}^1\Delta_g) + CO_2 \xrightarrow{k_{14}} O_2(X\,{}^3\Sigma_g^-) + CO_2 \tag{25}$$

The rate coefficient for this reaction is 1.5×10^{-20} cm^{-3} sec^{-1}. The equilibrium density of $O_2(a\,{}^1\Delta_g)$ states is

$$[O_2(a\,{}^1\Delta_g)] = \frac{J_3[O_3]}{A + k_{14}[CO_2]}, \tag{26}$$

and the emission rate E of the 1.27-μm airglow is

$$E = A[O_2(a\,{}^1\Delta_g)].$$

At high altitudes where the atmospheric density is sufficiently low so that $k_{14}[CO_2] \ll A$, the emission rate of 1.27-μm airglow is directly proportional to the O_3 photodissociation rate and to the O_3 density:

$$E = J_3[O_3].$$

At low altitudes where $k_{14}[CO_2] \gg A$, the 1.27-μm emission rate is proportional to the ratio of O_3 density to CO_2 density:

$$E = AJ_3[O_3]/k_{24}[CO_2]. \tag{27}$$

Calculations of the column emission rate of 1.27-μm airglow for a particular model atmosphere (Fig. 22) show that deactivation occurs for altitudes lower than 40 km. It is possible to measure the 1.27-μm emission in the Mars atmosphere by using large ground-based telescopes on Earth. Figure 23 shows the spectrum of the $O_2(a\,^1\Delta_g - X\,^3\Sigma_g^-)$ observed with a Fourier transform spectrometer on the 5-m Mt. Palomar telescope.

The abundance of O_3 on Mars is variable. It is variable as a function of time and location. Equation (18) shows that the O_3 density should vary as a function of the following quantities: (1) P(O), the photodissociation rate of

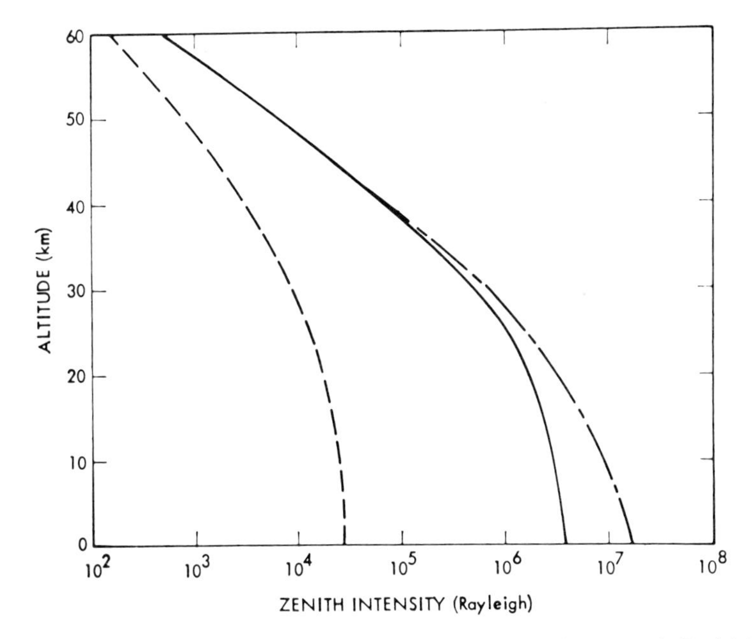

Fig. 22. Zenith intensities for the 1.27-μm (0,0) band of molecular oxygen in Rayleighs, for the ozone profile shown in Fig. 18. k_{32} ($cm^3\ sec^{-1}$): ---, 4×10^{-18}; ——, 1.5×10^{-20}; —·—, 0. From Kong and McElroy (1977).

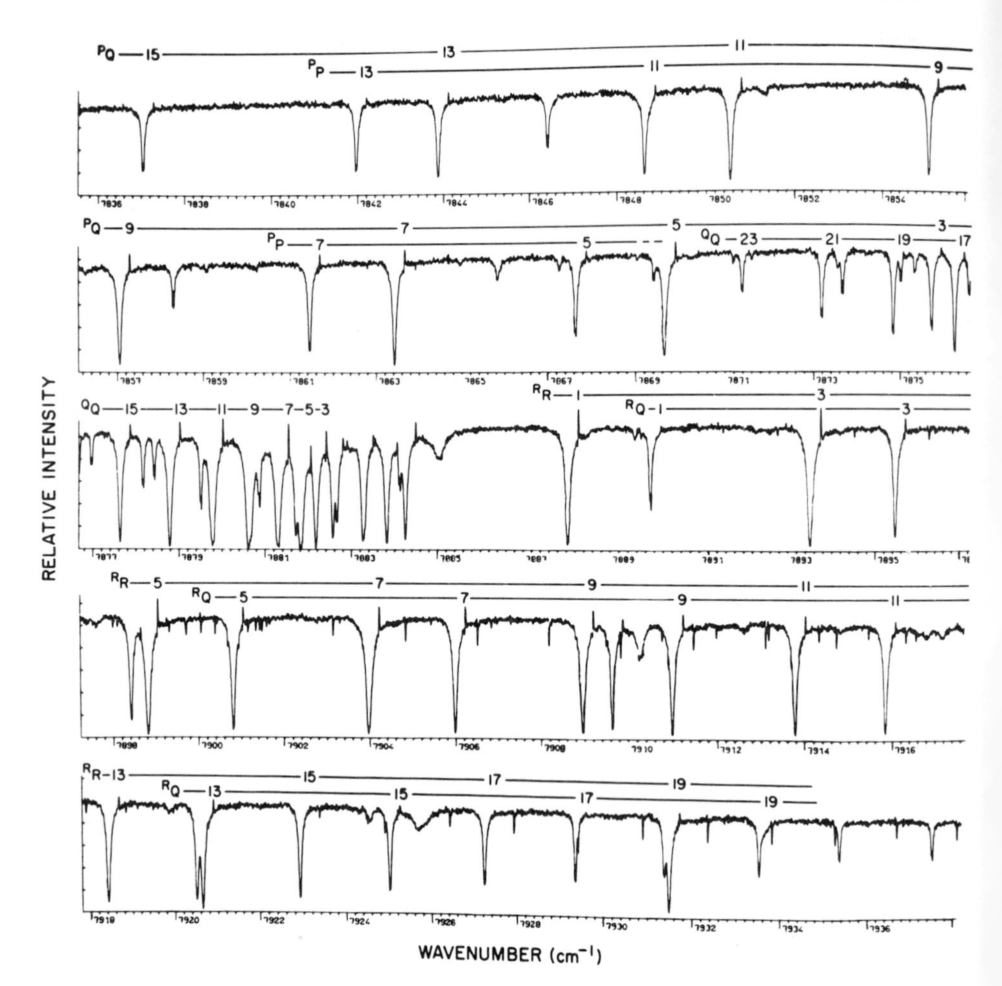

Fig. 23. Mount Palomar daytime observations of Mars in the 1.27-μm band, from May 1975. The sharp Martian molecular oxygen emission lines are Doppler-shifted by approximately $+0.2\ \mathrm{cm}^{-1}$ with respect to the corresponding stronger and broader terrestrial molecular oxygen absorption features; the lines are labeled by branch and lower-state quantum number K″. From Noxon *et al.* (1976). Reprinted courtesy of J. F. Noxon and *The Astrophysical Journal*, published by the University of Chicago Press. Copyright 1976 the American Astronomical Society.

CO_2, which varies as a function of season, latitude, and time of day because of its dependence on solar zenith angle, (2) six reaction rate coefficients, some of which have a strong dependence on temperature, (3) the O_2 and CO densities, (4) the CO_2 density, which varies as a function of location because of large altitude changes in the topography and as a function of time because of the seasonal change in pressure on Mars, and (5) the inverse dependence on the H_2O density in the atmosphere. It is the last quantity, the H_2O abundance, that produces the most dramatic variations in the O_3 abundance.

Ozone was first measured on Mars over the polar cap. The highly reflecting polar cap made it possible to measure the strong UV absorption continuum between 2000 and 3100 Å. The variation in O_3 over the north polar cap as a function of season is shown in Fig. 24. Starting in late winter and through the spring the amount of O_3 decreased monotonically. By the beginning of the summer, the O_3 abundance is below the detection limit of the *Mariner* UV spectrometer. Figure 25 shows the change in H_2O content of the Mars atmosphere as a function of season. At high northerly latitudes, the H_2O content increases during the northern spring, $L_s = 0$–$90°$. A comparison of the *Mariner* and the *Viking* data over the same seasonal period shows that as H_2O increases, O_3 decreases. Equation (18) shows the reason for this inverse relationship.

The *Mariner* UV spectrometer measured O_3 in the winter hemisphere in regions that were not over the polar cap. As shown in Fig. 26, the amount of O_3 was highly variable, but on any particular day in the winter the maximum amount of O_3 was found near 60°N latitude. The reason for this is that the cold, dry air from the polar regions meets the moist air from the equatorial regions in the latitude band 40–60°N. Equatorward of 40°N, there is sufficient H_2O in the atmosphere to supply enough odd-hydrogen to keep the O_3 content below the detection limit of the *Mariner* UV spectrometer, 3 μm atm. North of 60°N the solar UV radiation that produces the odd-oxygen is being attenuated by the increasing path through the atmosphere. The high variability of the O_3 content means simply that the H_2O content of the Mars atmosphere is highly variable.

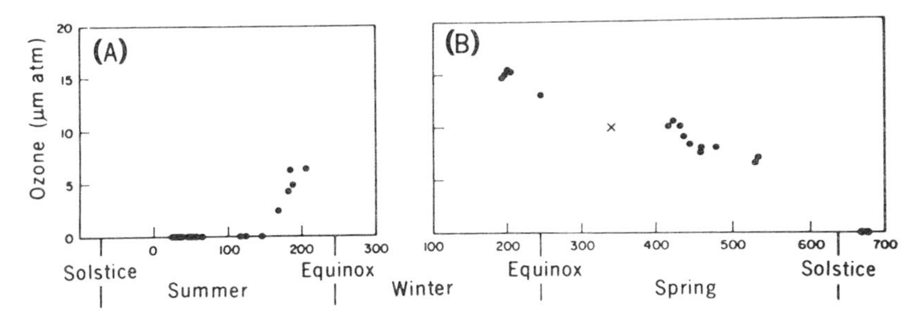

Fig. 24. (A) Amount of ozone observed over the south polar cap during the summer. (B) Amount of ozone observed over the north polar cap during the winter, spring, and beginning of summer. X, Refers to the *Mariner 7* measurement made over the south polar cap in 1969. From Barth *et al.* (1973). Copyright 1973 by the American Association for the Advancement of Science.

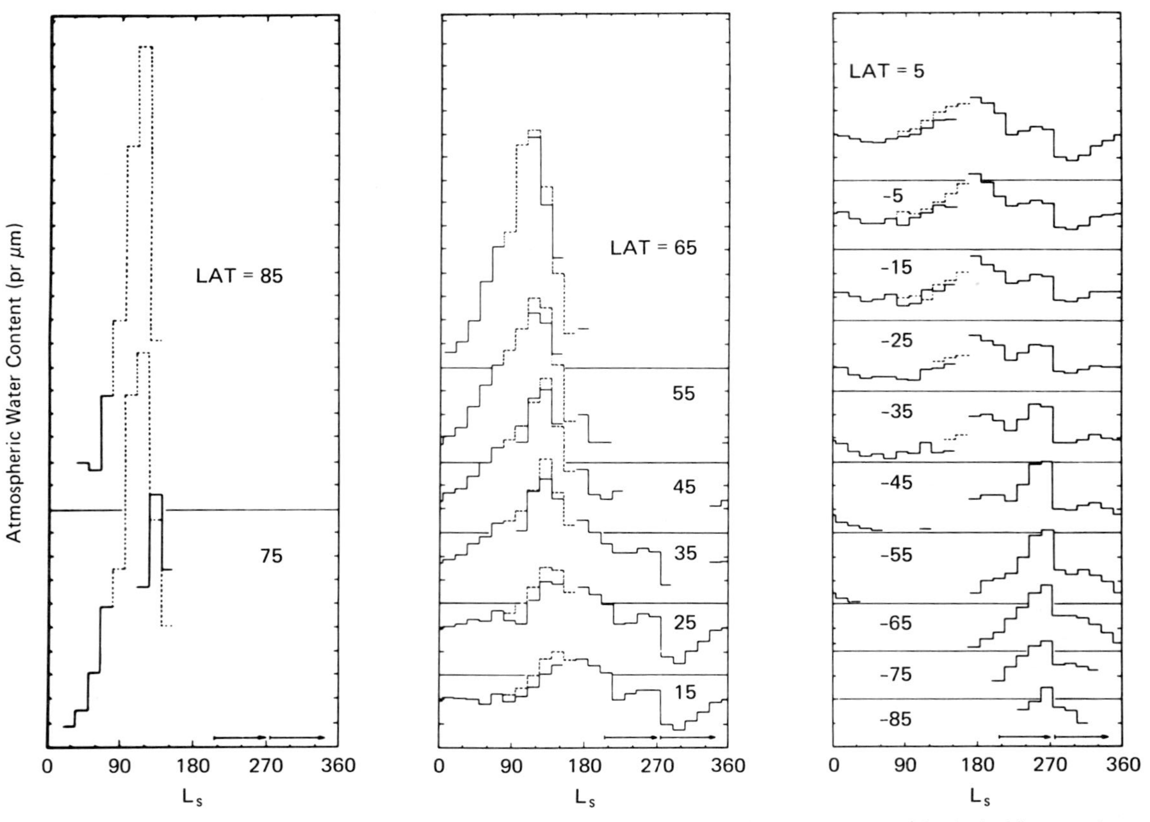

Fig. 25. Observed water vapor abundance in each 10° latitude strip as a function of Martian season. The dashed lines are data taken during the second year of *Viking* observations. The arrows mark the times of the global dust storms. The horizontal lines represent the base level for each plot, and the tick marks along the ordinate are spaced 5 pr μm apart. From Jakosky and Farmer (1982). Copyright 1982 the American Geophysical Union.

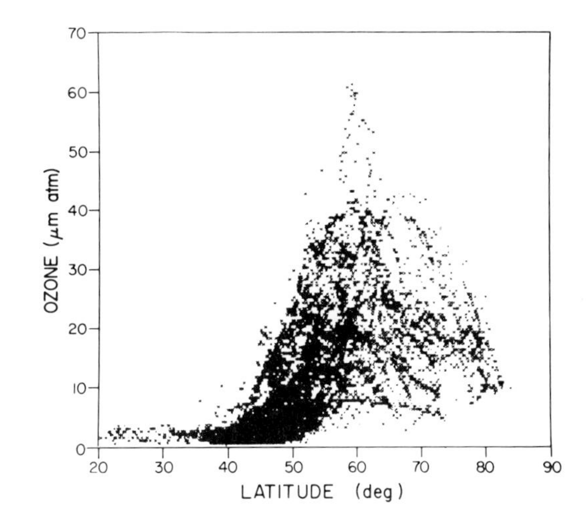

Fig. 26. *Mariner 9* measurements of ozone during the northern winter, $L_s = 330$–$360°$. The hundreds of individual measurements were made over a 100-day period as the observation track moved poleward. The amount of ozone varied widely from day to day. From Traub *et al.* (1979). Reprinted courtesy of W. A. Traub and *The Astrophysical Journal*, published by the University of Chicago Press. Copyright 1979 the American Astronomical Society.

In the upper atmosphere of Mars, CO_2 is ionized by solar UV radiation shortward of 900 Å:

$$CO_2 + h\nu \xrightarrow{J_6} CO_2^+ + e^- \qquad \lambda < 900\ \text{Å} \tag{28}$$

Most of the CO_2^+ that are formed are converted to O_2^+ by the ion–atom and ion–molecule reactions:

$$CO_2^+ + O \xrightarrow{k_{21}} O_2^+ + CO \tag{29}$$

$$CO_2^+ + O \xrightarrow{k_{22}} O^+ + CO_2 \tag{30}$$

$$O^+ + CO_2 \xrightarrow{k_{23}} O_2^+ + CO \tag{31}$$

All of these reactions are fast, and the first directly converts CO_2^+ to O_2^+. The second and third also convert CO_2^+ to O_2^+ through the intermediate ion O^+

The major loss process for O_2^+ is dissociative recombination:

$$O_2^+ + e^- \xrightarrow{k_{24}} 2\,O \tag{32}$$

Even though CO_2 is the major neutral constituent of the upper atmosphere of Mars, O_2^+ is the major ion in the Mars ionosphere.

Using this simplified reaction scheme, the equilibrium densities for the major ions are

$$\begin{aligned} [O_2^+] &= I_{CO_2}[CO_2]/k_{24}[e^-], \\ [CO_2^+] &= I_{CO_2}[CO_2]/(k_{21} + k_{22})[O], \end{aligned} \tag{33}$$

where

$$I_{CO_2} = \sum F_\lambda \sigma^+_{CO_2} \exp\left(-\sigma_{CO_2} \int_h^\infty [CO_2]\, dz\right), \tag{34}$$

where I is the ionization frequency, F_λ the solar flux, and $\sigma^+_{CO_2}$ the ionization cross section of CO_2. Both the O_2^+ and CO_2^+ densities reach a maximum near 130 km. This is the altitude where the rate of photoionization reaches a maximum. Above this altitude, the ionization rate may be eliminated from Eq. (33) to show that the ratio of CO_2^+ to O_2^+ is inversely proportional to the atomic oxygen density:

$$\frac{[CO_2^+]}{[O_2^+]} = \frac{k_{24}}{(k_{21} + k_{22})} \frac{[e^-]}{[O]}.$$

The measurement of these densities in the Mars ionosphere may be used to determine the atomic oxygen density:

$$[O] = \frac{k_{24}}{(k_{21} + k_{22})} \frac{[O_2^+]}{[CO_2^+]} [e^-].$$

For a CO_2^+ density of $\sim 10\%$ of the O_2^+ density at 130 km, the atomic oxygen makes up $\sim 2\%$ of the neutral atmosphere at this altitude.

There are additional ion reactions in the Mars ionosphere, and additional ions are formed. The ion flow diagram in Fig. 27 shows the reaction chain. The neutral species in the diagram are arranged according to ionization potential with N_2 which has the largest threshold, at the top and nitric oxide (NO), with the smallest, at the bottom. The production of ions from photoionization flows from the upper right-hand direction. Each of the molecular ions undergoes dissociative recombination. This loss process flows toward the lower left-hand direction:

$$N_2^+ + e^- \xrightarrow{k_{25}} N + N \tag{35}$$

$$CO_2^+ + e^- \xrightarrow{k_{26}} CO + O \tag{36}$$

$$O_2^+ + e^- \xrightarrow{k_{24}} 2\,O \tag{32}$$

$$NO^+ + e^- \xrightarrow{k_{27}} N + O \tag{37}$$

Other ion reactions cause one type of ion to change to another, flowing down the diagram from upper left to lower right. Ionized molecular nitrogen reacts with atomic oxygen to produce NO^+:

$$N_2^+ + O \xrightarrow{k_{28}} NO^+ + N \tag{38}$$

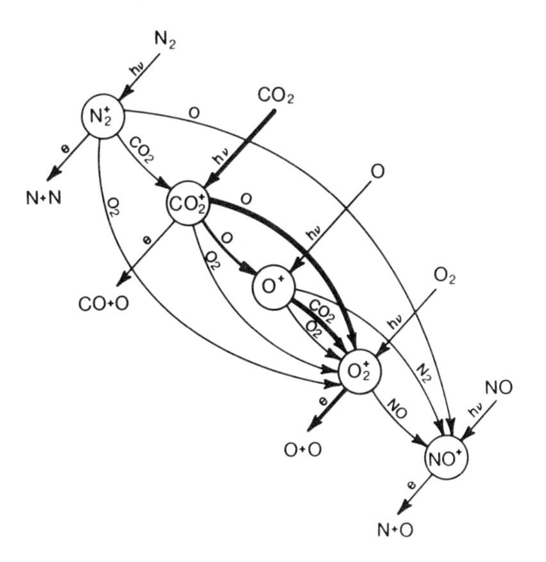

Fig. 27. Ion reactions in Mars ionosphere. Chemical symbols in circles represent the equilibrium density of the ions. Ions are arranged in order of decreasing ionization potential in moving from top to bottom. Production processes are shown by arrows pointing out of the circle. The chemical symbol along side the arrow indicates the neutral species reacting with the ion. The dark arrows show the major flow of ionization.

Molecular nitrogen ions also undergo charge exchange with CO_2 and O_2:

$$N_2{}^+ + CO_2 \xrightarrow{k_{29}} CO_2{}^+ + N_2 \tag{39}$$

$$N_2{}^+ + O_2 \xrightarrow{k_{30}} O_2{}^+ + N_2 \tag{40}$$

Ionized carbon dioxide reacts with O to produce $O_2{}^+$ and undergoes charge exchange with atomic and molecular oxygen:

$$CO_2{}^+ + O \xrightarrow{k_{21}} O_2{}^+ + CO \tag{29}$$

$$CO_2{}^+ + O \xrightarrow{k_{22}} O^+ + CO_2 \tag{30}$$

$$CO_2{}^+ + O_2 \xrightarrow{k_{31}} O_2{}^+ + CO_2 \tag{41}$$

Ionized atomic oxygen reacts with CO_2 to produce $O_2{}^+$ and with N_2 to produce NO^+. Charge exchange occurs with O_2:

$$O^+ + CO_2 \xrightarrow{k_{23}} O_2{}^+ + CO \tag{31}$$

$$O^+ + N_2 \xrightarrow{k_{32}} NO^+ + N \tag{42}$$

$$O^+ + O_2 \xrightarrow{k_{33}} O_2{}^+ + O \tag{43}$$

Nitric oxide ions at the bottom of the energy ladder are formed by the reactions of O^+ with N_2 and N_2^+ with O and by the charge exchange of NO with O_2^+:

$$O^+ + N_2 \xrightarrow{k_{32}} NO^+ + N \tag{42}$$

$$N_2^+ + O \xrightarrow{k_{28}} NO^+ + N \tag{38}$$

$$O_2^+ + NO \xrightarrow{k_{34}} NO^+ + O_2 \tag{44}$$

The equilibrium densities of all of the ions may be calculated with the help of Fig. 27 and Table IV. Under the assumption of photochemical equilibrium, the production rate of ionization is equal to the loss rate of each individual ion, X^+:

$$P_i = L_i[X^+].$$

The equilibrium density $[X^+]$ is equal to the production rate P_i divided by the loss frequency L_i. In Fig. 27, the processes that produce ions flow from the upper right and upper left, and the processes that lead to loss of the ions flow to the lower right and the lower left. An examination of Fig. 27 leads to the

Table IV

CHEMICAL REACTIONS IN THE MARS IONOSPHERE

Reaction number	Reaction	Rate coefficient, value ($cm^3\ sec^{-1}$)
(29)	$CO_2^+ + O \rightarrow CO + O_2^+$	k_{21}, 1.6×10^{-10}
(30)	$CO_2^+ + O \rightarrow CO_2 + O^+$	k_{22}, 1.0×10^{-10}
(31)	$O^+ + CO_2 \rightarrow O_2^+ + CO$	k_{23}, 9.4×10^{-10}
(32)	$O_2^+ + e^- \rightarrow O + O$	k_{24}, $1.6 \times 10^{-7}(300/T)^{0.5}$
(35)	$N_2^+ + e^- \rightarrow N + N$	k_{25}, $2.2 \times 10^{-7}(300/T)^{0.39}$
(36)	$CO_2^+ + e^- \rightarrow CO + O$	k_{26}, 3.8×10^{-7}
(37)	$NO^+ + e^- \rightarrow N + O$	k_{27}, $4.3 \times 10^{-7}(300/T)$
(38)	$N_2^+ + O \rightarrow NO^+ + N$	k_{28}, $1.4 \times 10^{-10}(300/T)$
(39)	$N_2^+ + CO_2 \rightarrow N_2 + CO_2^+$	k_{29}, 7.7×10^{-10}
(40)	$N_2^+ + O_2 \rightarrow O_2^+ + N_2$	k_{30}, $5.0 \times 10^{-11}(300/T)^{0.8}$
(41)	$CO_2^+ + O_2 \rightarrow CO_2 + O_2^+$	k_{31}, 5×10^{-11}
(42)	$O^+ + N_2 \rightarrow NO^+ + N$	k_{32}, 6.0×10^{-13}
(43)	$O^+ + O_2 \rightarrow O_2^+ + O$	k_{33}, $2.0 \times 10^{-11}(300/T)^{0.4}$
(44)	$O_2^+ + NO \rightarrow NO^+ + O_2$	k_{34}, 4.4×10^{-10}
(51)	$N(^2D) + CO_2 \rightarrow NO + CO$	k_{35}, 1.8×10^{-13}
(52)	$N(^4S) + NO \rightarrow N_2 + O$	k_{36}, 3.4×10^{-11}
(53)	$N(^2D) + NO \rightarrow N_2 + O$	k_{37}, 3.4×10^{-11}
(61)	$CO_2^+ + H_2 \rightarrow CO_2H^+ + H$	k_{12}, 1.4×10^{-9}
(62)	$CO_2H^+ + e^- \rightarrow CO_2 + H$	k_{13}, 3.5×10^{-7}

following equilibrium equations:

$$[N_2^+] = \frac{I_{N_2}[N_2]}{k_{28}[O] + k_{29}[CO_2] + k_{30}[O_2] + k_{25}[e^-]}, \tag{45}$$

$$[CO_2^+] = \frac{I_{CO_2}[CO_2] + k_{29}[N_2^+][CO_2]}{k_{21}[O] + k_{22}[O] + k_{31}[O_2] + k_{26}[e^-]}, \tag{46}$$

$$[O^+] = \frac{I_O[O] + k_{22}[CO_2^+][O]}{k_{23}[CO_2] + k_{32}[N_2] + k_{33}[O_2]}, \tag{47}$$

$$[O_2^+] = \frac{\begin{array}{l} I_{O_2}[O_2] + k_{21}[CO_2^+][O] + k_{23}[O^+][CO_2] \\ \quad + k_{33}[O^+][O_2] + k_{31}[CO_2^+][O_2] + k_{30}[N_2^+][O_2] \end{array}}{k_{34}[NO] + k_{24}[e^-]}, \tag{48}$$

$$[NO^+] = \frac{\begin{array}{l} I_{NO}[NO] + k_{28}[N_2^+][O] + k_{32}[O^+][N_2] \\ \quad + k_{34}[O_2^+][NO] \end{array}}{k_{27}[e^-]}, \tag{49}$$

$$[e^-] = [N_2^+] + [CO_2^+] + [O^+] + [O_2^+] + [NO^+]. \tag{50}$$

These six equations are solved simultaneously for a particular Mars model atmosphere. The calculations are carried out in an iterative manner, first assuming an electron distribution, then solving for the ion densities, and then recalculating the electron density. This procedure is repeated until consistent results are obtained.

The results of a photochemical equilibrium calculation of the Mars ion densities for conditions appropriate to the time of the *Viking* mission are shown in Fig. 28. The atomic oxygen density (1×10^9 atoms cm^{-3} at 130 km) has been adjusted to bring the calculated CO_2^+ densities into agreement with the ion densities measured by the *Viking 1* entry probe. The figure also shows the agreement between the calculated and measured O_2^+, indicating that the photoionization rate has been correctly calculated and that the scale height of CO_2 is also correct. The thermospheric temperature for this model is 225 K.

Photoelectrons are present in the Mars ionosphere as a result of the photoionization processes. Photoionization by solar UV photons of wavelengths shorter than the threshold for a particular ionization process produces electrons with excess kinetic energy. These photoelectrons form a population of energetic electrons that have sufficient energy to ionize, dissociate, and excite neutral species in the atmosphere.

When NO^+ undergoes dissociative recombination, both excited and ground-state nitrogen atoms are formed in the ratio of 3:1,

$$NO^+ + e^- \longrightarrow N(^2D) + O \qquad 0.75$$

$$NO^+ + e^- \longrightarrow N(^4S) + O \qquad 0.25$$

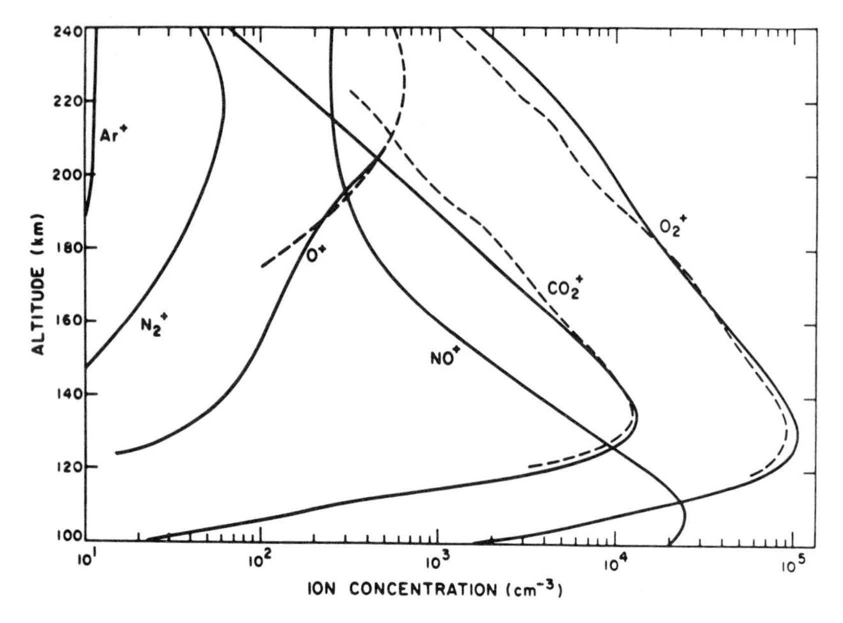

Fig. 28. Steady-state ion concentrations. The solid curves are calculated values. The dashed curves are those measured by *Viking 1*. From Fox and Dalgarno (1979). Copyright by the American Geophysical Union, 1979.

The branching ratio of those recombinations resulting in excited and ground-state atoms are listed to the right of the reaction. The excited nitrogen atoms react with CO_2 and produce NO:

$$N(^2D) + CO_2 \xrightarrow{k_{35}} NO + CO \tag{51}$$

Nitrogen atoms in excited states and the ground state react with NO, converting this odd-nitrogen species to N_2:

$$N(^4S) + NO \xrightarrow{k_{36}} N_2 + O \tag{52}$$

$$N(^2D) + NO \xrightarrow{k_{37}} N_2 + O \tag{53}$$

The mixing ratio of NO in the Mars upper atmosphere is directly proportional to the density of excited nitrogen atoms and inversely proportional to the total density of ground-state and excited nitrogen atoms:

$$[NO] = \frac{k_{35}[N(^2D)]}{\{k_{36}[N(^4S)] + (k_{37}[N(^2D)]\}}[CO_2]. \tag{54}$$

Other sources of excited nitrogen atoms are photodissociation and electron impact dissociation of N_2:

$$N_2 + h\nu \xrightarrow{J_7} N(^2D) + N(^4S) \qquad \lambda < 1020\ \text{Å}$$

$$N_2 + e^- \longrightarrow N(^2D) + N(^4S) \qquad E > 12.1\ \text{eV} \tag{55}$$

$$N_2 + e^- \longrightarrow N(^2D) + N(^2D) \qquad E > 14.5\ \text{eV}$$

The threshold wavelength and energies for these reactions are listed to the right of the reaction. Ionospheric reactions also produce excited and ground-state nitrogen atoms:

$$N_2^+ + O \xrightarrow{k_{28}} N(^2D) + NO^+ \tag{38}$$

$$O^+ + N_2 \xrightarrow{k_{32}} N(^4S) + NO^+ \tag{42}$$

$$N_2^+ + e^- \xrightarrow{k_{25}} N(^2D) + N(^4S) \tag{35}$$

Processes that produce a loss of excited nitrogen atoms, in addition to the reaction with CO_2, are deactivation by atomic oxygen and ionospheric electron collisions,

$$N(^2D) + O \longrightarrow N(^4S) + O$$

$$N(^2D) + e^- \longrightarrow N(^4S) + e^-$$

and by emission of a photon,

$$N(^2D) \longrightarrow N(^4S) + h\nu \qquad \lambda = 5200\ \text{Å}$$

Figure 29 shows altitude profiles for the production of $N(^2D)$ and $N(^4S)$ by these processes. The largest source of excited nitrogen atoms is the dissociative recombination of NO^+, a process that produces $N(^2D)$ in 75% of the reactions.

Figure 30 shows the result of model calculations of NO and N in which the fraction of the electron impact dissociation reactions producing $N(^4S)$ was varied. The result shows that a maximum amount of NO is produced if the electron impact reaction favors the production of $N(^2D)$. This figure also demonstrates that a larger density of NO leads to a smaller density of atomic nitrogen since the principal loss process for atomic nitrogen is the reaction with NO:

$$N + NO \xrightarrow{k_{36}} N_2 + O \tag{52}$$

Another extremely important loss reaction for NO is photodissociation, which proceeds through a predissociation process to produce ground-state nitrogen atoms:

$$NO + h\nu \xrightarrow{J_8} N(^4S) + O \qquad \lambda < 1910\ \text{Å} \tag{56}$$

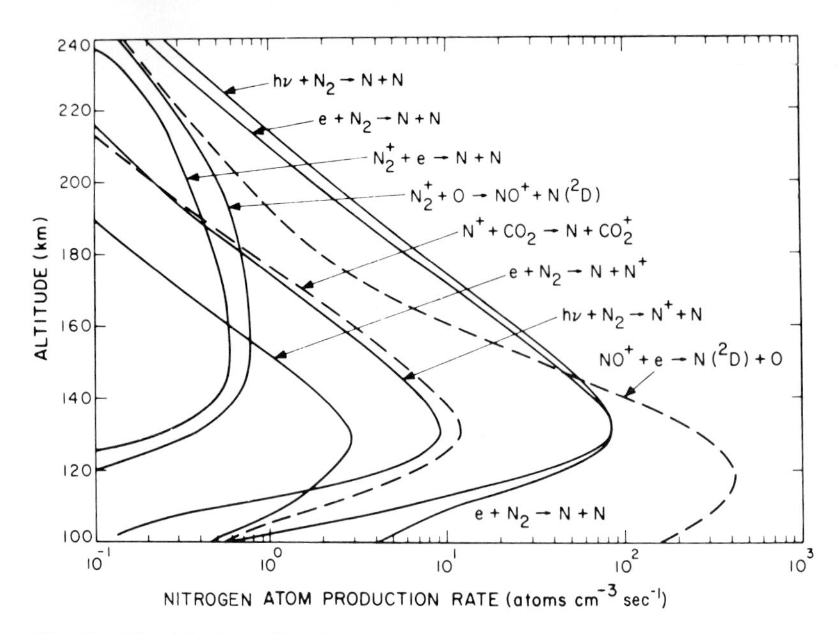

Fig. 29. The altitude profiles of the production rates of excited nitrogen atoms [N(^{4}S) and N(^{2}D)]. From Fox and Dalgarno (1980). Copyright 1980 Pergamon Press Ltd.

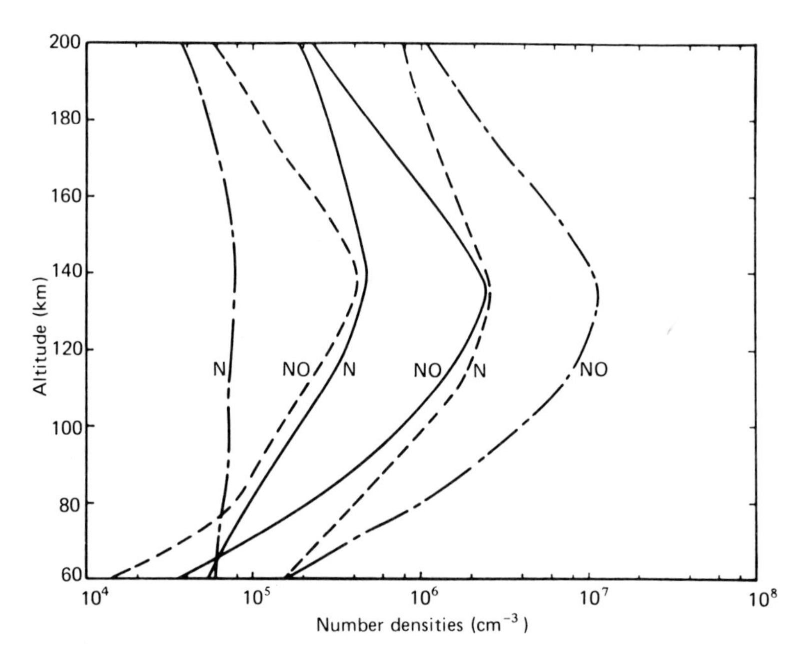

Fig. 30. Number densities of atomic nitrogen and nitric oxide. The three cases correspond to yields of N(^{4}S) due to electron impact dissociation of N_2 of 100 (---), 50 (—), and 0% (—·—), respectively. From Yung *et al.* (1977).

The cross section for this process is small, and the solar radiation penetrates deeply into the atmosphere below the altitudes where $N(^2D)$ is formed. The predissociation reaction limits the amount of NO that is transported into the lower atmosphere.

A flow diagram that illustrates the relationship of the reactions that determine the equilibrium density of NO and atomic nitrogen is shown in Fig. 31. The neutral constituents are arranged across the top in the order of increasing ionization energy. The key reactions are those flowing down the right hand side of the diagram that show the paths that produce excited nitrogen atoms. The bottom center of the diagram illustrates the processes that produce NO and the ultimate conversion of odd-nitrogen back to even-nitrogen.

Photoionization of CO_2 by solar UV radiation shortward of 686 Å produces ${CO_2}^+$ in an excited state:

$$CO_2 + h\nu \longrightarrow {CO_2}^+(B\,^2\Sigma_u^+) + e^- \qquad \lambda < 686\ \text{Å}$$

This excited ${CO_2}^+$ subsequently radiates, producing an airglow emission.

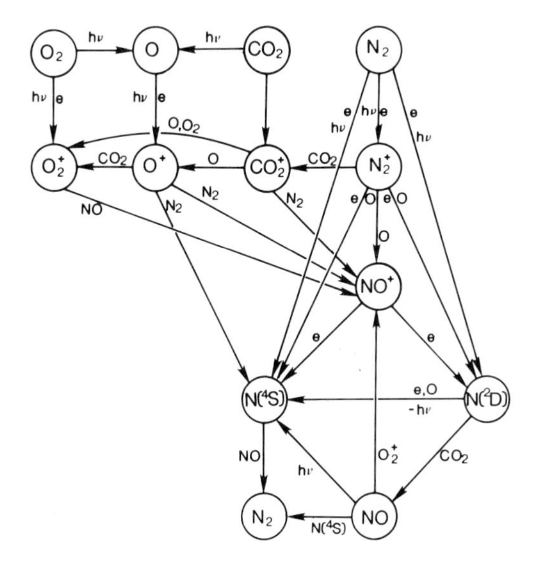

Fig. 31. Chemical reactions producing odd-nitrogen in the Mars upper atmosphere. Chemical symbols in the circles indicate the equilibrium density of molecules, atoms, and ions. The neutral species at the top are in the order of increasing ionization potential moving from left to right. Arrows show the path of the chemical reactions. Chemical symbols along side of the arrow indicate the chemical reaction. Even-nitrogen is at the top of the diagram, odd-nitrogen is at the bottom.

This particular transition is UV doublet at 2883 and 2896 Å:

$$CO_2{}^+(B\,{}^2\Sigma_u^+) \longrightarrow CO_2{}^+(X\,{}^2\Pi_g) + h\nu \qquad \lambda = 2883, 2896\ \text{Å}$$

This emission is present in the dayglow of Mars. Figure 32 shows the UV dayglow spectrum between 1900 and 4000 Å. The $CO_2{}^+$ UV doublet is the prominent emission in the center of the spectrum.

The emission rate of the UV doublet is calculated by summing the product of the solar flux and the ionization cross section over the appropriate wavelength interval, taking into account the attenuation of the solar radiation:

$$E(h) = \sum F_\lambda \sigma_\lambda [CO_2] \exp\left(-\sigma \int_h^\infty [CO_2]\, dz \right).$$

The results of such a calculation are shown in Fig. 33. Photoionization is the dominant process producing the UV doublet. Above 150 km, the emission rate is directly proportional to the CO_2 density. Below that altitude, attenuation takes place, and the emission rate reaches a maximum at 130 km. The measurement of the variation of the emission rate as a function of altitude is a method of measuring the density of CO_2 as a function of altitude and of determining the temperature of the atmosphere.

Photoionization of CO_2 by solar UV radiation shortward of 716 Å produces the Fox–Duffendack–Barker ($A\,{}^2\Pi_u$–$X\,{}^2\Pi_g$) bands of $CO_2{}^+$. These

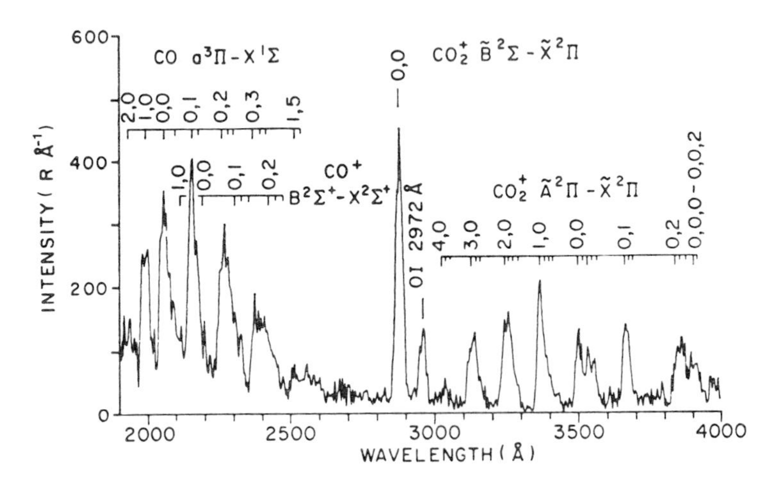

Fig. 32. Ultraviolet spectrum of the upper atmosphere of Mars, 1900–4000 Å, 20-Å resolution. The spectrum was obtained by observing the atmosphere tangentially at an altitude between 160 and 180 km. From Barth *et al.* (1971). Copyright 1971 the American Geophysical Union.

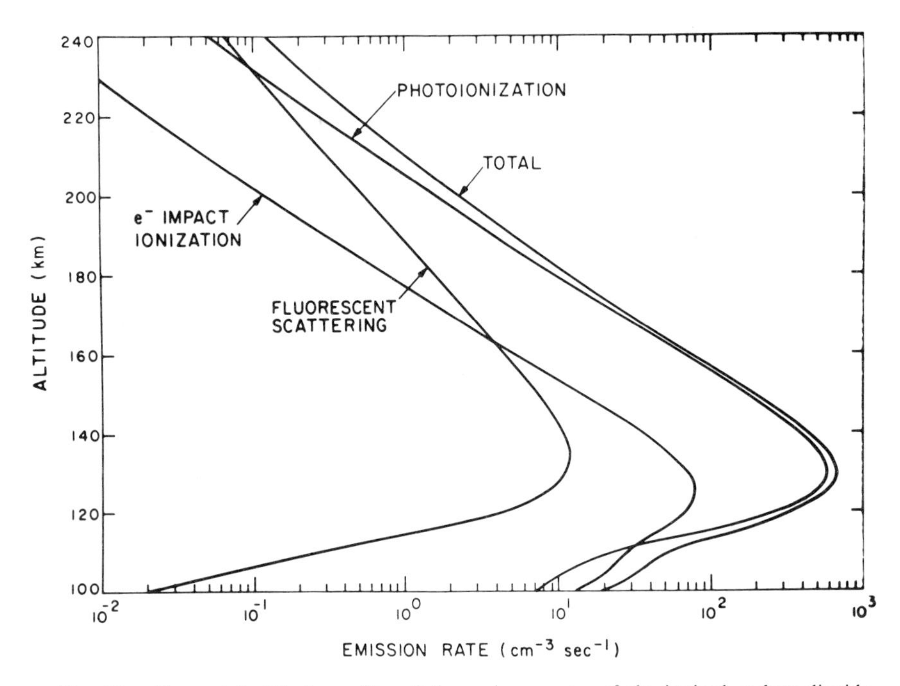

Fig. 33. Computed altitude profiles of the major sources of the ionized carbon dioxide ultraviolet doublet ($B\,^2\Sigma_u^+ - X\,^2\Pi_g$). From Fox and Dalgarno (1979). Copyright 1979 the American Geophysical Union.

bands also appear in the Mars UV dayglow in Fig. 32. The computed altitude profile of the emission rate of this band system is shown in Fig. 34. Photoionization is the dominant mechanism that produces this airglow emission below 150 km. Above this altitude, fluorescent scattering of solar UV radiation by $CO_2{}^+$ is a stronger source:

$$\begin{aligned} CO_2{}^+(X\,^2\Pi_g) + h\nu &\xrightarrow{J_9} CO_2{}^+(A\,^2\Pi_u) \\ CO_2{}^+(A\,^2\Pi_u) &\longrightarrow CO_2{}^+(X\,^2\Pi_g) + h\nu \end{aligned} \tag{57}$$

The emission rate of a fluorescent scattering airglow emission is calculated by the following equation:

$$E = F_\lambda \lambda^2(\pi e^2/mc^2) f [CO_2{}^+],$$

where F_λ is solar flux at wavelength λ, the wavelength where the absorption takes place. The oscillator strength of the transition is f, and $\pi e^2/mc^2$ is equal to 8.829×10^{-13} cm. The measurement of the Fox–Duffendack–Barker bands above 150 km is a method of measuring the density distribution of $CO_2{}^+$ in the Mars ionosphere. The altitude profile of the fluorescent

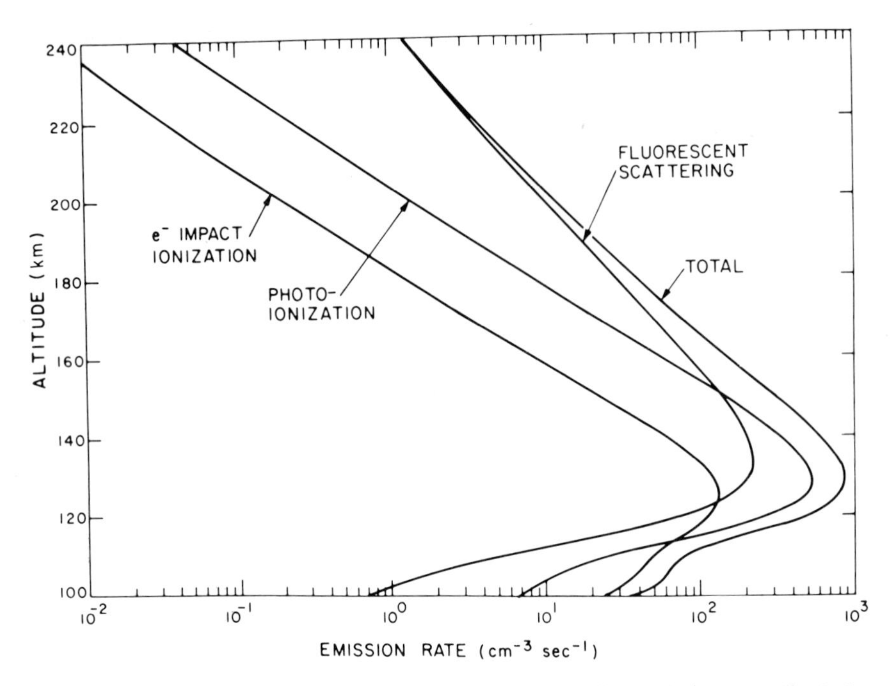

Fig. 34. Computed altitude profiles of the major sources of the CO_2^+ Fox–Duffendack–Barker bands ($A\,^2\Pi_u$–$X\,^2\Pi_g$). From Fox and Dalgarno (1979). Copyright 1979 the American Geophysical Union.

scattering emission rate follows the scale height of CO_2^+, which is larger than the scale height of the CO_2. The altitude profile of the photoionization emission rate follows the CO_2 scale height.

Photodissociation of CO_2 by solar UV radiation shortward of 1082 Å produces CO in an excited state:

$$CO_2 + h\nu \xrightarrow{J_{10}} CO(a\,^3\Pi) + O \qquad \lambda = 1082\ \text{Å} \tag{58}$$

The excited carbon monoxide molecule subsequently radiates the Cameron bands of CO:

$$CO(a\,^3\Pi) \longrightarrow CO(X\,^1\Sigma^+) + h\nu \qquad \lambda = 2063\ \text{Å}$$

The indicated wavelength corresponds to the (0,0) band, the transition between the lowest vibrational level of the excited state to the lowest vibrational level of the ground state. The Cameron bands are the most intense emission of the Mars UV airglow. The spectral features of this system may be seen in the 1900- to 2800-Å region of the spectrum in Fig. 32. The Cameron bands are also produced by photoelectron impact dissociation of CO_2:

$$CO_2 + e^- \longrightarrow CO(a\,^3\Pi) + O + e^- \qquad E > 11.5\ \text{eV}$$

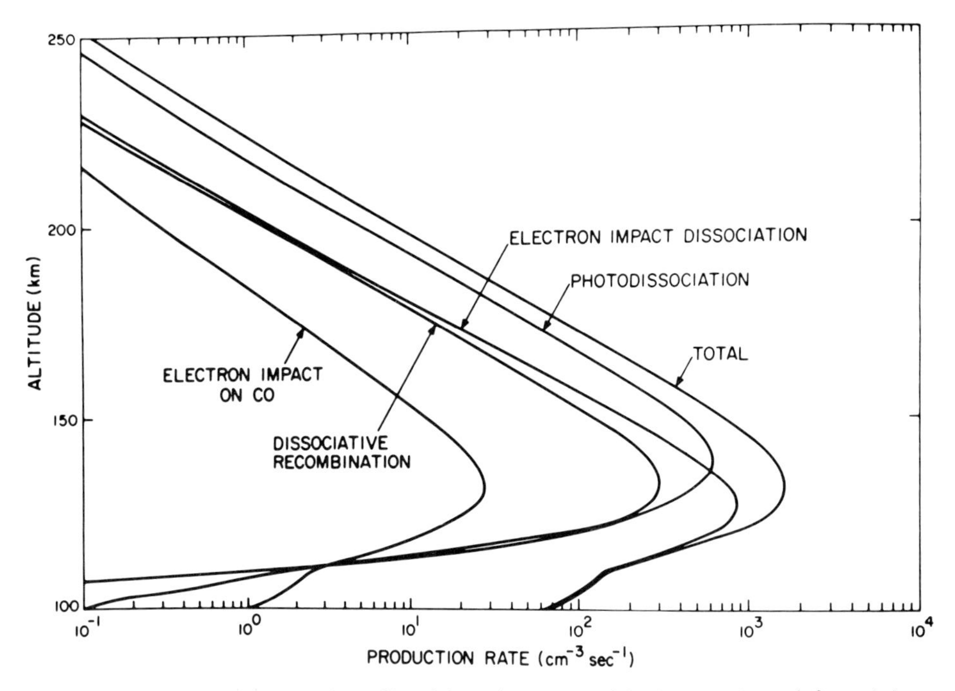

Fig. 35. Computed altitude profiles of the major sources of the Cameron bands (a $^3\Pi$–X $^1\Sigma$) of carbon monoxide. From Fox and Dalgarno (1979). Copyright 1979 Pergamon Press Ltd.

Figure 35 shows the computed altitude profiles for the emission rate of the Cameron bands from photodissociation and electron impact dissociation. These calculations show that the maximum emission rate occurs near 130 km, where photoelectron impact dissociation is the dominant mechanism. At higher altitudes, photodissociation is the larger contributor to the excitation. Above the maximum, for both mechanisms, the emission rate is directly proportional to the CO_2 density. The measurement of these bands is a method of determining the density distribution of CO_2 as a function of altitude and, from the scale height, determining the temperature of the atmosphere.

Photodissociation of CO_2 by solar UV radiation shortward of 1280 Å produces atomic oxygen in an excited state:

$$CO_2 + h\nu \xrightarrow{J_{11}} CO + O(^1S) \qquad \lambda < 1286\ \text{Å} \tag{59}$$

The excited atom subsequently radiates to produce the 2972- and 5577-Å lines of the Mars dayglow:

$$O(^1S) \longrightarrow O(^3P) + h\nu \qquad \lambda = 2972\ \text{Å}$$

$$O(^1S) \longrightarrow O(^1D) + h\nu \qquad \lambda = 5577\ \text{Å}$$

The 2972-Å line may be seen in the UV spectrum in Fig. 32. The calculation of the production rate of O(^{1}S) shown in Fig. 36 demonstrates that photodissociation is the dominant mechanism and that there are two maxima in the altitude profile. The lower maximum is the result of photodissociation of CO_2 by solar Lyα radiation, which penetrates to the 90-km level of the atmosphere. The emission rate of 2972 Å above the maximum in the altitude profile is directly proportional to the CO_2 density, and its measurement may be used to determine the scale height and temperature of the atmosphere.

The altitude profiles of these airglow emissions were measured by the UV spectrometer experiments on *Mariner* 6, 7, and 9. The results of these measurements are shown in Fig. 37. From the airglow measurements, the thermospheric temperatures were determined to be 350 K in 1969 and 325 K in 1971.

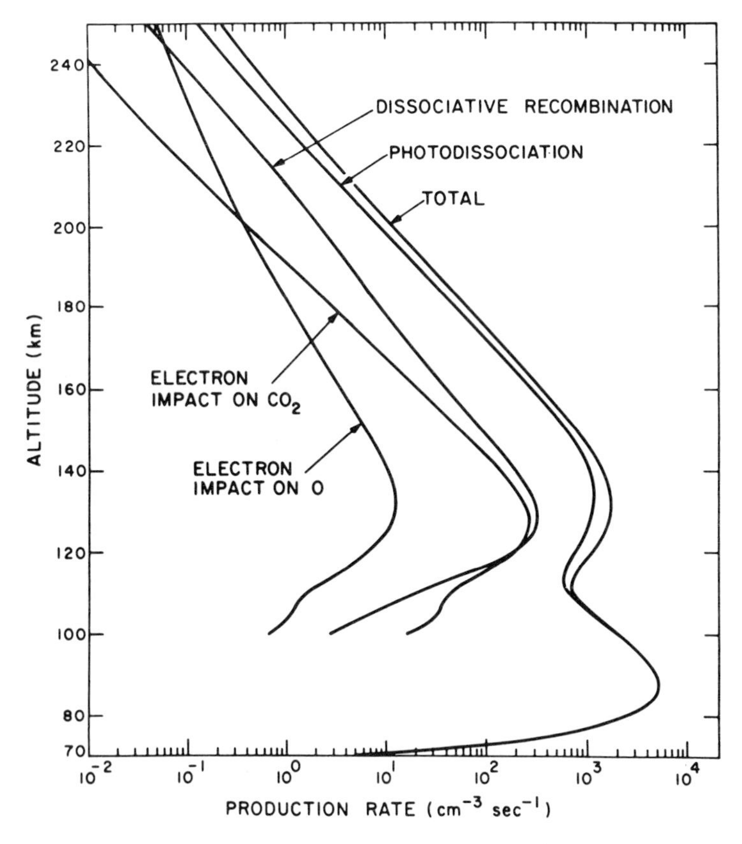

Fig. 36. Computed altitude profiles of the major sources of O(^{1}S). From Fox and Dalgarno (1979). Copyright 1979 Pergamon Press Ltd.

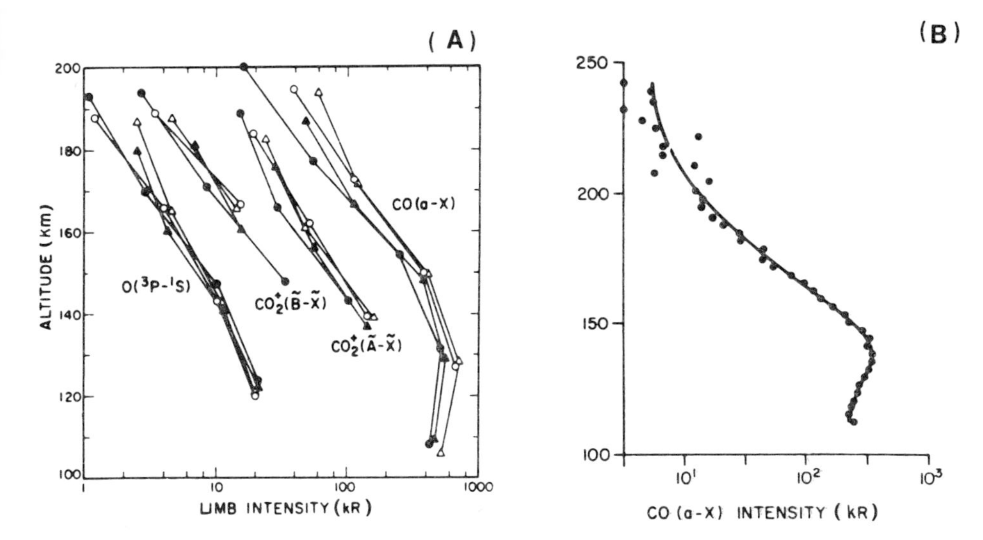

Fig. 37. (A) Limb intensity profiles of four airglow emissions observed by *Mariner 6* and *7* in 1969 (different data collections are indicated by open and filled symbols). A thermospheric temperature of 350 K was determined from the scale height of these observations. From Stewart (1972). Copyright 1972 the American Geophysical Union. (B) Limb intensity profile of carbon monoxide Cameron bands measured by *Mariner 9* in 1971. A thermospheric temperature of 325 K was determined from these observations. From Stewart *et al.* (1972). Copyright 1972 Pergamon Press Ltd.

The UV dayglow spectrum of Mars between 1100 and 1800 Å is shown in Fig. 38. In this region of the spectrum, the most prominent molecular emission is the fourth positive bands of CO ($A\,^1\Pi$–$X\,^1\Sigma^+$). In the Mars ionosphere, this airglow emission is produced by the following mechanisms:

1. The photodissociation of CO_2 by solar UV radiation shortward of 920 Å:

$$CO_2 + h\nu \xrightarrow{J_{12}} CO(A\,^1\Pi) + O \qquad \lambda < 920\ \text{Å} \tag{60}$$

2. The dissociative recombination of ${CO_2}^+$:

$${CO_2}^+ + e^- \longrightarrow CO(A\,^1\Pi) + O$$

3. The photoelectron impact dissociation of CO_2:

$$CO_2 + e^- \longrightarrow CO(A\,^1\Pi) + O + e^- \qquad E > 13.5\ \text{eV}$$

The atomic carbon lines are produced predominantly by processes that dissociate CO_2.

The airglow line at 1304 Å is produced by the resonant scattering of solar radiation by O in the Mars atmosphere. This spectral feature is actually a

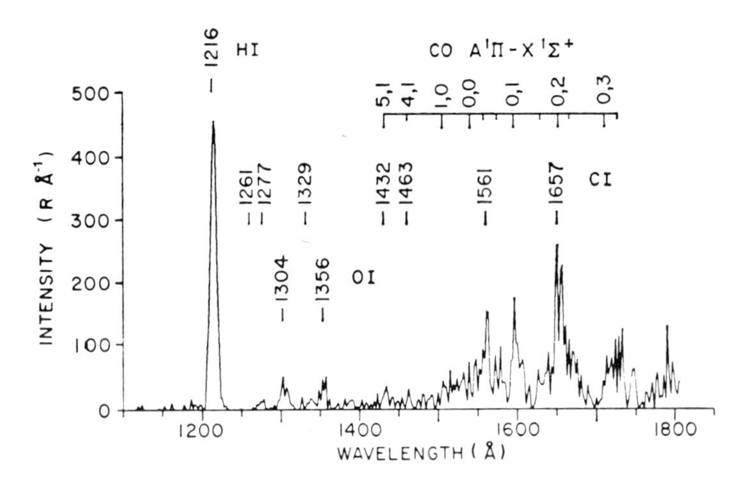

Fig. 38. Ultraviolet spectrum of the upper atmosphere of Mars, 1100–1800 Å, 10-Å resolution. The spectrum was obtained by observing the atmosphere tangentially at an altitude between 140 and 160 km. This spectrum was obtained from the sum of four individual observations. From Barth *et al.* (1971). Copyright 1971 the American Geophysical Union.

triplet at 1302, 1305, and 1306 Å. The solar lines are reversed, and the fluxes in the center of each solar line are nearly equal. The emission rate of airglow emission is calculated from the following expression:

$$E = F_\lambda \lambda^2 (\pi e^2/mc^2) f[\mathrm{O}],$$

where F_λ is flux in the center of the solar lines (3.9×10^9 photons $\mathrm{cm^{-1}\,sec^{-1}\,Å^{-1}}$), f the oscillator strength 0.048, and $\pi e^2/mc^2 = 8.83 \times 10^{-13}$ cm. The emission rate coefficient g calculated from these values is $2.8 \times 10^{-6}\ \mathrm{sec^{-1}}$, where g is defined by the relationship

$$E = g[\mathrm{O}].$$

Measurements of the 1304-Å atomic oxygen airglow emission gave the result in Fig. 8, a model with 1%.atomic oxygen at 135 km.

The strongest atomic emission in the Mars airglow is the atomic hydrogen line at 1216 Å. The line is produced by the resonant scattering of the solar Lyα line at 1216 Å. The solar line is broad (~ 1 Å wide) and reversed from radiative transfer effects in the solar atmosphere. Atomic hydrogen in the Mars atmosphere scatters photons from the center of the solar line. The emission rate of the Lyα line in the Mars airglow is calculated by the following equation:

$$E = F_\lambda \lambda^2 (\pi e^2/mc^2) f[\mathrm{H}],$$

where F_λ is the solar flux in the center of the line (4.0×10^{11}

photons cm^{-2} sec^{-1} Å^{-1} was measured from above the Earth's atmosphere in 1968; when adjusted to the orbit of Mars, the value is 1.7×10^{11} photons cm^{-2} sec^{-1} Å^{-1}), and f the oscillator strength of the Lyα transition, which is equal to 0.416. The emission rate coefficient g calculated from these values is 9.4×10^{-4} sec^{-1}. The H density determined from the *Mariner* measurements is 3×10^4 atoms cm^{-3} at an altitude of 220 km.

VI. Escape of Gases from the Mars Atmosphere

Hydrogen atoms escape from the top of the Mars atmosphere. The temperature of the atmosphere is sufficiently high and the gravitational energy sufficiently low that a significant number of the atoms have enough kinetic energy to escape from the gravitational influence of Mars. The escape flux is calculated from the following equation:

$$F_{\text{esc}} = [\text{H}]\frac{v_m}{2\pi^{1/2}}\left(\frac{GMm}{r_c kT} + 1\right)\exp\left(-\frac{GMm}{r_c kT}\right),$$

where v_m is equal to $(2kT/m)^{1/2}$, the most probable velocity of the hydrogen atom, GMm/r_c is the gravitational energy at r_c, the distance from the center of Mars to the top of the atmosphere, and kT, the product of the Boltzmann constant and the temperature, is proportional to the kinetic energy. The thermal escape flux is strongly controlled by the quantity in the exponent. For conditions at the time of the *Mariner 6* and *7* observations, the temperature is 350 K, and the exponential factor is calculated to be 4.4. The escape flux for these conditions is 1.2×10^8 atoms cm^{-2}. The ultimate source of the escaping atomic hydrogen is the dissociation of H_2O in the lower atmosphere. The amount of liquid water needed to supply this escape flux for 1 year is 5.7×10^{-4} μm. The average annual H_2O abundance in the lower atmosphere is 15 pr μm, many times more than ample to supply atomic hydrogen to the upper atmosphere. The path from the photochemistry of H_2O in the lower atmosphere to the escape of atomic hydrogen from the upper atmosphere, however, is indirect. The dissociation of H_2O produces odd hydrogen, in particular, atomic hydrogen and HO_2. These two species react to form H_2, which then flows upward to the upper atmosphere:

$$\text{H} + \text{HO}_2 \xrightarrow{k_{10}} \text{H}_2 + \text{O}_2 \tag{19}$$

In the ionosphere, H_2 reacts with CO_2^+ to produce atomic hydrogen:

$$\text{H}_2 + \text{CO}_2^+ \xrightarrow{k_{12}} \text{H} + \text{CO}_2\text{H}^+ \tag{61}$$

$$\text{CO}_2\text{H}^+ + e^- \xrightarrow{k_{13}} \text{H} + \text{CO}_2 \tag{62}$$

The result of these two reactions is to convert H_2 to hydrogen atoms while CO_2^+ recombines with an electron to form CO_2. The atomic hydrogen produced in the ionosphere flows upward to the top of the atmosphere, where it escapes. Figure 39 shows fluxes of H and H_2 in the Mars atmosphere. Molecular hydrogen is formed near 25 km, it flows upward to 125 km, where it is converted to atomic hydrogen by the ionospheric reactions. The H that is formed here flows upward to the top of the atmosphere. The fluxes in Fig. 38 are the result of a self-consistent model, which included 31 chemical reactions and eddy diffusion.

There are nonthermal escape mechanisms operating on Mars. Because of the smaller gravitational potential on Mars compared to the Earth, ionic reactions sometimes produce atoms with sufficient kinetic energy to escape from the gravitational field of Mars. In particular, the major ion in the ionosphere (O_2^+) produces oxygen atoms in a variety of states with varying

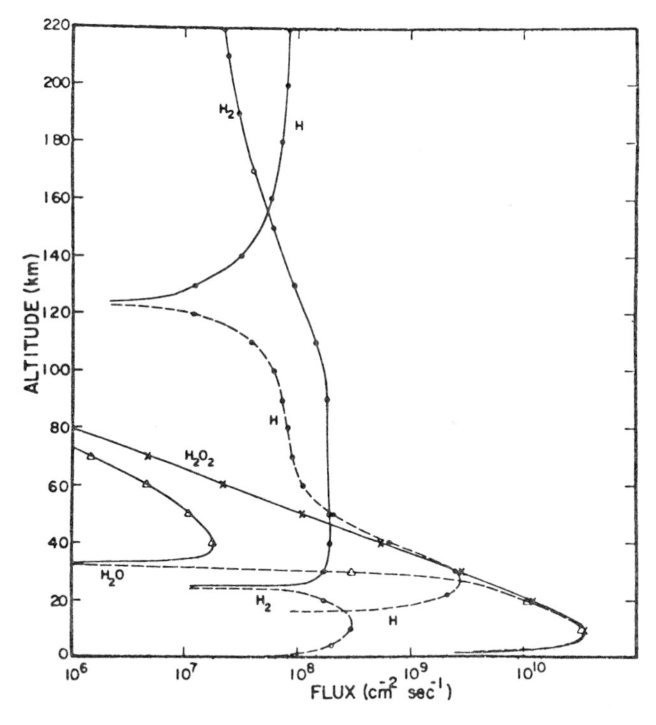

Fig. 39. Fluxes of odd-hydrogen (indicated by H), H_2, H_2O, and H_2O_2 for an O_2 mixing ratio of 1.3×10^{-3}. Solid curves indicate upward fluxes, and dashed curves indicate downward fluxes. From Liu and Donahue (1976).

amounts of kinetic energy when it undergoes dissociative recombination:

$$O_2^+ + e^- \longrightarrow O(^3P) + O(^3P) + 7.0 \text{ eV}$$

$$O_2^+ + e^- \longrightarrow O(^3P) + O(^1D) + 5.0 \text{ eV}$$

$$O_2^+ + e^- \longrightarrow O(^1D) + O(^1D) + 3.0 \text{ eV}$$

$$O_2^+ + e^- \longrightarrow O(^3P) + O(^1S) + 2.8 \text{ eV}$$

$$O_2^+ + e^- \longrightarrow O(^1D) + O(^1S) + 0.8 \text{ eV}$$

The first two excited states of atomic oxygen (^{1}D and ^{1}S) lie 2.0 and 4.2 eV above the ^{3}P ground state. The energy released in dissociative recombination is the difference between the ionization energy (12.1 eV) and the dissociation energy (5.1 eV) of O_2.

The energy required for an oxygen atom at the top of the Mars atmosphere to escape is 2.0 eV. This is calculated from the equation

$$E = GMm/r_c,$$

where G is the gravitational constant, M the mass of Mars, m the mass of the oxygen atom, and r_c the distance from the center of Mars to the top of the atmosphere. Since the excess kinetic energy in the dissociative recombination reactions is divided equally between the two oxygen atoms, the first two reactions produce oxygen atoms with sufficient energy to escape. One-half of these energetic atoms (i.e., those moving upward) should escape. *Pioneer Venus* observations of energetic oxygen atoms at the top of the atmosphere of Venus show that 22% of the O_2^+ dissociative recombinations result in two ^{3}P oxygen atoms and 55% produce one ^{3}P and one ^{1}D atom. Thus 77% of the dissociative recombinations lead to energetic oxygen atoms with sufficient energy to escape from Mars. Using the measured O_2^+ densities shown in Fig. 28, the flux of energetic oxygen atoms escaping from above the 190 km level is calculated from the following equation:

$$F_{\text{esc}} = 0.5 \times 0.77 \times k_{24} \int_h^\infty [O_2^+][e^-]\,dz,$$

where k_{24} is the dissociative recombination rate, $[O_2^+]$ and $[e^-]$ are the ion and electron densities, and h is equal to 190 km. The result of this calculation shows the flux of escaping energetic oxygen atoms to be approximately one-half the flux of escaping hydrogen atoms. This is a remarkable result since it suggests that for every two hydrogen atoms that escape, one oxygen atom escapes. This means that H_2O is escaping from the upper atmosphere of Mars in the form of hydrogen and oxygen atoms. If these escape processes, the thermal escape of atomic hydrogen and the energetic atom escape of atomic oxygen, have been operating for the lifetime of the solar system (4.5×10^9 years), then an amount of H_2O equivalent to 2.5 m deep has escaped from Mars.

Other ionospheric reactions produce atoms with sufficient energy to escape. The dissociative recombination of CO_2^+ produces CO and atomic oxygen, and excess kinetic energy equal to the difference between the ionization energy (13.7 eV) and the dissociative energy (5.4 eV) of CO_2:

$$CO_2^+ + e^- \longrightarrow CO + O + 8.3 \text{ eV}$$

The energy shared between the molecule and the atom is inversely proportional to the ratio of the masses: CO with 3.0 eV, and O with 5.3 eV. The atomic oxygen has sufficient energy to escape, but the CO does not.

Atomic nitrogen also escapes from the top of the Mars atmosphere in a nonthermal process. The energy required for atomic nitrogen escape is 1.74 eV. The processes that produce nitrogen atoms with excess kinetic energy are (see Fig. 29) the electron impact dissociation of N_2 by electrons with energy greater than 13.2 eV,

$$e^- + N_2 \longrightarrow N(^4S) + N(^4S) + e^- + 3.5 \text{ eV} \qquad E > 13.2 \text{ eV}$$

the dissociative recombination of N_2^+ (ionization energy 15.5 eV, dissociation energy 9.8 eV),

$$N_2^+ + e^- \longrightarrow N(^4S) + N(^4S) + 5.7 \text{ eV}$$

and photodissociation of N_2,

$$N_2 + h\nu \xrightarrow{J_{13}} N(^4S) + N(^4S) \qquad \lambda < 940 \text{ Å} \tag{63}$$

A model atmosphere calculation based on the atomic nitrogen production rates in Fig. 29 gives an escape rate for atomic nitrogen of 2.0×10^5 atoms cm^{-2} sec^{-1}. The total number of N_2 in the Mars atmosphere is 5.6×10^{21} molec cm^{-2}. At this escape rate, the present amount of nitrogen will be gone in 1.8×10^9 years. If this escape rate has been operating during the entire life of the solar system, then the original amount of nitrogen in the Mars atmosphere must have been at least two and a half times greater than it is today.

VII. Climatic Change on Mars

Astronomical changes in the Mars orbit parameters produce climatic changes on Mars. Gravitational perturbations by the eight other planets cause the inclination of the orbit to change. Solar torque on the equatorial bulge of Mars causes the spin axis to precess. The combination of these effects produces a precession of the angle of perihelion with a period of 51,000 years. The location of the time of perihelion compared with the time of southern summer solstice (Fig. 1) will change with this period. The pattern of dust storms on Mars will be different as perihelion moves away from summer solstice.

The obliquity of Mars (the angle between the plane of the equator and the plane of the orbit) changes with a period of 120,000 years. The present obliquity is 25.1°. In the past and in the future, the obliquity can be as large as 35° and as small as 15°. A calculation of the variation of the Mars obliquity over the past several million years is shown in Fig. 40. At 35°, the annual solar radiation falling on the polar regions would be 35% greater than its current value, while at 15°, the annual solar radiation would be only 60% of its current value. Under either of these conditions, the equilibrium temperature of the CO_2 polar cap would be different from what it is today. For example, if the temperature of the winter polar cap fell to 135 K, then additional CO_2 would precipitate out of the atmosphere and the atmospheric pressure over the entire planet would fall to less than 1 mbar. Conversely, if the equilibrium temperature of the winter polar cap were as high as 160 K, then additional CO_2 would sublime, and if there were a sufficient reservoir of CO_2, the atmospheric pressure would rise to more than 50 mbars. With a different total atmospheric pressure, the climate on Mars would be different from the current climate. At higher obliquity and higher pressures, more H_2O would come out of the polar caps. At lower obliquity and lower pressures, the permanent caps would collect more H_2O. With large changes in the pressure and in the H_2O content of the atmosphere, the photochemistry of Mars would be dramatically different than it is today.

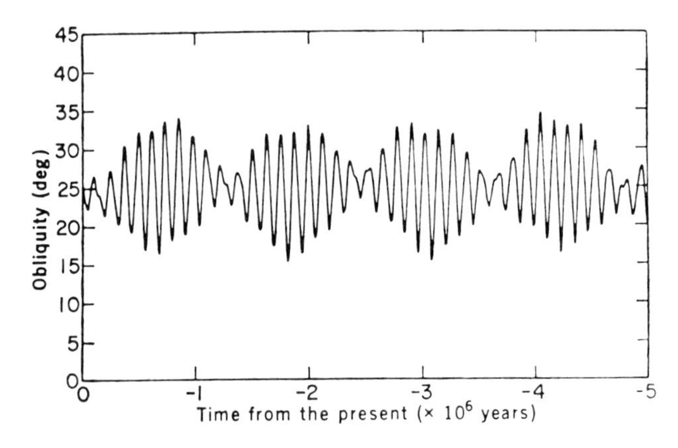

Fig. 40. Variations in the obliquity of Mars for the past 5 million years. From Ward (1973). Copyright 1973 the American Geophysical Union.

VIII. Directions for Future Research

The ultimate goal of research in the photochemistry of the Mars atmosphere is reaching an understanding of all of the photochemical processes that are occurring in the Mars atmosphere at the present time and discovery of what role photochemistry played in the past in the evolution of the Mars atmosphere. With this knowledge, it should be possible to predict the future course of photochemical change in the Mars atmosphere. Mars is an excellent laboratory to study the photochemistry of a CO_2 atmosphere. The chemical system is relatively simple, and changes do occur throughout the Mars year, particularly with the seasonal injection and removal of H_2O. Changes also occur in the photochemistry of the Mars atmosphere, as a result of changes in solar UV and particle radiation. These changes occur with both a 27-day and an 11-year cycle. To understand the intimate relationship between changes on the Sun and changes in the Mars atmosphere is an important goal.

To understand the influence of the Sun on the photochemistry of Mars and to understand the seasonal changes in the photochemistry of Mars, a photochemistry mission to Mars is needed. Such a spacecraft mission would measure all of the pertinent physical parameters and chemical species in the Mars atmosphere and also would measure the incoming radiation from the Sun. Such a mission would need to last for at least 2 Mars years, in order to identify clearly the seasonal changes in the photochemistry. To understand the influence of the Sun on the Mars photochemistry, this mission should be carried out during the period that the Sun is most active, so that a direct cause and effect can clearly be established.

To understand the past photochemical evolution of the Mars atmosphere, there is an ultimate mission to Mars. An automated laboratory placed on one of the residual polar caps would drill hundreds of meters deep. Core samples would be removed and analyzed for their chemical content. In the past, any condensible gas present in the atmosphere on a planetary scale would have condensed onto the winter polar cap, and therefore the entire photochemical history of Mars is recorded in the polar caps.

References

Barker, E. S. (1972). Detection of molecular oxygen in the Martian atmosphere. *Nature (London)* **238,** 447.

Barker, E. S., Schorn, R. A., Woszcyk, A., Tull, R. G., and Little, S. J. (1970). Mars: detection of atmospheric water vapor during the southern hemisphere spring and summer season. *Science* **170,** 1308.

Barth, C. A. (1974a). Free radicals in the atmospheres of Mars and Venus. *Ber. Bunsenges. Phys. Chem.* **78,** (2), 163.

Barth, C. A. (1974b). The atmosphere of Mars. *Annu. Rev. Earth Planet. Sci.* **2,** 333.

Barth, C. A., and Dick, M. L. (1974). Ozone and the polar hood of Mars. *Icarus* **22,** 205.

Barth, C. A., and Hord, C. W. (1971). *Mariner* ultraviolet spectrometer: topography and polar cap. *Science* **173,** 197.

Barth, C. A., Fastie, W. G., Hord, C. W., Pearce, J. B., Kelly, K. K., Stewart, A. I., Thomas, G. E., Anderson, G. P., and Raper, O. F. (1969). *Mariner 6*: ultraviolet spectrum of Mars upper atmosphere. *Science* **165,** 1004.

Barth, C. A., Hord, C. W., Pearce, J. B., Kelly, K. K., Anderson, G. P., and Stewart, A. I. (1971). *Mariner 6* and *7* ultraviolet spectrometer experiment: upper atmosphere data. *J. Geophys. Res.* **76,** 2213.

Barth, C. A., Stewart, A. I., Hord, C. W., and Lane A. L. (1972). *Mariner 9* ultraviolet spectrometer: Mars airglow spectroscopy and variations in Lyman alpha. *Icarus* **17,** 457.

Barth, C. A., Hord, C. W., Stewart, A. I., Lane, A. L., Dick, M. L., and Anderson, G. P. (1973). *Mariner 9* ultraviolet spectrometer experiment: seasonal variation of ozone on Mars. *Science* **179,** 795.

Belton, M. J. S., Broadfoot, A. L., and Hunten, D. M. (1968). Abundance temperature of CO_2 on Mars during the 1967 opposition. *J. Geophys. Res.* **73,** 4795.

Biemann, K., Owen, T., Rushneck, D. R., and Howarth, D. W. (1976). The atmosphere of Mars near the surface: isotope ratios and upper limits on noble gases. *Science* **194,** 76.

Biemann, K., Oro, J., Toulmin, P., III, Orgel, L. E., Nier, A. O., Anderson, D. M., Simmonds, P .G., Flory, D., Diaz, A. V., Rushneck, D. R., Biller, J. E., and Lafleur, A. L. (1977). The search for organic substances and inorganic volatile compounds in the surface of Mars. *JGR, J. Geophys. Res.* **82,** 4641.

Carleton, N. P., and Traub, W. A. (1972). Detection of molecular oxygen on Mars. *Science* **177,** 988.

Carr, M. H. (1981). "The Surface of Mars." Yale Univ. Press, New Haven, Connecticut.

Connes, J., Connes, P., and Maillard, J. P. (1969). "Atlas des spectres infrarouges de Venus, Mars, Jupiter et Saturne." CNRS, Paris.

Farmer, C. B., Davies, D. W., Holland, A. L., LaPorte, D. D., and Doms, P. E. (1977). Mars: water vapor observations from the *Viking* orbiters. *JGR, J. Geophys. Res.* **82,** 4225.

Fjeldbo, G., Sweetnam, D., Brenkle, J., Christensen, E., Farless, D., Mehta, J., Seidel, B., Michael, W., Jr., Wallio, A., and Grossi, M. (1977). *Viking* radio occultation measurements of the Martian atmosphere and topography: primary mission coverage. *JGR, J. Geophys. Res.* **82,** 4317.

Fox, J. L., and Dalgarno, A. (1979). Ionization, luminosity, and heating of the upper atmosphere of Mars. *JGR, J. Geophys. Res.* **84,** 7315.

Fox, J. L., and Dalgarno, A. (1980). The production of nitrogen atoms on Mars and their escape. *Planet. Space Sci.* **28,** 41.

Hanel, R. A., Conrath, B. J., Hovis, W. A., Kunde, V. G., Lowman, P. D., Maguire, W., Pearl, J., Pirraglia, J., Prabhakara, C., and Schalchman, B. (1972). Investigation of the Martian environment by infrared spectroscopy on *Mariner 9*. *Icarus* **17,** 423.

Hanson, W. B., Sanatani, S., and Zuccaro, D. R. (1977). The Martian ionosphere as observed by the *Viking* retarding potential analyzers. *JGR, J. Geophys. Res.* **82,** 4351.

Hess, S. L., Ryan, J. A., Tillman, J. E., Henry, R. M., and Leovy, C. B. (1980). The annual cycle of pressure on Mars measured by *Viking* landers 1 and 2. *Geophys. Res. Lett.* **7,** 197.

Horn, D., McAfee, J. M., Winer, A. M., Herr, K. C., and Pimentel, G. C. (1972). The composition of the Martian atmosphere: minor constituents. *Icarus* **16,** 543.

Hunten, D. M. (1974). Aeronomy of the lower atmosphere of Mars. *Rev. Geophys. Space Phys.* **12,** 529.

Jakosky, B. M., and Farmer, C. B. (1982). The seasonal and global behavior of water vapor in the

Mars atmosphere: complete global results of the *Viking* atmospheric water detector experiment. *JGR, J. Geophys. Res.* **87,** 2999.

Kieffer, H. H., Martin, T.Z., Peterfreund, A. R., Jakosky, B. M., Miner, E. D., and Palluconi, F. D. (1977). Thermal and albedo mapping of Mars during the *Viking* primary mission. *JGR, J. Geophys. Res.* **82,** 4249.

Kliore, A. J. (1982). "The Mars Reference Atmosphere." Pergamon, Oxford.

Kliore, A. J., Cain, D. L., Levy, G. S., Eshleman, V. R., Fjeldbo, G., and Drake, F. D. (1965). Occultation experiment: results of the first direct measurement of Mars' atmosphere and ionosphere. *Science* **149,** 1243.

Kong, T. Y., and McElroy, M. B. (1977). Photochemistry of the Martian atmosphere. *Icarus* **32,** 168.

Leighton, R. B., and Murray, B. C. (1966). Behavior of carbon dioxide and other volatiles on Mars. *Science* **153,** 136.

Liu, S. C., and Donahue, T. M. (1976). The regulation of hydrogen and oxygen escape from Mars. *Icarus* **28,** 231.

McElroy, M. B. (1973). Atomic and molecular processes in the Martian atmosphere. *Adv. Mol. Phys.* **9,**323.

McElroy, M. B., and McConnell, J. C. (1971). Dissociation of CO_2 in the Martian atmosphere. *J. Atmos. Sci.* **28,** 880.

McElroy, M. B., Kong, T. Y., and Yung, Y. L. (1977). Photochemistry and evolution of Mars' atmosphere: a *Viking* perspective. *JGR, J. Geophys. Res.* **82,** 4379.

Masursky, H. (1973). An overview of geologic results from *Mariner 9. J. Geophys. Res.* **78,** 4037.

Neugebauer, G., Munch, G., Kieffer, H., Chase, S. C., Jr., and Miner, E. (1971). Mariner 1969 infrared radiometer results: temperature and thermal properties of the Martian surface. *Astrophys. J.* **76,** 719.

Nier, A. O., and McElroy, M. B. (1977). Composition and structure of Mars' upper atmosphere: results from the neutral mass spectrometers on *Viking 1* and *2*. *JGR, J. Geophys. Res.* **82,** 4341.

Noxon, J. F., Traub, W. A., Carleton, N. P., and Connes, P. (1976). Detection of O_2 dayglow emission from Mars and the Martian ozone abundance. *Astrophys. J.* **207,** 1025.

Parkinson, T. D., and Hunten, D. M. (1972). Spectroscopy and aeronomy of O_2 on Mars. *J. Atmos. Sci.* **29,** 1380.

Seiff, A., and Kirk, D. B. (1977). Structure of the atmosphere of Mars in summer at mid-latitudes. *JGR, J. Geophys. Res.* **82,** 4364.

Spinrad, H., Munch, G., and Kaplan, L. D. (1963). The detection of water vapor on Mars. *Astrophys. J.* **137,** 1319.

Stewart, A. I. (1972). *Mariner 6* and *7* ultraviolet spectrometer experiment: implications of CO_2^+, CO, and O airglow. *J. Geophys. Res.* **77,** 54.

Stewart, A. I., Barth, C. A., Hord, C. W., and Lane, A. L. (1972). *Mariner 9* ultraviolet spectrometer experiment: structure of Mars' upper atmosphere. *Icarus* **17,** 469.

Strickland, D. J., Thomas, G. E., and Sparks, P. R. (1972). *Mariner 6* and *7* ultraviolet spectrometer experiment: analysis of the OI 1304- and 1356-angstrom emissions. *J. Geophys. Res.* **77,** 4052.

Strickland, D. J., Stewart, A. I., Barth, C. A., and Hord, C. W. (1973). *Mariner 9* ultraviolet spectrometer experiment: Mars atomic oxygen 1304-Å emission. *J. Geophys. Res.* **78,** 4547.

Traub, W. A., Carleton, N. P., Connes, P, and Noxon, J. F. (1979). The latitude variation of O_2 dayglow and O_3 abundance on Mars. *Astrophys. J.* **229,** 846.

Ward, W. R. (1973). Large-scale variations in the obliquity of Mars. *Science* **181,** 260.

Yung, Y. L., Strobel, D. F., Kong, T. Y., and McElroy, M. B. (1977). Photochemistry of nitrogen in the Martian atmosphere. *Icarus* **30,** 26.

8

The Photochemistry of the Atmospheres of the Outer Planets and Their Satellites

DARRELL F. STROBEL

Department of Earth and Planetary Sciences
The Johns Hopkins University
Baltimore, Maryland

ISBN 0-12-444920-4

I. Introduction

In sharp contrast to the terrestrial planets discussed in the previous chapters, the giant planets in the outer solar system are massive (15–320 Earth masses), larger (4–11 Earth radii), and possess multiple satellites and rings. In the case of Neptune we do not know whether it has a ring system, but its satellites have unusual orbits. Some physical properties of these planets are given in Table I. All have central core regions with 10 to 30 Earth masses of rock and/or ice surrounded by adiabatic hydrogen–helium envelopes. *Adiabatic envelope* denotes the fluid that surrounds the core and whose temperature and pressure decrease with height at the rate prescribed thermodynamically for isentropic displacements and indicative of convective energy transport. The fluid may consist of hydrogen and helium in various phases (e.g., gaseous and metallic hydrogen, gaseous and liquid helium). With the exception of Uranus, the giant planets emit twice as much radiation in the infrared (IR) as they absorb from the Sun in the visible. Their large masses enable them to gravitationally retain their primordial atmospheres and elemental composition.

For the largest planets, Jupiter (Fig. 1) and Saturn (Fig. 2), we expect their composition to be the same as the solar elemental composition. Reactive atoms such as carbon, nitrogen, and oxygen are primarily present in the form of saturated hydrides (CH_4, NH_3, and H_2O) at approximately the solar ratios of carbon, nitrogen, and oxygen. Condensable substances, for example, H_2O and NH_3, are expected to form clouds at appropriate levels in these atmospheres when their saturation vapor pressures are reached. A chemical equilibrium model constructed with an adiabatic temperature lapse rate, where the temperature decrease with height is equal to the gravitational acceleration divided by the specific heat at constant pressure, is widely regarded as a correct first-order description of the deep atmosphere. The presence of certain species, for example, C_2H_6 and C_2H_2, far in excess of chemical equilibrium expectations is indicative of photochemical processes occurring high in the atmosphere and is the subject matter of this chapter.

Uranus and Neptune are intermediate planets with significant amounts of ices relative to rock. A substantial fraction of these ices may be present in the adiabatic hydrogen–helium envelopes. Our knowledge of Uranus and Neptune is limited by the absence of intensive flyby observations; the *Voyager 2* spacecraft should remedy this in 1986 and 1989, respectively.

The outermost planet in the solar system is Pluto, whose orbit is sufficiently eccentric to pass closer to the Sun than Neptune's orbit. By virtue of Pluto's small mass, great orbital eccentricity, and great inclination, it is suspected that Pluto originally was a satellite of another planet and was forcibly ejected from its parent planet's gravitational field to orbit the Sun independently. The strongest candidate for the parent planet appears to be Uranus.

Table I

PLANETARY DATA

Planet	Semimajor axis of orbit (AU)[a]	Sidereal period (years)	Rotation period (hr)	Radius (km at the equator)	Density ($g\ cm^{-3}$)	Surface gravity ($cm\ sec^{-2}$ at the equator)
Earth	1	1	23.93	6,378	5.518	979
Jupiter	5.2	11.86	9.92	71,400	1.334	2365
Saturn	9.54	29.46	10.66	60,300	0.69	954
Uranus	19.2	84.01	16–24	24,500	1.26	840
Neptune	30.06	164.79	~18	25,100	1.67	1120
Pluto	39.44	247.7	~150	1,500	1.1	45.4

[a] $1\ AU \equiv 1.496 \times 10^8$ km.

Fig. 1. *Voyager 1* photo of Jupiter at a range of $\sim 33 \times 10^6$ km. Objects as small as 600 km can be seen in this image. The Great Red Spot is in the lower center of the image.

Methane gas (CH_4) has been detected on Pluto in the near IR. The observed amount ($\sim 7 \times 10^{22}$ cm^{-3} or 85 μbars) is equivalent to the vapor pressure of a CH_4 frost at ~ 57 K covering the surface. The atmospheric composition is consistent with models of Pluto's internal composition as being predominantly frozen volatiles with a mean density of ~ 1 g cm^{-3}. A CH_4 atmosphere is stable against hydrodynamic escape as there is insufficient energy at Pluto's orbital distance to supply heating to offset adiabatic cooling by the escaping gas.

Fig. 2. *Voyager 1* photo of Saturn which was enhanced to increase the visibility of large, bright features in Saturn's North Temperate Belt, which are believed to be gigantic convective storms. The right edge of the rings was "clipped" due to a slight spacecraft drift.

In the Jovian satellite system the Galilean satellites have a distinct density gradient from rock for Io ($\rho \sim 3.55$ g cm^{-3}, Table II) to ice for Callisto ($\rho \sim 1.83$ g cm^{-3}). Saturn's satellites, in contrast, display no such regularity in density. Only Titan (Fig. 3) has a substantial atmosphere. Its formation at low temperatures (~ 60 K) allowed the retention of volatile ices as clathrates (CH_4, NH_3). In contrast Ganymede and Callisto condensed at higher temperatures (> 100 K) and have comparable densities but no atmospheres (see Table II).

The massive N_2 atmosphere on Titan is consistent with the early, steady-state, thermal models developed by Lewis (1971) for icy satellites. He suggested a surface composed of a mixture of H_2O ice and solid hydrates of CH_4 and NH_3. Over the age of the solar system, photolysis of gaseous NH_3 in equilibrium with these hydrates produced enough N_2 to accumulate to the present observed abundance. Alternatively, the Saturnian nebula temperature

Table II

SATELLITE DATA

Planet	Satellite	Distance from planet (10^3 km)	Sidereal period (days)	Radius (km)	Density (g cm^{-3})	Surface gravity (cm sec^{-2})
Earth	Moon	384	27.3	1738	3.34	162
Jupiter	Io	422	1.77	1815	3.55	181
	Europa	671	3.55	1569	3.04	132
	Ganymede	1070	7.15	2631	1.93	144
	Callisto	1883	16.7	2400	1.83	125
Saturn	Titan	1222	16	2575	1.89	136
Neptune	Triton	355	6	~2500	~2.1	~150

Fig. 3. Layers of haze covering Titan as seen by *Voyager 1* at 22,000 km. The upper level of the thick aerosol appears brown-orange in color, and the divisions in the haze occur at 200, 375, and 500 km altitude above the limb.

at Titan's orbital distance was below 60 K and N_2 condensed preferentially as $N_2 \cdot 7H_2O$. Subsequently, Titan's surface warmed up sufficiently to evaporate N_2 as the dominant component of the atmosphere.

Tidal heating of Io's interior has generated repeated volcanic outgassing of volatiles (Fig. 4) with subsequent loss to the Jovian magnetosphere by processes not totally understood. Over the age of Io the chemical evolution of its volatile composition has been extensive. Almost all H_2O, N_2, and CO_2 have escaped, with the remaining volatile inventory now dominated primarily

Fig. 4. *Voyager 1* photo of volcanic eruption on Io taken at a distance of 500,000 km. The plume structure rises more than 100 km above the surface.

by sulfur compounds. Present-day volcanic activity probably maintains SO_2 as the dominant component of Io's atmosphere. Given its sporadic nature, the atmosphere could be transient.

The only other satellite with a possible atmosphere is Triton, the largest satellite of Neptune. The current upper limit on the CH_4 atmospheric abundance is 2.7×10^{21} cm^{-2}, but CH_4 has been detected on the surface. Also, an IR absorption feature at 2.15 μm is suggestive of liquid or solid nitrogen on the surface and would imply that some N_2 is in the atmosphere. There is considerable uncertainty on Triton's radius and hence density (Table II).

Wildt (1937) first discussed the photochemistry of planetary atmospheres in the outer solar system. His major concern was the maintenance of substantial amounts of CH_4 and NH_3 in their H_2-dominated, cold atmospheres despite fast dissociation by solar ultraviolet (UV) radiation. The generally accepted explanation for the stability of CH_4 and NH_3 was proposed by Strobel (1969), who concluded that photochemical products would be transported downward to the hot ($T > 1000$ K), dense ($P > 200$ bars) interior where thermal decomposition and subsequent reaction with H_2 would regenerate CH_4 and NH_3. Upward transport would replenish these species.

II. Observations

The composition and vertical temperature structure of a planetary atmosphere are fundamental quantities that are needed to acquire an understanding of photochemistry. These quantities are acquired from ground-based and spacecraft observations. The former were the only source of data until the early 1970s; continuous improvements in instrumentation have kept ground-based planetary spectroscopy competitive with spacecraft observations. It is beyond the scope of this chapter to provide a historical review of ground-based data, but the interested reader is referred to the Proceedings of the Third Arizona Conference on Planetary Atmospheres on the Atmospheres of the Jovian Planets (*J. Atmos. Sci.* **26,** 795–1001, 1969) and references therein.

A. *Pioneer 10* and *11*

Pioneer 10 flew by Jupiter at a distance of ~130,000 km on 4 December 1973 to begin a new era in discovery and exploration of the outer planets. The *Pioneer* mission had three objectives: (1) to investigate the interplanetary medium, (2) to study the asteroid belt, and (3) to make the first *in situ* measurements of the near environment of Jupiter. In essence the mission was more of a pioneering demonstration that spacecraft could be successfully sent to the outer solar system than a scientific investigation of the outer solar system. As a consequence of these objectives and its instrumentation, *Pioneer* scientific results on photochemistry were very limited. The principal results at Jupiter were the determination of the hydrogen to helium ratio, the detection of an ionosphere and hence atmosphere on Io, the detection of He 584-Å resonantly scattered radiation, and density measurements of the Jovian atmosphere and ionosphere.

Only *Pioneer 11* explored the Saturn system, with a close encounter on 1 September 1979. Atmospheric and ionospheric density measurements were obtained on Saturn, and a hydrogen torus was detected around Titan. Perhaps the major contribution of *Pioneer 11* spacecraft was a better overall understanding of Saturn's rings.

B. *Voyager 1* and *2*

The *Voyager 1* and *2* exploration of Jupiter and Saturn initiated a new era in outer solar system studies. *Voyager 1* encountered Jupiter on 5 March 1979, *Voyager 2* on 9 July 1979. The most significant discovery during the first encounter was volcanic activity on Io (see Fig. 4). Associated with Io's volcanic activity is a plasma torus with electron temperature ~ 6 eV, electron density $\sim 2 \times 10^3$ cm^{-3}, and the ions S^+, O^+, S^{2+}, S^{3+} preferentially confined to radial distances $(5\text{–}7)R_J$ with vertical thickness $\sim 2R_J$ ($1R_J$ = radius of Jupiter). These ions radiate extreme-UV radiation in excess of 10^{12} W. A ring of material was discovered around Jupiter; thus Neptune is the only giant planet without a confirmed ring system. Information on the composition and structure of Jupiter's atmosphere derived from *Voyager 1* and *2* measurements will be discussed in Section III.

Voyager 1 and *2* continued their tour of the outer solar system; the close encounters with Saturn occurred on 12 November 1980 and 26 August 1981, respectively. In terms of photochemistry, the major contribution of the *Voyager* mission was the first definitive knowledge of Titan's atmosphere (Fig. 3): the detection of a massive N_2 atmosphere (~ 1.5 bars) with a suite of organic molecules perhaps characteristic of the Earth's prebiotic atmosphere (Chapter 1). In addition there were major discoveries in Saturn's rings: spokes, kinks, density waves, eccentric ringlets, etc.

III. Composition and Thermal Structure

A. Thermal Structure

In Fig. 5 representative vertical temperature profiles as a function of pressure are given for the giant planets and Titan in regions where photochemical processes are known to be important. The profiles for Jupiter,

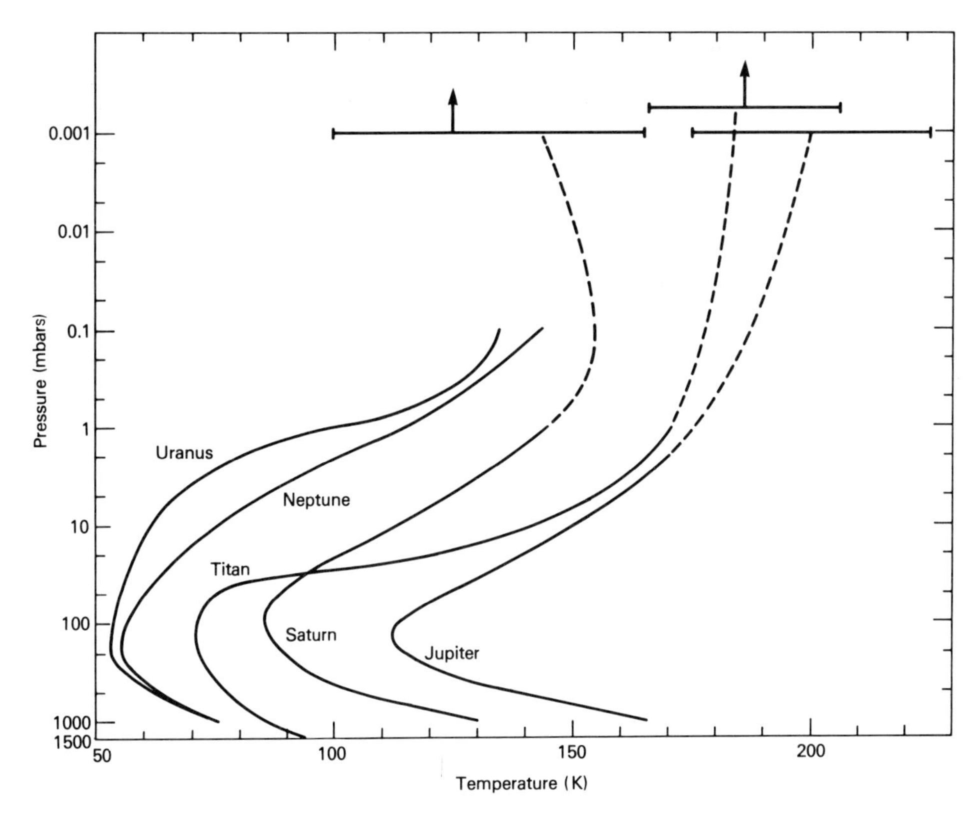

Fig. 5. Representative vertical temperature profiles: for Jupiter, Saturn, and Titan, based on *Voyager* data; for Uranus and Neptune, constructed from ground-based data and theoretical models. The number density (cm^{-3}) is obtained by $7.3 \times 10^{18}[P\ (\text{mbars})/T\ (\text{K})]$.

Saturn, and Titan are based on *Voyager* data; ground-based data and theoretical models were used to generate the Uranus and Neptune profiles. On the giant planets the temperature profile in the highest pressure region is convectively controlled, with an adiabatic lapse rate up to approximately the 800- to 900-mbar level, where the transition to radiative control occurs. The temperature minima at the tropopause for Jupiter, Saturn, Uranus, and Neptune are approximately 110, 85, 53, and 55 K, respectively. The tropopause on each planet will serve as an effective cold trap for NH_3 and a variety of photochemical products. On Jupiter the temperature rises above the 3×10^{-4}–mbar level to $\sim$1200 K at ionospheric heights. On Saturn the low-pressure temperature ($\sim 125^{+40}_{-25}$ K) with the arrow in Fig. 5 pertains to the 10^{-5}-mbar level, with the temperature eventually increasing to $\sim$420 K at ionospheric heights. Although Uranus is closer to the Sun than Neptune, the temperature inversion in the stratosphere is stronger on Neptune.

The surface pressure and temperature on Titan are 1.5 bars and 94 K. The tropopause on Titan is located at 130 mbars, where the temperature (71.4 K) is sufficient to condense CH_4 in the upper troposphere. Most photochemical products of N_2 and CH_4 are expected to precipitate out in the vicinity of this cold trap. The low-pressure temperature on Titan given in Fig. 5 with an arrow is applicable to the 7×10^{-9}–mbar level.

The other satellites with atmospheres, Io and Triton, have unknown thermal structures. In the case of Io, *Voyager* IR measurements suggest an average dayside surface temperature of ~130 K. Plasma temperatures inferred from the *Pioneer 10* radio occultation measurements of Io's ionosphere indicate a neutral exospheric temperature as high as ~1000 K. For Triton our knowledge is limited to a surface temperature of ~62 K, which is sufficient to freeze CH_4.

B. Composition

The composition of Jupiter's atmosphere given in Table III is almost entirely based on *Voyager* data. The hydrogen to helium ratio is in agreement

Table III

COMPOSITION OF JUPITER'S ATMOSPHERE

Constituent	Volume mixing ratio[a]
H_2	0.89
He	0.11
CH_4	0.00175
C_2H_2	0.02 ppm
C_2H_4[b]	7 ppb
C_2H_6	5 ppm
CH_3C_2H[b]	2.5 ppb
C_6H_6[b]	2 ppb
CH_3D	0.35 ppm
NH_3[c]	180 ppm
PH_3	0.6 ppm
H_2O[c]	1–30 ppm
GeH_4	0.7 ppb
CO	1–10 ppb
HCN	2 ppb

[a] ppm ≡ Parts per million; ppb ≡ parts per billion.
[b] Tentative identification, polar region.
[c] Value at 1 to 4 bars.

Table IV
COMPOSITION OF SATURN'S ATMOSPHERE

Constituent	Volume mixing ratio
H_2	0.94
He	0.06
CH_4	0.0045
C_2H_2	0.11 ppm
C_2H_6	4.8 ppm
CH_3C_2H[a]	No estimate
C_3H_8[a]	No estimate
CH_3D	0.23 ppm
PH_3	2 ppm

[a] Tentative identification.

with solar abundances. The CH_4/H_2 and NH_3/H_2 ratios, however, are enhanced a factor of 2 over solar ratios. In contrast, the predominant form of oxygen on Jupiter is H_2O, whose mixing ratio indicates a strong depletion of oxygen on Jupiter relative to the Sun (a factor of about 50 to 150). Similarly, the GeH_4 mixing ratio is also below the expected solar value by a factor of ~ 10. The C/H, N/H, Ge/H, and O/H ratios on Jupiter would suggest formation of Jupiter from accretion of planetesimals in the primordial solar nebula rather than condensation from a homogeneous nebula.

Although the abundance of constituents is given in terms of the mixing ratio, it should be noted that in most instances this is a quantity inferred from a spectroscopic measurement of column density, which is sensitive to temperature. For example, with a temperature profile 10 K warmer, the inferred C_2H_6 mixing ratio in Table III would decrease to ~ 1 ppm.

In Jupiter's stratosphere, hydrocarbons are seen in thermal IR emission, in particular C_2H_2 and C_2H_6. In the polar regions C_2H_4, CH_3C_2H, and C_6H_6 have been tentatively identified from *Voyager* data. Benzene (C_6H_6) is especially unexpected for reasons to be discussed below. The other species in Table III will be discussed in Section V.

The composition of Saturn's atmosphere is presented in Table IV. We note that the He/H_2 ratio is one-half of the Jupiter and solar values, which may indicate removal of helium from the atmosphere by condensation in the high-pressure interior and/or gravitational separation. The C/H ratio is enhanced on Saturn, as it is on Jupiter, relative to the Sun. Because NH_3 sublimates in the upper troposphere, an accurate N/H ratio cannot be inferred on Saturn. In Saturn's stratosphere the mixing ratios of C_2H_6 and C_2H_2 are comparable to Jovian values, and CH_3C_2H and C_3H_8 have been tentatively identified.

Table V

COMPOSITION OF TITAN'S ATMOSPHERE

Constituent	Volume mixing ratio		
N_2		$0.76–0.98^a$	
	Surface	Stratosphere	Thermosphere (3900 km)
CH_4	0.02–0.08	$\leqslant 0.026$	0.08 ± 0.03
Ar	<0.16		<0.06
Ne	<0.002		<0.01
CO	60 ppm		<0.05
H_2	0.002 ± 0.001		
C_2H_6		20 ppm	
C_3H_8		1–5 ppm	
C_2H_2		3 ppm	~0.0015 (3400 km)
C_2H_4		0.4 ppm	
HCN		0.2 ppm	<0.0005 (3500 km)
C_2N_2		0.01–0.1 ppm	
HC_3N		0.01–0.1 ppm	
C_4H_2		0.01–0.1 ppm	
CH_3C_2H		0.03 ppm	
CO_2		1–5 ppb	

[a] Preferred value.

The atmospheric compositions of Uranus and Neptune are considerably less certain than those Jupiter and Saturn. The consensus is that CH_4/H_2 is enhanced over the solar value in the deep atmosphere, but there is not unanimity on its value. There is also some uncertainty about the stratospheric value of the CH_4 mixing ratio, particularly for Neptune, where the CH_4 mixing ratio appears to be at the saturation value of the local temperature rather than the cold trap, tropopause temperature (Fig. 5). The latter would suggest strong convective overshooting in the tropopause region. The N/H ratios deep in the atmospheres of Uranus and Neptune are less than the solar value. Accurate values for the He/H_2 ratio are not available. Ethane (C_2H_6) has been detected only on Neptune; its mixing ratio is $\sim 10^{-6}$.

The combined *Voyager* IR, UV, and radio occultation data yield the bulk of the composition entries in Table V for Titan's atmosphere. The dominant constituent is N_2, with a preferred mixing ratio of ~ 0.98. The preferred interpretation of the radio occultation data does not require any ^{36}Ar (the only likely constituent heavier than N_2 that does not freeze out at 94 K) to obtain the observed temperature to mean molecular mass ratio. Titan has many important organic molecules seen in thermal emission, which is discussed in Section VI. In addition, the detections of CO_2 at the part per billion level and CO at the 60-ppm level suggest meteoritic infall of water-bearing material or initial outgassing of CO from a hydrate in the crust.

The only constituent observed in Io's atmosphere by *Voyager* was SO_2. Over a region where the surface temperature was $\gtrsim 130$ K, the surface density and pressure were 5×10^{12} cm^{-3} and 1×10^{-7} bar, respectively, which is consistent with the vapor pressure of SO_2 at 130 K. *Voyager* IR observations suggest that the SO_2 atmosphere is in equilibrium with the surface temperature; for example, at night when the surface cools down the SO_2 pressure is substantially less. No other constituents have been detected.

The other Galilean satellites have no detectable atmospheres. In the case of Ganymede, a UV stellar occultation by *Voyager 1* yielded an upper limit density and pressure of $\sim 10^9$ cm^{-3} and 10^{-11} bar, respectively, for any constituent with a 10^{-17} cm^2 absorption cross section in the 910- to 1700-Å region (e.g., O_2, H_2O, CO_2, CH_4). Thus only an exosphere could exist on Ganymede. For Europa and Callisto no equivalent measurements are available; it is probable that their atmospheres are only exospheres, as Ganymede, with H_2O and its photochemical products O, H, O_2.

For Triton only upper limits on CH_4 are available; no atmosphere has been detected, although CH_4 and/or N_2 are expected.

IV. Some Basic Principles

Before discussing the photochemistry of these atmospheres, we review some basic principles introduced in Chapter 1. The globally averaged density profiles are solutions of the continuity equations

$$d\phi_i/dz = P_i - L_i n_i, \tag{1}$$

where ϕ_i is the vertical flux, P_i the production rate per unit volume, L_i the chemical loss rate, n_i the number density of constituent i, and z the altitude. As discussed in Chapter 1, the vertical flux for an isothermal atmosphere is written

$$\phi_i = -D_i(dn_i/dz + n_i/H_i) - K(dn_i/dz + n_i/H_a), \tag{2}$$

where T is temperature, $H_i = kT/m_i g$ the scale height of the ith constituent, $H_a = kT/m_a g$ is the scale of the background atmosphere, g the gravitational acceleration, m_a the mean molecular mass of the atmosphere, m_i the constituent's mass, K the eddy diffusion coefficient, and D_i the average molecular diffusion coefficient. The homopause is defined as the level where $D_i = K$.

For constituents that do not readily escape the gravitational field of the planet or satellite, $\phi_i = 0$ at the exobase, and in the absence of chemistry ($P_i = L_i = 0$), the density varies above the homopause as

$$n_i \propto \exp(-z/H_i), \tag{3}$$

which is very sensitive to the constituent's mass. Below the homopause the density behaves as

$$n_i \propto \exp(-z/H_a), \tag{4}$$

the same as the background atmospheric density $[n_a = n_0 \exp(-z/H_a)]$, and the mixing ratio, $f_i = n_i/n_a$, is a constant. Extending these arguments to a family of species whose net chemistry satisfies

$$\sum_i (P_i - L_i n_i) = 0, \tag{5}$$

we find

$$\sum_i \phi_i = \text{const} = 0 \tag{6}$$

when loss to the surface or interior and thermal escape are negligible. But $\phi_i = -Kn_a\, df_i/dz$, and Eq. (6) reduces to

$$\sum_i f_i = \text{const.} \tag{7}$$

Hence the mixing ratio of an element is conserved in all forms below the homopause.

Another useful solution to Eqs. (1) and (2) in the homosphere, where $D_i \ll K$, is when Eq. (5) holds with $\phi_i \neq 0$. Then

$$n_i = \frac{\phi_0}{K}\left(\frac{1}{H_a} - \frac{1}{H_K}\right)^{-1} + c\exp(-z/H_a), \tag{8}$$

where ϕ_0 is the downward flux, equal to the net column production rate high in the atmosphere, and $K \propto \exp(z/H_K)$, $H_K > H_a$. When $\phi_0 = 0$, we recover the previous solution $f_i = c = \text{const}$. When the integration constant c is zero, then the constituent is being transported downward at its maximum velocity $K(1/H_a - 1/H_K)$ and $n_i \propto 1/K$. It follows that n_i is a maximum where K is a minimum and illustrative of why pollutants accumulate in a region of slow mixing such as the Earth's stratosphere (see Chapter 3). Conversely, if an element in all forms is concentrated preferentially in a region where K is a minimum (e.g., the tropopause), it implies that $\phi_0 > 0$ and that an external source for the element exists.

V. Photochemistry of the Outer Planets

A. Hydrogen and Helium

Molecular hydrogen is the dominant constituent in the atmospheres of the outer planets. It has an ionization continuum below 804 Å and a dissociation continuum below 845 Å:

$$H_2 + h\nu \longrightarrow H + H(2p) \quad \lambda < 845 \text{ Å} \tag{9}$$

$$H_2 + h\nu \longrightarrow H_2^+ + e^- \quad \lambda < 804 \text{ Å} \tag{10}$$

$$\longrightarrow H^+ + H + e^- \tag{11}$$

In addition, it is possible to dissociate H_2 by fluorescence after discrete absorption in the Lyman and Werner bands. For example,

$$H_2(X\,^1\Sigma_g^+)_{(v=0)} + h\nu\,(\lambda < 1109 \text{ Å}) \longrightarrow H_2(B\,^1\Sigma_u^+)_{(v=v')} \longrightarrow H_2(X)_{(v=v'')} + h\nu \tag{12}$$

in the Lyman bands, and if $v'' > 14$, H_2 will dissociate in the vibrational continuum of the ground state. This process is particularly important in the interstellar medium as a mechanism for dissociating H_2 by starlight.

At least two hydrogen atoms are eventually produced as a result of H_2 ionization, since

$$H_2^+ + H_2 \longrightarrow H_3^+ + H \tag{13}$$

$$H_3^+ + e^- \longrightarrow H_2 + H \text{ or } 3\,H \tag{14}$$

Similarly, ionization of He leads to formation of hydrogen atoms by

$$He^+ + H_2 \longrightarrow H_2^+ + He \tag{15}$$

followed by reactions (13) and (14). At ionospheric pressures recombination of H is extremely improbable, and a net downward flux of H from the ionosphere to the lower atmosphere occurs on the outer planets. On Jupiter the magnitude of this flux is $\sim 10^9$ cm^{-2} sec^{-1} (Strobel, 1975; Yung and Strobel, 1980). By scaling the solar flux, we estimate this flux to be 3×10^8, 7×10^7, and 3×10^7 cm^{-2} sec^{-1} on Saturn, Uranus, and Neptune, respectively.

Electron and ion densities depend critically on whether the major ion is atomic or molecular because of the great disparity in recombination rates. Radiative recombination of H^+,

$$H^+ + e^- \longrightarrow H + h\nu \tag{16}$$

proceeds at a rate of $\sim 3.5 \times 10^{-12}\,(T_e/300)^{-0.7}$ cm^3 sec^{-1} (T_e, electron temperature), whereas H_3^+ dissociatively recombines [reaction (14)] at a rate of $2.3 \times 10^{-7}\,(T_e/300)^{-1/2}$. Since He^+ reacts rapidly with H_2 ($k_{15} \sim 1 \times$

10^{-13} cm^3 sec^{-1}), H^+ is the only important atomic ion. In photochemical equilibrium at high altitudes,

$$[H^+] = P(H^+)/k_{16}[e^-], \tag{17}$$

$$[H_3{}^+] = P(H_3{}^+)/k_{14}[e^-], \tag{18}$$

where brackets denote number density and $P(x)$ denotes production rate of the indicated ion. For equal production rates, $P(H^+) = P(H_3{}^+)$, at $T_e =$ 1000 K, the density ratio $[H^+]/[H_3{}^+]$ would be 8×10^4 and illustrates the importance of H^+ sources for the ionosphere. At high altitudes where, based on *Pioneer* and *Voyager* measurements, $[H^+] = [e^-] < 10^6$ cm^{-3}, the recombination lifetime exceeds 10^6 sec, and ionospheric plasma is susceptible to significant transport by neutral winds and electric fields.

In addition to radiative recombination, protons may react with vibrationally hot H_2:

$$H^+ + H_{2(v \gtrsim 4)} \longrightarrow H_2{}^+ + H \tag{19}$$

as the reaction is exothermic only if $v \geqslant 4$. At higher pressures H^+ may undergo recombination:

$$H^+ + H_2 + H_2 \longrightarrow H_3{}^+ + H_2 \tag{20}$$

In the vicinity of the homopause, hydrocarbon reactions become important:

$$\begin{aligned} H^+ + CH_4 &\longrightarrow CH_3{}^+ + H_2 \\ &\longrightarrow CH_4{}^+ + H \\ CH_4{}^+ + H_2 &\longrightarrow CH_5{}^+ + H \\ CH_3{}^+ + CH_4 &\longrightarrow C_2H_5{}^+ + H_2 \end{aligned} \tag{21}$$

The presence of C_2H_2, C_2H_4, and C_2H_6 results in additional reactions to form more complex $C_xH_y{}^+$ ions, which upon recombination lead to nonmethane hydrocarbons. This is a minor pathway for producing complex hydrocarbons in comparison to neutral photochemistry discussed in Section V,B.

In summary, H^+ is expected to be the dominant ion in the upper ionosphere when H_2 is not too vibrationally hot. Given the long recombination time constant, substantial transport by wind and electric field systems is expected. Our ignorance of these winds and electric fields makes theoretical predictions very uncertain. We expect $H_3{}^+$ and hydrocarbon ions to be dominant in the lower ionosphere and the electron density to be significantly smaller as a result of fast recombination.

In the polar regions of Jupiter energetic ions from the magnetosphere precipitate and lose most of their energy above the homopause. Formation of hydrocarbon ions is expected to be small. The energy loss of these precipitating ions is important in the thermal balance of Jupiter's thermosphere, the production of hydrogen atoms, and the production of global wind systems by auroral heating. Similar remarks apply to Saturn, but the auroral energy loss is two orders of magnitude less than that of Jupiter, where the power dissipation is $\gtrsim 2 \times 10^{13}$ W. In the vicinity of the tropopause on the outer planets galactic cosmic rays are the principal ionization sources at high magnetic latitudes. On Uranus and Neptune, cosmic rays may be the dominant ionization source. Recent UV observations of Uranus suggest of auroral emission and a magnetospheric energy source for its atmosphere.

B. Hydrocarbons

The photochemistry of CH_4 is driven by absorption of intense solar Lyman α radiation, because CH_4 absorbs only UV radiation below ~ 1450 Å. At Lyα the primary dissociation paths are

$$
\begin{aligned}
CH_4 + h\nu &\longrightarrow {}^1CH_2 + H_2 \\
&\longrightarrow {}^{1\text{ or }3}CH_2 + 2\,H \\
&\longrightarrow CH + H + H_2
\end{aligned}
\tag{22}
$$

where the first two have comparable probability and the latter a yield of ~ 0.08. $^3CH_2$ is the ground state form of CH_2 and is generally unreactive with stable molecules. The net CH_4 photochemistry in an H_2 atmosphere can be summarized by

$$
\begin{aligned}
2(CH_4 + h\nu &\longrightarrow {}^1CH_2 + H_2) \\
2({}^1CH_2 + H_2 &\longrightarrow CH_3 + H) \\
2\,CH_3 + M &\longrightarrow C_2H_6 + M \\
\hline
2\,CH_4 &\longrightarrow C_2H_6 + 2\,H
\end{aligned}
\tag{23}
$$

$$
\begin{aligned}
CH_4 + h\nu &\longrightarrow {}^1CH_2 + H_2(\text{or } 2\,H) \\
CH_4 + h\nu &\longrightarrow {}^3CH_2 + 2\,H \\
{}^1CH_2 + H_2 &\longrightarrow CH_3 + H \\
{}^3CH_2 + CH_3 &\longrightarrow C_2H_4 + H \\
\hline
2\,CH_4 &\longrightarrow C_2H_4 + 4\,H
\end{aligned}
\tag{24}
$$

$$
\begin{aligned}
CH_4 + h\nu &\longrightarrow CH + H + H_2 \\
CH + CH_4 &\longrightarrow C_2H_4 + H \\
\hline
2\,CH_4 &\longrightarrow C_2H_4 + H_2 + 2\,H
\end{aligned}
\tag{25}
$$

Rapid photolysis of C_2H_4 leads to C_2H_2 formation:

$$\begin{aligned} C_2H_4 + h\nu &\longrightarrow C_2H_2 + H_2 \\ &\longrightarrow C_2H_2 + 2\,H \end{aligned} \tag{26}$$

and C_2H_2 photolysis in the upper atmosphere yields a net dissociation of H_2:

$$\begin{aligned} C_2H_2 + h\nu &\longrightarrow C_2H + H \\ &\longrightarrow C_2 + H_2 \\ C_2H + H_2 &\longrightarrow C_2H_2 + H \\ C_2 + H_2 &\longrightarrow C_2H + H \\ C_2H + H_2 &\longrightarrow C_2H_2 + H \\ \hline 2\,H_2 &\longrightarrow 4\,H \end{aligned} \tag{27}$$

The net photolysis of C_2H_6,

$$\begin{aligned} C_2H_6 + h\nu &\longrightarrow C_2H_4 + H_2 \\ &\longrightarrow C_2H_4 + 2\,H \\ &\longrightarrow C_2H_2 + 2\,H_2 \\ &\longrightarrow CH_4 + {}^1CH_2 \end{aligned} \tag{28}$$

yields C_2H_2 and CH_4 after rapid photolysis of C_2H_4 [reactions (26)]. (A minor path of 2 CH_3 is omitted.) A comparison of the time constant of C_2H_6 photolysis with the time constant for vertical transport (Fig. 6), however, implies that the bulk of C_2H_6 will be transported down and accumulate in the stagnant lower stratosphere. A similar conclusion holds for C_2H_2 as a consequence of the "do nothing" chemical cycle in reactions (27).

Methane is recycled by the reactions

$$H + CH_3 + M \longrightarrow CH_4 + M \tag{29}$$

and

$$\begin{aligned} H + C_2H_4 + M &\longrightarrow C_2H_5 + M \\ H + C_2H_5 &\longrightarrow 2\,CH_3 \\ 2(CH_3 + H + M &\longrightarrow CH_4 + M) \\ \hline C_2H_4 + 4\,H &\longrightarrow 2\,CH_4 \end{aligned} \tag{30}$$

Hydrogen atoms may be catalytically recombined by

$$\begin{aligned} H + C_2H_2 + M &\longrightarrow C_2H_3 + M \\ H + C_2H_3 &\longrightarrow C_2H_2 + H_2 \\ \hline 2\,H &\longrightarrow H_2 \end{aligned} \tag{31}$$

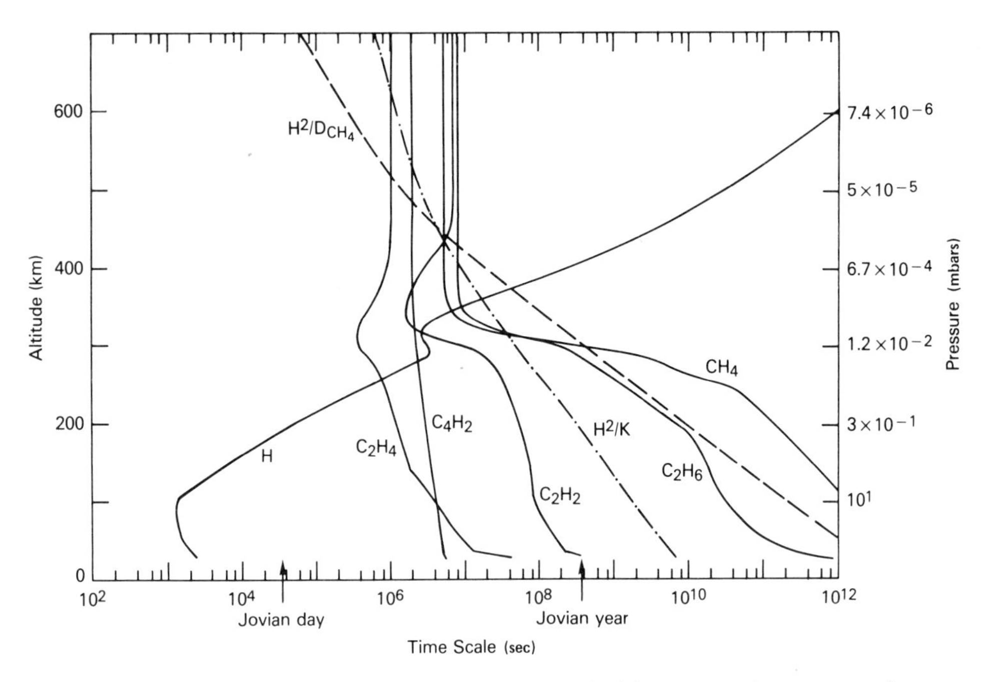

Fig. 6. Photochemical loss time constants compared with transport time constants for molecular diffusion (H^2/D_{CH_4}) and eddy diffusion (H^2/K) on Jupiter, where $K = 1.3 \times 10^6$ $(2.17 \times 10^{13}/[M])^{0.6}$ $cm^2\ sec^{-1}$. From Gladstone (1982).

Up-to-date calculations of pure hydrocarbon photochemistry in the atmosphere of Jupiter were given by Gladstone (1982). He found that ~20% of the photons absorbed by CH_4 produce C_2H_6 and ~10% yield C_2H_4 (and thus C_2H_2 by photolysis), with the remaining 70% initiating "do nothing" cycles that terminate by reactions (29) and (30). We can use expression (8) with $c = 0$ to estimate the total C_2H_6 and C_2H_2 density at the tropopause, where $K \sim 10^3\ cm^2\ sec^{-1}$, $H_K = 1.3\ H_a$, and $H_a \sim 20$ km (Yung and Strobel, 1980). The globally averaged Lyα column dissociation rate of CH_4 is $\sim 5 \times 10^9\ cm^{-2}\ sec^{-1}$. With an efficiency of 30%, this results in a ϕ_0 value of $\sim 7.5 \times 10^8\ cm^{-2}\ sec^{-1}$ for C_2 hydrocarbons. Thus

$$n_i \sim 6.5 \times 10^{12} \quad cm^{-3}$$

for a mixing ratio of $\sim 1 \times 10^{-6}$ at the tropopause (~100 mbars in Fig. 5), in reasonably good agreement with the measurements of Kostiuk *et al.* (1983) but not the *Voyager* IR results in Table III.

Detailed hydrocarbon mixing ratios calculated by Gladstone (1982) are shown in Fig. 7, where the downward flux of C_2H_6 into troposphere was severely constrained at the lower boundary. Good agreement with the

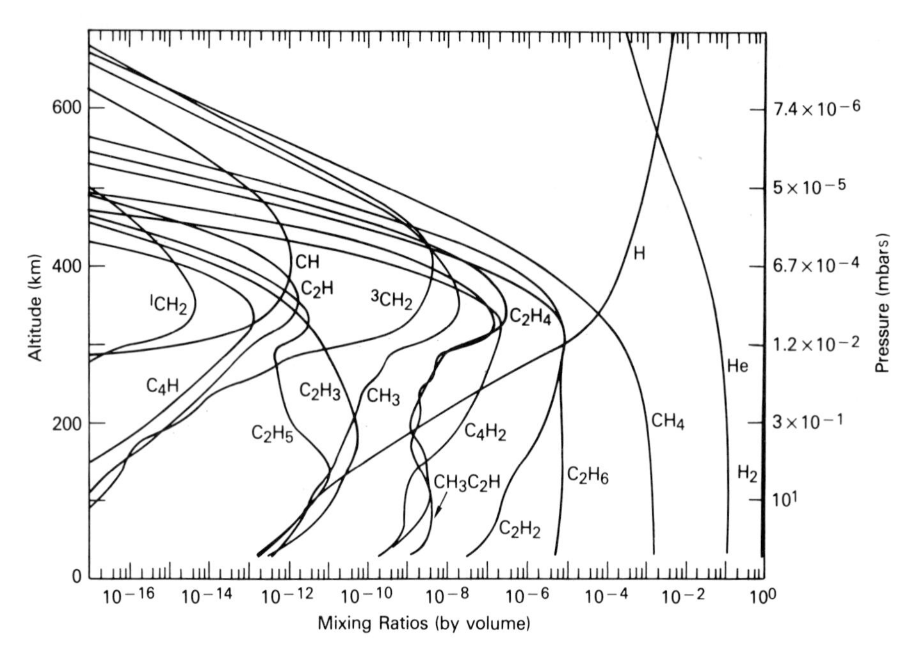

Fig. 7. Altitude profiles of the hydrocarbon mixing ratios computed by Gladstone for Jupiter. From Gladstone (1982).

available data base (thermal emission in the IR (the principal source of Table III), UV reflection spectra from *International Ultraviolet Explorer* (*IUE*), and *Voyager* ultraviolet spectrometer (UVS) occultation data) was obtained. Thermochemical conditions in the deep troposphere, however, would seem to require a large downward flux of C_2H_6 close to the maximum permitted flux. This would lower the C_2H_6 mixing ratio to $\sim 10^{-6}$ in agreement with Kostiuk *et al.* (1983), without appreciably altering the good agreement obtained for other species such as C_2H_2, with a mixing ratio of $\sim 2 \times 10^{-8}$ at the tropopause.

Gladstone's (1982) predicted abundance of C_2H_4 ($\lesssim 3 \times 10^{-9}$, Fig. 7) is consistent with the observational upper limit in the equatorial regions of Jupiter. The detection of C_2H_4, CH_3C_2H, and C_6H_6 in the polar regions of Jupiter strongly suggest an origin related to energetic electron and ion precipitation. But the *Voyager* measured energies of electrons and ions in the magnetosphere are not sufficiently high to allow precipitation and penetration to stratospheric depths. The absence of rapid photolysis of C_2H_4 may explain why it is detected only at high latitudes. It is extremely difficult to suggest a viable mechanism for C_6H_6. Formation by reaction of a metastable (triplet) state of C_2H_2 produced in C_2H_2 photolysis,

$$C_2H_2^{**} + 2C_2H_2 \longrightarrow C_6H_6 \tag{31a}$$

is extremely unlikely due to probable deactivation by collisions with H_2.

To first order, Gladstone's calculations can be scaled to Saturn, although the temperature there is lower and vertical mixing more vigorous. On Uranus and Neptune, where the $[CH_4]/[H_2]$ value may be different from Jupiter, the applicability of his quantitative results is uncertain. Qualitatively, we would expect C_2H_6 and C_2H_2 to be the major photochemical products.

C. Ammonia and Phosphine

We know from UV and IR data that NH_3 and PH_3 are absent in detectable abundances above the lower stratospheres on the outer planets. This is due, in part, to sublimation of NH_3 by the tropopause cold traps and to photochemical destruction of these constituents in the tropopause regions.

The column densities of CH_4, C_2H_6, and H_2 are sufficient to absorb most solar photons below 1650 Å before they reach the lower stratosphere. Photolysis of NH_3 is restricted to two paths when $\lambda > 1650$ Å:

$$\begin{aligned} NH_3 + h\nu &\longrightarrow NH_2(\tilde{X}\,^2B_1) + H \\ &\longrightarrow NH(a\,^1\Delta) + H_2 \end{aligned} \tag{32}$$

The latter path is a minor one, and the reaction

$$NH(a\,^1\Delta) + H_2 \longrightarrow NH_2 + H \text{ or } NH_3 \tag{33}$$

is probably very fast, hence NH_2 and H are the sole products of NH_3 photolysis. Because NH_2 reactions with stable molecules have high activation energies, NH_2 reacts preferentially with H and itself:

$$NH_2 + NH_2 + M \longrightarrow N_2H_4 + M \tag{34}$$

$$NH_2 + H + M \longrightarrow NH_3 + M \tag{35}$$

At the cold temperatures of the outer planets, the most probable fate of N_2H_4 is sublimation,

$$(N_2H_4)_g \longrightarrow (N_2H_4)_s$$

The few N_2H_4 molecules that remain in gas phase are susceptible to photolysis and hydrogen atom attack,

$$N_2H_4 + h\nu \longrightarrow N_2H_3 + H \tag{36}$$

$$N_2H_4 + H \longrightarrow N_2H_3 + H_2 \tag{37}$$

followed by

$$H + N_2H_3 \longrightarrow N_2H_2 + H_2 \tag{38a}$$

$$\longrightarrow 2\,NH_2 \tag{38b}$$

$$N_2H_3 + N_2H_3 \longrightarrow N_2H_4 + N_2H_2 \tag{39a}$$

$$\longrightarrow 2\,NH_3 + N_2 \tag{39b}$$

$$N_2H_2 \longrightarrow N_2 + H_2 \tag{40}$$

where paths (38a) and (39a) are dominant and yield N_2 as the ultimate NH_3 photolysis product. Note that reaction (40) represents spontaneous decomposition.

After sublimation the solid N_2H_4 particles descend in the troposphere, evaporate in a region where dissociating solar UV radiation has been attenuated by NH_3 absorption and H_2 Rayleigh scattering, and undergo no further photochemistry. The crucial step in the photochemistry of NH_3 is the fate of the NH_2 radical [reactions (34) and (35)], as once N_2H_4 is formed, NH_3 is lost from the tropopause. Of course, deep in the interior, thermal decomposition of N_2H_4 and reaction with H_2 will recycle NH_3, but at the tropopause, sublimation of N_2H_4 leads effectively to irreversible NH_3 photolysis. The continuity equation for NH_3 may be written as

$$d\phi_{NH_3}/dz = -\epsilon J[NH_3], \tag{41}$$

where ϵ is the fraction of dissociation events that are irreversible and J the NH_3 dissociation rate. An approximate solution for the NH_3 density profile is

$$[NH_3] \propto \exp(-z/H_m), \tag{42}$$

when $H_m \ll H_a$ and where $H_m = \sqrt{K/\epsilon J}$ is the "photomechanical" scale height. On Jupiter an analysis of the UV albedo implies $H_m \sim 3$ km, with $\epsilon \sim 0.25$, $K \sim 2 \times 10^4$ cm^2 sec^{-1}, and where $H_a \sim 15$ km. Accordingly, the NH_3 mixing ratio decreases rapidly with height (Fig. 8).

The above simple description of NH_3 photochemistry is complicated by the presence of PH_3. Although one would expect PH_3 and NH_3 photochemistries to be similar, Norrish and Oldershaw (1961) interpreted their laboratory work otherwise. They suggested that the reaction $PH_2 + PH_2$ yielded $PH + PH_3$, not P_2H_4, followed by $PH + PH \rightarrow P_2 + H_2$ and $P_2 + P_2 \rightarrow P_4$. Prinn and Lewis (1975) argued that P_4 (red phosphorus) was the end product of PH_3 photochemistry on Jupiter and could account for the red coloring material observed on Jupiter, in particular the Great Red Spot. (It should be pointed out that red phosphorus has an ill-defined structure.) Experimental work confirms that the photochemistry of PH_3 is similar to NH_3 (Ferris and Benson, 1981), however, with P_2H_4 as the product of PH_2 recombination.

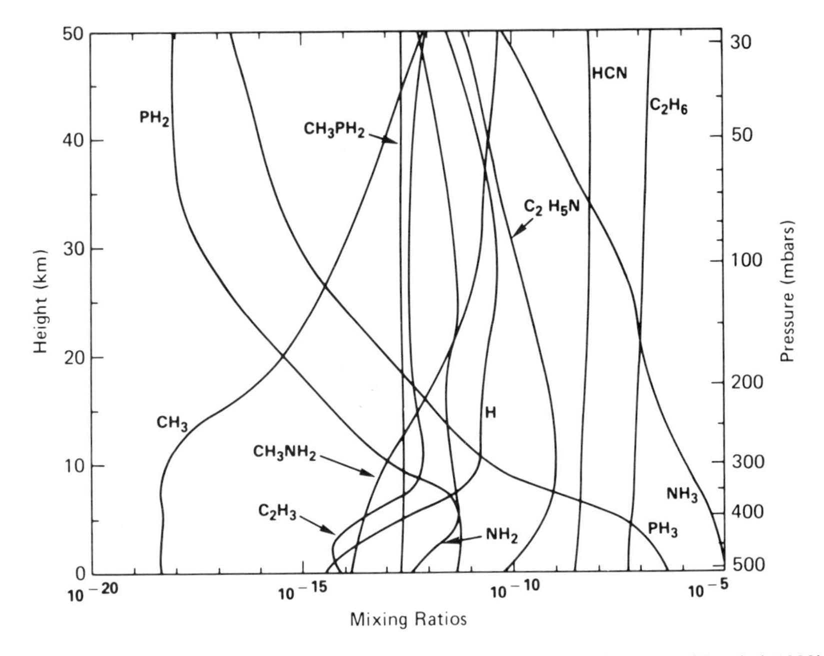

Fig. 8. Calculated mixing ratios from the photochemical model of Kaye and Strobel (1983b) for Jupiter.

On the basis of more recent calculations (Kaye and Strobel, 1983a,b,c) a first-order description of NH_3 and PH_3 photochemistry is given by

$$NH_3 + h\nu \longrightarrow NH_2 + H \tag{32}$$

$$PH_3 + h\nu \longrightarrow PH_2 + H \tag{43}$$

$$PH_2 + H \longrightarrow PH_3 \tag{44}$$

$$NH_2 + H \longrightarrow NH_3 \tag{35}$$

$$NH_2 + NH_2 \longrightarrow (N_2H_4)_s \tag{34}$$

$$PH_2 + PH_2 \longrightarrow (P_2H_4)_s \tag{45}$$

$$NH_2 + PH_2 \longrightarrow (H_2NPH_2)_s \tag{46}$$

$$H + PH_3 \longrightarrow PH_2 + H_2 \tag{47}$$

$$H + PH_2 \longrightarrow PH + H_2 \tag{48}$$

$$H + PH \longrightarrow P + H_2 \tag{49}$$

$$P + PH \longrightarrow P_2 + H \tag{50}$$

$$P_2 + P_2 \longrightarrow P_4 \tag{51}$$

$$P + H + M \longrightarrow PH + M \tag{52}$$

$$PH + H_2 + M \longrightarrow PH_3 + M \tag{53}$$

$$CH_4 + C_2H_2 + h\nu \longrightarrow CH_3 + H + C_2H_2 \tag{54}$$

$$CH_3 + NH_2 \longrightarrow CH_3NH_2 \tag{55}$$

$$CH_3 + PH_2 \longrightarrow CH_3PH_2 \tag{56}$$

$$CH_3 + H \longrightarrow CH_4 \tag{29}$$

Vapor pressure curves as a function of temperature are not available to determine how important reaction (45) is in removing P_2H_4 and ensuring irreversible photolysis of PH_3 on Jupiter. With a melting point of 174 K, P_2H_4 certainly sublimes on Saturn, Uranus, and Neptune. Gas-phase P_2H_4 is subject to photolysis and hydrogen atom reactions:

$$P_2H_4 + h\nu \longrightarrow P_2H_3 + H \tag{57}$$

$$H + P_2H_4 \longrightarrow P_2H_3 + H_2 \tag{58}$$

$$\begin{aligned} H + P_2H_3 &\longrightarrow P_2H_2 + H_2 \\ &\longrightarrow 2\,PH_2 \end{aligned} \tag{59}$$

$$\begin{aligned} P_2H_3 + P_2H_3 &\longrightarrow P_2H_2 + P_2H_4 \\ &\longrightarrow 2\,PH_3 + P_2 \end{aligned} \tag{60}$$

$$P_2H_2 \longrightarrow P_2 + H_2 \tag{61}$$

with P_2, P_x as the ultimate products. In reaction (61) P_2H_2 spontaneously decomposes.

The CH_3 source represented reaction by reactions (61a) is the C_2H_2-catalyzed photodissociation of CH_4 (Allen *et al.*, 1980):

$$\begin{aligned} C_2H_2 + h\nu &\longrightarrow C_2H + H \\ &\longrightarrow C_2 + H_2 \\ C_2H + CH_4 &\longrightarrow CH_3 + C_2H_2 \\ C_2 + CH_4 &\longrightarrow {}^1CH_2 + C_2H_2 \\ {}^1CH_2 + H_2 &\longrightarrow CH_3 + H \\ \hline CH_4 + h\nu &\longrightarrow CH_3 + H \end{aligned} \tag{61a}$$

Although reaction sequence (61a) is the most important source of CH_3 in the lower stratosphere, subsequent recombination of CH_3 is an insignificant C_2H_6 source in comparison to direct CH_4 photolysis.

On Jupiter the hydrogen atoms liberated from NH_3 photolysis [reactions (32)] accelerates the photochemical destruction of PH_3 by reaction (47) and produces the rapid PH_3 mixing ratio fall-off above 20 km shown in Fig. 8. In this figure the respective photochemical scale heights of NH_3 and PH_3 are approximately 3 and 0.5 km [cf. Eq. (42)] and are consistent with the absence of a significant UV signature in albedo data; UV photons sample only the stratosphere because H_2 Rayleigh scattering prevents deeper penetration. Below 25 km the NH_3 mixing ratio is controlled by the temperature profile, as NH_3 is saturated. The most abundant radical is atomic hydrogen, as a result of its slow recombination and potent photolytic sources.

On Saturn the roles of NH_3 and PH_3 are reversed, as PH_3 is more abundant than NH_3 in the photochemical region. At the lower boundary in Fig. 9 the respective mixing ratios of PH_3 and NH_3 are 2×10^{-6} and 6×10^{-9}. The concentration of PH_3 is rapidly decreased with altitude with a scale height of ~ 3.5 km; $[NH_3]$ is below the saturation level in much of Fig. 9 as a result of photochemistry. The combination of reactions (45) and (53) prevents the buildup of any significant abundance of P_2, the precursor of red phosphorus. In the case of reaction (53), a spin-forbidden reaction, even with a rate coefficient as small as 10^{-41} cm^6 sec^{-1}, the model does not predict spectroscopically detectable amounts of P_2 (or P_4). Potentially observable amounts of the organophosphorus compounds HCP and CH_3PH_2, $>5 \times$

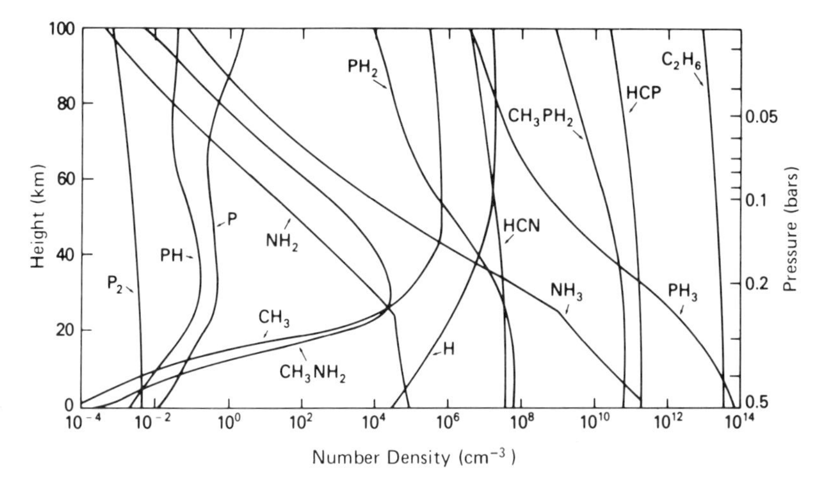

Fig. 9. Calculated mixing ratios from the photochemical model of Kaye and Strobel (1983c) for Saturn.

10^{17} cm^{-2}, however, were predicted by Kaye and Strobel (1983c). Considerable uncertainty must be attached to these results since the relevant heats of sublimation are not known.

Calculations for Uranus and Neptune, such as those shown in Figs. 8 and 9, cannot realistically be carried out without adequate vapor pressure data, as the tropopause temperatures dip to ~ 55 K on these planets. As noted before, even CH_4 sublimates at these temperatures.

A particularly interesting product of NH_3 photochemistry is HCN, a precursor of α-amino acids and nucleic acid bases. The role of HCN in the photochemistry of reduced atmospheres and in the chemical evolution of the Earth's prebiotic atmosphere has been of substantial interest (Chapter 1). On Jupiter, Tokunaga *et al.* (1981) observed by IR absorption an HCN column density of $\sim 1.3 \times 10^{17}$ cm^{-2} or mixing ratio of $\sim 2 \times 10^{-9}$. No thermal IR emission was detected; hence HCN must be confined to the vicinity of the tropopause or below. Only an upper limit of 6.5×10^{17} cm^{-2} was obtained for HCN on Saturn. This result is not surprising; the most probable local source of nitrogen atoms is NH_3, which has a low abundance (<6 ppb).

Although it is fashionable to invoke lightning discharges and subsequent equilibrium chemistry at high temperatures in the shocked atmosphere as a principal source of HCN, the absolute upper limit mixing ratio from this source is $\sim 3 \times 10^{-12}$, and with probable energy efficiencies the actual HCN mixing ratio is $\sim 10^{-16}$. Also, thermochemically formed HCN cannot be convected out of the interior with a mixing ratio of $\sim 10^{-9}$.

One probable HCN source is the UV photolysis of C_2H_5N, a plausible high-pressure product of the recombination of NH_2 and C_2H_3, which originate from NH_3 photolysis [reaction (32)] and addition of hydrogen atoms to acetylene [reactions (31)], respectively,

$$NH_2 + C_2H_3 + M \longrightarrow C_2H_5N + M \tag{62}$$

$$C_2H_5N + h\nu \longrightarrow HCN + CH_3 + H \tag{63}$$

Dissociation path (63) applies to the isomers cyclic aziridine and vinylamine ($H_2C{=}CH{-}NH_2$). A two-step process,

$$C_2H_5N + h\nu \longrightarrow H_2CN + CH_3 \tag{64}$$

$$H + H_2CN \longrightarrow H_2 + HCN \tag{65}$$

is more likely for the isomers ethylidenimine ($H_3C{-}CH{=}NH$) and *N*-methylmethyleneimine ($H_2C{=}N{-}CH_3$). With plausible yields of 0.1 for C_2H_5N in reaction (62) and 0.33 for either reaction (63) or (64), an HCN column density of $\sim 2 \times 10^{17}$ cm^{-2} is obtained (Fig. 8), which is in reasonably good agreement with observations.

An additional source of HCN in the vicinity of Jupiter's tropopause involves hot hydrogen atom (H*) chemistry,

$$
\begin{aligned}
NH_3 + h\nu &\longrightarrow NH_2 + H^* \\
H^* + CH_4 &\longrightarrow H_2 + CH_3 \\
CH_3 + NH_2 &\longrightarrow CH_3NH_2 \\
CH_3NH_2 + h\nu &\longrightarrow CH_3NH^{**} + H \\
&\longrightarrow H_2C{=}NH^{**} + 2\,H \\
&\longrightarrow HCN + 2\,H + H_2 \\
CH_3NH^{**} &\longrightarrow H_2C{=}NH + H \\
H_2C{=}NH^{**} &\longrightarrow HCN + 2\,H \\
CH_3NH + H &\longrightarrow CH_3NH_2
\end{aligned}
\tag{66}
$$

with a predicted HCN column density of $<4 \times 10^{16}$ cm^{-2}, somewhat less than the C_2H_5N source (Kaye and Strobel, 1983b). The double asterisks in reactions (66) indicate sufficient internal energy to undergo further decomposition. In both cases slow mixing ($K \sim 10^4$ cm^2 sec^{-1}) above the NH_3 cloud tops is required to accumulate HCN preferentially in the tropopause region (Fig. 8). This condition is also required to explain the ortho and para hydrogen equilibration in the same region. Below the NH_3 cloud tops more rapid mixing in the convection zone will cause a sharp decrease in the HCN mixing ratio, whereas HCN photolysis,

$$HCN + h\nu \longrightarrow H + CN \tag{67}$$

followed by

$$CN + C_2H_2 \longrightarrow HC_3N + H \tag{68}$$

will deplete HCN in the upper stratosphere.

D. Carbon Monoxide

Most oxygen on the outer planets should be in the form of H_2O, which has been detected only on Jupiter. The discovery of CO on Jupiter with a column density of $\sim 6 \times 10^{17}$ cm^{-2} was unexpected. One possible source is rapid outward convection of CO from the hot interior with a mixing ratio of $\sim 10^{-9}$ in the troposphere. The observational inference of a low CO rotational temperature suggested that CO accumulates in the Jovian tropopause region and must have an extraplanetary source as discussed in Section IV. To produce CO, two sources for oxygen have been proposed:

1. Ablation of meteoroidal material that contains substantial amounts of H_2O (Prather *et al.*, 1978)
2. Infall of material from the Galilean satellites of Jupiter, principally Io and Europa, via the Jovian magnetosphere (Strobel and Yung, 1979)

The flux of H_2O associated with meteoroidal material has been estimated at $(1–200) \times 10^7$ cm^{-2} sec^{-1}, whereas the oxygen flux from the Galilean satellites into Jupiter's atmosphere on the basis of *Voyager* data is about $(0.3–5) \times 10^7$ cm^{-2} sec^{-1}. From Eq. (8) with $c = 0$, $H_K = 1.3H_a$ as in Section V,B,

$$n_i = (1.3H_a/0.3K_0)\phi_0 \exp(-z/1.3H_a), \tag{69}$$

where K_0 is the minimun value of K, which occurs just above the ammonia cloud tops and where z is defined to be zero. The column density, whose principal contribution is above the height where K is a minimum, is

$$N \simeq \int_0^\infty n_i\,dz = \frac{5.6H_a^{\,2}}{K_0}\phi_0. \tag{70}$$

Since the observed N approximates 6×10^{17} cm^{-2} and H_a approximates 2×10^6 cm, then

$$\phi_0/K_0 \sim 2.7 \times 10^4 \quad \text{cm}^{-4}.$$

For constituents with slow mixing (e.g., C_2H_6) the relevant K_0 approximates 10^3 cm^2 sec^{-1} in contrast to NH_3 with rapid chemistry, where $K_0 \sim 10^4$ cm^2 sec^{-1} is inferred. Thus what is required is

$$\phi_0 \sim 2.7 \times 10^7 \quad \text{cm}^{-2}\,\text{sec}^{-1}, \tag{71}$$

which is consistent with both extraplanetary sources.

There are observational tests to distinguish these two sources. Io's volcanoes, atmosphere, and torus contain substantial quantities of sulfur in addition to oxygen. By analogy with CO and the chemical similarity of sulfur and oxygen, the infall of sulfur should lead to observable amounts of CS. Meteoroidal material contains sufficient amounts of silicon to produce observable amounts of SiO in Jupiter's atmosphere. Neither species has been detected.

Ultimately, oxygen in any form (e.g., O^+, O, H_2O) will be converted to CO. The details remain uncertain, but in Fig. 10 the essential reactions are schematically presented. Note that $O^+ + H_2$ represents a sequence of reactions:

$$\begin{aligned} O^+ + H_2 &\longrightarrow OH^+ + H \\ OH^+ + H_2 &\longrightarrow H_2O^+ + H \\ H_2O^+ + CH_4 &\longrightarrow H_3O^+ + CH_3 \\ H_3O^+ + e^- &\longrightarrow H_2O + H \text{ or } OH + H_2 \end{aligned} \tag{72}$$

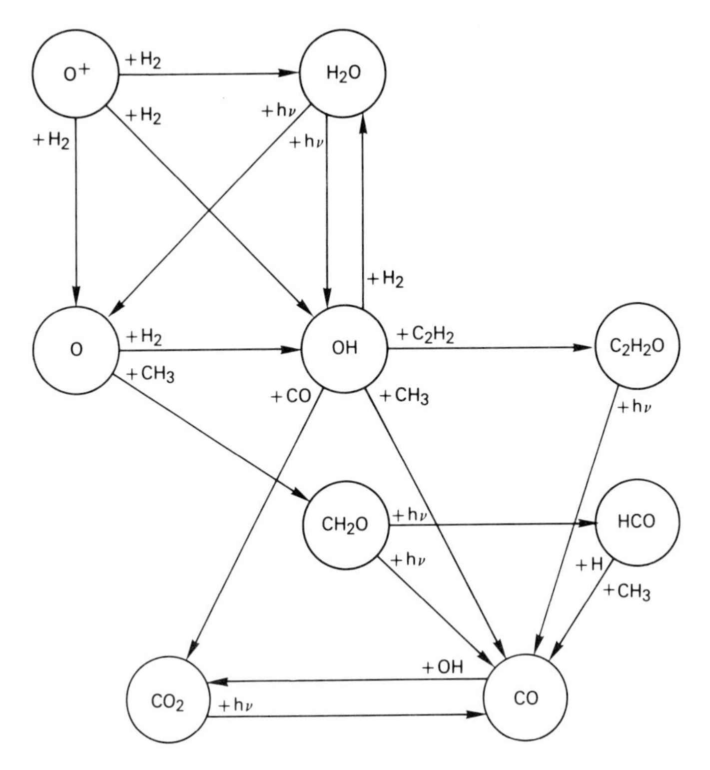

Fig. 10. Schematic diagram of oxygen photochemistry on Jupiter. The oxygen sources are a flux of water vapor in the form of meteoroidal material and energetic oxygen ions from the magnetosphere. The sinks are sublimation of water vapor and a downward flux carbon monoxide to the troposphere. From Strobel (1983).

In the Jovian stratosphere CO_2 and H_2O undergo rapid photolysis, whereas CO cannot in the presence of solar radiation longward of 1650 Å. CO is not susceptible to chemical attack by OH in the lower stratosphere because sublimation of H_2O removes the hydroxyl source. In the thermal inversion region OH must react only once with C_2H_2 (or C_2H_4) in a cycle of repeated H_2O dissociation, as oxygen is mixed down to the tropopause to form CO irreversibly. The relevant time constants (Chapter 1) are 10^7 sec for H_2O photolysis and 10^9 sec for vertical transport.

On the other outer planets ablation of meteoroidal material is expected to be important. In addition, on Saturn the rings, which are mostly water ice, must be included as an oxygen source. If the mass of Saturn's rings has not decreased significantly over the age of the Saturn system, then the infall of H_2O must be $< 5 \times 10^6$ cm^{-2} sec^{-1}, at least a factor of 5 less than required on Jupiter to account for CO.

E. Lightning

The production of organic matter in laboratory experiments by electrical discharges has stimulated debate on the importance of Jupiter lightning as a source of organic matter. Lightning and thunderstorms occur in convective regions of the atmosphere which on Jupiter would be below the ammonia cloud tops and most plausibly concentrated at the 6-bar, 300-K level of the atmosphere. The production of organic matter in the high-temperature, shocked atmosphere is inefficient because the ratio of energy dissipated by electrical storms to convective energy flux is only $\sim 4 \times 10^{-5}$ (Borucki *et al.*, 1982). In addition, the large abundance of H_2 ensures that hydrogen atoms are the most abundant radicals in the shocked gas. Radicals such as CH_3, NH_2, OH, SH, etc. are likely to recombine with H rather than with each other. As a result of these inefficiencies, lightning discharges can produce and maintain carbon species at a maximum mixing ratio of $\sim 10^{-14}$ in the water clouds, of which 98% is CO (Lewis, 1980). The next most abundant carbon compounds are HCN and C_2H_2, with respective mixing ratios of $\sim 10^{-16}$ and $\sim 4 \times 10^{-17}$.

These compounds are produced in the convective regions of the upper troposphere where vertical mixing is large ($K_t \sim 10^7$–10^9 cm^2 sec^{-1}). They cannot be transported up to the tropopause region where K_0 approximates 10^3–10^4 cm^2 sec^{-1} and increase their mixing ratios. This can be demonstrated from Eq. (8) where, in terms of mixing ratio, the tropospheric solution, f_t, with constant K_t is

$$f_t = c_t - (|\phi_0| H_a / K_t n_0) \exp(z/H_a), \tag{73}$$

where the upward flux to the stratosphere is $|\phi_0|$, and the background neutral density n_a is $n_0 \exp(-z/H_a)$, with $z = 0$ at the interface of the discontinuity in vertical mixing above the NH_3 cloud tops. Note that an upward flux causes the mixing ratio to decrease with increasing height as $|\phi_0| = -K n_a (df/dz)$ must be a constant and positive. At the interface the flux must be continuous, thus

$$|\phi_0| = -K_t n_a (df_t/dz) = -K_0 n_a (df_s/dz), \tag{74}$$

where K_0 is the minimum value of K and f_s the mixing ratio above the interface. Clearly

$$\frac{df_s}{dz} = \frac{K_t}{K_0} \frac{df_t}{dz} \sim 10^4 \frac{df_t}{dz} < 0, \tag{75}$$

and $f_t > f_s$. Thus the lightning-produced mixing ratios of 10^{-16} and 4×10^{-17} for HCN and C_2H_2, respectively, cannot account for the tropopause values of 2×10^{-9} and 2×10^{-8} (Table III).

F. Aerosol and Haze Layers

The stratospheres of Jupiter, Saturn, and possibly, Uranus and Neptune along with Titan share in common aerosol or haze layers, which are referred to as Axel or Danielson dust. These aerosols are fine particles that absorb UV and visible light. Because of their small size they are poor emitters and must heat up and collisionally transfer their energy to atmospheric molecules. The origin of these particles is widely held to be photochemical products of CH_4. In the case of Titan, as discussed in Section VI,B, heavy hydrocarbons readily sublimate in the lower stratosphere and are the most likely source of haze seen in *Voyager* images (Fig. 3). Complex molecules with conjugated bonds (e.g., polyacetylenes) and nitriles have the requisite absorption properties on Titan.

On Jupiter and Saturn the large H_2 abundance and substantial concentrations of hydrogen atoms make hydrogen atom cracking reactions efficient and prevent significant production of complex hydrocarbons. Simple hydrocarbons such as C_2H_6 and C_2H_2 do not accumulate in sufficient concentrations to sublimate. Furthermore, their optical properties are inconsistent with observations. NH_3 and PH_3 photochemistry produces condensates such as N_2H_4, P_2H_4, and H_2NPH_2. Liquid N_2H_4, which absorbs out to ~3500 Å, could partially account for the observed optical properties. Inorganic chemistry on the outer planets is driven by a larger number of solar photons than is organic chemistry and may make a greater contribution to aerosol formation.

VI. Photochemistry of Titan

Titan's unique atmosphere is particularly conducive to chemical evolution. With an extended atmosphere (exobase at 0.6 of a Titan radius from the surface), weak gravitational acceleration ($g \sim 55$ cm sec^{-2} at the exobase), and warm exosphere ($T > 186$ K), hydrogen in atomic and molecular form escapes efficiently by thermal processses at the exobase. This ensures that CH_4 photolysis produces a suite of heavy hydrocarbons whose ultimate fate is to accumulate on the "surface" and produce an ocean composed mostly of C_2H_6 with dissolved CH_4, N_2, and other trace species. The extensive atomic hydrogen torus mapped by the *Voyager* UV spectrometer around Saturn between 8 and 25 R_S (R_S = Saturn's radius) is the signature of hydrogen escape. In contrast to the outer planets and as a result of escape, Titan's atmosphere has only trace amounts of H_2.

A. Nitrogen

The N_2 bond is extremely difficult to break. Only energetic magnetospheric electrons interacting with Titan's thermosphere and exosphere, solar radiation below 1000 Å, and secondary electrons generated by cosmic rays in the 5- to 50-mbar region are capable of dissociating appreciable amounts of N_2. Most important is precipitation of magnetospheric electrons. Dissociation of N_2 by electron impact yields

$$e^{-*} + N_2 \longrightarrow N(^2D) + N(^4S) + e^- \quad (76)$$

followed by

$$N(^2D) + CH_4 \longrightarrow NH + CH_3 \quad (77)$$

If $N(^4S)$ reacts with NH,

$$N(^4S) + NH \longrightarrow N_2 + H \quad (78)$$

then N_2 is regenerated, and CH_4 is catalytically dissociated to CH_3 plus H. Alternatively, if $N(^4S)$ reacts with reactive radicals,

$$\begin{aligned} N(^4S) + {}^3CH_2 &\longrightarrow HCN + H \\ N(^4S) + CH_3 &\longrightarrow HCN + 2\,H \text{ or } H_2CN + H \\ H + H_2CN &\longrightarrow HCN + H_2 \end{aligned} \quad (79)$$

and

$$NH + H \longrightarrow N(^4S) + H_2 \quad (80)$$

then HCN is formed. For example,

$$\begin{aligned} 2[e^{-*} + N_2 &\longrightarrow N(^4S) + N(^2D) + e^-] && (76) \\ 2[N(^2D) + CH_4 &\longrightarrow NH + CH_3] && (77) \\ 2[N(^4S) + CH_3 &\longrightarrow HCN + H_2(\text{or } 2\,H)] && (79) \\ NH + N(^4S) &\longrightarrow N_2 + H && (78) \\ NH + H &\longrightarrow N(^4S) + H_2 && (80) \\ \hline N_2 + 2\,CH_4 &\longrightarrow 2\,HCN + 3\,H_2(\text{or } H_2 + 4\,H) && (81) \end{aligned}$$

Dissociation of CH_4 is a by-product of N_2 dissociation. The HCN yield from reactions (76)–(81) is ~10% (Yung *et al.*, 1984).

Nitrile compounds are produced by

$$HCN + h\nu \longrightarrow CN + H \quad (82)$$

followed by

$$CN + C_2H_2 \longrightarrow HC_3N + H \quad (83)$$

$$CN + C_2H_4 \longrightarrow C_2H_3CN + H \quad (84)$$

$$CN + HCN \longrightarrow (CN)_2 + H \quad (85)$$

HC_3N and $(CN)_2$ have been detected on Titan (Table V and Fig. 11). CH_3CN has not been observed as CH_3CN is not dominant product of

$$CN + CH_4 \longrightarrow HCN + CH_3 \tag{86}$$

because abstraction rather than insertion into an alkane is faster.

The ion chemistry of $N_2{}^+$ preserves the N_2 bond in the upper atmosphere, but dissociative ionization of N_2 by electron impact is followed by CH_4 reactions:

$$\begin{aligned} N^+ + CH_4 &\longrightarrow CH_3{}^+ + N + H \\ &\longrightarrow CH_4{}^+ + N \\ &\longrightarrow H_2CN^+ + H + H \\ &\longrightarrow HCN^+ + H_2 + H \end{aligned} \tag{87}$$

$$\begin{aligned} HCN^+ + CH_4 &\longrightarrow H_2CN^+ + CH_3 \\ &\longrightarrow C_2H_3{}^+ + NH_2 \end{aligned} \tag{88}$$

$$H_2CN^+ + e^- \longrightarrow HCN + H \tag{89}$$

for a net yield of HCN and N. Although the dissociative ionization rate of N_2 is ~ 0.1 of the N_2 dissociation rate by electron impact, the HCN yield is ~ 0.25, as compared to the 0.1 yield from N_2 dissociation.

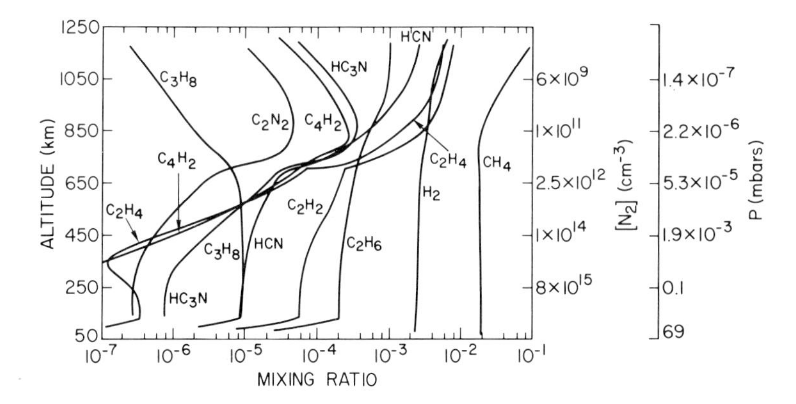

Fig. 11. Volume mixing ratios of selected species based on the calculations of Yung *et al.* (1984) as a function of altitude above the surface of Titan, pressure, and N_2 number density. To compare these ratios with Table V they must be averaged above 50 km by using the ratio of column density to N_2 column density. The CH_3C_2H mixing ratio (not shown) is approximately three times the HC_3N mixing ratio. Adapted from Yung *et al.* (1984).

B. Hydrocarbons

The photolysis of CH_4 has been discussed in Section V,B in the context of H_2 atmospheres. The N_2 atmosphere on Titan requires some significant changes in that discussion, as $^1CH_2$ is rapidly quenched by N_2:

$$^1CH_2 + N_2 \longrightarrow {}^3CH_2 + N_2 \tag{90}$$

which severely limits the production of methyl radicals by direct CH_4 photolysis. The radicals produced by CH_4 dissociation react to form C_2H_2, C_2H_4, and C_3H_4, prominent minor constituents on Titan (Table V), but not C_2H_6, the principal product of CH_4 photolysis on the outer planets:

$$^3CH_2 + {}^3CH_2 \longrightarrow C_2H_2 + H_2 \tag{91}$$

$$^3CH_2 + C_2H_2 + M \longrightarrow C_3H_4 + M \tag{92}$$

$$CH + CH_4 \longrightarrow C_2H_4 + H \tag{93}$$

$$^1CH_2 + CH_4 \longrightarrow 2\,CH_3 \tag{94}$$

$$^3CH_2 + CH_3 \longrightarrow C_2H_4 + H \tag{95}$$

$$C_2H_4 + h\nu \longrightarrow C_2H_2 + H_2(\text{or } 2\text{ H}) \tag{26}$$

Only the isomer CH_3C_2H of C_3H_4 has been detected.

Methyl radicals are produced by catalytic dissociation of CH_4 (Allen *et al.*, 1980):

$$\begin{array}{l} C_{2n}H_2 \longrightarrow C_{2n}H + H \qquad n = 1, 2, 3 \\ C_{2n}H + CH_4 \longrightarrow C_{2n}H_2 + CH_3 \\ \hline CH_4 \longrightarrow CH_3 + H \end{array} \tag{96}$$

which has the distinct advantage of polyacetylenes absorbing solar radiation longward of 3000 Å, in contrast to $\lambda < 1450$ Å for CH_4. Formation of polyacetylenes occurs by

$$C_{2n}H + C_{2m}H_2 \longrightarrow C_{2(n+m)}H_2 + H \tag{97}$$

The methyl radicals mostly recombine to yield

$$CH_3 + CH_3 + M \longrightarrow C_2H_6 + M \tag{98}$$

but some react with H to recycle CH_4,

$$CH_3 + H + M \longrightarrow CH_4 + M \tag{29}$$

Ethane may be attacked by the ethynl radical (C_2H),

$$C_2H + C_2H_6 \longrightarrow C_2H_5 + C_2H_2 \tag{99}$$

to yield subsequently propane,

$$CH_3 + C_2H_5 + M \longrightarrow C_3H_8 + M \tag{100}$$

and butane,

$$2\,C_2H_5 + M \longrightarrow C_4H_{10} + M \tag{101}$$

of which only propane has been observed on Titan (Table V).

All of these complex hydrocarbons sublimate in the lower stratosphere, where the temperature is as low as 71 K. The extensive haze layers observed by the *Voyager* imaging system (Fig. 3) is most likely constituted of large condensed organic molecules.

Many complex molecules formed on Titan are susceptible to hydrogen atom cracking, for example, in the case of polyacetylenes,

$$\begin{aligned} H + C_{2n}H_2 + M &\longrightarrow C_{2n}H_3 + M \qquad n > 1 \\ H + C_{2n}H_3 &\longrightarrow C_2H_2 + C_{2n-2}H_2 \end{aligned} \tag{102}$$

if the hydrogen atom concentration is appreciable. To overcome this problem, Yung *et al.* (1984) proposed a faster branch to reactions (102)

$$H + C_{2n}H_3 \longrightarrow C_{2n}H_2 + H_2 \tag{103}$$

for a net yield of reactions (102) and (103) of $2\,H \rightarrow H_2$. The recombination of hydrogen atoms is essential to understanding Titan's photochemistry. Theoretical mixing ratios of some observed species are shown in Fig. 11 as a function of altitude from the model of Yung *et al.* (1984). With the faster branch [reaction (103)] they obtain reasonably good agreement with observations.

The ultimate fate of these heavy hydrocarbons is deposition on Titan's surface because carbon cannot escape Titan. The predominant photochemical product is C_2H_6, which forms an ocean with dissolved CH_4, N_2, and other trace species. Solid C_2H_2 accumulates at the bottom of the ocean.

C. Carbon Monoxide and Carbon Dioxide

The discoveries of CO and CO_2 in the atmosphere of Titan adds complexity to its photochemistry. The two sources of oxygen are substantial amounts of CO when the atmosphere was initially formed and ablation of meteoroidal material. The latter is considered to be more probable.

In an atmosphere with appreciable CO, magnetospheric electrons initiate the chemical destruction of CO:

$$e^{-*} + CO \longrightarrow C + O(^1D) + e^- \tag{104}$$

$$e^{-*} + N_2 \longrightarrow N_2(A\,^3\Sigma) + e^- \tag{105}$$

The energy of $N_2(A\,^3\Sigma)$ can be transferred almost resonantly to CO,

$$N_2(A\,^3\Sigma) + CO \longrightarrow CO(a\,^3\Pi) + N_2 \qquad (106)$$

followed by

$$CO(a\,^3\Pi) + CO \longrightarrow CO_2 + C \qquad (107)$$

The $O(^1D)$ reacts with CH_4,

$$O(^1D) + CH_4 \longrightarrow OH + CH_3 \qquad (108)$$

followed by

$$CO + OH \longrightarrow CO_2 + H \qquad (109)$$

to also convert CO to CO_2. Sublimation of CO_2 terminates the chain of reactions and produces a 1-m layer of dry ice on the surface over geological time given an initial CO mixing ratio of 0.1 (Samuelson *et al.*, 1983).

Photolysis of meteoroidally derived H_2O,

$$\begin{aligned} H_2O + h\nu &\longrightarrow OH + H \\ &\longrightarrow H_2 + O(^1D, ^1S) \end{aligned} \qquad (110)$$

and reaction of OH with hydrocarbon radicals,

$$OH + CH_3 \longrightarrow CO + 2\,H_2 \qquad (111)$$

$$OH + {}^3CH_2 \longrightarrow CO + H_2 + H \qquad (112)$$

yields CO and subsequently CO_2,

$$CO + OH \longrightarrow CO_2 + H \qquad (109)$$

CO_2 can be readily converted back to CO by

$$CO_2 + h\nu \longrightarrow CO + O(^1D, ^3P) \qquad (113)$$

$${}^3CH_2 + CO_2 \longrightarrow H_2CO + CO \qquad (114)$$

$$\begin{aligned} H_2CO + h\nu &\longrightarrow H_2 + CO \\ &\longrightarrow H + HCO \end{aligned} \qquad (115)$$

$$HCO + h\nu \longrightarrow H + CO \qquad (116)$$

$$H + HCO \longrightarrow H_2 + CO \qquad (117)$$

The ultimate fate of oxygen is sublimation of CO_2. A meteoroidal H_2O flux of 6×10^5 cm^{-2} sec^{-1} maintains CO at an abundance of 1.1×10^{-4} and CO_2 at 1.5×10^{-9} according to Yung *et al.* (1984), which is in reasonably good agreement with observations (Table V). In their model [CO] is proportional to $[CH_3]$, and $[CO_2]$ is proportional to the flux of H_2O.

VII. Photochemistry of Io

The SO_2 atmosphere on Io is substantially different from the atmospheres discussed previously. Photolysis of SO_2,

$$\begin{aligned} SO_2 + h\nu &\longrightarrow SO + O \qquad \lambda < 2210\ \text{Å} \\ SO_2 + h\nu &\longrightarrow S + O_2 \qquad \lambda < 2070\ \text{Å} \\ S + O_2 &\longrightarrow SO + O \\ SO + SO &\longrightarrow SO_2 + S \\ \hline SO_2 &\longrightarrow S + 2\,O \end{aligned} \tag{118}$$

leads to the production of sulfur and oxygen atoms; the latter readily escape Io by thermal processes. Recombination occurs through reactions

$$SO + SO \longrightarrow SO_2 + S \tag{119}$$

$$SO + O_2 \longrightarrow SO_2 + O \tag{120}$$

According to calculations by Kumar (1982), atomic oxygen emerges as the dominant species at the exobase with significant mixing ratios ($\sim 10^{-3}$–10^{-1}) of the photochemical products S, O, SO, and O_2 in the lower atmosphere. Sulfur dioxide is ultimately lost from the atmosphere by thermal Jeans escape (mostly oxygen atoms), ionization and magnetospheric pickup (S^+, O^+, SO_2^+), and sputtering from the atmosphere. The slow thermal escape rate of sulfur atoms in comparison to oxygen atoms may lead to a net accumulation of sulfur in the atmosphere and at the surface. Also, the ratio of oxygen to sulfur supplied to the plasma torus may exceed the 2:1 ratio expected with SO_2 as the parent molecule.

The ionosphere on Io detected by *Pioneer 10* is produced by solar radiation and magnetospheric electrons:

$$\begin{aligned} SO_2 + h\nu\ (\lambda < 1005\ \text{Å}) &\longrightarrow SO_2^+ + e^- \\ e^{-*} + SO_2 &\longrightarrow SO_2^+ + 2\,e^- \\ S + h\nu &\longrightarrow S^+ + e^- \\ O + h\nu &\longrightarrow O^+ + e^- \end{aligned} \tag{121}$$

and recombines dissociatively by reactions such as

$$SO_2^+ + e^- \longrightarrow SO + O \tag{122}$$

Given the uncertain knowledge of Io's atmospheric composition, thermal structure, and day–night distribution, it is difficult to discuss its photochemistry in depth. The atmosphere probably plays a fundamental role in transferring volcanic gases to the Jovian magnetosphere. Analyses of the *Voyager* data from the Io plasma torus indicates that Io must supply $\sim 10^{28}$ ions sec^{-1} to the torus to account for its UV radiative loss and plasma densities.

VIII. Directions for Future Research

In early 1986 *Voyager 2* will fly by Uranus for the first close encounter with this planet. If the data return is comparable to that obtained on Jupiter and Saturn, it will lead to an explosive growth in our knowledge of this intermediate planet. Interpretation of ground-based data has been difficult, and considerable uncertainty on the composition and thermal structure of Uranus remains. It is anticipated that *Voyager* observations will rectify this situation. Of particular importance is the mixing ratio of CH_4 in the vicinity of the tropopause and in the thermal inversion region.

In 1989, provided the spacecraft is still operating, *Voyager 2* will fly by Neptune to explore the other intermediate planet and complete the grand tour of the outer planets with the exception of Pluto. Items of key interest are whether Neptune has a ring system, the CH_4 mixing ratio profile, and the reason why Neptune has a stronger temperature inversion than Uranus does. It is hoped that *Voyager 2* will provide us with comprehensive measurements of Triton, as *Voyager 1* did of Titan.

Late in this decade (1988) the *Galileo* spacecraft will conduct in-depth investigations of the Jovian system as an orbiter. An entry probe with a neutral mass spectrometer will descend into the atmosphere of Jupiter and provide the greatest advance in knowledge since *Voyager*. It is hoped that noble gases, H_2, CH_4, NH_3, N_2, and other gases will be measured, down to the parts per million range. For H_2O, only 100 ppm may be attainable. In addition, isotopic ratios will be obtained.

The Galilean satellites will be observed at distances of $0(\sim 1000)$ km. Composition, surface density, and pressure will be accurately measured or low upper limits will be obtained. It is hoped that the atmospheric composition of Io and the exospheric composition of Europa will be determined.

In the laboratory much work remains to be done. The photolysis of C_2H_2, an important constituent in these atmospheres, has only minor dissociation paths to C_2H and C_2 for a total yield of ~ 0.2. What is the fate of the photon energy for the other 0.8 yield? Is the vinylidene radical ($H_2C{=}C$) formed, and if so, what is its ultimate fate in H_2 and N_2 atmospheres? Accurate measurements are needed for many reactions at low temperatures, in particular NH_2, PH_2, and PH reactions. Vapor pressure as a function of low temperature are also required for many constituents, with P_2H_4 the most critical.

As spectroscopy improves in sensitivity and detection of constituents at less than the parts per billion level becomes possible, the demands for accurate laboratory studies and photochemical models will increase significantly. While it is possible to get lost in the details, it should be remembered that one of the most important goals of this research is to understand the Earth's primitive atmosphere and the origin of life. The striking similarity of Titan's atmosphere with the mildly reducing, prebiotic terrestrial atmosphere (Chapter 1; N_2 as the dominant constituent and trace amounts of CH_4, H_2O, H_2, CO, CO_2 and other hydrocarbons) suggests that this effort in our philosophical quest for understanding our origin will ultimately be rewarding.

Acknowledgment

This work was supported in part by NASA's Office of Planetary Atmospheres.

References

Allen, M., Pinto, J. P., and Yung, Y. L. (1980). Titan: aerosol photochemistry and variations related to the sunspot cycle. *Astrophys. J.* **242,** L125–L128.

Borucki, W. J., Bar-Nun, A., Scarf, F. L., Cook, A. F., II, and Hunt, G. E. (1982). Lightning activity on Jupiter. *Icarus* **52,** 492–502.

Ferris, J. P., and Benson, R. (1981). An investigation of the mechanism of phosphine photolysis. *J. Am. Chem. Soc.* **103,** 1922–1927.

Gladstone, G. R. (1982). Radiative transfer and photochemistry in the upper atmosphere of Jupiter. Ph.D. Thesis, California Institute of Technology, Pasadena.

Kaye, J. A., and Strobel, D. F. (1983a). HCN formation on Jupiter: the coupled photochemistry of ammonia and acetylene. *Icarus* **54,** 417–433.

Kaye, J. A., and Strobel, D. F. (1983b). Formation and photochemistry of methylamine in Jupiter's atmosphere. *Icarus* **55,** 399–419.

Kaye, J. A., and Strobel, D. F. (1983c). Phosphine photochemistry in Saturn's atmosphere. *Geophys. Res. Lett.* **10,** 957–960.

Kostiuk, T., Mumma, M. J., Espenak, F., Deming, D., Jennings, D. E., Maguire, W., and Zipoy, D. (1983). Measurements of stratospheric ethane in the Jovian south polar region from infrared heterodyne spectroscopy of the ν_9 band near 12μm. *Astrophys. J.* **265,** 564–569.

Kumar, S. (1982). Photochemistry of SO_2 in the atmosphere of Io and implications on atmospheric escape. J. Geophys. Res. **87,** 1677–1684.

Lewis, J. S. (1971). Satellites of the outer planets: their physical and chemical nature. *Icarus* **15,** 174–185.

Lewis, J. S. (1980). Lightning synthesis of organic compounds on Jupiter. *Icarus* **43,** 85–95.

Norrish, R., and Oldershaw, G. (1961). The flash photolysis of phosphine. *Proc. R. Soc. London Ser. A* **262,** 1–9.

Prather, M. J., Logan, J. A., and McElroy, M. B. (1978). Carbon monoxide in Jupiter's upper atmosphere: an extraplanetary source. *Astrophys. J.* **223,** 1072–1081.

Prinn, R. G., and Lewis, J. S. (1975). Phosphine on Jupiter and implications for the great red spot. *Science* **190,** 274–276.

Samuelson, R. E., Maguire, W. C., Hanel, R. A., Kunde, V. G., Jennings, D. E., Yung, Y. L., and Aikin, A. C. (1983). CO_2 on Titan. *JGR, J. Geophys. Res.* **88,** 8709–8715.

Strobel, D. F. (1969). The photochemistry of methane in the Jovian atmosphere. *J. Atmos. Sci.* **26,** 906–911.

Strobel, D. F. (1975). Aeronomy of the major planets: photochemistry of ammonia and hydrocarbons. *Rev. Geophys. Space Phys.* **13,** 372–382.

Strobel, D. F. (1983). Photochemistry of the reducing atmosphere of Jupiter, Saturn, and Titan. *Int. Rev. Phys. Chem.* **3,** 145–176.

Strobel, D. F., and Yung, Y. L. (1979). The Galilean satellites as a source of CO in the Jovian upper atmosphere. *Icarus* **37,** 256–263.

Tokunaga, A. T., Beck, S. C., Geballe, T. R., Lacy, J. H., and Serabyn, E. (1981). The detection of HCN on Jupiter. *Icarus* **48,** 283–289.

Wildt, R. (1937). Photochemistry of planetary atmospheres. *Astrophys. J.* **86,** 321–336.

Yung, Y. L., and Strobel D. F. (1980). Hydrocarbon photochemistry and Lyman alpha albedo of Jupiter. *Astrophys. J.* **239,** 395–402.

Yung, Y. L., Allen, M., and Pinto, J. P. (1984). Photochemistry of the atmosphere of Titan: comparison between model and observations. *Astrophys. J., Suppl. Ser.* **55,** 465–506.

III

Comets

9

The Photochemistry of Comets

WALTER F. HUEBNER

Theoretical Division
Los Alamos National Laboratory
Los Alamos, New Mexico

ISBN 0-12-444920-4

I. Introduction

Until we have a spacecraft mission to a comet, the physical and chemical nature of its nucleus can only be inferred from observations of the comet atmosphere (coma). Photodissociation and photoionization, however, destroy the mother molecules evaporating* from the nucleus. To extract the primary information, the chemical kinetics and physics of the coma must be modeled with a computer and the results compared with coma observations. Such models are important tools in support of comet missions (such as the *Giotto* mission to Halley's comet), in the analysis and interpretation of comet mission data and of ground-based and Earth-satellite-based observations, and in the understanding of the earliest history of our solar system.

II. Classification

Comets differ significantly from planets and their satellites in size, appearance, atmosphere, and orbital motion. In this section we first define what a comet is; then, in order to understand better their physics and chemistry, we classify them.

* To make a clear distinction between phase changes from solid to gas and from gas to solid, both called sublimation, we shall call the former vaporization and the latter condensation. But it must be remembered that the liquid phase implied by these terms does not have any meaning in the context used here.

A. Definition of a Comet

A comet is a minor body of the solar system containing a chemically rich mixture of volatile materials and refractory grains. This mixture, the icy conglomerate nucleus (Whipple, 1950, 1951), is the source for a comet's diffusely appearing atmosphere (coma) and two morphologically different tails. Coma and tails develop before and subside and disappear after perihelion passage. Since the nucleus is only a few kilometers in size, it cannot bind its atmosphere gravitationally, as is typically the case for planets. Since the mass of a comet is less than 10^{-10} the mass of the Earth, planets can severely perturb comet orbits, but the reverse effect is negligible. The volatile materials in the nucleus are primordial ices (frozen gases) that condensed before or during formation of the solar system; in the surface layer of the nucleus they have been postprocessed by ultraviolet (UV) radiation and cosmic rays. Although comets probably also exist in interstellar space, none have been observed with certainty. The word comet originates from the Greek *kometes*, meaning longhaired, because of their appearance as hairy stars or tail stars.

Comets have many properties in common, but they differ from each other in details and sometimes even in some of their main features. Typically, when a comet in its highly eccentric orbit around the Sun has a heliocentric distance less than about 1.5 AU, it consists of a head (composed of a nucleus surrounded by a nearly spherical coma), a plasma tail, and a dust tail. In dust-poor comets the dust tail may be faint or absent, while in CO^+-poor comets the plasma tail may be so faint that it is not visible. Comets that have experienced many solar passages, as for example Comet Encke, have their most volatile ice components depleted to such a degree that they only feature a faint coma and no tail. The brightness of a comet can vary by several magnitudes from day to day. Flaring, jet-, fountain-, and envelope-like structures are observed in active comet comae.

The longest plasma tails are several times 10^7 km long in visible fluorescent bands of CO^+ and shorter in N_2^+ and H_2O^+. Plasma tails are straight, show activity and structures such as knots, kinks, disconnections, and outward spreading flares (the "swan" feature in Comet Kohoutek), and point (within a few degrees) radially away from the Sun. From the coma shorter, spike-like features (streamers) emanate at small angles to the plasma tail and tend to close on it as time progresses while new streamers appear at the birthplace of the old ones.

The dust tail shows curvature, is smooth, and near the head it points away from the Sun. Yet, exceptions to smoothness exist. For example, the nearly parallel bands (striae) that point almost radially away from the Sun but

Fig. 1. Comet Mrkos (1957 V), on 15 August 1957 at 0245 UT. The head of the comet is at the left of the picture, the plasma tail is the faint, narrow, straight feature pointing toward the upper right corner. The dust tail is bright near the comet head; it fades as it curves toward the right, then toward the bottom right. The faint straight features emanating from the dust tail at small but increasing angles with respect to the plasma tail are the striae. The photograph was obtained with a $f/1.5$ Schmidt camera near Fort Worth, Texas (courtesy of J. A. Farrell, Los Alamos).

originate near the concave edge of the dust tail, as observed in Comets Mrkos (1957 V)* and West (1976 VI). Some of the discussed features of comets are illustrated in Fig. 1.

A large reservoir of comets composes the Oort cloud, a spherical shell around the Sun with outer radius of $\sim 50{,}000$ AU. The orbits of these comets differ from those of the planets and the asteroids in that they are not confined to the region of the ecliptic. Aside from some clustering, which can be explained by perturbations from stars passing through the Oort cloud (Biermann *et al.*, 1983), the aphelia distribution on the sky of these comets is isotropic. Another group of comets, the periodic comets (this refers to comets with periods $P < 200$ years), have orbits that predominantly lie close to the ecliptic. With this preliminary grouping we have arrived at the classification of comets.

* When a comet is discovered, it is named after the discoverer and labeled by the year, followed by a small letter indicating the sequence of discovery in that year. Later, these comets are relabeled with the year and a roman numeral that indicates the sequence of perihelion passage of that comet in that year.

B. Classes of Comets

Marsden *et al.* (1978) analyzed the nearly parabolic orbits of comets and classified them according to their accuracy. The accuracy depends on observational circumstances such as number and time interval of observations, on brightness of the comet, and on nongravitational effects (the recoil on the nucleus from the asymetrically evaporating gas). Gravitational perturbations of the planets were taken into account in the analysis by Marsden *et al.* (1978).

Considering only the comets with the most accurately known orbits, another classification emerges from the analysis of Marsden *et al.*: the semimajor axis a of a comet relates to its total mechanical energy ($\sim 1/a$) in the solar system and to the radius of the Oort cloud (since a comet's aphelion distance $Q \simeq 2a$ for nearly parabolic orbits). The analysis shows a gap in the reciprocal of the "original" semimajor axes $[(1/a)_{\text{orig}}]$ between about 100×10^{-6} and 200×10^{-6} AU^{-1}. This gap suggests a boundary in the Oort cloud and implies that comets with $(1/a)_{\text{orig}} < 100 \times 10^{-6}$ AU^{-1} are "new" comets that probably make their first passage through the inner part of the solar system. Orbits with the smallest error appear to cluster about the value 37.6×10^{-6} AU^{-1} (Biermann, 1982). Although a few comets show small (in magnitude) negative values in $(1/a)_{\text{orig}}$, their corresponding mean error is always large enough so that they cannot be classified as hyperbolic or interstellar comets. They belong to the group of "new" comets.

From the average change in $1/a$ ot only those comets whose orbits were made more elliptical during one revolution around the Sun, Marsden *et al.* defined "old" comets to have $(1/a)_{\text{orig}}$ more than five times larger than the average change. This average perturbation, which is $\sim 400 \times 10^{-6}$ AU^{-1}, decreases somewhat with increasing perihelion distance q. All comets lying between the "old" and "new" comet groups we shall call "intermediate" comets.

Many of the parabolic comets have orbits that are not known to sufficient accuracy to be classified as "new," "intermediate," or "old"; we shall simply call them parabolic comets. "Parabolic" is only a label indicating that the highly eccentric elliptic orbits are nearly parabolic.

Some recent "new" comets include 1953 II, 1957 III, 1959 I, 1960 II, 1962 III, 1973 XII, 1975 II, 1975 V, and 1975 XI. To the group of "intermediate" comets belong 1908 III, 1957 V, 1959 IV, 1969 IX, 1970 III, 1974 III, and 1975 IX. Among the "old" comets are 1961 VIII, 1962 VIII, 1964 VIII, 1970 II, and the Kreutz group of Sun-grazers: 1882 II, 1963 V, and 1965 VIII.

Marsden (1983) labeled the next major group as periodic comets; by definition they have a period less than 200 years. About two-thirds of these comets have been observed at two or more perihelion passages. Periodic comets are often indicated by "P/" preceding their name. To this group belong

P/Encke, which has been observed at over 50 perihelion passages and has the shortest period $P = 3.3$ years, and P/Halley, which has been observed during 28 of its 29 perihelion passages since 240 BC (as recorded by the Chinese) and has a period $P = 76$ years. (Although there was a favorable apparition in 164 BC, it was not recorded.) The last apparition of P/Halley was in 1910; the next will be in 1986. A special section is devoted to this comet.

In summary, we have defined a comet to be a minor body of the solar system. Comets contain varying amounts of grains and frozen gases, that is, primordial ices, that vaporize when heated by the Sun. The classification of comets bears on the composition and volatility of the frozen gases in the surface layer of the nucleus.

III. Composition and Structure

The nucleus of a comet is too small to be observed visually from Earth, but it has recently been detected by radar. Even at large heliocentric distances, when the coma is weak or absent, the nucleus may be surrounded by a small but compact dust cloud, giving it a starlike appearance. The optical (photometric) center of the coma, by some observers called the nucleus, must not be confused with the real icy conglomerate nucleus. Until a spacecraft visits a comet, the composition and structure of the nucleus must be inferred from observations of the coma and to a lesser degree from observations of the tail.

A. Development of Coma and Tails

To understand better the structure of a comet and its changes and stages of development, we can go through the following *Gedanken* observation as a comet approaches the Sun. At large heliocentric distance, ~ 10 AU (orbital distance of Saturn), the nucleus slowly vaporizes the most volatile icy components from its dayside surface layer. Even if the nucleus is rotating rapidly, there will be insignificant evaporation on its nightside, since the amount of heat stored in its surface layer from the daytime insolation is still too small.

As the comet approaches the orbital distance of Jupiter (~ 5 AU) its rate of vaporization increases rapidly and may extend to the nightside, if the nucleus is rotating. Typical rotation periods are about 6–20 hours. Since they are lighter than dust grains, small icy grains of micrometer or submicrometer size and of a less volatile component (e.g., water, clathrates, or hydrates) will be carried along by the escaping gases evaporating from the more volatile components of the icy conglomerate. A coma developes with a very small ice halo. The ice grains in this halo evaporate under insolation as they move away from the nucleus. Further on its way toward the Sun these icy grains evaporate more rapidly, the ice halo shrinks, and the gas production increases further.

The gas now carries small dust particles with it. The density of the outward streaming gas is so low that the collision mean free path in it is large compared to the coma, and free molecular flow applies. The gas production increases rapidly as the comet passes the distance corresponding to the asteroid belt and approaches that of the orbit of Mars (~ 1.5 AU). During this phase the gas production becomes high enough to make the collision mean free path of the molecules small compared to the range of the visual coma ($\sim 10^4$ km) so that the gas acts like a fluid. This is also about the place where most comets develop a tail and the chance of discovering the comet becomes large. The size of the dust grains in the coma has increased with the increasing gas production rate. In the neighborhood of the Earth's orbit, the tail is well developed (sometimes two tails, a plasma and a dust tail, develop), and the visual coma reaches its maximum size ($\sim 10^5$ km).

In comets with heliocentric distance less than 0.7 AU (orbit of Venus) the sodium D lines are observed in the head and extend into the tail at still smaller heliocentric distances. Typical examples of such comets are 1882 I, 1882 II, 1910 I, 1911 V, 1927 IX, 1941 I, 1947 XII, 1948 XI, 1957 III, 1957 V, and 1965 VIII (Ikeya–Seki). If the comet is of the Sun-grazing type, then the length of the sodium component in the tail reaches its maximum when the comet is at a heliocentric distance $r \sim 0.1$ or 0.2 AU, where also the lines of potassium appear. At still smaller distances the forbidden O lines, Ca^+, V, Cr, Mn, Fe, Co, Ni, and Cu lines are observed. Examples where these elements have been identified in the spectra are the comets 1882 II and Ikeya–Seki. The forbidden O lines can, however, also be observed in large comets at larger heliocentric distances. The appearance of the metal spectra indicates that the dust must be evaporating. At less than $\sim 4R_\odot$ (solar radii) (i.e., <0.02 AU) from the center of the Sun, the dust coma will essentially vaporize completely, and the dust tail will turn into a pure plasma tail. Although no spectra were taken of comet Ikeya–Seki at $r < 4R_\odot$ to prove this, the analysis of the direction and curvature of that comet's tail by Lüst and Schmidt (1968) indicates its plasma nature.

After perihelion passage the processes discussed above will be in reverse order, but the gas production rate, in general, decreases faster on recession from the Sun than it increased during approach. Unless the nucleus splits into two or more large parts (e.g., as happened in Comet West), thereby increasing the surface area and exposing virgin ices to the Sun's radiation, the surface area will be reduced in comparison to what it was before perihelion passage, thus reducing the total gas output. A large comet with $q \sim 1$ AU will lose $\sim 10^7$ tons of its mass, or a few decimeters of its surface layer, during perihelion passage. This corresponds to $\sim 10^{-4}$ of its total mass. The gas production rate per unit area will, however, also decrease faster with time because the insolation decreases with increasing distance, allowing meteoritic material (the refractory grains in the icy conglomerate nucleus) that the evaporating ices expose to remain on the surface to form an insulating layer. Before perihelion such a layer could not form so easily since the dust was dragged into the atmosphere as the gas production increased with decreasing solar distance. The third reason for a faster reduction of the gas yield is that the more volatile ices have been depleted to a large extent by the time the comet recedes from the Sun.

B. Dust and Plasma Tails

The dust tail is composed of micrometer- and submicrometer-sized particles. Pressure exerted by solar radiation on the grains accelerates them radially away from the Sun. The smaller particles are accelerated more easily; the larger particles remain closer to the nucleus of the comet. The acceleration is proportional to the geometric cross section of the grains and to the Mie efficiency factor for radiation pressure. This factor and the intensity of the solar radiation decrease in such a way that particles smaller than 0.1 μm receive accelerations less than the gravitational acceleration exerted by the Sun. Since the comet's velocity perpendicular to the Sun–comet direction is not negligible compared to the radial velocity component of the accelerating dust, the dust tail lags the comet's motion.

Because of its stability, CO^+ dominates the plasma tail. With the exception of N_2^+, the other ions listed in Table I are much less stable against photodissociation and photoionization; their range in the tail is much shorter, sometimes so insignificant that they cannot be detected at all. Close to the head of the comet, the ions in the tail move with a speed of 20 to 40 km sec^{-1} as determined from Doppler shifts in the spectrum of H_2O^+, while at larger distances from the head, the ion speed is $\sim$100 km sec^{-1} as determined from the motion of structures such as knots (Lüst, 1962). This latter measurement assumes bulk motion of the structures. The rapid accelerations are caused by coupling to the solar wind. It is this acceleration together with the activity of the plasma tail and the relationship of its small deviation from solar radial direction discovered by Hoffmeister in 1943 that prompted Biermann (1951) to postulate the solar wind as a corpuscular stream emanating from the Sun, many years before it was verified by direct measurements. At solar minimum the solar wind has an average speed of $\sim$450 km sec^{-1} and a density of $\sim$10 protons cm^{-3} at 1 AU heliocentric distance. This corresponds to a solar mass loss of more than 10^{12} g sec^{-1}, but, if constant, this is still a negligible fraction

Table I

IONS DETECTED IN COMETS

CO^+	H_2O^+	CO_2^+	N_2^+	CH^+	CN^+	OH^+	C^+	SH^+

of the Sun's total mass over its lifetime. Yet comets show a visible effect by moving through this extended atmosphere of the Sun.

Since only a few percent of the gas produced by a comet is ionized in the head (i.e., $R < 10^5$ km) and because of the rapid acceleration of the ions, the density in the far reaches of the plasma tail is less than 1 ion cm^{-3}. Aside from further photodissociation and photoionization of the tail ions and destruction of grains (formation of striae), no chemistry takes place in the plasma or dust tails; the densities are too low, and the collision mean free paths are too large.

Table II

NEUTRAL SPECIES IDENTIFIED IN COMETS

H	C	O	Na	S	CH	NH	OH	C_2	CN	CO	CS	S_2	NH_2	H_2O	HCN	C_3	NH_3	CH_3CN

C. Coma and Nucleus

A coma spectrum typically consists of many emission bands from different molecular species superimposed on a continuum of sunlight reflected by the dust. The excitation of the emission bands is primarily by fluorescence, as has been beautifully illustrated by Arpigny (1976), who showed that particular transitions in CN would be only weakly excited when their Doppler-shifted wavelengths happen to coincide with (solar) Fraunhofer (absorption) lines. This is known as the Swings effect; the Doppler shift is caused by the comet's radial velocity component relative to the Sun. Another Doppler shift superimposed on that from the comet motion, but caused by the outward streaming gas relative to the nucleus in the subsolar and antisolar directions, gives rise to different degrees of excitation in different regions of the comet; this is known as the Greenstein effect. Collisional excitation or creation of excited states as a result of photodissociation contribute only in a minor way. In addition to the ions listed in Table I, the atoms, radicals, and molecules given in Table II have been identified in comet comae. While atomic hydrogen, carbon, oxygen, and sulfur have been detected by their resonance lines in the UV spectrum, atomic nitrogen has not been detected. Although nitrogen is expected to be present, its resonance line is masked by the dominating hydrogen Lyα line, which makes detection of nitrogen very difficult. Some species have been detected in microwave emissions in only one comet [HCN and CH_3CN in Comet Kohoutek (1973 XII) and NH_3 in Comet *IRAS*–Araki–Alcock (1983 d)*]. Cometary spectra vary considerably from comet to comet, from place to place within the coma of a particular comet, and with

* See footnote in Section II,A.

distance. Figure 2 shows some cometary spectra; note the differences in CO^+ and dust.

Generally, the molecular species observed in comet spectra are not those that exist in the nucleus, rather they are the products of photochemical reactions of the evaporated mother molecules. Nevertheless, these observed species give clues about the elemental abundance and chemical composition of the frozen gases in the nucleus (e.g., see Delsemme, 1982). Interpreting these clues is one of the major objectives of cometary spectroscopy and chemical modeling. Present indications are that the frozen gases in the nucleus are composed of carbon, nitrogen, oxygen, and sulfur combined with hydrogen,

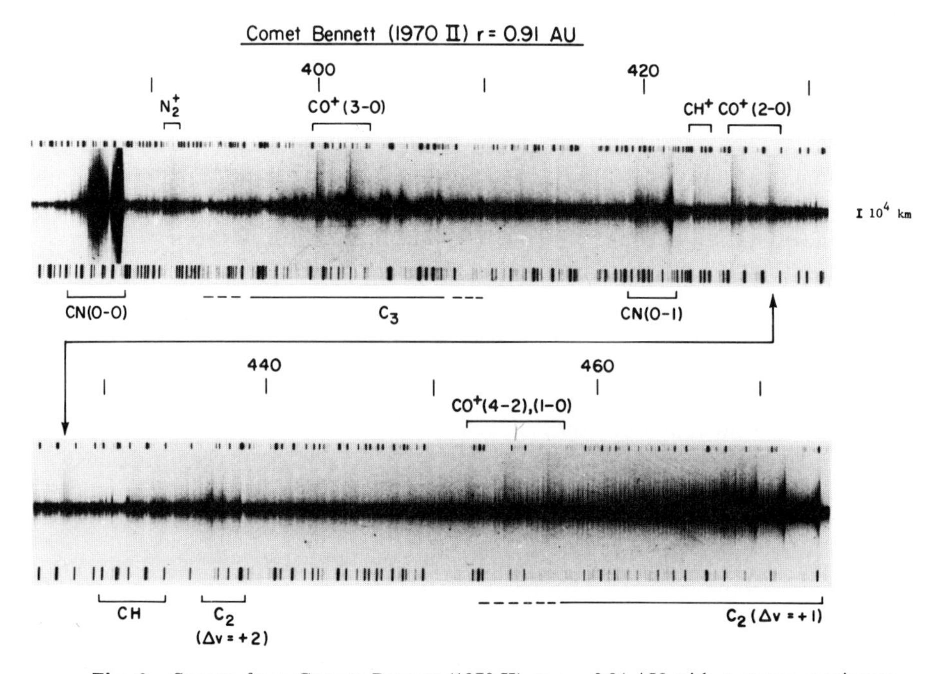

Fig. 2. Spectra from Comets Bennett (1970 II) at $r = 0.91$ AU with a strong continuum, Tago–Sato–Kosaka (1969 IX) at $r = 1.00$ AU with a medium continuum, and Ikeya (1963 I) at $r = 0.72$ AU with a weak continuum. Note the emissions of the ions CO^+, $N_2{}^+$, and CH^+ extend in the tail direction (up, in Bennett), while the neutral species CN, C_2, C_3, CH, and OH are nearly symmetric in the coma. Although the neutral species in Tago–Sato–Kosaka and Ikeya are strong, there is no CO^+. Also, CN has a large range, while CH has a very short range in the coma. The continuum is sunlight reflected by dust. The spectra were obtained with Coudé spectrographs, in the case of Bennett and Tago–Sato–Kosaka with the 200-inch telescope of the Hale Observatory on Palomar Mountain, and in case of Ikeya with the 1.5-m telescope of the Haute-Provence Observatory of the French Centre National de la Recherche Scientifique (courtesy C. Arpigny, Liège, Belgium).

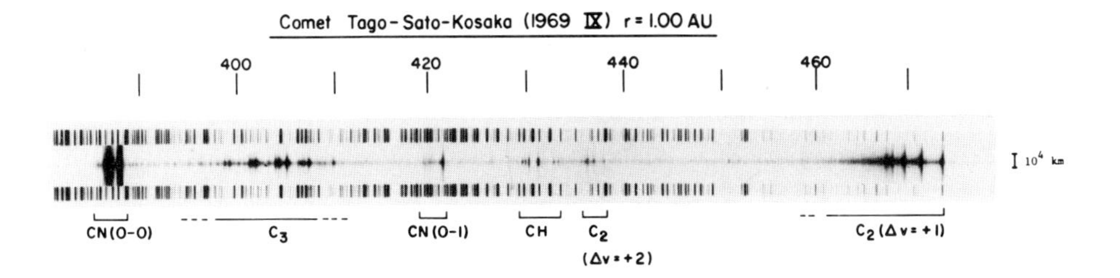

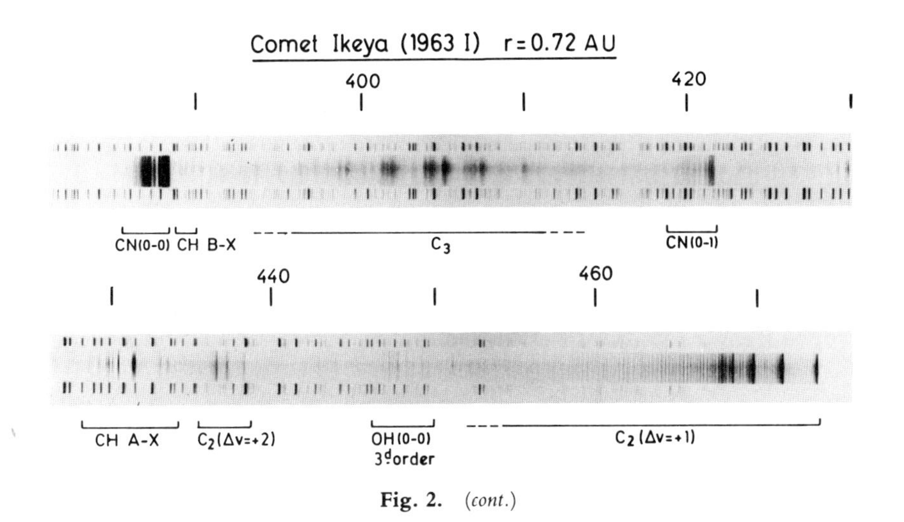

Fig. 2. *(cont.)*

but not enough hydrogen to fully saturate the chemical bonds in all molecules. Water (H_2O) is present in large quantities; H_2 is too volatile to be present in any measurable quantity.

Spectroscopy and narrow-band photometry have provided much information about the physical conditions in the coma and about the identity and spatial distribution of the constituents in the coma and tail. But because of calibration problems, the variability of the coma, and the masking of the spectra by the dust continuum, abundances of the various species are still very uncertain. Exceptions to this are the relative and absolute abundances of H and OH; the ratio for these is close to 2, indicating that their parent is most likely H_2O and is a major constituent. The observations of H and OH are made by rockets and Earth satellites in the UV range of the spectrum. The hydrogen coma has a typical radius of more than 10^6 km at 1 AU heliocentric distance; it is larger than the Sun! The hydrogen expands with at least two major velocity components of about 8 and 20 km sec^{-1}, while the velocity of the bulk gas in the visual coma is only 1 km sec^{-1} on the average. The hydrogen coma was predicted by Biermann (1968) several years before it was detected in Lyα radiation.

D. The Difference between Comet Comae and Planetary Atmospheres

In general, planetary atmospheres are gravitationally bound. Exceptions are Mercury, where the gas density at the surface is less than 10^8 molec cm^{-3}, and the tops of the atmospheres, where the lightest constituents escape by thermal motion. In comets the atmosphere is generated by evaporation from the nucleus, has a density of about 10^{14} molec cm^{-3} at the surface (at $r \simeq$ 1 AU), and escapes almost instantaneously at supersonic speed. The density in a planetary atmosphere has an approximately exponential behavior with height and a small range compared to the radius of the planet, while in a comet the coma density varies approximately with the square of the reciprocal of the radius and (the collision-dominated part) is $\sim 10^4$ times larger than the nucleus. A comet is practically all atmosphere. Planetary atmospheres exhibit tangential winds and turbulence created through contact of the atmosphere with the planet's surface. In comets the dominant velocity is radially outward, in the nearly uniform bulk motion as well as in the jets and envelopes; an exception to this is the much smaller plasma component that deviates from the radial outward motion when it couples with the solar wind outside of the contact surface on the subsolar hemisphere and moves into the plasma tail. A comet coma depends on the volatility of the frozen gases stored in the nucleus and on the heliocentric distance of the comet, which is a function of time. Compared to this a planet's atmosphere is constant.

For these reasons, the chemistry in planetary atmospheres is nearly in a steady state and is dominated by radical reactions, while in comets, chemistry is time dependent (in a radially moving volume element) and based on ion gas-phase reactions akin to the chemistry in interstellar clouds. But it is primarily the dilution of the gas, as it streams outward, that makes chemical kinetics necessary in modeling a comet coma.

E. Observational Results

Table III is a partial list of observed comet species, with an indication of their ranges (projected on the sky) into the coma and the heliocentric distances at which they become observable in large, bright comets as they approach the Sun. The table is based primarily on homogeneous evaluations taken from the *Atlas of Cometary Spectra* (Swings and Haser, 1956) for visual emissions and

Table III

PARAMETERS OF SOME OBSERVED SPECIES

Species	Detectable at r (AU)	Range into coma ρ (km)	Shape or range into plasma tail
H	2	10^7	Oblong in direction of tail
O			Extends into tail
Na	0.7		Extends into tail
CH	1.5	10^4	
NH	1.7	$10^{5.5}$	
OH	1.8	$10^{5.7}$	
C_2	1.8	$10^{5.5}$	Circular, defines visual size and color
CN	3	10^6	Circular, defines photographic size
CS		$10^{2.7}$	
S_2		$10^{2.7}$	
NH_2	2	$10^{4.5}$	
C_3	2.2	$10^{3.7}$	
CH^+			Short tail
OH^+			Short tail
CO^+		$10^{4.7}$	Very long tail
$N_2{}^+$			Long tail
H_2O^+			Long tail
$CO_2{}^+$			Medium long tail

heliocentric from Keller (1976) for UV emissions, supplemented with more recent data for sulfur compounds. Detectability as a function of heliocentric distance and range into the coma are averages; deviations from these can be expected for individual comets. Beside species density, other factors, such as the strength of radiative transitions and the instrumentation, will affect the observed range of a species. The main purpose of the table is to facilitate comparison between species. Also, any model for a bright comet must reproduce these features, at least qualitatively.

F. Inference about the Composition and Structure of the Nucleus

The structure and composition of the nucleus must be inferred from modeling and analysis of coma observations. In models the nucleus is always assumed to be spherical and homogeneous, although in reality it is probably highly irregular and inhomogeneous. Modeling has already led to one important conclusion: The composition must contain some molecules of interstellar origin. But coma modeling alone cannot resolve the question about chemical composition. Modeling the formation of comet nuclei in the presolar or solar nebula or in a companion fragment of the presolar nebula also gives clues about the composition. Unfortunately, such modeling depends on the place of formation of comet nuclei, and at this time, their place of origin is as uncertain as their composition. Therefore the more common sequence in analysis is: coma observations give clues about nucleus composition which in turn gives clues about places of origin. Nevertheless, a model for the formation of comet nuclei based on this sequence must result in a composition consistent with coma models. Some models for condensation on grains in dense (10^5 molec cm^{-3}) interstellar clouds lead to inconsistencies: The heavy molecules condense in a very short time ($\sim 10^5$ years) compared to the real age of such clouds. Yet these molecules have been detected in dense clouds. Therefore, the formation of heavy molecules in these clouds must be faster than present knowledge predicts. The application of the Clausius–Clapeyron equation to interstellar cloud models to predict the composition of icy mantles on grains is not valid, because equilibrium (which the Clausius–Clapeyron equation implies) does not exist in these clouds.

Table IV summarizes attempts to intelligently guess the composition of the frozen gases in the nucleus. Sulfur is not considered because it is a minor elemental constituent and because sulfur compounds have been detected only recently; they have little effect on CNO chemistry. They will be considered in the future. The first five compositions have their origin in the suggestions made by Wurm (1943) and are valid if comets have their origin in the neighborhood of the giant planets. Composition 1 corresponds to relative solar abundance between carbon, nitrogen, and oxygen with as much H as can be bound to saturate the chemical bonds of these atoms, consistent with low temperature chemical equilibrium. The next four compositions are "near equilibrium" compositions, assuming that some of the C is bound in CO_2 or CO. Model calculations based on some of these compositions produce enough CN in agreement with observations at 1 AU, but they fail to produce the observed abundances of C_2 and C_3. They also fail in reproducing the observed heliocentric relationships summarized in Table III (Giguere and Huebner, 1978; Huebner and Giguere, 1980).

Compositions 6 and 7 (Biermann *et al.*, 1982) and 8 (Mitchell *et al.*, 1981) were the first containing interstellar molecules representing attempts to reproduce relative abundances of CN, C_2, and C_3 and their behavior as a function of heliocentric distance as presented in Table III. Composition 8 was also a crude attempt to link the molecular abundances in comets directly to those in interstellar clouds.

Delsemme (1982) has devoted much effort to determine the chemistry of the nucleus. He is a proponent for clathrates in the nucleus and has suggested composition 9. This and composition 10 (Yamamoto *et al.*, 1983) are attempts to analyze comet compositions in the light of possible interstellar abundances.

Table IV

SOME ASSUMED COMPOSITIONS FOR THE FROZEN GASES IN THE NUCLEUS (%)

	Composition number									
Species	1	2	3	4	5	6	7	8	9	10
H_2O	61.11	55.56	53.34	48.89	48.89	43.004	33.004	61.127	74.036	62.433
CH_4	30.45	—	2.22	11.11	11.11	13.464	1.650	5.094	—	0.001
NH_3	8.44	11.11	11.11	11.11	11.11	0.080	0.094	0.007	—	0.002
CO_2	—	33.33	33.33	28.89	—	12.057	—	1.698	7.404	24.886
CO	—	—	—	—	28.89	2.814	—	13.583	4.970	12.428
H_2CO	—	—	—	—	—	22.105	25.931	0.082	4.462	—
N_2	—	—	—	—	—	5.225	6.129	15.282	—	0.039
HCN	—	—	—	—	—	0.482	0.566	0.102	2.941	0.012
CH_3CN	—	—	—	—	—	0.402	0.471	—	1.420	0.012
NH_2CH_3	—	—	—	—	—	0.201	0.236	—	—	—
$H_2C_3H_2$	—	—	—	—	—	0.005	0.236	0.020	—	0.125
C_2H_2	—	—	—	—	—	0.161	0.094	0.071	2.231	0.031
HCOOH	—	—	—	—	—	—	17.445	—	—	—
CH_3OH	—	—	—	—	—	—	14.144	0.815	—	—
NO	—	—	—	—	—	—	—	1.019	—	—
O_2	—	—	—	—	—	—	—	1.019	—	—
C_2H_4	—	—	—	—	—	—	—	0.061	—	0.031
HC_3N	—	—	—	—	—	—	—	0.020	—	—
$H_2N_2H_2$	—	—	—	—	—	—	—	—	1.826	—
HC_4H	—	—	—	—	—	—	—	—	0.710	—

These have not been analyzed sufficiently in coma models, mainly because the reaction paths and rate coefficients are not known well enough. Table IV is interesting because it shows a variety of independent attempts for intelligently guessing the composition of the frozen gases in the nucleus and the general conclusion that interstellar molecules must be present, even if in some cases only as minor or trace quantities. It also illustrates the uncertainty in current thinking about the details of the interstellar components.

In summary, until we get a direct sample from a comet nucleus, the composition must be determined from coma observations and coma models supplemented with models for nucleus formation that must be consistent with the place of origin and chemical postprocessing in the nucleus surface layer by UV and cosmic radiation.

IV. Origin of Comets

The time and place of the origin of comet nuclei is not known. Before we can discuss the alternatives, whether comets form in the region of the giant planets, in the outer parts of the solar nebula, or in a companion fragment of this nebula, we must describe what is known about the Oort cloud.

A. The Oort Cloud

Orbital analysis of all parabolic comets indicates that a reservoir of $\sim 2 \times 10^{11}$ comet nuclei exists in a shell with outer radius somewhat over 50,000 AU from the Sun. This shell is known as the Oort cloud, named after its proposer (Oort, 1950). These nuclei have a total mass close to that of the Earth. Most "new" comets come from a rather narrow outer region of the Oort cloud. Their aphelion distances Q are typically between about 48,000 and 56,000 AU. The latter value represents more precisely the outer radius of the cloud. The inner radius demarcates the efficiency of passing starts to perturb enough comets in the cloud to bring them into the inner solar system and make them visible to us. Whether the cloud extends to much smaller radii (sometimes refered to as the inner cloud) cannot be proved directly. Detections of small comets by the *Infrared Astronomical Satellite* (*IRAS*) suggest that the number of comets in Oort's cloud may be significantly larger.

B. Formation in the Outer Planetary Region?

The older hypothesis that comet nuclei (in this context also called cometesimals) formed in the region of the giant planets, but before the planets formed, has now lost most of its appeal. The attraction for this hypothesis was that at this place the solar nebula would be dense enough to allow condensation on grains and accretion into small bodies that were the forerunners (planetesimals) for planet formation. Gravitational instabilities would cause coagulation and aggregation of ice-covered grains in eddies and turbulent regions in the solar nebula. After the planets formed, the remaining cometesimals would be thrown into the Oort cloud through gravitational perturbations by the planets. During this process many comets would be lost to interstellar space and many would be brought into the inner solar system where they would eventually disintegrate. The major difficulty with this hypothesis is that computer modeling of comet orbits by Everhart (1973) showed that it is not possible by this process to form a homogeneous Oort cloud as we know it.

Hills (1982) proposed that cometesimals form by radiation pressure in the outer proto-Sun at radii of a few times 10^3 AU. According to this, radiation pressure forces grains away from the radiation source. At places where the dust density is high, grains cause self-shielding, making radiation pressure less effective. Thus grains in dilute regions can move to dense regions that then become denser, and grains can coagulate and aggregate into cometesimals. This model favors the existence of an inner Oort cloud.

C. Formation in the Outer Solar Nebula or in a Companion Fragment Nebula?

Biermann and Michel (1978) proposed that cometesimals formed through gravitational instabilities of the dust layer in the equatorial plane of the outer presolar nebula. They derived a relationship between the nebula temperature and an upper limit to the cometesimal mass, applicable to a period long before the Sun developed, when approximate equilibrium between gravity, centrifugal force, and pressure gradient existed. For an assumed temperature of 15 to 20 K they obtained as an upper limit for a cometesimal mass $\sim 10^{15}$ kg, which corresponds to a radius of ~ 10 km. This radius is consistent with observed gas production rates of large comets. Similar results are obtained for comet formation in a companion fragment of the solar nebula. If the original mass of the nebula was 2 to 3 $M_\odot$ (solar masses), then the dispersive action of the new Sun removing those parts of the nebula that did not become part of the Sun, planets, planetesimals, or cometesimals reduces the gravitational binding of the cometesimals, and their orbits expand. Stellar perturbations can then easily produce the cometesimal distribution in real and velocity space as it is now observed in the Oort cloud.

In summary, the tendency of modern hypotheses on the origin of comets reflects the generally accepted conclusion that they must form further out in the presolar nebula then had previously been assumed. The further out they form, the easier it is for passing stars or stellar systems to perturb them into the Oort cloud and the more likely they will contain interstellar molecules. This would be particularly true if comets had their origin in a companion fragment of the solar nebula.

V. Measurements

Two types of measurements must be considered: measurements on comets and laboratory measurements of properties relevant to analyzing and modeling comets.

A. Measurements on Comets

Several types of measurements on comets are relevant to our discussion. From a series of determinations of comet positions on the sky, orbital parameters and their accuracy can be obtained, from which comets can be classified. If the comet is "new" then it can be expected to contain large amounts of volatile material, but if it is a short-period comet it will have lost most of its volatile materials. All other comets will fall between these extremes.

Some radar measurements have been made at Arecibo to obtain effective reflection areas for the nucleus. Corresponding mean values for the nuclear radius R_0 are 0.6 km for Comet P/Encke, 0.8 km for Comet *IRAS*–Araki–Alcock, and 0.11 km for Comet Sugano–Saigusa–Fujikawa. The latter two comets passed the Earth at $\Delta = 0.03$ and 0.04 AU in 1983.

Photometric and spectral measurements and their variation with heliocentric distance tell us about the volatility, composition, and size of the comet. The relative intensity of emission features tell us about the relative abundances of various species. The profile of C_2 emissions has been found to be quite flat as a function of distance into the coma compared to emissions from other molecules. The averaged C_2 column density in the coma increases more rapidly than that of other species, notably that of OH, with decreasing heliocentric distance of the comet. These different behaviors of C_2 tell us that it is not a primary decay product. The scattered radiation intensity of Lyα from the hydrogen cloud provides very accurate production rates. For a number of comets the production rate at $r \simeq 1$ AU is nearly the same, about 2 to 4×10^{29} molec sec^{-1}, and typically varies with $r^{-2.3}$.

Infrared (IR) emission from the dust gives information about the temperature of the dust and the size of the dust grains. Dust temperatures are higher than would be expected from absorption of sunlight and blackbody reemission. Grains appear to be of submicrometer size, they cannot reemit the long wavelength IR radiation and therefore heat up. Silicate features have been detected. The mass ratio of dust to gas shows large variations. For a dust-rich comet it is about 0.7–0.8, as determined from comets Arend–Roland (1957 III) and Bennett (1970 II) (see also Fig. 2).

The ratio of CO^+ to H_2O^+ has been measured only once (in the near tail of Comet Kohoutek). Model calculations cannot reproduce the observed high ratio $CO^+/H_2O^+ \simeq 30$.

Measurement of the angle of the plasma tail relative to the radius vector from the Sun gives information about the solar wind. Thus comets can be used as solar wind probes, particularly in regions away from the ecliptic where it is expensive to use satellites. These measurements have been extensively studied and analyzed by Brandt (1982). The curvature of the dust tail provides clues about the composition (conducting or dielectric material) and size of the dust grains in the tail.

Most important for coma modeling are column densities corrected for aperture size of the telescope. Conversion to production rates, a practice often exercised by observers, is less useful for modeling, since it already involves application of some coma model.

B. Laboratory Measurements

The most important laboratory measurements for coma modeling are determinations of rate coefficients of chemical reactions and photolytic cross sections, including the branching ratios for the products. A table of solar photo rate coefficients and associated excess energies is presented in Appendix I. Chemical reactions with rate coefficients can be found in the papers by Giguere and Huebner (1978), Huebner and Giguere (1980), and Mitchell *et al.* (1981). Most of these reactions have not been included in Appendix II because the list is too long. Heterogeneous (grain surface) reactions have not been considered in coma chemistry models, because very little is known about them. This could be a fertile field for laboratory investigations. Measurements of oscillator strengths are needed to convert observed intensity profiles to species column densities for comparison with model calculations.

Laboratory determinations of physicochemical properties of frozen gases, mixtures of frozen gases, and mixtures of frozen gases and grains are needed to model the vaporization of comet nuclei. Space Shuttle experiments exposing artificially made icy conglomerate materials directly to solar radiation could be particularly revealing.

In summary, more comet and laboratory measurements are needed to improve our understanding of the physics and chemistry of comets. Specifically, photo cross sections and electron dissociative recombination branching ratios for complex molecules and molecular ions are needed.

VI. Photochemistry of the Coma

Modeling a comet coma must take into account the physics and chemistry of the (neutral) coma gas from the point of production on the nucleus surface to the outer hydrogen coma, the production of the plasma component and its streaming until it interacts with the solar wind and moves into the plasma tail, the chemical interaction of the plasma component with the neutral gas, and the entrainment of the dust by the coma gas and the solar radiation pressure that moves the dust into the tail. Such a model involves collision-dominated flow in the inner coma, transition to free molecular flow, multifluid flow (the lighter hydrogen decouples from the bulk of the heavier gas), counterstreaming of the plasma relative to the neutral gas in the solar wind interaction region between the contact surface and the bow shock, heating and cooling, and photochemical reactions. No model exists that takes all these processes into account. In the following sections a dust-free model is described without solar wind interaction.

A. Physics of the Collision-Dominated Coma

It is not possible to completely separate the chemistry from the physics of the coma. Some aspects of chemistry influence the physics; they are described in this section. The main photolytic and chemical reaction processes are described in the next section.

The frozen gases in an icy conglomerate nucleus will evaporate under the influence of solar radiation. The gas so produced will be in equilibrium with the surface. Energy conservation requires that the energy of insolation, corrected for reflection from the nucleus, must equal the energy that is reradiated plus the energy used for evaporation, if we assume that the energy conducted into the interior is negligible and (since only a dust-free coma is considered) that reflection and reradiation by dust onto the nucleus are ignored. Mathematically expressed (for uniform distribution of energy over the nucleus surface), this becomes

$$(F_0/r^2)\pi R_0^2/(4\pi R_0^2)e^{-\tau}(1 - A) = \epsilon\sigma_0 T_0^4 + ZL/N_0, \tag{1}$$

where F_0 is the solar energy flux at $r = 1$ AU heliocentric distance, R_0 the radius of the comet nucleus, τ the mean (visual) optical depth in the coma, A the mean (visual) albedo of the nucleus, ϵ its mean IR emissivity, σ_0 the Stefan–Boltzmann constant, T_0 the equilibrium temperature between gas and icy conglomerate on the surface, Z the molecular production rate per unit surface area, L the mean latent heat of sublimation (vaporization) per mole, and N_0 Avogadro's number. The quantities R_0, τ, A, ϵ, and L are parameters. This leaves two unknowns: Z and T_0.

To solve for Z and T_0, an additional equation is needed. This equation is obtained by combining the equations of state,

$$p_0 = p_r \exp[(T_r^{-1} - T_0^{-1})L/(kN_0)], \qquad \text{Clausius–Clapeyron,} \tag{2}$$

$$n_0 = p_0/(kT_0), \qquad \text{ideal gas,} \tag{3}$$

and the relationship between gas density and production rate per unit area,

$$Z = n_0 v_s. \tag{4}$$

In these equations p_r is the pressure at a reference temperature T_r, p_0 and n_0 are the gas pressure and gas number density just above the icy conglomerate surface at radius R_0, k is the Boltzmann constant, and

$$v_s = (\gamma N_0 k T_0/M)^{1/2} \tag{5}$$

is the sound speed, where the mean of the ratio of specific heats is

$$\gamma = \sum_i n_i \gamma_i \Big/ \sum_i n_i, \tag{6}$$

and the mean molecular weight is

$$M = \sum_i n_i M_i \Big/ \sum_i n_i. \tag{7}$$

Here γ_i, M_i, and n_i are the ratio of specific heats, the gram molecular weight, and the number density of chemical species i. In practical applications of these equations, the mean quantities L, p_r, and L/T_r in Eq. (2) are defined in a way analogous to Eq. (6) or (7), but they can also take into account clathrates.

Simultaneous solution of these equations for heliocentric distance $r \simeq 1$ AU yields a gas temperature $T_0 \simeq 150$ K, gas production rate per unit surface area $Z \simeq 10^{17}$ molec cm^{-2} sec^{-1} and number density $n_0 \simeq 10^{13}$ to 10^{14} molec cm^{-3}. These values are sufficiently large to provide collision energies and collision frequencies between molecules to cause exothermic reactions for some distance (typically several 10^3 to 10^4 km) into the coma. The extent of these reactions and the question whether some endothermic reactions could also take place depend on density, temperature, and flow speed profiles. These, particularly the temperature, depend on the photolytic and chemical reactions.

The brightness of a comet depends on the type, abundance, and total number of molecules in the coma. (Some abundant molecules like H_2O do not fluoresce in the visual spectrum; only minor constituents like CN, C_2, and C_3 contribute significantly.) For dusty comets, the brightness depends also on the amount of dust. But since the dust is entrained by the coma gas, the brightness is proportional to the number of coma molecules. As is apparent from the above equations, at small r, Z is proportional to r^{-2}; at large r, Z decreases about exponentially with increasing r. For spinning, water-dominated comets a sharp knee in the functional behavior of Z occurs at $r \sim 1.4$ AU. For more volatile material (smaller L), the knee occurs at larger r. A comet that shows a coma, tail, or strong coma and tail activity at $r > 1.4$ AU is therefore not likely to be water dominated. Some of these unusual comets are described in a separate section.

Only one-dimensional coma calculations will be described here. Two-dimensional UV optical depth effects (discussed in the next section) have been considered; they show deviations from a one-dimensional calculation only in a narrow cone in the tail direction of the comet. Therefore, for the inner coma a one-dimensional calculation is sufficient, but in the outer coma, where solar wind interactions occur, two-dimensional calculations will be absolutely essential for the ions. The inner coma models are based on the processes occurring in a shell of coma gas as it expands and moves radially outward with velocity v.

To obtain the density, temperature, and flow speed profiles, the fluid dynamic conservation equations must be solved. Ideally they should be solved for each species with collisional coupling between the species and the chemical kinetics as part of the source and sink terms. This would require three equations for each species, or over 300 equations for a typical coma calculation with over 100 species. So many differential equations are an unncessary complexity for solving this problem. The usual approach taken is to solve the chemical reaction network with a stiff differential equation solver method on a fine mesh and to solve the conservation equations for mass (rather than atomic elements), momentum, and energy for the bulk motion of all molecules in a Lagrangian coordinate system, but corrected for the fast nonthermalized hydrogen species:

$$d(\rho v R^2) = \mu R^2\, dR, \qquad \text{mass conservation,} \tag{8}$$

$$d(\rho v^2 R^2) + dp = \Pi R^2\, dR, \qquad \text{momentum conservation,} \tag{9}$$

$$d[\rho v^3 R^2/2 + \gamma p v R^2/(\gamma - 1)] = \epsilon R^2\, dR, \qquad \text{energy conservation.} \tag{10}$$

In these equations ρ, v, and p are the mass density, velocity, and pressure of the gas at a distance R from the center of the nucleus. The terms μ, Π, and ϵ are the sources and sinks for mass, momentum, and energy, each quantity per unit volume and per unit time. Contributions to these terms from fast hydrogens are discussed below. Effects on these terms from gas–dust collisions have been investigated by Marconi and Mendis (1982). But effects from gas–dust collisions, solar wind interaction, and radiation pressure on the chemistry have never been investigated in a comprehensive model.

Endothermic reactions and escaping radiation, emitted from collisionally excited molecular species, are energy sinks. In addition, expansion of the coma gas into a near vacuum is a cooling process. Excess energy imparted to products of exothermic chemical reactions or of photodissociations are energy sources if the product particles can be thermalized collisionally. If such a product particle has a speed much higher than the bulk speed of the fluid, then this fast particle is likely to leave the shell (energy sink for that shell) but can be collisionally thermalized in another shell (energy source for that shell). The probability of leaving one shell and being thermalized in another depends on the thermalization mean free path of the fast particle.

The excess photon energy above the dissociation threshold, if it is not too large, goes mostly into kinetic energy of the dissociation products. This has been verified experimentally for water dissociation at low excess photon energy. There is no verification for high excess energy, but then the photon energy is far above the threshold energy where the cross section is small, and the process contributes relatively little to the overall energy budget of the coma. Excess kinetic energy in dissociation and kinetic energy of exothermic reaction products are the main sources for the pressure increase relative to the density decrease and the related temperature rise in the region $R \simeq 10^2$ to 10^4 km (see Fig. 4). Solar photo rate coefficients and excess energies have been compiled in Appendix I.

Because of momentum conservation, the particle with smaller mass m_s of two photofragments receives the larger fraction of the excess energy in the ratio $m_l/(m_l + m_s)$, where m_l is the fragment with the larger mass. For example, in the photodissociation $H_2O + h\nu \rightarrow OH + H$, for which the excess energy in the solar radiation field at 1 AU heliocentric distance $E_x = 3.4$ eV, the kinetic energy of H is 3.4 eV $\times \frac{17}{18} \simeq 3.2$ eV, while that for OH is 3.4 eV $\times \frac{1}{18} \simeq 0.2$ eV. The 3.2-eV kinetic energy of H corresponds to a speed of ~ 25 km sec^{-1} relative to the ~ 1 km sec^{-1} average speed of the streaming bulk gas, approximately in agreement with observation. On the other hand, the 0.2-eV kinetic energy of OH corresponds to only ~ 1.5 km sec^{-1} relative to the bulk gas, and this energy is rapidly shared with it. In photoionization the electron receives practically all of the excess energy.

Product molecules with molecular weights nearly equal to that of the bulk fluid in which they are immersed are almost thermalized in one collision; their thermalization mean free path is about equal to their collisional mean free path, $\Lambda_{th} \simeq \Lambda_{coll}$. Molecules much lighter than the bulk fluid's mean value require many collisions before they are thermalized. For example, fast hydrogen atoms in water vapor require about 9 or 10 random walk collisions to lose half their energy (we shall call that thermalized), $\Lambda_{th} \simeq 9.5^{1/2}\Lambda_{coll}$.

In the photodissociation of H_2O, hydrogen receives most of the excess energy but tends to carry it away instead of sharing it with the bulk of the coma gas. The coma gas will therefore not be heated effectively by this process. This also holds true for any dissociation or reaction producing H or H_2, and to a lesser degree it also holds for atomic carbon, nitrogen, and oxygen products and their hydrides.

We shall assume that the fluid is composed of three components: one for fast atomic hydrogen H(f), one for fast molecular hydrogen H_2(f), and one for the rest of the bulk fluid including the thermalized atomic and molecular hydrogen. If the cross-section radius $a_b = 4.4$ Å for collisions between molecules of the bulk fluid and $a_f = 3.3$ Å for both of the fast hydrogen components H(f) and H_2(f) colliding with fluid molecules, then the collision mean free path between bulk fluid molecules is $\Lambda_b = (\sqrt{2}n\pi a_b^2)^{-1} \simeq 1.16 \times 10^{14}/n$ cm and for fast hydrogens [H(f) and H_2(f)] it is $\Lambda_f \simeq 2.07 \times 10^{14}/n$ cm, where n is the fluid number density in cm^{-3}. The mean free path concept is most important near the diffuse collision zone boundary, where fluid flow changes to free molecular flow. Here the average random walk model of Combi and Delsemme (1980) provides a formalism for an effective mean free path for purely radial outflow models, such as the Haser model. This is unnecessary if the first flight functions discussed below are used, since they contain angle integrations, where we define $\Lambda = 9.5^{1/2}\Lambda_f$ and $\Lambda' = \Lambda_b$.

Mean free paths depend on the local number density n and therefore on the radial distance R. The radius at which fluid flow gradually changes to free molecular flow is the critical radius R_c, defined as the place in the coma where the local mean free path equals the radius $R_c \equiv \Lambda(R)|_{R=R_c}$. Different species have different critical radii, for fast hydrogen $R_c \simeq 3 \times 10^3$ km at 1 AU heliocentric distance. Once the critical radius and the production of fast hydrogen $P_H = \sum_i k_i^H n_i^H$ are known, the fast particle first flight flux $\Phi(R)$ and first flight current density (first angular moment of flux) $j(R)$ for an R^{-2} density distribution (Huebner and Keady, 1984) can be used to obtain the fast particle density and the inhomogeneous terms for the conservation equations, where k_i^H is the rate coefficient for production of hydrogen from species i, and n_i^H is the number density of molecular species i that produces hydrogen. Similar terms apply for the production of H_2(f). The fast particle first flight flux is

$$\Phi(R) = P_f R^2 \Phi_1(R/R_c)/R_c, \tag{11}$$

and the fast particle first flight current density is

$$j(R) = P_f R^2 j_1(R/R_c)/R_c, \tag{12}$$

where the subscript f signifies reactions producing fast species H or H_2. The functions $\Phi_1(R/R_c)$ and $j_1(R/R_c)$ are shown in Fig. 3.

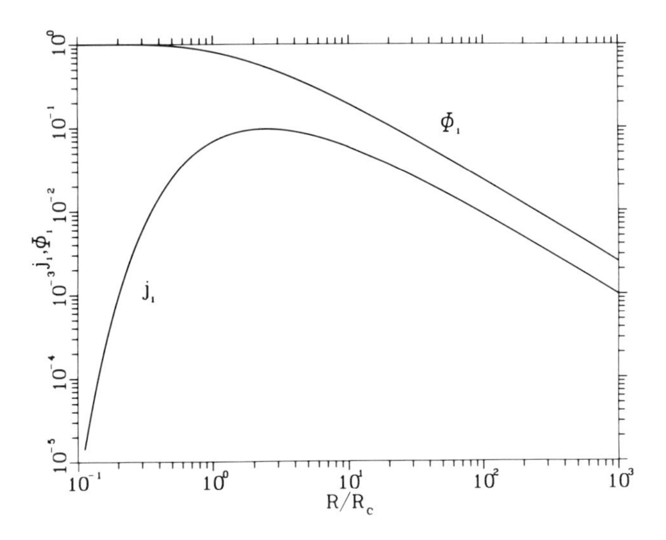

Fig. 3. First flight emitted particle flux Φ_1 and current density j_1 for fast particles as function of R/R_c.

The inhomogeneous terms of the fluid conservation equations can be written

$$\mu(R)R^2\,dR = -d[R^2 j(R)]m[v_f/(v_f + v)], \tag{13}$$

$$\Pi(R)R^2\,dR = d[R^2 j(R)]|mv_f|[v_f/(v_f + v)], \tag{14}$$

$$\epsilon(R)R^2\,dR = E\{P_f R^2\,dR - d[R^2 j(R)]\}[v_f/(v_f + v)] + E'\{PR^2\,dR - d[R^2 j'(R)]\} - \epsilon_{rad} R^2\,dR, \tag{15}$$

where the right-hand sides are summed over particles, E is the fraction of excess energy imparted to fast hydrogens and E' the fraction of excess energy imparted to products other than fast hydrogens ($E_x = E + E'$), P_f and P are the production rates ($cm^{-3}\ sec^{-1}$) of fast hydrogens and all other products, respectively, $j'(R)$ is the first flight current density of particles other than fast hydrogens with $R_c' = \Lambda'(R)|_{R=R_c'}$ for the bulk fluid, and v stands for the speed of the bulk fluid, v_f for that of the fast hydrogens relative to the bulk fluid. Since the function $j(R)$ considers the birth and death of the fast particles at a cost to the bulk fluid in Eq. (13), negative values of $j(R)$ take into account the addition of mass to the bulk fluid. The energy gain of the fast particles in Eq. (15) is not at a cost to the bulk fluid but comes from the photon field and must therefore be added separately. The additional factor, involving only velocities, corrects for the motion of the fluid in which the fast particles are produced. The transit time of fast hydrogen relative to the transit time of the bulk fluid through a given distance is equal to the reciprocal of the corresponding speeds, $v/(v_f + v)$. One minus this fraction, namely, $v_f/(v_f + v)$, determines the loss of fast hydrogen from the moving shell. If the fast hydrogen were produced with $v_f = 0$, then this "fast" hydrogen would never leave the shell, regardless how large its mean free path is. On the other hand, if $v_f \gg v$, then the loss of fast hydrogen from the fluid shell is determined entirely by its mean free path.

The density of fast particles at point R is obtained by dividing $\Phi(R)$ by the speed of the fast particles at R. As Huebner and Keady have shown, the functions $\Phi_1(R)$ and $j_1(R)$ are related to the Feautrier formalism in radiative transfer. Thus the procedure outlined here is suitable for including higher-order collisional effects without using the more costly Monte Carlo procedure.

Radiative cooling is designated by ϵ_{rad}; the main contributor to it is H_2O. A semiempirical equation for it has been given by Shimizu (1976), but it does not include consideration for radiation trapping in the coma. An optical depth calculation is sufficient to approximate the radiation trapping. The optical depth is

$$\tau' = \int_R^\infty n\sigma\, dR \simeq n_0 v_0 R_0^2 \sigma \int_R^\infty \frac{dR}{vR^2} \simeq \frac{0.4 n_0 R_0^2 \sigma}{R}, \tag{16}$$

where a constant value for $v = 2.5\, v_0$ was assumed for the inner coma to account approximately for the rapid increase of v near the nucleus. For IR H_2O radiation $\sigma = 4 \times 10^{-15}$ cm^2, $n_0 \simeq 10^{13}$ cm^{-3} (for $r \simeq 1$ AU), and $R_0 =$ 2.5 km, $\tau' \simeq 10^4/R$, with R in kilometers. This means that radiative cooling is not important before the onset of dissociative and chemical heating, unless the comet is very small or is at a large heliocentric distance r so that its gas production and therefore its density n are very low. Shimizu's formula for radiative cooling, modified by the above optical depth, is

$$\epsilon_{rad} = \frac{8.5 \times 10^{-19}\, T^2 n^2}{n + 2.7 \times 10^7\, T} \exp \tau' \quad \text{erg cm}^{-3}\ \text{sec}^{-1}, \tag{17}$$

where n is the R-dependent number density of H_2O.

Equations (8)–(10) can now be solved for ρ, p, and v. The local number density n can then be obtained using the mean molecular weight for the local gas composition, and from this and the ideal gas law the local temperature T can be obtained. Model results for density ρ, pressure p, temperature T, and velocities v_s, v, $v_{H(f)}$, and $v_{H_2(f)}$ are shown in Fig. 4. These results are influenced by chemistry that will be discussed below.

B. Photochemistry

Chemical reactions between evaporated mother molecules are very slow and of secondary importance only, even in large comets, but dissociation and ionization by solar UV radiation produce highly reactive radicals and ions. While it is mostly the visual part of the solar spectrum that causes production of gas from the nucleus, it is the UV spectrum that is responsible for initiating chemical reactions. Near the nucleus surface, however, the coma is opaque to UV radiation except for small comets and comets at large heliocentric distance. From the relationship between density n and distance R in the coma, from the photo cross sections of the coma constituents, and from their relative abundances, the attenuation of the solar UV radiation can be calculated. This attenuation depends not only on the gas density and distance above the nucleus, but also on wavelength and angle of incidence of the sunlight. Calculations show, however, that the attenuation varies only weakly with angle as measured from the comet–Sun axis except for a narrow cone in the antisolar direction. Therefore only one dimension is considered here.

Mother molecules dominate the composition in the region where attenuation is largest. Assuming an R^{-2} dependence for the density n, which is a good approximation except very close to the nucleus, the UV optical depth is

$$\tau''(\lambda) \simeq -\int_{\infty}^{R} \sum_i n_i \sigma_i(\lambda)\, dR = \sum_i n_i \sigma_i(\lambda) R. \tag{18}$$

The summation over i is effectively only over mother molecules. The contribution from radicals is very small and can be calculated only after a complete chemistry calculation of the coma. A second iteration of the coma calculation can be carried out in which the contribution from the radicals and a more exact density profile are taken into account.

The effective photo rate coefficients can be calculated from

$$k_i = \sum_j k_i(\lambda_j) e^{-\tau''(\lambda_j)}, \tag{19}$$

where the $k_i(\lambda_j)$ are the photo rate coefficients for species i in the wavelength bin λ_j. In the optically thin case the values for the k_i are consistent with the tabular entries in Appendix I.

An alternative to producing radicals by dissociation of mother molecules in the solar UV radiation field is the decomposition of clathrates containing radicals in visible sunlight. In that case chemical reactions could take place even in the innermost coma where UV radiation is strongly attenuated. So far, clathrates have not been identified in comets.

Another very important process is photodissociative ionization (PDI), a process in which an extreme-UV photon causes a molecule to be ionized into an excited state from which the molecular ion decays by dissociation. An example is

$$CO_2 + h\nu \longrightarrow O + CO^+ + e^- \tag{20}$$

Even though the rate coefficients for PDI are small, the hard photons

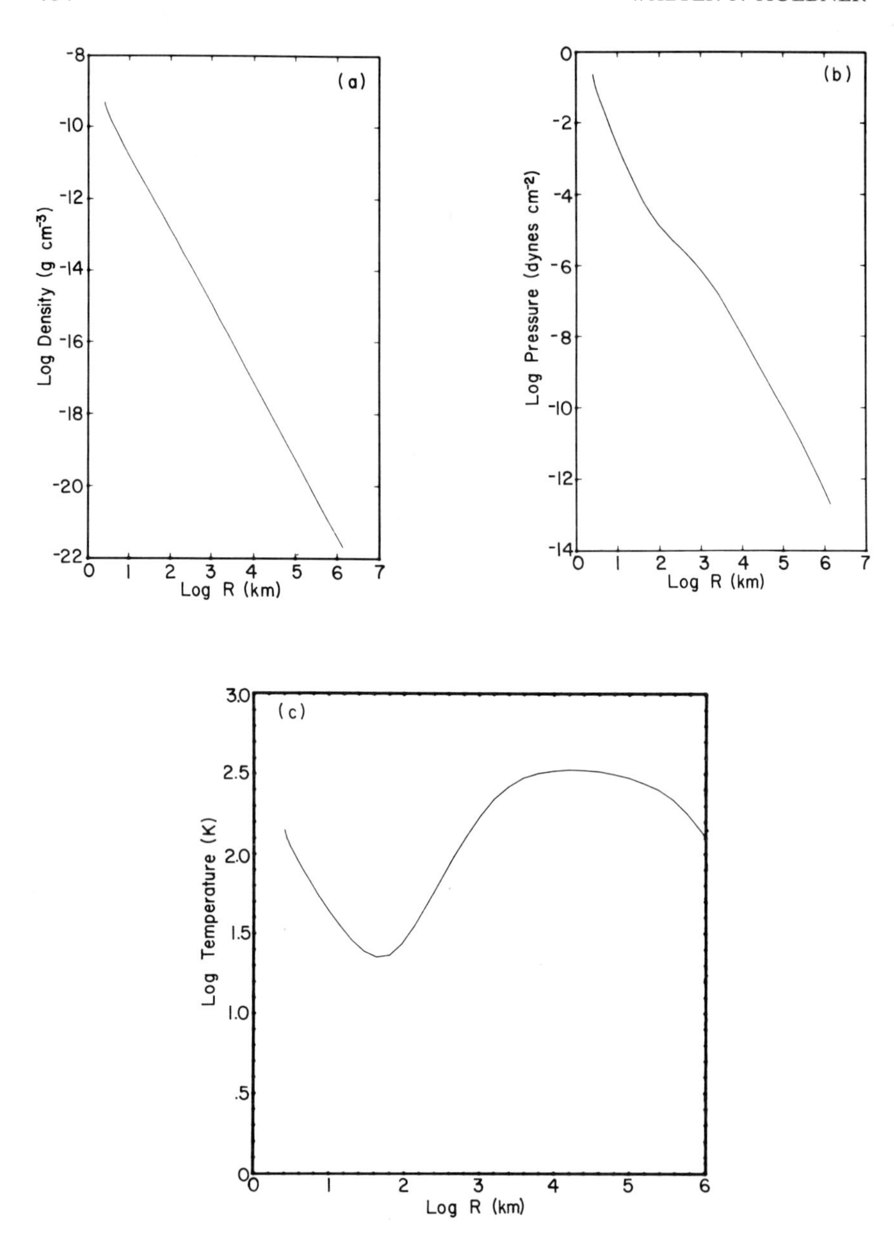

Fig. 4. Model results for density ρ (a), pressure p (b), temperature T (c), and velocities (d) plotted as function of distance into the coma, log$R$. The bottom curve represents the sound speed v_s, the next curve up is the bulk speed v, this is followed by the fast molecular hydrogen speed $v_{H_2(f)}$, and finally the fast atomic hydrogen speed $v_{H(f)}$. Results correspond to composition 6 in Table IV, for $r = 1$ AU.

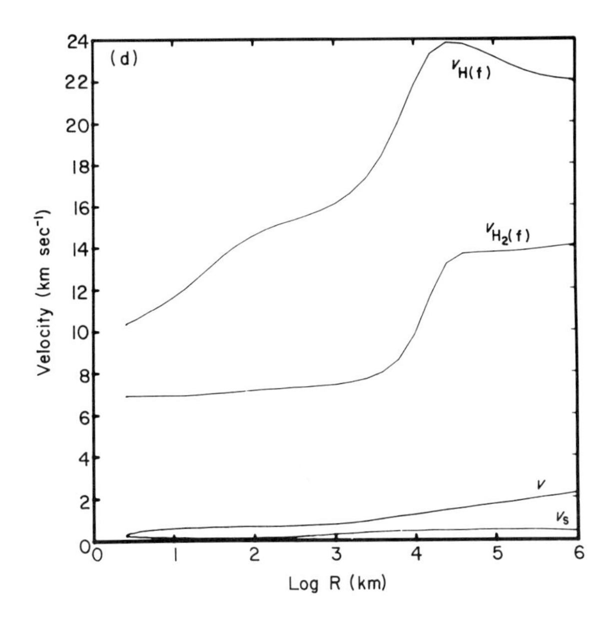

Fig. 4. *(cont.)*

necessary, for PDI penetrate deeply into the coma where they are deposited in a relatively thin shell with high density *n*. Therefore, this process is important for the production of CO^+ deep in the coma, at $\sim 10^3$ km from the nucleus, where CO^+ is observed. Photodissociative ionization of other CO-bearing molecules gives similar results.

Photon energies in excess of the ionization potential produce energetic electrons (often $\geqslant 10$ eV; see Appendix I) that can cause further ionization and dissociation by impact. This is a secondary process, with effects on the coma that have not been investigated in detail but are expected to increase the fraction of ions.

Electron-dissociative recombination is another very important process in which an electron recombines with a molecular ion and simultaneously dissociates that ion. Because of the Coulomb interaction between the electron and the ion, this process has a very large cross section and therefore a large rate coefficient. The cross section increases inversely proportionally to the electron velocity relative to the ion, or proportional to $T^{-1/2}$.

Ion gas-phase reactions including positive ion–atom interchange and positive ion charge transfer compose a large part of the chemical reaction network that must be considered. Examples of these two reactions are

$$CO^+ + H_2O \longrightarrow HCO^+ + OH \tag{21}$$

$$CO^+ + H_2O \longrightarrow H_2O^+ + CO \tag{22}$$

Also important is radiative electronic state deexcitation, such as

$$O(^1D) \longrightarrow O(^3P) + h\nu \tag{23}$$

which is one of the forbidden O transitions.

Of lesser importance are neutral rearrangements, radiation-stabilized neutral recombinations, radiation-stabilized positive ion–neutral associations, neutral–neutral associative ionizations, radiative recombinations, three-body neutral recombinations, and three-body positive ion–neutral associations. Negative ions play no role in comet comae because the densities are too low. Heterogeneous (grain surface) reactions may be important but have not been considered because virtually no rate coefficients are available.

Because of the expansion of the coma gas as it moves radially outward, chemical reactions take place in a continually diluting gas exposed to ever increasing solar UV radiation. For this reason chemical kinetics must be considered instead of a steady state. Some chemical reactions proceed very fast compared to the fluid dynamic dilution, while others are slow. In the practical application of solving a network with several hundred photochemical reactions, the time step for the chemistry needs to be much smaller than the fluid dynamic time step at which the density, temperature, and attenuated solar flux are recalculated. Fluid dynamic time steps are spaced approximately logarithmically. Only at large distance from the nucleus does the chemical time step approach the fluid dynamic time step. A stiff differential equation solver technique is used to solve the chemical reaction network. Although some very fast chemical reactions may reach a steady state in any one fluid dynamic time step, many reactions will not, and some species will "freeze-in" as the gas expands and reactions terminate.

Since different reactions dominate in the same comet at different heliocentric distances and at different distances into the coma, a network of several hundred reactions is required. Competing processes leading to the same product or destroying the same reactant can change their relative importance. For example, at heliocentric distance $r \simeq 1$ AU, in the innermost coma, CO^+ is produced by photodissociative ionization of CO-bearing molecules, while in the intermediate coma it is produced by photoionization of CO. The crossover of these two processes occurs within $\sim 10^3$ km, while at $r \simeq 3$ AU this crossover is at $\sim 5 \times 10^3$ km. At $r \simeq 1$ AU, destruction of CO^+ in the inner coma occurs by positive ion–atom interchange [e.g., reaction (21)] and charge exchange [e.g., reaction (22)], while in the outer coma electron dissociative recombination is the dominant destruction mechanism of CO^+. The crossover occurs within $\sim 10^4$ km, while at $r \simeq 3$ AU it is at $\sim 2 \times 10^4$ km. Another example is the production of H_3O^+. For composition 6 in Table IV, at heliocentric distance $r \simeq 3$ AU, the process

$$H_2O^+ + H_2O \longrightarrow H_3O^+ + OH \tag{24}$$

dominates in the entire coma out to 10^5 km. At $r \simeq 1$ AU, it dominates out to $\sim 3 \times 10^4$ km; beyond this the reaction

$$H_2^+ + H_2O \longrightarrow H_3O^+ + H \tag{25}$$

dominates. At $r \simeq 0.6$ AU, reaction (24) dominates out to $\sim 10^4$ km, at $\sim 3 \times 10^4$ km the reaction

$$HCO^+ + H_2O \longrightarrow H_3O^+ + CO \tag{26}$$

dominates, while at $\sim 10^5$ km,

$$H_2O^+ + H_2 \longrightarrow H_3O^+ + H \tag{27}$$

dominates. Different initial compositions may cause different reactions to dominate. Competing, parallel processes also lead to an unexpected benefit: they tend to minimize the effects from uncertainties in the rate coefficients.

Figure 5 illustrates results for number densities from a calculation using composition 6. Predicted, but as yet not detected, species include H_3O^+ with certainty, HCO^+, C_2H_2, C_2H_4, and possibly HCO_2^+ and CH_2OH^+ if a hydrocarbon precursor exists in the nucleus with sufficient abundance, NH_4^+ if NH_3 exists even with small abundance, and NO and H_2CN^+. (Even though C_2H_2 is a common result of coma chemistry, in composition 6 it was also a mother molecule.)

C. Nongravitational Forces

Equation (1) expresses the energy balance at the nucleus surface as an angular average. A more detailed examination shows that the incident solar radiation is highest per unit surface area at the subsolar point on the nucleus. This means that vaporization occurs predominantly on one side of the nucleus, but not necessarily in the subsolar direction; thermal inertia can shift the most active region to the afternoon side of the nucleus. The effective velocity component for this mass loss is ~ 100 m sec^{-1}. With a mass loss of $\sim 10^6$ g sec^{-1}, a nongravitational acceleration of $\sim 10^{-5}$ cm sec^{-2} results at $r \simeq 1$ AU. This compares to a solar gravitational acceleration of ~ 0.7 cm sec^{-2} at 1 AU. Although about five orders of magnitude smaller, the nongravitational acceleration causes measurable perturbation on the orbit of comets that have small (typically < 1.5 AU) perihelion distances.

These nongravitational forces were recognized by Whipple (1950, 1951) and supported his reasons for proposing the icy conglomerate nucleus. Only a solid nucleus, as opposed to a loose assembly of particles, can react as a unit to the recoil from asymmetrically evaporating gases. Many comet orbits have been analyzed by Marsden and co-workers for the effects of nongravitational forces.

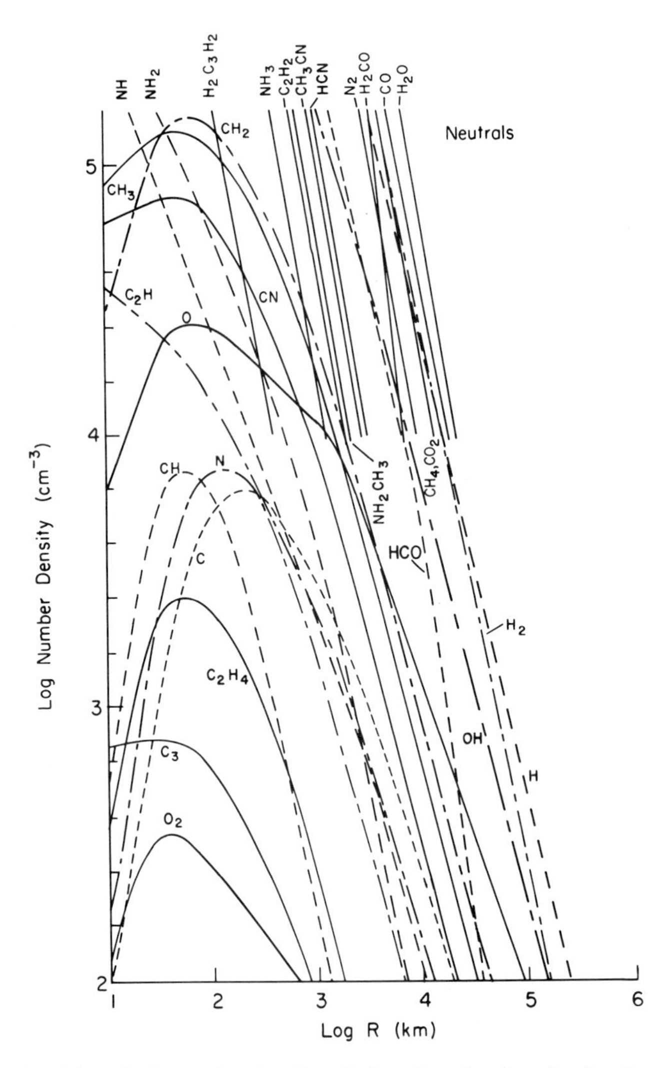

Fig. 5. Model results for number densities of selected species plotted as function of distance into the coma, log$R$. Results correspond to composition six in Table IV, for $r = 1$ AU. Mother molecules are not shown below a density of 10^4 cm^{-3} in order to keep the graph readable. Also, the ion species are plotted separately for the purpose of clarity.

D. Transition to Free Molecular Flow and the Outer Coma

Outside the critical radius for the bulk fluid, collisions become rare, chemical reactions (except those involving Coulomb interactions) cease, dissociation and ionization products stop sharing their excess energy, and

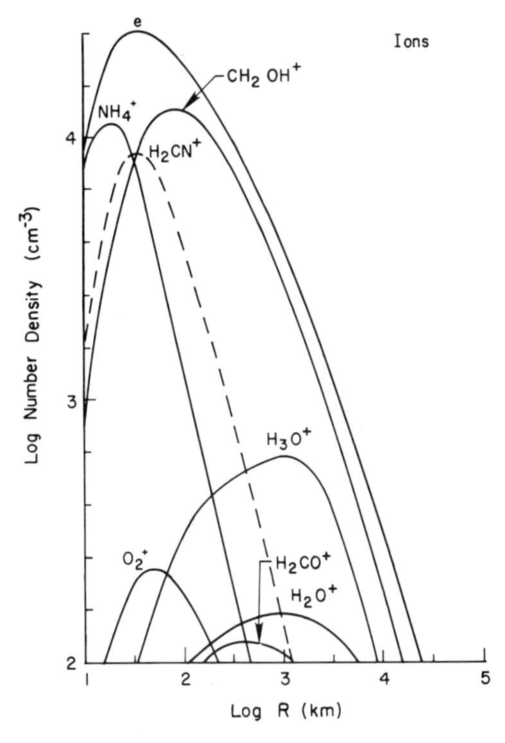

Fig. 5. *(cont.)*

continue on their path as determined by their initial velocity. (We ignore here further complications caused by radiation pressure on neutrals, solar wind interaction, and gyration of ions and electrons in magnetic fields that cause counter streaming and two-stream instabilities.) The average random walk model of Combi and Delsemme (1980) becomes an important tool for modeling, [but see comment above Eq. (11)] and brightness profile analysis. Their model is a good, simple approximation to the actual random walk model as simulated by Monte Carlo calculations. It takes into account the random ejection direction of product particles and removes the often-discussed discrepancies between observed and theoretical scale lengths. It assumes that the dissociation product receives its ejection velocity, on the average, perpendicular to the velocity of its mother molecule in that molecule's frame of reference.

Of particular interest and importance in the free molecular flow region is the hydrogen coma. The velocity profile of atomic hydrogen can be determined from the Doppler shift of solar Lyα radiation scattered by the hydrogen coma. Analysis has shown that at least two major velocity components exist: one at

~ 20 km sec^{-1}, the other at ~ 8 km sec^{-1}. It is believed that the first of these components results from the excess energy imparted to H in the dissociation of H_2O. The second component may have its origin in the dissociation of OH, which itself is a dissociation product of H_2O. The dissociation cross section of OH is only approximately known; predissociation and superposition of solar Fraunhofer lines on the predissociation lines play an important role.

From the column density profiles of H and OH a production rate of H_2O can be determined. For a large number of bright comets this H_2O production rate is $\sim 10^{29}$ molec sec^{-1}. Assuming $Z \simeq 10^{17}$ molec cm^{-2} sec^{-1}, this yields a nucleus radius of a few kilometers.

Radiation pressure of solar Lyα on the hydrogen coma causes deviations from spherical symmetry. On the subsolar side, the coma intensity (and number density) decreases more rapidly than in the direction perpendicular toward the Sun and it extends further in the tail direction.

E. Solar Wind Interaction

The neutral component of the coma gas streams almost radially outward, essentially unimpeded by the solar wind. Small deviations from this are caused by radiation pressure on the neutrals and near the contact surface (the boundary of solar wind penetration into the coma) where some collisions occur between neutrals and nonradially moving ion streams. Radiation pressure is most effective on species with strong resonance transitions in the visible region of the spectrum. Just as in the hydrogen coma, radiation pressure causes an asymmetry of the coma in the tail direction.

On the other hand, coma ions move radially outward until they encounter the contact surface. At that point they experience strong interactions with the solar wind ions. Between the contact surface (probably several times 10^3 km from the nucleus in the subsolar direction) and the bow shock ($\sim 10^6$ km from the nucleus) the solar wind is loaded down with coma ions. Solar wind electrons impacting with some of the coma neutrals may cause some of these ions. The ion-loaded solar wind slows down as it penetrates inward from the bow shock, and finally stagnates at the subsolar point on the contact surface. In this region cometary ions couple to the solar wind and are swept tailwards. The solar wind interaction with the coma has been modeled extensively by Schmidt and Wegmann (1982). They have also taken into account the Alfvén critical velocity effect. This effect is a collective process, caused by plasma instabilities, in which electrons produced by collisonal ionization are heated so that they cause further ionization when a plasma with a magnetic field collides with a neutral gas and the relative kinetic energy between plasma (ion) and gas (molecule) is larger than the ionization energy of a gas molecule.

In the region between the contact surface and outer shock the outward streaming coma plasma and the solar wind can be treated as fluids since

Coulomb interactions result in large collision cross sections. Photolytic processes continue in this region and further complicate the situation. Fortunately, at least in first approximation, this region can be treated semiindependently from the region inside the contact surface. But multifluid flow must be considered.

F. Comparisons of Models with Observations

No direct comparisons of the model density, pressure, and temperature profiles shown in Fig. 4 are possible. Comparisons with the model velocity profiles, however, can be made. Observations indicate a mean bulk velocity in the visual coma (out to a few 10^4 km) of ~ 1 km sec^{-1}. This is in good agreement with model results. The velocity profile for the fast atomic hydrogen H(f) in Fig. 4 is an average from many different dissociations, but H_2O dissociation is an important contributor to it. In the outer coma, beyond several 10^4 km, the model indicates a velocity of ~ 20 km sec^{-1} or slightly higher, in agreement with the observed high velocity component. The low velocity component has not been separated out in the model calculations. H_2(f) cannot be observed; the abundance in the coma is too low for present instrumentation.

Model column densities for selected species are shown in Fig. 6. The atomic hydrogen density is split into two parts: the thermalized hydrogen, labeled H, and the fast hydrogen, labeled H(f). The total column density is the sum of these two components and gives a profile (not shown) that increases somewhat toward the center of the coma. This is a feature that qualitatively agrees with observations.

The C_2 column density in Fig. 6 is much flatter than that of any other neutral species. This is in agreement with observations. Many competing reaction paths lead to the production of C_2, giving it this profile. Comparing the ranges of various species taking into account their relative oscillator strengths (e.g., a factor of ~ 40 for C_3/C_2), we note qualitative agreement with the observed ranges listed in Table III.

Reactions (21) and (22) are main reasons to keep the abundance ratio of CO^+/H_2O^+ in model calculations from reaching values much larger than one. This is in disagreement with the only measured value which is ~ 30.

If the total observed flux F_λ at wavelength λ from an optically thin coma is contained within the solid angle determined by the aperture of an instrument, and since the aperture's projected size on the coma is proportional to the geocentric distance Δ of the comet the number of molecules N emitting at this wavelength is

$$N = 4\pi\Delta^2 F_\lambda / g_\lambda, \tag{28}$$

where Δ is in AU and g_λ is the fluorescence rate factor for scattering a solar

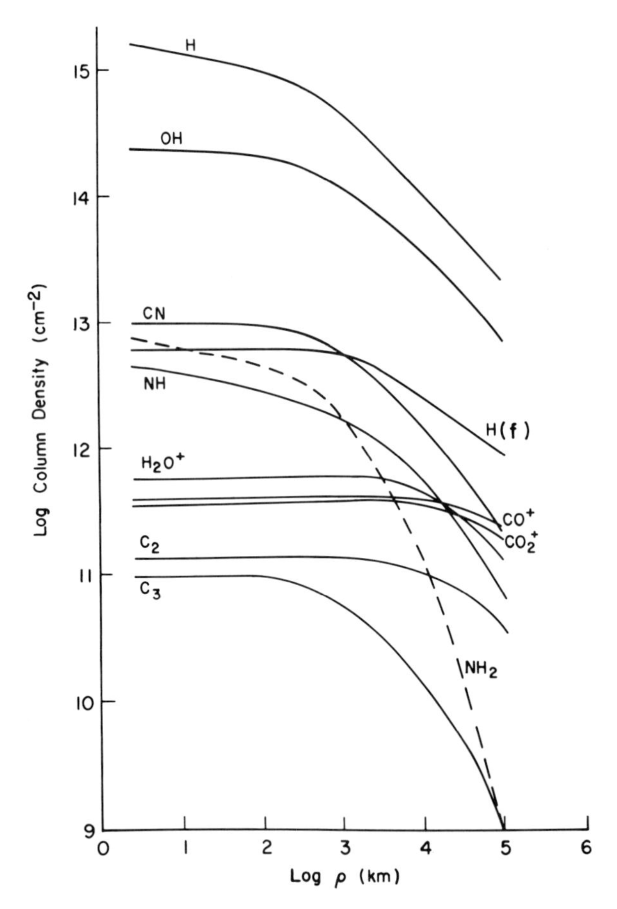

Fig. 6. Model results for column densities of selected species plotted as function of projected distance on the sky, logρ. Thermal hydrogen is indicated by H, while fast hydrogen is labeled H(f). Note that the column density for C_2 is flat compared to that of the other neutral species. Results correspond to composition six in Table IV for $r = 1$ AU.

photon ($\sec^{-1}$),

$$g_\lambda = (\pi e^2/mc) f_\lambda F_\odot(\lambda)/r^2, \tag{29}$$

where e and m are the charge and mass of the electron, c is the speed of light, f_λ the oscillator strength for the radiative transition, $F_\odot(\lambda)$ the solar flux (photons cm^{-2} sec^{-1} Hz^{-1}) at wavelength λ, and r the heliocentric distance of the comet. Some observers convert this to a particle production rate Q, by assuming a steady state,

$$Q = N/\tau. \tag{30}$$

The lifetime τ of the species depends not only on its radiative destruction time but also on chemical reactions (including solar wind charge exchange for ions) and therefore on a coma model.

In summary, dust-free models for parabolic comets give reasonable agreement with overall, averaged observations. Although some models have been developed with dust or solar wind interaction, they are preliminary and lack internal consistency. Sulfur chemistry and observations of sulfur compounds may become probes for the most central part of the coma. We believe to understand the overall features in the physics and chemistry, but lack most details.

VII. Halley's Comet and Other Unusual Comets

The most famous of all comets is Comet P/Halley. Here we describe some of its features as well as those of some other unusual comets.

A. Historical Record of P/Halley

We already mentioned in Section II that Halley's comet has a period of about 76 years and that its early appearances were recorded by Chinese as early as 240 BC. Most comets are named after their discoverer. This is not the case for this comet. Edmond Halley, an English astronomer, used positional data from the apparitions of AD 1531, 1607, and 1682 and deduced that they defined the orbit of the same comet. He published his findings in 1705, predicting Halley's return in 1758 to 1759. His prediction proved to be correct, and the comet was named in honor of him.

Repeatedly Halley's comet has come very close to the Earth. In AD 837 the minimum Earth–comet separation was 0.04 AU (less than 100 Earth radii). In AD 374 and 607 the comet came as close as 0.09 AU, and in AD 141 it was at $\Delta = 0.17$ AU from the Earth. If the backwards calculations of the orbits are correct, then the comet was only 0.1 AU from the Earth in 164 BC, the year in which it was not recorded by the Chinese, a circumstance that is difficult to understand.

B. The Apparition of Halley's Comet in 1910

Halley's comet was studied extensively through photographs and spectra in the period from 1909 to 1911. An analysis of the observations was given by Bobrovnikoff (1927). A brief summary is given in Table V. In particular, the species CN, C_2, and C_3 are traced from $r \simeq 3$ AU preperihelion to $r = 1.4$ AU postperihelion. The range ρ (projected on the sky) of the coma species, in units

Table V

EXTENT AND RELATIVE INTENSITY OF SPECTRAL IMAGES OF HALLEY (1910 II)

		CN		C_2		C_3		
r (AU)	Date	ρ (10^4 km)	log I	ρ (10^4 km)	log I	ρ (10^4 km)	log I	Comments
2.95	22 Oct 09	Faint uniform continuum; CN and C_2 suspect (faint), C_3 (4050 Å) observed faintly						Doubtful that C_2 could be seen at this distance
~2.5	18 Dec 09	7.19	0.25	4.95	0.05	4.96	—	
~2.3	30 Dec 09	9.52	0.35	7.46	0.2	7.71	—	
~2.1	09 Jan 10	13.5	0.45	9.7	0.08	6.49	—	Maximum range of CN
~2.0	15 Jan 10	9.0	0.25	8.45	0.17	6.45	—	
~1.65	04 Feb 10	6.52	0.4	5.77	0.3	4.20	—	
1.6	10 Feb 10	6.25	0.25	6.9	0.15	4.28	—	CN and C_2 comparable strength
~1.15	08 Mar 10	6.91	0.05	10.66	0.05	~3.5	—	Reported spectral change; maximum range of C_2
0.6	21 Apr 10	~3.8	0.65	4.61	0.8	~2.0	—	Near perihelion (19 Apr 1910)
0.7	05 May 10	2.12	0.55	4.05	1.0	0.80	—	Primarily in reflected light, fluctuations; throughout May envelopes at minimum range
~0.85	27 May 10	1.54	1.0	1.43	1.4	0.94	—	
1.0	06 Jun 10	2.32	1.3	3.73	1.4	1.67	—	
1.4	26 Jun 10	5.97	0.15	4.81	0.1	4.65	—	CN bands dominate; C_3 and CN exhibit similar behavior

of 10^4 km, and the relative logarithmic intensities determined from photographic densities (Bobrovnikoff, 1927) are presented. Since the observational quality that existed at that time was much more limited compared to present capabilities, the values presented are not absolute. The proper behavior is intact, however; this is what is important for comparison.

General observational features include an apparent spectral change [solar ($\lambda \sim 4700$ Å) to violet ($\lambda \sim 4000$ Å) and back] occurring at $r \simeq 1.2$ AU pre- and postperihelion. Comparison with the cometary average at $r \simeq 0.8$ AU indicates that this is possibly an important effect. But the reality of this spectral change is somewhat in doubt. It is felt by some that in the preperihelion phase (constituting a violet to solar change), the decrease in brightness of CN, that is, the violet spectrum, was only apparent; CN and C_3 actually increase but are overshadowed by the reflection spectrum and not by a significant increase in the C_2/CN and C_2/C_3 ratios. Others feel that the spectral change can be attributed to "bad procedures," namely, poor quality of observations as well as fallacious corrections. Whatever the reason may be, the face value of the observations shows a spectral change that must be accounted for.

Among other observational features are the following: sodium D lines were observed in envelopes and extended into the tail during perihelion passage; CH and various bands of C_2, C_3, and CN were positively identified; no CO^+ was found in envelopes but was definitely identified in the tail together with $N_2{}^+$. The envelopes were essentially uniform in mixture of CN and C_2.

The Earth passed through the tail of the comet. Comet P/Halley has a dust tail and a plasma tail and is one of the most active periodic comets. The plasma tail showed a disconnection in 1910. The nucleus is believed to have a diameter of 5 to 10 km.

C. The *Giotto* Mission and Other Comet Missions in "1986"

Comet Halley was rediscovered on 16 October 1982 on its way for the 1986 apparition. At the time of rediscovery the comet had a visual brightness $m_v = 24.2$ and was beyond the distance of Saturn, at $r = 11.05$ AU, $\Delta = 10.94$ AU, within an arcsecond of its predicted position.

Six probes will investigate the comet and its environment. The *International Sun–Earth Explorer (ISEE) 3* satellite will "fly" through the tail of Halley's comet after first passing through the tail of Comet P/Giacobini–Zinner. It will collect data on the solar wind interaction with the plasma tail and will pass $\sim 3 \times 10^3$ km from the nucleus of P/Giacobini–Zinner on 11 September 1985, with a relative speed of about 21 km $\sec^{-1}$. In contrast, it will pass $\sim 3 \times 10^7$ km from P/Halley. Two Japanese instrument probes will make similar measurements. One will pass in front of Halley, measuring the unperturbed solar wind, and one will pass through the tail, measuring the solar wind interaction with the plasma tail. The hydrogen coma will be observed with a Lyα camera.

Two Soviet instrument probes are scheduled to make measurements. One will pass within 10^4 km of the nucleus on 8 March 1986, about a month after Halley has passed perihelion. The spacecraft will first "fly" to Venus and release the Venus lander, then continue for another 175 days toward Halley. Launch time is December 1984.

The best equipped spacecraft to Halleys comet is the European Space Agency's (ESA) *Giotto* mission. It is planned to pass the nucleus at $\sim 10^3$ km on 13 March 1986. The *Giotto* spacecraft will carry a camera for imaging the inner coma and the nucleus, neutral, ion, and dust mass spectrometers to measure coma composition, a dust impact detector to study the dust environment, various plasma analyzers, an electron analyzer, an energetic-particle experiment, a magnetometer to study plasma physical processes, and an optical probe for *in situ* measurements of the cometary dust and gas environment.

Halley missions are particularly difficult because of the retrograde motion of the comet, that is, the comet motion, when projected on the ecliptic, is in the opposite direction to the motion of the planets. Therefore, the spacecraft will have a flyby velocity of 68 km sec^{-1}. The comet will be at $r \simeq 0.89$ AU and a distance $\Delta \simeq 0.98$ AU from the Earth. The *Giotto* spacecraft will approach the comet from the nightside, which is favorable for the spacecraft's survival since the large dust particles are predominantly injected into the sunward hemisphere of the coma. Target delivery capability is about as good as the uncertainty of the position of the nucleus in the coma, which is ~ 500 km. This can be improved to ~ 100 km using data from the first Soviet probe.

D. Other Unusual Comets

Two comets that are always quoted as unusual are Morehouse (1908 III) and Humason (1962 VIII). Morehouse was an "intermediate" comet with perihelion distance $q = 0.945$ AU, and Humason was an "old" comet with $q = 2.133$ AU. Both comets were dominated by their CO^+ tails and their unusual activities during the entire observing periods; they were dust-poor. Comet Humason had a CO^+ tail and brightness outbursts at $r > 6$ AU.

Another unusual comet is Schwassmann–Wachmann 1 (1925 II). With an eccentricity of $e = 0.105$, it has an almost circular orbit with $q \simeq 5.4$ AU and $Q \simeq 6.7$ AU. It is observed to brighten sometimes by six to nine magnitudes over a period of 1 day and subside to its quiescent condition over a period of ~ 1 week. This happens about once a year. During a recent outburst CO^+ was detected.

A less well known, but unusual comet is Stearns (1927 IV); it had a perihelion distance $q = 3.68$ AU. The comet showed C_2 Swan bands on a continuous spectrum, a bright stubby tail, and a narrow tail. Presumably the continuous spectrum was sunlight reflected by dust in the coma, the stubby tail was a dust tail, and the narrow one a plasma tail. What is remarkable about the comet is that it was observable for 4 years, almost all of this time after perihelion (only a few days before perihelion), out to a distance of $r = 11.5$ AU and $\Delta = 11.9$ AU. It belongs to the group of "intermediate" comets.

Comet Schuster (1975 II) behaved similarly. It had a perihelion distance $q = 6.88$ AU, but was not discovered until a year after perihelion passage in February 1976. It was then observed for nearly 2 years as it receded from the Sun. It belongs to the group of "new" comets.

A recent, very active comet is Comet Bowell (1980 b). It showed strong continuum (dust) activity at heliocentric distances $r > 4$ AU.

Obviously in these comets the volatiles were not depleted, nor were they caged in clathrates. In the first three, the presence of CO^+ suggests that the volatile component might have been CO. No tail spectra are available for Comet Stearns. The large amount of volatiles (perhaps only in pockets in the nucleus of Schwassmann–Wachmann 1) suggests that these comets were formed in a different place in the presolar nebula or at a different time. Many parabolic comets are bright at large distances before perihelion and then fade or do not increase their brightness as rapidly as might be expected from the observations at discovery time. It has been suggested that in those comets a volatile frosting may cover their surface but is depleted before perihelion passage.

A completely different set of unusual comets is the Kreutz group. This is a group of Sun-grazers with perihelion distance $q < 0.009$ AU. (For comparison, the solar radius is ~ 0.0047 AU.) To this group belong the Great March Comet (1843 I), the Great Southern Comet of 1880 (1880 I), the Great September Comet (1882 II), the Great Southern Comet of 1887 (1887 I), and the Comets du Toit (1945 VII), Pereyra (1963 V), Ikeya–Seki (the Day-Light Comet) (1965 VIII), White–Ortiz–Bolelli (1970 VI), and Howard–Koomen–Michels (the comet that fell into the Sun) (1979 XI). In 1982 two more comets were reported that probably belong to the Kreutz group and that fell into the Sun. The last three comets were discovered with the U.S. Naval Research Laboratory's Earth-orbiting *Solwind* coronograph. One other Sun-grazing comet (1680) has $q = 0.00622$ AU but different orbital elements and does not belong to the Kreutz group. Three of these comets (1882 II, 1963 V, and 1965 VIII) belong to the class of "old" comets. For the others the orbital elements are probably not good enough to classify them. Apparently the Kreutz group comets are decay products of a large comet that in a previous passage, more than 10^4 years ago, split into many parts, some of which are now observed over a period of a few hundred years, as they again come into the inner solar system.

Presumably the three comets that fell into the Sun are very small since they were not observable at a larger distance from the Sun like the other Sungrazers. It is likely that there are many very small comets that we fail to detect unless the circumstances are very favorable. One such favorable circumstance was the discovery recently of Comet *IRAS*–Araki–Alcock (1983d), detected with the *Infrared Astronomical Satellite* (*IRAS*).The comet came very close to the Earth on 11 May 1983, when its geocentric distance was $\Delta = 0.03$ AU. It must have been small, or poor on volatiles, because it was observable only for a very short time before and after Earth passage. A comet that comes that close to the Earth moves over the sky at a rate of $\sim 2^\circ \text{ hr}^{-1}$.

Another very small comet (if it was a comet) collided with the Earth in 1908 in Siberia. This collision is known as the Tunguska event.

E. The Current Model for P/Halley

Based on the 1910 apparition of P/Halley and on knowledge obtained from observations of other "similar" comets, the parameters listed in Table VI have been assembled. These values are based on conclusions of the joint NASA/ESA working group, with some modifications by the author. Undoubtedly further modifications will be made during the coming years as we learn more about comets from model studies, comet observations in general, and observations of Comet P/Halley as it approaches the inner solar system.

It will be interesting to compare models after the Halley mission results have been analyzed. Undoubtedly many of our prejudices will have to be changed.

We have summarized what is known about Comet P/Halley and the prospects for gaining confirmation of this from comet missions. We have also

Table VI

PARAMETERS FOR COMET P/HALLEY

Nucleus radius	4 km
Density	1 g cm^{-3}
Mass	3×10^{17} g
Visual albedo	0.1
Amount of H_2O	83%
Mean latent heat of sublimation	10 kcal mol^{-1}
Total gas production rate at 1 AU	$5 \times 10^{29} \text{ molec sec}^{-1}$
Mass ratio of dust to gas	0.5

given some indications that some comets are quite different from others in size and in activity and therefore in composition and in origin.

VIII. Directions for Future Research

In the next few years it is very likely that observational data for comet comae and tails will be vastly increased as new techniques in radio, IR, visible, UV, and *in situ* measurements by space probes mature. Infrared molecular spectroscopy should soon start to produce very important data. This flood of new data will not by itself solve the comet problems, but in context with coma models this goal does become realistic.

A. Observations and Space Missions

High spatial resolution of coma activity, such as jets, envelopes, and fans can be obtained through use of an image-dissector scanner. Such instruments have gone into operation only relatively recently.

Only very few high-resolution spectra of the coma are available. They reveal many weak, unidentified lines. More spectra of this type are needed and will be made.

Radio observations will reveal new species and reveal mechanisms for UV pumping to excite molecules and populate levels that emit in the radio range. *IRAS* has already revealed its usefulness for comet observations. With IR observations it will be possible to identify molecules that otherwise escape detection.

The recent detection of S_2 in Comet *IRAS*–Araki–Alcock shows that UV observations despite their enormous successes in the last decade have still not exhausted their potential for comet observations. Much more can be expected from them.

We have already described plans for space missions to Comets P/Giacobini–Zinner and P/Halley. Undoubtedly space missions to other periodic comets will be planned in the future. Missions to parabolic comets are much more difficult, because their appearance and orbits do not allow for enough time between launch and intercept. A sample return or a physical probing of the nucleus will be very revealing and rewarding experiments.

But there are also other possibilities: creation of artificial comets; not on a large scale with an "ice cube," but with chemical releases from satellites. The objective is to study chemical reactions that are thought to occur in cometary or planetary atmospheres and in interstellar clouds. Optical, radar, radio, and *in situ* diagnostics can be employed to measure chemical composition, plasma concentrations, velocities, and temperatures.

B. Improvement of Models

Models for comet comae as well as models for the formation of comets must be improved. For the latter a better understanding of the chemistry of formation of heavy molecules in dense interstellar clouds is needed. Heterogeneous processes may play an important role.

Improved coma models must take into account, in an internally consistent way, the coma–solar wind interaction, sulfur chemistry, affects of the dust on the coma environment (including surface grain chemistry), two-dimensional streaming of the plasma component, and inhomogeneities in the coma flow such as envelopes and fountains. It is not known if icy grain halos really exist. They have been proposed several times, but observational verification is very difficult. Model calculations with and without extended sources (icy grains) may provide some clues useful for further observations.

To summarize, new and not-so-new observational techniques will provide a wealth of new information. Space and laboratory experiments can provide new insight and new data for comet analyses and comet models. Much has been learned, much still needs to be learned.

Acknowledgments

I wish to thank Drs. C. Arpigny, Rh. Lüst, H. U. Schmidt, and D. G. Torr for valuable comments on the manuscript.

References

Arpigny, C. (1976). Interpretation of comet spectra. *In* "The Study of Comets" (B. Donn, M. Mumma, W. Jackson, M. A'Hearn, and R. Harrington, eds.), NASA SP-393, p. 797. Natl. Tech. Inf. Serv., Springfield, Illinois.

Biermann, L. (1951). Kometenschweife und solare Korpuskularstrahlung. *Z. Astrophys.* **29,** 274.

Biermann, L. (1968). On the emission of atomic hydrogen in comets. *Rep.—J. Inst. Lab. Astrophys.* **93.**

Biermann, L. (1982). On the global size of Oort's cloud of cometary nuclei and their total number. *In* "Compendium in Astronomy" (E. G. Mariolopoulos, P. S. Theocaris, and L. N. Mavridis, eds.), p. 183. Reidel Publ., Dordrecht, Netherlands.

Biermann, L., and Michel, K. W. (1978). On the origin of cometary nuclei in the presolar nebula. *Moon Planets* **18,** 447.

Biermann, L., Giguere, P. T., and Huebner, W. F. (1982). A model of a comet coma with interstellar molecules in the nucleus. *Astron. Astrophys.* **108,** 221.

Biermann, L., Huebner, W. F., and Lüst, Rh. (1983). Aphelion clustering of "new" comets: star tracks through Oort's cloud. *Proc. Natl. Acad. Sci. U.S.A.* **80,** 5151.

Bobrovnikoff, N. T. (1927). On the spectrum of Halley's comet. *Astrophys. J.* **66,** 145.

Brandt, J. C. (1982). Observations and dynamics of plasma tails. *In* "Comets" (L. L. Wilkening, ed.), p. 519. Univ. of Arizona Press, Tucson.

Combi, M. R., and Delsemme, A. H. (1980). Neutral cometary atmospheres. I. An average random walk model for photodissociation in comets. *Astron. J.* **237,** 663.
Delsemme, A. H. (1982). Chemical composition of cometary nuclei. *In* "Comets" (L. L. Wilkening, ed.), p. 85. Univ. of Arizona Press, Tucson.
Everhart, E. (1973). Examination of several ideas of comet origins. *Astron. J.* **78,** 329.
Giguere, P. T., and Huebner, W. F. (1978). A model of comet comae. I. Gas-phase chemistry in one dimension. *Astrophys. J.* **223,** 638.
Hills, J. (1982). The formation of comets by radiation pressure in the outer protosun. *Astron. J.* **87,** 906.
Huebner, W. F., and Giguere, P. T. (1980). A model of comet comae. II. Effects of solar photodissociative ionization. *Astrophys. J.* **238,** 753.
Huebner, W. F., and Keady, J. J. (1984). First-flight escape from spheres with R^{-2} density distribution. *Astron. Astrophys.* **135,** 177.
Keller, H. U. (1976). The interpretations of ultraviolet observations of comets. *Space Sci. Rev.* **18,** 641.
Lüst, Rh. (1962). Bewegung von Struckturen in der Koma und im Schweif des Kometen Morehouse. *Z. Astrophys.* **65,** 236.
Lüst, Rh., and Schmidt, H.-U. (1968). The influence of the solar corona on the tail of comet Ikeya–Seki 1965 VIII. *Mitt. Astron. Ges.* **25,** 211.
Marconi, M. L., and Mendis, D. A. (1982). The photochemistry and dynamics of a dusty cometary atmosphere. *Moon Planets* **27,** 27.
Marsden, B. G. (1983). "Catalogue of Cometary Orbits," 4th ed. Smithsonian Astrophysical Observatory, Cambridge, Massachusetts.
Marsden, B. G., Sekanina, Z., and Everhart, E. (1978). New osculating orbits for 110 comets and analysis of original orbits for 200 comets. *Astron. J.* **83,** 64.
Mitchell, G. F., Prasad, S. S., and Huntress, W. T. (1981). Chemical model calculations of C_2, C_3, CH, CN, OH, and NH_2 abundances in cometary comae. *Astrophys. J.* **244,** 1087.
Oort, J. H. (1950). The structure of the cloud of comets surrounding the solar system and a hypothesis concerning its origin. *Bull. Astron. Inst. Neth.* **11,** 91.
Schmidt, H.-U., and Wegmann, R. (1982). Plasma flow and magnetic fields in comets. *In* "Comets" (L. L. Wilkening, ed.), p. 538. Univ. of Arizona Press, Tucson.
Shimizu M. (1976). The structure of cometary atmospheres. I: temperature distribution. *Astrophys. Space Sci.* **40,** 149.
Swings, P., and Haser, L. (1956). "Atlas of Representative Cometary Spectra." Inst. d'Astrophys., Liege.
Whipple, F. L. (1950). A comet model. Part I. The acceleration of comet Encke. *Astrophys. J.* **111,** 375.
Whipple, F. L. (1951). A comet model. Part II. Physical relations for comets and meteors. *Astrophys. J.* **113,** 464.
Wurm, K. (1943). Die Natur der Kometen. *Mitt. Ham. Sternwarte* **8,** No. 51.
Yamamoto, T., Nakagawa, N., and Fukui, Y. (1983). The chemical composition and thermal history of the ice of a cometary nucleus. *Astron. Astrophys.* **122,** 171.

IV

Appendixes

I

Unattenuated Solar Photo Rate Coefficients at 1 AU Heliocentric Distance

$\sum_i Z_i$	Mother species	Photolysis products	Binding energy equivalent wavelength (Å)	Rate coefficient (10^{-6} sec^{-1})	Total rate coefficient (10^{-6} sec^{-1})	Photolysis products excess energy (eV)	Chapter and (equation)[a]
Monatomics							
1	H	$H^+ + e^-$	911.75	0.073	0.073	3.5	
2	He	$He^+ + e^-$	504.27	0.052	0.052	16	5(137)
6	$C(^3P)$	$C^+ + e^-$	1101.07	0.41	0.41	5.9	
6	$C(^1D)$	$C^+ + e^-$	1240.27	3.6	3.6	1.0	
6	$C(^1S)$	$C^+ + e^-$	1445.66	4.3	4.3	2.1	

(*cont.*)

Appendix I (*cont.*)

$\sum_i Z_i$	Mother species	Photolysis products	Binding energy equivalent wavelength (Å)	Rate coefficient (10^{-6} sec^{-1})	Total rate coefficient (10^{-6} sec^{-1})	Photolysis products excess energy (eV)	Chapter and (equation)[a]
Monatomics							
7	N	$N^+ + e^-$	853.06	0.19		15	5(131)
					0.19		
8	$O(^3P)$	$O^+ + e^-$	910.44	0.21		22	5(56), 6(18), 8(121)
					0.21		
8	$O(^1D)$	$O^+ + e^-$	827.9	0.18		22	5(56), 6(18), 8(121)
					0.18		
8	$O(^1S)$	$O^+ + e^-$	858.3	0.20		19	5(56), 6(18), 8(121)
					0.20		
16	$S(^3P)$	$S^+ + e^-$	1196.75	1.1		6.3	8(121)
					1.1		
16	$S(^1D)$	$S^+ + e^-$	1121.43	1.1		6.2	8(121)
					1.1		
16	$S(^1S)$	$S^+ + e^-$	1164.12	1.0		5.4	8(121)
					1.0		
Diatomics							
2	H_2	$H(1s) + H(2s,2p)$	844.79	0.044		2.2	8(9)
		$H_2^+ + e^-$	803.67	0.054		6.6	8(10)
		$H + H^+ + e^-$	685.8	0.0095		25	8(11)
					0.11		
7	CH	$C + H$	3589.9	12,000		0.48	
		$CH^+ + e^-$	1170	0.76		6.4	
					12,000		
10	HF	$H + F$	2110	1.8		8.1	
					1.8		
12	C_2	$C + C$	2030	0.17		12	
					0.17		

14	$CO(X\,^1\Sigma^+)$	C + O	1117.8	0.28		2.6	
		$C(^1D) + O(^1D)$	863.4	0.042		6.0	
		$CO^+ + e^-$	884.79	0.31		10	
		$O + C^+ + e^-$	554.7	0.0079		29	
		$C + O^+ + e^-$	501.8	0.0060		31	
					0.65		
14	$CO(a\,^3\Pi)$	C + O	2431.8	75		2.2	
		$CO^+ + e^-$	1549.1	8.6		2.0	
		$O + C^+ + e^-$	758.3	0.011		29	
		$C + O^+ + e^-$	662.7	0.0084		33	
					84		
14	N_2	N + N	1270.4	0.66		3.4	5(401), 7(55), 7(63)
		$N_2^+ + e^-$	796	0.35		18	5(58)
		$N + N^+ + e^-$	510.4	0.015		29	5(58), 5(92), 5(127)
					1.0		
15	NO	N + O	1910	2.2		1.8	3(70), 5(409), 7(56)
		$NO^+ + e^-$	1340	1.3		8.2	
		$N + O^+ + e^-$	616.2	0.0018		19	
		$O + N^+ + e^-$	589.3	0.032		25	
					3.5		
16	O_2	$O(^3P) + O(^3P)$	2423.7	0.060		8.3	1(24), 3(1), 3(14), 3(15), 5(4)
		$O(^3P) + O(^1D)$	1759	4.2		1.3	3(13), 5(3), 5(377)
		$O(^1S) + O(^1S)$	923	0.041		0.82	5(381)
		$O_2^+ + e^-$	1027.8	0.51		17	5(57), 5(118)
		$O + O^+ + e^-$	585	0.051		26	5(57)
					4.9		
18	HCl	H + Cl	2798	7.2		4.4	3(119), 6(20)
					7.2		
18	F_2	F + F	7950	570		1.8	
					570		

(cont.)

Appendix I *(cont.)*

$\sum_i Z_i$	Mother species	Photolysis products	Binding energy equivalent wavelength (Å)	Rate coefficient (10^{-6} sec^{-1})	Total rate coefficient (10^{-6} sec^{-1})	Photolysis products excess energy (eV)	Chapter and (equation)[a]
Diatomics							
34	Cl_2	Cl + Cl	4950	5100		2.1	
					5100		
43	BrO	Br + O(^{3}P)	5150	~4800		~1.3	
		Br + O(^{1}D)	2830	~5.8		~1.1	
					~4800		
Triatomics							
9	NH_2	NH + H	1970.5	1.8		1.6	
					1.8		
10	H_2O	H + OH	2424.6	10		3.4	1(18), 3(27), 5(26), 7(7), 8(110)
		H_2 + O(^{1}D)	1770	1.4		3.5	8(10)
		$H_2O^+ + e^-$	984	0.33		12	
		$H + OH^+ + e^-$	684.4	0.055		19	
		$H_2 + O^+ + e^-$	664.8	0.0058		37	
		$OH + H^+ + e^-$	662.3	0.013		25	
					12		
14	HCN	H + CN(A $^2\Pi_i$)	1950	13		4.3	8(67), 8(82)
					13		
17	HO_2	OH + O	4395	6600		1.0	
					6600		
18	H_2S	HS + H	3160	3300		0.77	6(40)
		$H_2S^+ + e^-$	1185.25	0.56		2.2	
		$H_2 + S^+ + e^-$	927	0.15		6.9	
		$H + HS^+ + e^-$	867	0.073		12	
					3300		

22	CO_2	$CO(X\,^1\Sigma^+) + O(^3P)$	2275	0.017		1.7	1(21), 6(4), 7(1), 8(113)
		$CO(X\,^1\Sigma^+) + O(^1D)$	1671	0.92		4.3	6(4), 8(113)
		$CO(a\,^3\Pi) + O$	1082	0.28		2.0	7(58)
		$CO_2^+ + e^-$	899.22	0.66		17	7(28), 6(15)
		$O + CO^+ + e^-$	636.93	0.050		27	9(20)
		$CO + O^+ + e^-$	650.26	0.064		28	
		$O_2 + C^+ + e^-$	546.55	0.029		30	
					2.0		
22	N_2O	$N_2 + O(^1D)$	3407	1.0		2.8	3(55)
		$N_2 + O(^1S)$	2115	4.9		6.8	
					5.9		
23	NO_2	$NO + O(^3P)$	3978	8600		0.34	2(5), 3(59), 4(25)
		$NO + O(^1D)$	2439	42		1.3	
		$NO_2^+ + e^-$	1270	1.3		13	
					8600		
24	O_3	$O + O_2$	11,790	1100		2.8	1(32), 3(3), 3(16), 3(17), 4(27), 5(9)
		$O(^1D) + O_2(a\,^1\Delta_g)$	3100	9100		0.67	2(8), 2(31a), 3(18), 3(19), 4(27), 5(8), 5(413), 7(12), 7(24)
					10,000		
26	HOCl	OH + Cl	5130	560		1.7	3(131)
					560		
30	OCS	$CO + S(^3P)$	3973	560		0.92	3(134), 6(25), 6(40)
		$CO + S(^1S)$	3369	88		3.4	
		CS + O	1820	1.1		0.35	
		$OCS^+ + e^-$	1120	0.14		1.9	
		$CO + S^+ + e^-$	907	0.0087		58	
		$S + CO^+ + e^-$	704	0.0020		62	
		$O + CS^+ + e^-$	654	0.00027		57	
		$CS + O^+ + e^-$	629	0.00019		62	
		$SO + C^+ + e^-$	564	0.00056		61	
					~650		

(cont.)

Appendix I (*cont.*)

$\sum_i Z_i$	Mother species	Photolysis products	Binding energy equivalent wavelength (Å)	Rate coefficient (10^{-6} sec^{-1})	Total rate coefficient (10^{-6} sec^{-1})	Photolysis products excess energy (eV)	Chapter and (equation)[a]
Triatomics							
32	SO_2	$SO + O$	2210	190		0.49	3(147), 6(35), 8(118)
		$S + O_2$	2070	58		0.68	8(118)
		$SO_2^+ + e^-$	1005	1.1		12	8(121)
					250		
32	ClNO	$Cl + NO$	7748	6300		2.7	
					6300		
33	ClOO	$ClO + O$	4960	9000		2.2	
					9000		
38	CS_2	$CS + S$	2792	~3500		0.92	
		$CS_2^+ + e^-$	1230	0.55		2.4	
		$CS + S^+ + e^-$	834	0.012		53	
		$S + CS^+ + e^-$	775	0.0078		51	
		$C + S_2^+ + e^-$	731	0.00035		48	
		$S_2 + C^+ + e^-$	656	0.0012		50	
					~3500		
44	HOBr	$OH + Br$	5180	~1800		~1.3	
					~1800		
62	HOI	$OH + I$	~5200	~870		~1.6	
					~870		
68	INO	$I + NO$	16,500	110,000		>1.6	
					110,000		
Tetratomics							
10	NH_3	$NH_2 + H$	2798	170		1.9	1(14), 8(32), 8(66)

		$NH(a\,^1\Delta) + H_2$	2240	3.7		1.6	8(32)
		$NH_3^+ + e^-$	1220	0.61		5.8	
		$H + NH_2^+ + e^-$	786.2	0.18		11	
		$H_2 + NH^+ + e^-$	~775	0.0069		26	
		$H_2 + H + N^+ + e^-$	~560	0.0033		30	
		$NH_2 + H^+ + e^-$	~387	0.0033		20	
					180		
14	C_2H_2	$H + C_2H$	2306	10		3.2	8(27)
		$H_2 + C_2$	2006	2.7		3.1	8(27)
		$C_2H_2^+ + e^-$	1086	0.78		5.1	
		$H + C_2H^+ + e^-$	697	0.074		16	
					14		
16	H_2CO	$H_2 + CO$	>7000	160		>2.1	3(101), 8(115)
		$H + HCO$	3340	84		0.37	2(10), 3(100), 8(115)
		$H + H + CO$	2750	32		3.0	
		$H_2CO^+ + e^-$	1141.6	0.40		3.2	
		$H + HCO^+ + e^-$	1043	0.20		7.3	
		$H_2 + CO^+ + e^-$	882	0.12		29	
					280		
18	PH_3	$PH_2 + H$	~3650	61		3.5	8(43)
					61		
18	H_2O_2	$OH + OH$	5765	140		3.0	2(14), 2(29), 3(46), 7(23)
					140		
22	HNCO	$NH(c\,^1\Pi) + CO$	3540	15		5.1	
		$H + NCO(A\,^2\Sigma)$	2530	14		4.1	
					29		
24	HONO	$OH + NO$	5930	2300		1.7	3(75)
					2300		
31	NO_3	$NO + O_2$	90,000	170,000		1.8	3(65)
		$NO_2 + O$	5800	110,000		0.28	3(64)
					290,000		

(cont.)

Appendix I (*cont.*)

$\sum_i Z_i$	Mother species	Photolysis products	Binding energy equivalent wavelength (Å)	Rate coefficient (10^{-6} sec^{-1})	Total rate coefficient (10^{-6} sec^{-1})	Photolysis products excess energy (eV)	Chapter and (equation)[a]
Tetratomics							
32	COF_2	$COF + F$	2235	23	23	4.0	3(156)
40	$ClNO_2$	$Cl + NO_2$	~8800	2600	2600	>2.7	
40	ClONO	$Cl + NO_2$	17,000	11,000	11,000	>2.4	
40	COFCl	COF + Cl	3160	~73	~73	2.9	
41	ClO_3	$ClO_2 + O$?	~3500	~26,000	~26,000	~0.59	
48	$COCl_2$	COCl + Cl	3690	160	160	3.3	
Pentatomics							
10	CH_4	$CH_3 + H$	2770	1.2		5.9	
		$CH_2(\tilde{a}\,^1A_1) + H_2$	2373	5.5		5.2	1(15), 8(22)
		$CH + H_2 + H$	1370	0.50		1.9	8(22)
		$CH_4^+ + e^-$	945	0.36		5.5	
		$H + CH_3^+ + e^-$	866	0.20		8.0	
		$H_2 + CH_2^+ + e^-$	822	0.021		20	
		$CH_3 + H^+ + e^-$	686	0.0091		27	
		$H_2 + H + CH^+ + e^-$	545	0.0042	7.8	25	
24	HCOOH	$CO_2 + H_2$	~7000	320		4.8	
		HCO + OH	~2600	560		1.8	

		$HCOOH^+ + e^-$	1094.4	0.91		3.9	
		$OH + HCO^+ + e^-$	902	0.28		21	
					880		
26	CH_3Cl	$CH_3 + Cl$	2260	25		3.5	3(108)
					25		
26	HC_3N	$HC_2 + CN$	>1632	28		>2.1	
					28		
32	HNO_3	$OH + NO_2$	5980	210		4.3	3(79)
					210		
42	CHF_2Cl	$CHF_2 + Cl$	2040	19		4.6	
					19		
48	$ClONO_2$	$Cl + NO_3$	7210	~ 220		2.5	3(128)
		$ClONO + O$	3910	~ 790		2.4	3(129)
					1000		
58	CF_2Cl_2	$CF_2Cl + Cl$	3460	12		5.4	3(109)
		$CF_2 + Cl_2$	2160	20		4.6	
					32		
66	$CFCl_3$	$CFCl_2 + Cl$	3860	56		5.5	3(110)
					56		
66	$BrONO_2$	$BrO + NO_2$	8660	~ 4900		~ 1.8	
					~ 4900		
74	CCl_4	$CCl_3 + Cl$	4070	44		3.8	3(111)
		$CCl_2 + Cl + Cl$	2090	56		3.0	
					100		
84	$IONO_2$	$IO + NO_2$	~ 7850	~ 5900		~ 1.9	
					~ 5900		
Hexatomics							
16	C_2H_4	$C_2H_2 + H_2$	7200	24		6.2	8(26)
		$C_2H_2 + H + H$	1960	23		1.7	8(26)
		$C_2H_4{}^+ + e^-$	1180	0.58		7.3	
		$H_2 + C_2H_2{}^+ + e^-$	945	0.20		12	
		$H + C_2H_3{}^+ + e^-$	898	0.23		13	
					48		

(cont.)

Appendix I (*cont.*)

$\sum_i Z_i$	Mother species	Photolysis products	Binding energy equivalent wavelength (Å)	Rate coefficient (10^{-6} sec^{-1})	Total rate coefficient (10^{-6} sec^{-1})	Photolysis products excess energy (eV)	Chapter and (equation)[a]
Hexatomics							
18	CH_3OH	$H_2CO + H_2$	~7000	250		6.5	
		$CH_3 + OH$	~3150	13		4.4	
		$CH_3OH^+ + e^-$	1143	0.50		2.6	
		$H + CH_3O^+ + e^-$	1006	0.12		3.5	
		$H_2 + H_2CO^+ + e^-$	976	0.12		4.4	
					270		
40	HO_2NO_2	$HO_2 + NO_2$	13,400	~330		~4.2	3(84)
		$OH + NO_3$	7300	~330		~3.4	
					660		
Septatomics							
24	CH_3CHO	$CH_4 + CO$	~7000	~9.8		~3.7	
		$CH_3 + HCO$	3501	24		0.87	
		$CH_3CO + H$	3375	22		0.43	
					~55		
26	CH_3OOH	$CH_3O + OH$	6470	440		5.5	3(98)
					440		
54	N_2O_5	$NO_2 + NO_2 + O$	4055	850		2.3	3(68)
					850		
Octatomics							
18	C_2H_6	$C_2H_4 + H_2$	8743	3.7		9.0	8(28)
		$CH_3 + CH_3$	3220	0.88		7.4	
		$C_2H_5 + H$	2900	3.3		6.8	
		$CH_4 + {}^1CH_2$	2726	2.2		6.1	8(28)

		$C_2H_6^+ + e^-$	1064	0.49		10
					11	
66	CH_3CCl_3	$CH_3 + CCl_3$	~2400	87		3.1
					87	
Supraoctatomics						
48	$CH_3O_2NO_2$	$CH_3O_2 + NO_2$	14,930	~740		~3.3
		$CH_3O + NO_3$	9560	~740		~2.8
					~1500	
50	$CH_3S_2CH_3$	$CH_3SS + CH_3$	4940	~1400		2.0
		$CH_3S + CH_3S$	3870	~1200		1.6
					~2500	
62	$CH_3CO_3NO_2$	$CH_3CO_3 + NO_2$	10,860	~800		3.4
		$CH_3CO_2 + NO_3$	9140	~800		3.2
					~1600	

[a] For example, 5(137) ≡ Chapter 5, Eq. (137).

II

Chemical Reaction Rates

Reaction	Rate[a]	Source[b]	Chapter and (equation)[c]
Oxygen species			
$O + O + M \rightarrow O_2 + M$	$4.7 \times 10^{-33}(300/T)^2$	a	1(22), 3(5), 5(5), 6(6)
$O + O_2 + M \rightarrow O_3 + M$	$k_0 = 6.0 \times 10^{-34}(T/300)^{-2.3}$	b	1(31), 2(6), 3(2), 4(26), 5(6), 7(15)
$O + O_3 \rightarrow 2\,O_2$	$8.0 \times 10^{-12}\exp(-2060/T)$	b	1(33), 3(4), 5(7)
$O(^1D) + M \rightarrow O + M$	$3.2 \times 10^{-11}\exp(+67/T)$	c	3(20)
Hydrogen species			
$OH + OH \rightarrow O + H_2O$	4.2×10^{-12}	b	1(20), 3(41), 5(29)
$H + HO_2 \rightarrow O + H_2O$	$\leqslant 9.4 \times 10^{-13}$	c	1(26), 5(30a)
$H + HO_2 \rightarrow O_2 + H_2$	7.4×10^{-11}	b	3(43), 5(30b), 7(19),
$HO_2 + HO_2 \rightarrow O_2 + H_2O_2$	$2.3 \times 10^{-13}\exp(+590/T)$	b	2(13), 3(45)
$OH + HO_2 \rightarrow O_2 + H_2O$	$(7 + 4P_{atm}) \times 10^{-11}$	b	3(42), 5(28), 7(9)
$OH + H_2O_2 \rightarrow HO_2 + H_2O$	$3.1 \times 10^{-12}\exp(-187/T)$	b	3(47)

(*cont.*)

Appendix II *(cont.)*

Reaction	Rate[a]	Source[b]	Chapter and (equation)[c]
Nitrogen species			
$NO + NH_2O_2 \rightarrow NO_2 + NH_2O$	6.2×10^{-13}	d	2(26c)
$NO_2 + NO_3 + M \rightarrow N_2O_5 + M$	$k_0 = 2.2 \times 10^{-30}(T/300)^{-2.8}$	b	3(66)
	$k_\infty = 1.0 \times 10^{-12}$	b	
$N_2O_5 + M \rightarrow NO_2 + NO_3 + M$	$k_0 = 2.2 \times 10^{-3}(T/300)^{-4.4}\exp(-11{,}080/T)$	c	3(67)
	$k_\infty = 9.7 \times 10^{14}(T/300)^{+0.1}\exp(-11{,}080/T)$	c	
$N(^4S) + NO \rightarrow N_2 + O$	3.4×10^{-11}	b	3(71), 5(407), 6(13), 7(52)
$N(^2D) + NO \rightarrow N_2 + O$	3.4×10^{-11}	b	5(418), 7(53)
$N(^2D) + M \rightarrow N(^4S) + M$	2×10^{-12} for M = O	aa	6(11)
	2.5×10^{-12} for M = CO	aa	6(11)
$N(^4S) + NH \rightarrow N_2 + H$	$1.1 \times 10^{-11}\ T^{0.5}$	e	8(78)
Carbon species			
$RO_2 + RO'_2 \rightarrow ROOR' + O_2$	General form of reaction		2(18)
$CH_3 + CH_3 + M \rightarrow C_2H_6 + M$	$k_0 = 1.3 \times 10^{-23}$	e	8(98)
	$k_\infty = 5.5 \times 10^{-11}$	e	
$^3CH_2 + CH_3 \rightarrow C_2H_4 + H$	7×10^{-11}	e	8(24), 8(95)
$CH + CH_4 \rightarrow C_2H_4 + H$	1×10^{-10}	e	8(25), 8(93)
$C_2H_2^{**} + 2\,C_2H_2 \rightarrow C_6H_6$	?	m	8(31a)
$C_2H + CH_4 \rightarrow CH_3 + C_2H_2$	$2.8 \times 10^{-12}\exp(-250/T)$	e	8(54)
$C_2 + CH_4 \rightarrow C_2H_2 + {}^1CH_2$	1.9×10^{-11}	e	8(54)
$^3CH_2 + {}^3CH_2 \rightarrow C_2H_2 + H_2$	5.3×10^{-11}	e	8(91)
$^3CH_2 + C_2H_2 + M \rightarrow C_3H_4 + M$	$k_0 = 7.6 \times 10^{-25}$	e	8(92)
	$k_\infty = 5.9 \times 10^{-12}$	e	
$^1CH_2 + CH_4 \rightarrow 2\,CH_3$	1.9×10^{-12}	e	8(94)
$C_2H + C_2H_6 \rightarrow C_2H_5 + C_2H_2$	6.5×10^{-12}	e	8(99)
$CH_3 + C_2H_5 + M \rightarrow C_3H_8 + M$	$k_0 = 1.3 \times 10^{-22}$	e	8(100)
	$k_\infty = 4.2 \times 10^{-11}\exp(-200/T)$	e	

$2\,C_2H_5 + M \rightarrow C_4H_{10} + M$	$k_0 = 2.9 \times 10^{-21}$	e	8(101)
	$k_\infty = 1.3 \times 10^{-11} \exp(-95/T)$	e	
$CO(a\,^3\pi) + CO \rightarrow CO_2 + C$	1.4×10^{-10}	n	8(107)
$^3CH_2 + CO_2 \rightarrow H_2CO + CO$	3.9×10^{-14}	e	8(114)
Oxygen–hydrogen species			
$O + OH \rightarrow O_2 + H$	$2.2 \times 10^{-11} \exp(+117/T)$	b	1(23), 3(32), 5(17), 5(416)
$O_2 + 2\,H_2 \rightarrow 2\,H_2O$	Net reaction cycle		1(25), 3(44)
$O_2 + H + M \rightarrow HO_2 + M$	$k_0 = 5.5 \times 10^{-32} (T/300)^{-1.6}$	b	1(27), 2(11), 3(30), 5(15), 6(35), 7(4)
$O_3 + OH \rightarrow HO_2 + O_2$	$1.6 \times 10^{-12} \exp(-940/T)$	b	1(34), 3(34), 5(19)
$O_3 + HO_2 \rightarrow OH + 2\,O_2$	$1.4 \times 10^{-14} \exp(-580/T)$	b	1(35), 2(30), 3(35), 5(20)
$O(^1D) + H_2O \rightarrow 2\,OH$	2.2×10^{-10}	b	1(19), 2(9), 2(31b), 3(28), 4(28), 5(27), 7(8)
$O(^1D) + H_2 \rightarrow H + OH$	1.0×10^{-10}	b	3(29), 7(20)
$O + HO_2 \rightarrow OH + O_2$	3.0×10^{-11}	b	3(31), 3(50), 5(18), 7(2)
$O_3 + H \rightarrow OH + O_2$	$1.4 \times 10^{-10} \exp(-470/T)$	b	3(33), 3(49), 5(16)
$O + H_2O_2 \rightarrow OH + HO_2$	$1.4 \times 10^{-12} \exp(-2000/T)$	b	3(48)
Oxygen–nitrogen species			
$O_3 + NO \rightarrow NO_2 + O_2$	$1.8 \times 10^{-12} \exp(-1370/T)$	b	1(37), 2(7), 3(58), 3(61)
$O + NO_2 \rightarrow NO + O_2$	9.3×10^{-12}	b	1(38), 3(60), 3(61)
$O_2 + N_2 \xrightarrow{\Delta} 2\,NO$	High-temperature reaction (combustion and lightning)		2(2)
$NH_2 + O_2 + M \rightarrow NH_2O_2 + M$	$<3 \times 10^{-18}$	b	2(26d)
$O_2 + NH_2O \rightarrow HNO + HO_2$	2.7×10^{-18}	d	2(26b)
$O_2 + HNO \rightarrow NO + HO_2$	1.3×10^{-19}	d	2(26e)
$O(^1D) + N_2O \rightarrow 2\,NO$	6.7×10^{-11}	b	3(56)
$O(^1D) + N_2O \rightarrow N_2 + O_2$	4.9×10^{-11}	b	3(57)
$O_3 + NO_2 \rightarrow NO_3 + O_2$	$1.2 \times 10^{-13} \exp(-2450/T)$	b	3(63)
$O_2 + N(^4S) \rightarrow NO + O$	$4.4 \times 10^{-12} \exp(-3220/T)$	hh	3(72), 5(406)
$O_2 + N(^2D) \rightarrow NO + O$	$4.4 \times 10^{-12} \exp(-3220/T)$	c	3(72), 5(404)
$O_2 + N(^2D) \rightarrow NO + O(^1D)$	$\simeq 5 \times 10^{-11}$	ee	5(378)

(*cont.*)

Appendix II *(cont.)*

Reaction	Rate[a]	Source[b]	Chapter and (equation)[c]
Oxygen–nitrogen species			
$O + HNO_2 \rightarrow OH + NO_2$	$<1 \times 10^{-15}$	l	3(77)
$O + HNO_3 \rightarrow OH + NO_3$	$<3.0 \times 10^{-17}$	b	3(81)
$O + HO_2NO_2 \rightarrow OH + NO_2 + O_2$	$7.0 \times 10^{-11} \exp(-3370/T)$	b	3(87)
$O(^1D) + N_2 \rightarrow O(^3P) + N_2$	2.3×10^{-11}	ff	5(372), 5(419)
$O + N(^2D) \rightarrow N(^4S) + O$	$\sim 7 \times 10^{-13}$	gg	5(403)
Oxygen–carbon species			
$\left(\frac{x+y}{4}\right) O_2 + C_xH_y \rightarrow xCO_2 + \frac{y}{2} H_2O$	High-temperature reaction		2(1)
$O_2 + CHO \rightarrow HO_2 + CO$	5.0×10^{-12}	d	2(12), 3(102)
$O_2 + R + M \rightarrow RO_2 + M$	General form of reaction		2(17)
$O_2 + RO \rightarrow R_{-1}CHO + HO_2$	General form of reaction		2(21)
$O(^1D) + CH_4 \rightarrow OH + CH_3$	1.4×10^{-10}	b	3(91), 8(108)
$OH + CH_4 \rightarrow H_2 + H_2CO$	1.4×10^{-11}	b	3(92)
$O_2 + CH_3 + M \rightarrow CH_3O_2 + M$	$k_0 = 6.0 \times 10^{-31} (T/300)^{-2}$	b	3(94)
	$k_\infty = 2.0 \times 10^{-12}$	b	
$O_2 + CH_3O \rightarrow H_2CO + HO_2$	$1.3 \times 10^{-13} \exp(-1350/T)$	c	3(99)
$O_2 + CHO \rightarrow HO_2 + CO$	$3.5 \times 10^{-12} \exp(+140/T)$	b	3(102)
$O(^1D) + CO_2 \rightarrow O + CO_2$	$7.4 \times 10^{-11} \exp(+117/T)$	b	7(13)
Hydrogen–nitrogen species			
$OH + NH_3 \rightarrow H_2O + NH_2$	$3.3 \times 10^{-12} \exp(-900/T)$	b	1(16), 2(26a)
$HO_2 + NO \rightarrow OH + NO_2$	$3.7 \times 10^{-12} \exp(+240/T)$	b	2(15), 3(73), 4(24), 6(37)
$OH + NO_2 + M \rightarrow HNO_3 + M$	$k_0 = 2.6 \times 10^{-30} (T/300)^{-3.2}$	b	2(22), 3(78)
	$k_\infty = 2.4 \times 10^{-11}$	b	
$OH(A\,^2\pi) + N_2 \rightarrow H + N_2O$	?		3(54)
$OH + NO + M \rightarrow HNO_2 + M$	$k_0 = 7.0 \times 10^{-31} (T/300)^{-2.6}$	b	3(74)
	$k_\infty = 1.5 \times 10^{-11}$	b	

$OH + HNO_2 \rightarrow H_2O + NO_2$	6.6×10^{-12}	d	3(76)
$OH + HNO_3 \rightarrow H_2O + NO_3$	$9.4 \times 10^{-15} \exp(+778/T)$	b	3(80)
$HO_2 + NO_2 + M \rightarrow HO_2NO_2 + M$	$2.3 \times 10^{-31} (T/300)^{-4.6}$	b	3(83)
$HO_2NO_2 + M \rightarrow HO_2 + NO_2 + M$	$k_0 = 5 \times 10^{-6} \exp(-1000/T)[N_2]$	c	3(85)
	$k_\infty = 3.4 \times 10^{14} \exp(-10420/T)$	c	
$OH + HO_2NO_2 \rightarrow NO_2 + H_2O + O_2$	$1.3 \times 10^{-12} \exp(+380/T)$	b	3(86)
$NH(a\,^1\Delta) + H_2 \rightarrow NH_2 + H$	? (probably fast)		8(33)
$NH(a\,^1\Delta) + H_2 \rightarrow NH_3$	? (probably fast)		8(33)
$NH_2 + NH_2 + M \rightarrow N_2H_4 + M$	$1.6 \times 10^{-28} [M]/(1 + 10^{-17} [M])$	g	8(34)
$NH_2 + H + M \rightarrow NH_3 + M$	$3.4 \times 10^{-30} [M]/(1 + 1.2 \times 10^{-19} [M])$	g	8(35)
$N_2H_4 + H \rightarrow N_2H_3 + H_2$	$9.9 \times 10^{-12} \exp(-1200/T)$	bb	8(37)
$N_2H_3 + H \rightarrow N_2H_2 + H_2$	$>2.7 \times 10^{-12}$		8(38a)
$N_2H_3 + H \rightarrow 2\,NH_2$	2.7×10^{-12}	cc	8(38b)
$N_2H_3 + N_2H_3 \rightarrow N_2H_4 + N_2H_2$	6.0×10^{-11}	cc	8(39a)
$N_2H_3 + N_2H_3 \rightarrow 2NH_3 + N_2$	$<6.0 \times 10^{-11}$	cc	8(39b)
$N_2H_2 \rightarrow N_2 + H_2$	Spontaneous decomposition		8(40)
$NH + H \rightarrow N(^4S) + H_2$	$1.7 \times 10^{-12} T^{0.68} \exp(-950/T)$	dd	8(80)
Hydrogen–carbon species			
$OH + CH_4 \rightarrow H_2O + CH_3$	$2.4 \times 10^{-12} \exp(-1710/T)$	b	1(17), 3(90), 4(22)
$H + CO + M \rightarrow HCO + M$	$2.0 \times 10^{-33} \exp(-850/T)$	e	1(28)
$HCO + HCO \rightarrow H_2CO + CO$	6.3×10^{-11}	e	1(29)
$OH + RH \rightarrow H_2O + R$	General form of reaction		2(16)
$HO_2 + CH_3O_2 \rightarrow CH_3OOH + O_2$	$7.7 \times 10^{-14} \exp(+1300/T)$	c	3(96)
$OH + CH_3OOH \rightarrow CH_3O_2 + H_2O$	1.0×10^{-11}	b	3(97)
$OH + CO \rightarrow H + CO_2$	$1.5 \times 10^{-13} (1 + 0.6P_{atm})$	b	3(103), 4(23), 6(28), 6(35), 6(57), 7(3), 8(109)
$H_2O + CO \rightarrow CO_2 + H_2$	?	z	6(31)
$H_2 + {}^1CH_2 \rightarrow CH_3 + H$	7.0×10^{-12}	e	8(23), 8(54)
$H_2 + C_2H \rightarrow C_2H_2 + H$	$1.9 \times 10^{-11} \exp(-1450/T)$	e	8(27)
$H_2 + C_2 \rightarrow C_2H + H$	1.4×10^{-12}	e	8(27)
$H + CH_3 + M \rightarrow CH_4 + M$	$k_0 = 1.7 \times 10^{-27}$	e	8(29)
	$k_\infty = 1.5 \times 10^{-10}$	e	

(cont.)

Appendix II *(cont.)*

Reaction	Rate[a]	Source[b]	Chapter and (equation)[c]
Hydrogen–carbon species			
$H + C_2H_4 + M \rightarrow C_2H_5 + M$	$k_0 = 1.1 \times 10^{-23} T^{-2} \exp(-1040/T)$	e	8(30)
	$k_\infty = 3.7 \times 10^{-11} \exp(-1040/T)$	e	
$H + C_2H_5 \rightarrow 2\,CH_3$	$1.9 \times 10^{-10} \exp(-440/T)$	e	8(30)
$H + C_2H_2 + M \rightarrow C_2H_3 + M$	$k_0 = 6.4 \times 10^{-25} T^{-2} \exp(-1200/T)$	e	8(31)
	$k_\infty = 9.2 \times 10^{-12} \exp(-1200/T)$	e	
$H + C_2H_3 \rightarrow C_2H_2 + H_2$	1.5×10^{-11}	e	8(31)
$H^* + CH_4 \rightarrow CH_3 + H_2$	?	g	8(66)
$OH + CH_3 \rightarrow CO + 2\,H_2$	6.7×10^{-12}	e	8(111)
$OH + {}^3CH_2 \rightarrow CO + H_2 + H$	5.0×10^{-12}	e	8(112)
$H + HCO \rightarrow H_2 + CO$	3.0×10^{-10}	e	8(117)
Hydrogen–nitrogen–carbon species			
$NO + RO_2 \rightarrow NO_2 + RO$	General form of reaction		2(20)
$NO + CH_3O_2 \rightarrow NO_2 + CH_3O$	$7.2 \times 10^{-12} \exp(+180/T)$	b	3(95)
$NO_2 + CH_3O_2 + M \rightarrow CH_3O_2NO_2 + M$	$k_0 = 1.5 \times 10^{-30} (T/300)^{-4.0}$	b	3(104)
	$k_\infty = 6.5 \times 10^{-12}$	b	
$CH_3O_2NO_2 + M \rightarrow NO_2 + CH_3O_2 + M$	$5.0 \times 10^{14} \exp(-9965/T)\,sec^{-1}$	c	3(105)
$N(^2D) + CO_2 \rightarrow NO + CO$	6.8×10^{-13}	aa	7(51)
$NH_2 + CH_3 \rightarrow CH_3NH_2$	$k_\infty = 8.7 \times 10^{-11} \exp(-35/T)$	f	8(55), 8(66)
$NH_2 + C_2H_3 + M \rightarrow C_2H_5N + M$	$\sim 4.2 \times 10^{-11}$	f	8(62)
$H_2CN + H \rightarrow H_2 + HCN$	$\sim 1.5 \times 10^{-11}$	e	8(65)
$CH_3NH^{**} \rightarrow H_2C{=}NH + H$	?	o	8(66)
$H_2C{=}NH^{**} \rightarrow HCN + 2\,H$	?	o	8(66)
$CH_3NH + H \rightarrow CH_3NH_2$	?	g	8(66)
$CN + C_2H_2 \rightarrow HC_3N + H$	5.0×10^{-11}	e	8(68), 8(83)
$N(^2D) + CH_4 \rightarrow NH + CH_3$	3.0×10^{-12}	e	8(77)
$N(^4S) + {}^3CH_2 \rightarrow HCN + H$	$\sim 5.0 \times 10^{-11} \exp(-250/T)$	e	8(79)

$N(^4S) + CH_3 \rightarrow HCN + 2H$	$\sim 5 \times 10^{-11} \exp(-250/T)$	e	8(79)
$N(^4S) + CH_3 \rightarrow H_2CN + H$			8(79)
$CN + C_2H_4 \rightarrow C_2H_3CN + H$	$\sim 3 \times 10^{-11}$ (estimate)		8(84)
$CN + HCN \rightarrow (CN)_2 + H$	$\sim 3.1 \times 10^{-11}$	e	8(85)
$CN + CH_4 \rightarrow HCN + CH_3$	$1.0 \times 10^{-11} \exp(-857/T)$	e	8(86)
Sulfur species			
$SO_2 + OH + M \rightarrow HSO_3 + M$	$k_0 = 3.0 \times 10^{-31} (T/300)^{-3.4}$	b	2(23a), 3(138), 4(30)
	$k_\infty = 2.0 \times 10^{-12}$	b	
$HSO_3 + OH \rightarrow H_2SO_4$	1.0×10^{-11}	d	2(23b)
$H_2S + OH \rightarrow HS + H_2O$	$1.1 \times 10^{-11} \exp(-225/T)$	b	2(24a), 3(144)
$HS + O_2 + M \rightarrow HSO_2 + M$	$\sim 1.0 \times 10^{-16}$	s	2(24b)
$HSO_2 + O_2 \rightarrow SO_2 + HO_2$	?	t	2(24c)
$CH_3SCH_3 + OH \rightarrow CH_3SCH_2 + H_2O$	$5.5 \times 10^{-12} \exp(+150/T)$	c	2(25a)
$CH_3SCH_2 + O_2 + M \rightarrow CH_3SCH_2O_2 + M$	?	u	2(25b)
$CH_3SCH_2O_2 + NO \rightarrow CH_3SCH_2O + NO_2$	?	u	2(25c)
$CH_3SCH_2O \rightarrow CH_3S + H_2CO$	?	u	2(25d)
$CH_3S + O_2 + M \rightarrow CH_3SO_2 + M$	?	u	2(25e)
$S + O_2 \rightarrow SO + O$	2.3×10^{-12}	b	3(135), 6(21), 6(34), 6(42), 8(118)
$OCS + O \rightarrow SO + CO$	$2.1 \times 10^{-11} \exp(-2200/T)$	b	3(136)
$SO + O_2 \rightarrow SO_2 + O$	$2.4 \times 10^{-13} \exp(-2370/T)$	b	3(137), 8(120)
$SO_2 + O + M \rightarrow SO_3 + M$	$4.0 \times 10^{-32} \exp(-1000/T)[N_2]$	c	3(139), 4(29)
$HSO_3 + OH \rightarrow SO_3 + H_2O$	$\sim 1.0 \times 10^{-11}$ (estimate)	i	3(140)
$HSO_3 + O \rightarrow SO_3 + OH$	$\sim 1.0 \times 10^{-11}$ (estimate)	i	3(141)
$HSO_3 + O_2 \rightarrow HO_2 + SO_3$	?		3(142)
$SO_3 + H_2O \rightarrow H_2SO_4$	1.0×10^{-12}	d	3(143)
$HS + O \rightarrow SO + H$	?		3(145)
$HS + O_2 \rightarrow SO + OH$	?		3(146)
$SO_3 + CO \rightarrow SO_2 + CO_2$	?		6(22), 6(48)
$SO + HO_2 \rightarrow SO_2 + OH$	9.0×10^{-16}	d	6(33), 6(35)
$HS + HS \rightarrow S + H_2S$	1.2×10^{-11}	bb	6(40)

(cont.)

Appendix II *(cont.)*

Reaction	Rate[a]	Source[b]	Chapter and (equation)[c]
Sulfur species			
$S + OCS \rightarrow S_2 + CO$	$2.8 \times 10^{-12} \exp(-2050/T)$	bb	6(41), 6(56)
$OCS + SO \rightarrow CO_2 + S_2$	?		6(43)
$OCS + S_2 \rightarrow CO + S_3$	?		6(44), 6(56)
$OCS + S_3 \rightarrow CO + S_4$	?		6(44), 6(56)
$S_2 + S_2 + M \rightarrow S_4 + M$	?		6(44)
$S^* + HCl \rightarrow SH + Cl$	?		6(57)
$H + SH \rightarrow H_2 + S$	2.5×10^{-11}	bb	6(57)
$S + S_2 + M \rightarrow S_3 + M$	?		6(57)
$SO + SO \rightarrow SO_2 + S$	8.3×10^{-15}	q	8(118)
Chlorine species			
$Cl + O_3 \rightarrow ClO + O_2$	$2.8 \times 10^{-11} \exp(-257/T)$	b	1(40), 3(123)
$ClO + O \rightarrow Cl + O_2$	$7.5 \times 10^{-11} \exp(-120/T)$	c	1(41), 3(124)
$Cl + CH_4 \rightarrow HCl + CH_3$	$9.6 \times 10^{-12} \exp(-1350/T)$	b	3(93), 3(120), 6(32)
$CH_3Cl + OH \rightarrow CH_2Cl + H_2O$	$1.8 \times 10^{-12} \exp(-1120/T)$	b	3(112)
$CHClF_2 + OH \rightarrow CClF_2 + H_2O$	$1.1 \times 10^{-12} \exp(-1620/T)$	c	3(113)
$CF_2Cl_2 + O(^1D) \rightarrow CF_2Cl + ClO$	1.4×10^{-10}	b	3(114)
$CFCl_3 + O(^1D) \rightarrow CFCl_2 + ClO$	2.3×10^{-10}	b	3(115)
$HCl + OH \rightarrow Cl + H_2O$	$2.8 \times 10^{-12} \exp(-425/T)$	b	3(116), 6(29)
$HCl + O \rightarrow Cl + OH$	$1.0 \times 10^{-11} \exp(-3340/T)$	b	3(117)
$HCl + O(^1D) \rightarrow Cl + OH$	1.4×10^{-10}	b	3(118)
$Cl + H_2 \rightarrow HCl + H$	$3.7 \times 10^{-11} \exp(-2300/T)$	b	3(121), 6(20), 6(30)
$Cl + HO_2 \rightarrow HCl + O_2$	$1.8 \times 10^{-11} \exp(+170/T)$	b	3(122), 6(23)
$Cl + HO_2 \rightarrow ClO + OH$	$4.1 \times 10^{-11} \exp(-450/T)$	b	6(24)
$ClO + NO \rightarrow Cl + NO_2$	$6.2 \times 10^{-12} \exp(+294/T)$	b	3(126)
$ClO + NO_2 + M \rightarrow ClONO_2 + M$	$k_0 = 1.8 \times 10^{-31} (T/300)^{-3.4}$	b	3(127)
	$k_\infty = 1.5 \times 10^{-11}$	b	

$ClO + HO_2 \rightarrow HOCl + O_2$	$4.6 \times 10^{-13} \exp(+710/T)$	b	3(130)
$HOCl + OH \rightarrow ClO + H_2O$	$3 \times 10^{-12} \exp(-150/T)$	b	3(132)
$HOCl + O \rightarrow ClO + OH$	$1 \times 10^{-11} \exp(-2200/T)$	c	3(133)
$ClO + CO \rightarrow Cl + CO_2$	$\sim 1.0 \times 10^{-12} \exp(\sim -3700/T)$	b	6(26)
$COCl + O_2 \rightarrow CO_2 + ClO$	2×10^{-14}	z	6(27)
$COCl + O_2 + M \rightarrow ClCO_3 + M$	$k_0 = 3.0 \times 10^{-31}$	z	6(38)
	$k_\infty = 5.7 \times 10^{-15}$	z	
	$\times \exp(+500/T)/(1 \times 10^{17} + 0.05\ M)$		
$ClCO_3 + Cl \rightarrow CO_2 + ClO + Cl$	1×10^{-11}	z	6(38)
$ClCO_3 + O \rightarrow CO_2 + ClO + O$	?	z	6(38)
$ClCO_3 + O \rightarrow CO_2 + Cl + O_2$	1×10^{-11}	z	6(38)
$Cl + H_2O \rightarrow OH + HCl$	?		6(57)
Fluorine species			
$F + O_3 \rightarrow FO + O_2$	$2.8 \times 10^{-11} \exp(-226/T)$	b	3(157)
$FO + O \rightarrow F + O_2$	5×10^{-11}	b	3(158)
$FO + NO \rightarrow F + NO_2$	2.6×10^{-11}	b	3(159)
$F + H_2O \rightarrow HF + OH$	$2.2 \times 10^{-11} \exp(-200/T)$	b	3(160)
$F + CH_4 \rightarrow HF + CH_3$	$3.0 \times 10^{-10} \exp(-400/T)$	b	3(161)
Bromine species			
$CH_3Br + OH \rightarrow CH_2Br + H_2O$	$7.6 \times 10^{-13} \exp(-890/T)$	c	3(162)
$Br + O_3 \rightarrow BrO + O_2$	$1.4 \times 10^{-11} \exp(-755/T)$	b	3(163)
$BrO + O \rightarrow Br + O_2$	3×10^{-11}	b	3(164)
$BrO + NO \rightarrow Br + NO_2$	$8.7 \times 10^{-12} \exp(+265/T)$	b	3(165)
$BrO + BrO \rightarrow 2Br + O_2$	$1.4 \times 10^{-12} \exp(+150/T)$	b	3(166)
$BrO + ClO \rightarrow Br + Cl + O_2$	6.7×10^{-12}	b	3(168)
$Br + Cl + O_2 \rightarrow Br + OClO$	6.7×10^{-12}	b	3(169)
Phosphine species			
$PH_2 + H \rightarrow PH_3$	$k_\infty = 3.7 \times 10^{-10} \exp(-340/T)$	f	8(44)
$PH_2 + PH_2 \rightarrow (P_2H_4)_s$	$k_\infty\ 2.8 \times 10^{-11} \exp(-30/T)$	f	8(45)
$PH_2 + NH_2 \rightarrow (H_2NPH_2)_s$	$k_\infty\ 1.0 \times 10^{-10} \exp(-18/T)$	f	8(46)

(cont.)

Appendix II *(cont.)*

Reaction	Rate[a]	Source[b]	Chapter and (equation)[c]
Phosphine species			
$PH_3 + H \rightarrow PH_2 + H_2$	$4.5 \times 10^{-11} \exp(-735/T)$	f	8(47)
$PH_2 + H \rightarrow PH + H_2$	$6.2 \times 10^{-11} \exp(-318/T)$	f	8(48)
$PH + H \rightarrow P + H_2$	$1.5 \times 10^{-10} \exp(-416/T)$	f	8(49)
$P + PH \rightarrow P_2 + H$	$5.0 \times 10^{-11} \exp(-400/T)$	f	8(50)
$P_2 + P_2 \rightarrow P_4$	?	r	8(51)
$P + H + M \rightarrow PH + M$	$3.4 \times 10^{-33} \exp(173/T)[M]$	f	8(52)
$PH + H_2 + M \rightarrow PH_3 + M$	$3.0 \times 10^{-36} [M]$	f	8(53)
$PH_2 + CH_3 \rightarrow CH_3PH_2$	$k_\infty = 1.2 \times 10^{-10} \exp(-37/T)$	f	8(56)
$P_2H_4 + H \rightarrow P_2H_3 + H_2$	$4.52 \times 10^{-11} \exp(-740/T)$	p	8(58)
$P_2H_3 + H \rightarrow P_2H_2 + H_2$	?	r	8(59)
$P_2H_3 + H \rightarrow 2\,PH_2$	?	r	8(59)
$P_2H_3 + P_2H_3 \rightarrow P_2H_2 + P_2H_4$	?	r	8(60)
$P_2H_3 + P_2H_3 \rightarrow 2\,PH_3 + P_2$	?	r	8(60)
$P_2H_2 \rightarrow P_2 + H_2$	?	r	8(61)
Photochemical smog species			
$H_2C{=}CHCH{=}CH_2 + OH \rightarrow H_2C{=}CHCHCH_2OH$	7.7×10^{-11}	v	2(37a)
$H_2C{=}CHCHCH_2OH + O_2 + M \rightarrow H_2C{=}CHC(O_2)HCH_2OH + M$	?		2(37b)
$H_2C{=}CHC(O_2)HCH_2OH + NO \rightarrow CHC(O)HCH_2OH + NO_2$	?		2(37c)
$H_2C{=}CHC(O)HCH_2OH \rightarrow H_2C{=}CHCHO + CH_2OH$	?		2(37d)
$CH_3CH_3 \rightarrow\rightarrow\rightarrow CH_3CHO$	Net reaction cycle		2(38a)
$CH_3CHO + OH \rightarrow CH_3C(O) + H_2O$	1.5×10^{-11}	d	2(38b)
$CH_3C(O) + O_2 \rightarrow CH_3C(O)O_2$	6.7×10^{-12}	d	2(38c)
$CH_3C(O)O_2 + NO_2 + M \rightarrow CH_3C(O)O_2NO_2 + M$	6×10^{-12}	b	2(38d)

Reaction	Rate	Source	Ref.
Aqueous/aerosol species[d]			
$H_2SO_4 \rightleftarrows H^+ + HSO_4^-$	$k_f = >5 \times 10^{10}$ sec^{-1}	h	2(27a)
	$k_r = \sim 5 \times 10^{10}$ M^{-1} sec^{-1}	h	
$HSO_4^- \rightleftarrows H^+ + SO_4^{2-}$	$k_f = 1 \times 10^9$ sec^{-1}	h	2(27b)
	$k_r = 1 \times 10^{11}$ M^{-1} sec^{-1}	h	
$HNO_3 \rightleftarrows H^+ + NO_3^-$	$k_f = >5 \times 10^{10}$ sec^{-1}	h	2(28b)
	$k_r = \sim 5 \times 10^{10}$ M^{-1} sec^{-1}	h	
$HSO_3^- + H_2O_2 \xrightarrow{H^+} HSO_4^- + H_2O$	$8.03 \times 10^{10}[H^+]$ M^{-1} sec^{-1}	w	2(32), 3(155)
$HSO_3^- + O_3 \rightarrow HSO_4^-$	$3.8 \times 10^5 + 1.05 \times 10^{16}[OH^-]$ M^{-1} sec^{-1}	y	2(33)
$RC(O)OH \rightleftarrows H^+ + RC(O)O^-$	$k_f = 1.4 \times 10^4$ M^{-1} sec^{-1}	x	2(34)
	$k_r = 4.5 \times 10^{10}$ M^{-1} sec^{-1}	x	
$N_2O_5 + H_2O \rightarrow HNO_3 + HNO_3$	?		3(150)
$ClONO_2 + H_2O \rightarrow HOCl + HNO_3$	?		3(151)
$H_2O_2 \xrightarrow{solution} H_2O_2$	$4.5 \times 10^{-6} \exp(7000/T)$ M/atm[e]	j	3(152)
$SO_2 \xrightarrow{solution} H_2SO_3$	$3.0 \times 10^{-5} \exp(3165/T)$ M/atm[e]	k	3(153)
$H_2SO_3 \rightleftarrows H^+ + HSO_3^-$	$1.8 \times 10^{-5} \exp(1960/T)$ M[f]	k	3(154)
$H_2SO_4 + 2\,H_2O \rightleftarrows SO_3 + 3\,H_2O$	?		6(39), 6(48)

[a] Reaction rates: For second-order reactions, the temperature dependence (if known) of the rate constant has the form $k = A\exp(E/RT)$, where A is the Arrhenius factor and E/R the activation temperature. If no temperature dependence is specified, the rate constant is usually for a temperature of 300 K. The units for second-order reaction rates are cm^3 molec^{-1} sec^{-1}. For third-order reactions, two rate constants are given: the low-pressure limiting rate constant and the high-pressure rate constant. The low-pressure limiting rate constant has the form $k_0(T) = k_0^{300}(T/300)^{-n}$, where the equation is suitable for air for the third body M, unless molecular nitrogen $[N_2]$ is specified as the third body. The units for the third-order low-pressure limiting rate constant are cm^6 molec^{-1} sec^{-1}. The high-pressure limiting rate constant has the form $k_\infty(T) = k_\infty^{300}(T/300)^{-m}$. The units for the third-order high-pressure limiting rate constant are cm^3 molec^{-1} sec^{-1}.

[b] Sources: **a,** Campbell, I. M., and Gray, C. N. (1973). *Chem. Phys. Lett.* **18,** 607; **b,** Demore, W. B., Molina, M. J., Watson, R. T., Golden, D. M., Hampson, R. F., Kurylo, M. J., Howard, C. J., and Ravishankara, A. R. (1983). *JPL Publ.* **83–62,** 1–210; **c,** Baulch, D. L., Cox, R. A., Crutzen, P. J., Hampson, R. F., Kerr, J. A., Troe, J., and Watson, R. F. (1982). *J. Phys. Chem. Ref. Data* **11,** 327–496; **d,** Graedel, T. E. (1979). *JGR, J. Geophys. Res.* **84,** 273–286; **e,** Yung, Y. L., Allen, M., and Pinto, J. P. (1984). *Astrophys. J., Suppl. Ser.* **55,** 465–506; **f,** Kaye, J. A., and Strobel, D. F. (1983). *Geophys. Res. Lett.* **10,** 957–960; **g,** Kaye, J. A., and Strobel, D. F. (1983). *Icarus* **55,** 399–419; **h,** Graedel, T. E., and Weschler, C. J. (1981). *Rev. Geophys. Space Phys.* **19,** 505–539; **i,** Turco, R. P., Whitten, R. C., and Toon, O. B. (1982). *Rev. Geophys. Space Phys.* **20,** 233–279; **j,** Martin,

Appendix II *(cont.)*

L. R., and Damschen, D. E. (1981). *Atmos. Environ.* **15,** 1615–1621; **k,** Maahs, H. G. (1982). *Geophys. Monogr., Am. Geophys. Union* **26,** 187–195; **l,** Hampson, R. F. (1980). "Chemical Kinetic and Photochemical Data Sheets for Atmospheric Reactions," FAA Rep. EE-80-17, Federal Aviation Administration, Washington, D.C. 20591 (available from National Technical Information Service, Springfield, Virginia 22151); **m,** Okabe, H. (1983). *J. Chem. Phys.* **78,** 1312; **n,** Taylor, G. W., and Setser, D. W. (1973). *J. Chem. Phys.* **58,** 4840; **o,** Gardner, E. P., and McNesby, J. R. (1982). *J. Phys. Chem.* **86,** 2646; **p,** Lee, J. H., Michael, J. V., Payne, W. A., Whytock, D. A., and Stief, L. J. (1976). *J. Chem. Phys.* **65,** 3280; **q,** Herron, J. T., and Huie, R. E. (1980). *Chem. Phys. Lett.* **76,** 322; **r,** Ferris, J. P., and Benson, R. (1981). *J. Am. Chem. Soc.* **103,** 1922; **s,** Sprung, J. L. (1977). *Adv. Environ. Sci. Technol.* **1,** 263; **t,** Thiemens, M. W., and Schwartz, S. E. (1978). *Inf. Conf. Photochem., 13th, Jan 1978,* Extended Abstracts; **u,** Niki, H., Maker, P. D., Savage, C. M., and Breitenbach, L. P. (1983). *Int. J. Chem. Kinet.* **15,** 647; **v,** Lloyd, A. C., Darnall, K. R., Winer, A. M., and Pitts, J. N. (1976). *J. Phys. Chem.* **80,** 789; **w,** Kunen, S. M., Lazrus, A. L., Kok, G. L., and Heikes, B. G. (1983). *JGR, J. Geophys. Res.* **88,** 3671; **x,** Eigen, M., Krause, W., Maass, G., and DeMaeyer, L. (1964). *Prog. React. Kinet.* **2,** Chapter 7; **y,** Maahs, H. G. (1983). *JGR, J. Geophys. Res.* **88,** 10721; **z,** Yung, Y. L., and DeMore, W. B. (1982). *Icarus* **51,** 199; **aa,** Fox, J. L. (1982). *Icarus* **51,** 248; **bb,** Hampson R. F., and Garvin, D. (1978). *NBS Spec. Publ. (U.S.)* **513**; **cc,** Atreya, S. K., Donahue, T. M., and Kuhn, W. R. (1977). *Icarus* **31,** 348.

[c] For example, 1(22) ≡ Chapter 1, Eq (22).

[d] k_f ≡ rate constant for forward reaction; k_r ≡ rate constant for reverse reaction.

[e] Units are moles per liter per atmosphere.

[f] Units are moles per liter.

Index

N

O

P

W

X